MECHANICS OF ELASTIC BIOMOLECULES

Edited by

WOLFGANG A. LINKE
University of Heidelberg, Heidelberg, Germany

HENK GRANZIER
Washington State University, Pullman, WA, USA

and

MIKLÓS S. Z. KELLERMAYER
University of Pécs, Pécs, Hungary

Reprinted from *Journal of Muscle Research and Cell Motility*, Volume 23: 5–6, 2002.

SPRINGER-SCIENCE+BUSINESS MEDIA, B.V.

A C.I.P. Catalogue record for this book is available from the Library of Congress

ISBN 978-94-010-3971-0   ISBN 978-94-010-0147-2 (eBook)
DOI 10.1007/978-94-010-0147-2

Printed on acid-free paper

Mechanics of Elastic Biomolecules

Edited by Wolfgang A. Linke, Henk Granzier and Miklós S.Z. Kellermayer

Extracellular Matrix Proteins

List of contributors

J.L. Ashworth
School of Biological Sciences, University of Manchester, 2.205 Stopford Building, Oxford Road, Manchester M13 9PT, UK

M.L. Bennink
Biophysical Techniques, Department of Applied Physics and MESA$^+$ Research Institute, University of Twente, PO Box 217, 7500 AE Enschede, The Netherlands

M. Berri
University of Wisconsin-Madison, Madison, Wisconsin 53706, USA

B. Bullard
European Molecular Biology Laboratory, D-69012 Heidelberg

V. Daggett
Department of Medicinal Chemistry, University of Washington, Seattle, Washington 98195-7610, USA

H.P. Erickson
Department of Cell Biology, Box 3079, Duke University Medical Center, Durham, NC 27710, USA

J.M. Fernandez
Department of Biological Sciences, Columbia University, New York, NY, USA

A. Fukuzawa
Department of Biology, Faculty of Science, Chiba University, Chiba, 263-8522, Japan

S. Furuike
Materials Science, Graduate School of Science and Engineering, Shizuoka 422-8529

M. Gao
Department of Physics and Beckman Institute, University of Illinois at Urbana-Champaign, IL 61801, USA

L. Grama
Department of Biophysics, University of Pécs, Medical School Pécs, Szigeti ut 12, Pécs H-7624 Hungary

W. Grange
Institute of Physics, NCCR Nanoscale Science, University of Basel, Klingelbergstrasse 82, CH-4056 Basel, Switzerland

H. Granzier
Department VCAPP, Washington State University, Pullman, WA 99164-6520, USA

M.L. Greaser
University of Wisconsin-Madison, Madison, Wisconsin 53706, USA

J. Greve
Biophysical Techniques, Department of Applied Physics and MESA$^+$ Research Institute, University of Twente, PO Box 217, 7500 AE Enschede, The Netherlands

L. Haston
School of Biological Sciences, University of Stirling FK9 4LA, UK

M. Hegner
Institute of Physics, NCCR Nanoscale Science, University of Basel, Klingelbergstrasse 82, CH-4056 Basel, Switzerland

M. Hiroshima
Structural Biology Center, National Institute of Genetics, Mishima, Shizuoka, 411-8540, Japan; Research Center for Allergy and Immunology, RIKEN, Yokohama, Kanagawa, 230-0045, Japan

T. Ito
Department of Biophysics, Graduate School of Science, Kyoto University, Kyoto 606-8502, Japan

M.S.Z. Kellermayer
Department of Biophysics, University of Pécs, Medical School Pécs, Szigeti ut 12, Pécs H-7624 Hungary

C.M. Kielty
School of Medicine, University of Manchester, 2.205 Stopford Building, Oxford Road, Manchester, M193 9PT, UK

S. Kimmura
Department of Biology, Faculty of Science, Chiba University, Chiba, 263-8522, Japan

D. Labeit
Anesthesiology and Intensive Operative Medicine, University Hospital Mannheim, 68135, Germany

S. Labeit
Anesthesiology and Intensive Operative Medicine, University Hospital Mannheim, 68135, Germany

K. Leonard
European Molecular Biology Laboratory, D-69012 Heidelberg

S.H. Leuba
Department of Cell Biology and Physiology, University of Pittsburgh School of Medicine, Hillman Cancer Center, UPCI Research Pavilion, Pittsburgh, PA 15213, USA

B. Li
Department of Medicinal Chemistry, University of Washington, Seattle, Washington 98195-7610, USA

W.A. Linke
Institute of Physiology and Pathophysiology, University of Heidelberg, Im Neuenheimer Feld 326, D-69120 Heidelberg, Germany

H. Lu
Department of Bioengineering, University of Illinois at Chicago, IL 60607, USA

J.F. Marko
Department of Physics, University of Illinois at Chicago, 845 West Taylor Street, Chicago IL 60607-7059, USA; Department of Bioengineering, University of Illinois at Chicago, 851 South Morgan Street, Chicago IL 60607-7052, USA

K. Maruyama
Department of Biology, Faculty of Science, Chiba University, Chiba, 263-8522, Japan

P.E. Mozdziak
University of Wisconsin-Madison, Madison, Wisconsin 53706, USA

T.M. Parker
Bioelastics Research Ltd., 2800 Milan Court, Suite 386, Birmingham, AL, 35211-6918, USA

M.G. Poirier
Department of Physics, University of Illinois at Chicago, 845 West Taylor Street, Chicago IL 60607-7059, USA

L.H. Pope
Biophysical Techniques, Department of Applied Physics and MESA^{+} Research Institute, University of Twente, PO Box 217, 7500 AE Enschede, The Netherlands

K. Schulten
Department of Physics and Beckman Institute, University of Illinois at Urbana-Champaign, IL 61801, USA

M.J. Sherratt
School of Biological Sciences, University of Manchester, 2.205 Stopford Building, Oxford Road, Manchester M13 9PT, UK

C.A. Shuttleworth
School of Biological Sciences, University of Manchester, 2.205 Stopford Building, Oxford Road, Manchester M13 9PT, UK

D. Stamenovic
Department of Biomedical Engineering, Boston University, 44 Cummington St., Boston, MA 02215, USA

M. Tokunaga
Structural Biology Center, National Institute of Genetics, Mishima, Shizuoka, 411-8540, Japan; Research Center for Allergy and Immunology, RIKEN, Yokohama, Kanagawa, 230-0045, Japan; Department of Genetics, Graduate University for Advanced Studies, Mishima, Shizuoka, 411-8540, Japan

D.W. Urry
University of Minnesota, Twin Cities Campus, Bio Technology Institute, 1479 Gortner Avenue, St. Paul, MN 55108-6106; Bioelastics Research Ltd., 2800 Milan Court, Suite 386, Birmingham, AL, 35211-6918, USA

N. Wang
Physiology Program, Harvard School of Public Health, 665 Huntington Ave., Boston MA 02115, USA

C.M. Warren
University of Wisconsin-Madison, Madison, Wisconsin 53706, USA

T.J. Wess
School of Biological Sciences, University of Stirling FK9 4LA, UK

Y. Wu
Department VCAPP, Washington State University, Pullman, WA 99164-6520, USA

M. Yamazaki
Materials Science, Graduate School of Science and Engineering, Shizuoka 422-8529; Department of Physics, Faculty of Science, Shizuoka University, Shizuoka 422-8529

N. Yonezawa
Department of Biology, Faculty of Science, Chiba University, Chiba, 263-8522, Japan

J. Zlatanova
Department of Chemistry and Chemical Engineering, Polytechnic University, Brooklyn, NY 11201

Preface

Biomolecular mechanics – the study of mechanical events and mechanical properties associated with biological molecules – is an exciting and rapidly expanding field. The advances made in recent years have greatly improved our understanding of the role of mechanical forces in cellular functions. We have all witnessed the great achievements in the mechanics of motor proteins and their associated cytoskeletal elements, actin filaments and microtubules. More recently, however, significant progress has been made also in a relatively novel area of biomolecular mechanics, the analysis of 'elastic biomolecules'. The new technologies that allow one to grab and stretch individual molecules and observe yet unseen phenomena captured the imagination of many. Thus, it appears timely to assemble a special overview volume on the mechanics of elastic biomolecules.

When it comes to defining what 'elastic biomolecules' are, naturally, opinions may vary. In this volume we included molecules whose elastic, viscous or viscoelastic properties are known to play a significant role in biological functions such as in the maintenance of cell and tissue mechanical integrity, sensing and/or relaying mechanical information and the modulation of organelle (e.g., chromosome) compactness. The list of included elastic biomolecules is by no means comprehensive. Several 'classical' elastic proteins are discussed such as elastin, titin, and fibrillin, whereas other well-known elastomeric proteins like resilin, collagen fibers, spider dragline silk, or byssal fiber were omitted. Also excluded are recombinant proteins like artificial silks or other synthetic polymers. Our major criterion for inclusion of a particular biomolecule in this volume was the availability of data from mechanical experiments at the molecular scale. By covering structurally quite diverse elastic molecules in a single volume we hope to provide the interested readership with a useful reference work; similarities and differences between molecules might be identified and some common, biologically relevant mechanisms of elasticity may be deduced.

The issue combines 17 contributions from leading research groups. The volume begins with a chapter on the elastic properties of DNA, chromatin and chromosomes – research that gave the initial impetus to the development of the field. The first four articles describe impressive findings of molecular-mechanics studies at the different organizational levels of the genetic material. These papers take the opportunity to describe the state-of-the-art methodology, such as single-molecule atomic force microscopy or optical tweezers, often utilized in the study of biomolecular mechanics. The following two contributions introduce the elastic proteins in invertebrate muscles. These molecular springs have only recently been shown to exhibit highly variable extensibility, depending on the specific requirements of the muscles. The next five papers address the molecular events underlying the elasticity of the giant muscle protein titin. New sequence information is provided and sophisticated single-molecule experiments are described. The results are also discussed in the context of the elasticity of whole muscle tissue. Another article gives a compelling example of how the powerful method of steered molecular dynamics simulations can be used to study the mechanical behavior of titin immunogobulin domains. The cytoskeleton contains many other extensible structures whose mechanical properties at the single-molecule level are only beginning to be understood. Here, two reports are included discussing pioneering mechanical work on the role of filamin A and vimentin in force transmission. Finally, a set of four articles is dedicated to the elasticity of extracellular matrix proteins. Although the elastic behavior of proteins like elastin, fibronectin, and fibrillin was recognized many years ago, it is only now that the elaborate mechanisms of their extensibility and elasticity can be studied directly by single-molecule mechanics, novel imaging methods, or molecular dynamics simulations.

It is difficult to cast a vision of the future for this newly sprung field. Undoubtedly, as the field rapidly progresses, more elastic biomolecules might be discovered and analyzed using single-molecule approaches. Certainly, the physiological mechanisms by which biomolecular elasticity is modulated will be addressed. Ultimately, the mechanisms by which biological systems sense, transmit and process mechanical information may be explored and better understood. We hope that such ventures will be facilitated by this volume.

We would like to express our gratitude to the authors for providing excellent contributions, descriptive illustrations and for meeting deadlines. We reserve special thanks to Myriam Poort at Kluwer Academic Publishers for the enthusiasm and encouragement in the starting of this project.

Wolfgang A. Linke
Heidelberg, Germany
Henk Granzier
Pullman, WA, USA
Miklós S.Z. Kellermayer
Pécs, Hungary

DNA, Chromatin and Chromosomes

Journal of Muscle Research and Cell Motility **23**: 367–375, 2002.

Mechanics and imaging of single DNA molecules

M. HEGNER* and W. GRANGE
Institute of Physics, NCCR Nanoscale Science, University of Basel, Klingelbergstrasse 82, CH-4056 Basel, Switzerland

Abstract

We review recent experiments that have revealed mechanical properties of single DNA molecules using advanced manipulation and force sensing techniques (scanning force microscopy (SFM), optical or magnetic tweezers, microneedles). From such measurements, intrinsic relevant parameters (persistence length, stretch modulus) as well as their dependence on external parameters (non-physiological conditions, coating with binding agents or proteins) are obtained on a single-molecule level. In addition, imaging of DNA molecules using SFM is presented.

Introduction

The study of nucleic-acid structure has experienced a fundamental shift toward methods that are capable of studying the physical properties of single molecules. During the last 10 years several force-sensitive experiments have been used to stretch and manipulate single DNA molecules. The application of hydrodynamic drag (Bensimon *et al.*, 1995; Perkins *et al.*, 1997; Otobe and Ohani, 2001), magnetic tweezers (Smith *et al.*, 1992; Strick *et al.*, 2000a), glass needles (Leger *et al.*, 1999), optical tweezers (OT) (Smith *et al.*, 1996; Bouchiat *et al.*, 1999; Hegner *et al.*, 1999) and scanning force microscopy (SFM) (Rief *et al.*, 1999) allowed to investigate the mechanical response of single DNA molecules. For all of these force-sensitive experiments, the following requirements have to be fulfilled. First, the molecule has to be tethered in-between two microscopic objects whose position can be precisely defined (nm accuracy). Second, the force acting on moving parts has to be quantified with force resolution in the picoNewton (pN) regime. Force spectroscopy on single molecules is thus technically most demanding. One can select a method of generating tension that is tailored to a specific application for a specific range of forces. Today, the magnetic beads and the OT technique are best suited to measure the entropic and enthalpic force arising in single-molecule DNA experiments. In particular, magnetic tweezers are sensitive in the range of 0.01–10 pN and OTs have sensitivity in a range of 0.1–200 pN (Grange *et al.*, 2002). The force limit (0.1 pN) of OT is two orders of magnitudes lower than in conventional SFM based techniques. The largest stretching forces (10–10,000 pN) are generated by SFM, in which the molecule of interest is stretched between a surface and a cantilever. For each of these methods the applied force can be continuously and accurately varied; thus, the energies of various nucleic acid and nucleic acid – protein structural transitions can be quantified.

* To whom correspondence should be addressed

In this article we focus on recent experiments performed on single DNA molecules using the techniques mentioned above. In particular, we discuss mechanical properties and stress-induced transitions of a DNA single molecule when stretched beyond its entropic regime under both physiological and non-physiological conditions.

This review is organized as follows. In the first part we briefly mention simple theoretical models often used to describe the behavior of a polymer either at low (entropic regime) or high (enthalpic regime) force. Anchoring the molecules is certainly a key point in single-molecule measurements, which determines the success of the experiment. This is discussed in details in a second part. Then, we present typical force vs. extension curves obtained either on dsDNA, ssDNA, and discuss how external parameters (salt, surrounding proteins, etc.) can affect the observed mechanical properties. We also briefly mention experiments performed to twist, unzip or unbind single DNA molecules. Finally, recent measurements on DNA imaging are highlighted.

Models for DNA elasticity

The relevance of single-molecule force experiments lies in the fact that important parameters that describe the stiffness of the polymer (e.g. DNA) can be extracted. Such parameters are not accessible when measurements are performed on an ensemble, simply because this would need the synchronization of all stretching events. Again, we emphasize that such stretching experiments (force vs. distance) need to have a very high-force sensitivity. To give order of magnitudes, the characteristic force required to align a polymer made of identical rigid units of length b is $\sim kT/b$ and is only about 0.1 pN for dsDNA.

Entropic elasticity

Depending on the polymer different models are applied to correctly describe its entropic elastic behavior

(Merkel, 2001). The freely jointed chain (FJC) model (Flory,1969) considers a chain of identical N rigid subunits that are freely able to rotate (i.e. the angle between the different segments is not fixed):

$$\frac{x}{L} = \coth\left(\frac{Fb}{k_B T}\right) - \frac{k_B T}{Fb} \tag{1}$$

where x is end-to-end length, L is the contour length of the polymer ($L=Nb$), and b denotes the Kuhn length. Alternatively, the wormlike chain model (WLC) can be used (Bustamante *et al.*, 1994; Marko and Siggia, 1995; Rivetti *et al.*, 1998). In the WLC model, the polymer is treated as an uniform flexible rod and bending costs energy. In this case, the force-extension relation reads:

$$\frac{FA}{k_B T} = \frac{1}{4}\left(1 - \frac{x}{L}\right)^{-2} + \frac{x}{L} - \frac{1}{4} \tag{2}$$

where A is known as the persistence length (the distance along which the molecule can be considered as rigid). Note that (i) Equation (2) is only an approximation to the WLC model and additional correction terms should be added for a more rigorous treatment (Bouchiat *et al.*, 1999) (ii) the persistence length A is defined to be half of the Kuhn length b.

Enthalpic elasticity

At higher forces (above 5–10 pN for DNA), the elastic response of the polymer under an applied tension is not more purely entropic. In this case, enthaplic contributions have to be taken into account (Odijk, 1995) and experimental data are usually fitted using an extensible WLC model (we consider that – due to some stretchability – the contour length linearly increases with the applied force) (Smith *et al.*, 1996):

$$\frac{x}{L} = 1 - \frac{1}{2}\left(\frac{k_B T}{FA}\right)^{1/2} + \frac{F}{S} \tag{3}$$

where S denotes the stretch modulus of the polymer. Alternatively, models were the Kuhn segments are able to stretched have been proposed (Smith *et al.*, 1996).

Immobilization of molecules to surfaces

A challenge in single-molecule experiments is to specifically and strongly attach molecules onto surfaces. In this review, we focus on DNA single molecules although the techniques presented in this section might be applied to other biological samples. Different procedures exist depending on the requirements whether (i) a covalent, a non-covalent, or a non-specific coupling has to be achieved (ii), or a SFM or an optical (magnetic) tweezers experimental setup is used.

Covalent coupling – although it involves more steps and is somehow more difficult to achieve – is always best suited due to its high strength. In SFM experiments dealing with short molecules, single thiol-modified ssDNA molecules were covalently attached to an activated glass surface (silanized with an aminopropyltriethoxysilane) through a poly(ethylene glycol)-α-maleimide-ω-*N*-hydroxysuccinimide-ester spacer (Chrisey *et al.*, 1996; Strunz *et al.*, 1999). Both the glass and the tip surfaces were treated in parallel except that different ssDNA (i.e. complementary ssDNA) were attached on the AFM tip and the glass surface, respectively. The experiment then just consists of approaching the AFM to the surface till a bond (due to hybridization) forms between the complementary ssDNA strands (Figure 1A., upper panel). Note that the use of spacers in this case (i.e. short molecules) is of prime importance because this reduces non-specific interactions. For OT experiments, Hegner (Hegner, 2000) developed a procedure for covalent binding of DNA molecules to polystyrene microspheres. Briefly, it consists (i) to chemically activate microspheres and covalently attach a thiol- or amino-modified 5-end of a dsDNA to the surface of the bead (ii) to attach a ligand (e.g. biotin) to the opposite 3-end. The ligand can then interact with receptors immobilized on a second set of beads, as shown in Figure 1B. (upper panel). This procedure allows preparation of DNA-microspheres in advance and storage of these beads for months. Moreover, high forces can be reached (200 pN) (Hegner *et al.*, 1999; Wuite *et al.*, 2000). This is not possible when using digoxygenin–antidigoxygenin recognition (i.e. non-covalent coupling) (Leger *et al.*, 1998; Merkel, 2001). Notice that a similar procedure was applied in magnetic tweezers experiments (Haber and Wirtz, 2000).

Non-covalent coupling is the most widely used method in OT experiments, certainly due to the ease of preparation. Many groups have worked on lambda DNA or its linearized fragments, which can be labeled at both its 5-ends with biotinylated nucleotides (Figure 1B., lower panel) using nucleotide incorporation with DNA polymerase enzymes. In contrast to covalent coupling, both ends have the same bioreactive group (biotin) and no beads stock solution can be prepared (Davenport *et al.*, 2000).

Finally, DNA can be immobilized between a surface and a tip of an AFM cantilever using non-specific interactions (Rief *et al.*, 1999) (Figure 1A., lower panel).

Overview on single-DNA experiments

Mechanical properties of bare DNA under physiological conditions

Low-force regime of dsDNA

Figure 2 shows a typical dsDNA force vs. fractional extension (x/L) curve obtained with a modified linearized pTYB1 plasmid (7477 bp). For this experiment, a 20-bp thiol-labeled dsDNA linker was ligated to the 5-end of the fragment and the single strand overhang of the 3-end was biotinylated with DNA polymerase

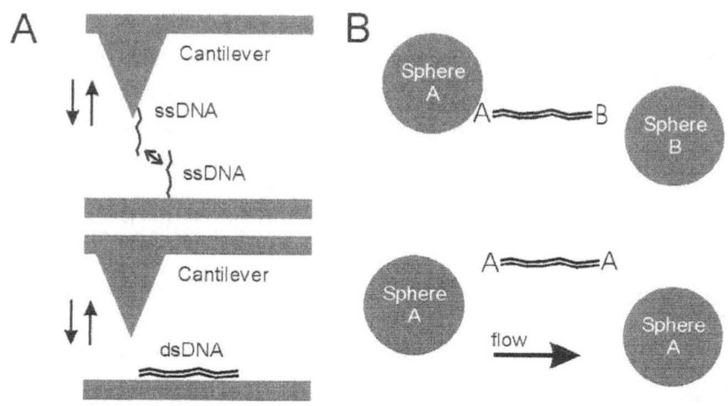

Fig. 1. Possible experimental arrangements for stretching single dsDNA molecules using either a SFM (A) or OT (B) in liquid and room temperature. A. Top panel: complementary ssDNA molecules are covalently attached to a treated AFM tip and a treated glass (Strunz *et al.*, 1999) or gold surface (Boland and Ratner, 1995). When approaching the tip to the surface, a bond may form between the two complementary ssDNAs (DNA hybridization). Bottom panel: an untreated tip is pressed onto an untreated glass surface. Due to non-specific interactions, dsDNA can be picked up and subsequently stretched (the tip has to be pressed to the surface and large forces (up to a few nN) have to be applied). B. Top panel: a dsDNA molecule is covalently attached to a chemically modified polystyrene microsphere (bond A), which is held by the optical trap. The free end of the dsDNA molecule has been labeled with a ligand that can interact with receptors immobilized on a sphere (held by suction on a moveable micropipette) (bond B). This bond (B) can be a weak non-covalent bond (e.g. biotin–streptavidin) (Hegner, 2000). Bottom panel: two microspheres are coated with identical receptors (e.g. streptavidin). One of the beads is held by the optical trap, while the second is held by suction on a moveable micropipette. A dsDNA, which is modified at its opposite 5′-ends with biotin ligands, is injected in the chamber and a continuous flow is applied. Upon attachment of the DNA to the bead in the optical trap (bond A), the measured drag force on the bead increases. The pipette is then moved in the vicinity of the optical trap to attach the free end of the DNA to the pipette bead (bond A). For this experiment, continuous flow is required to prevent looping (Davenport *et al.*, 2000).

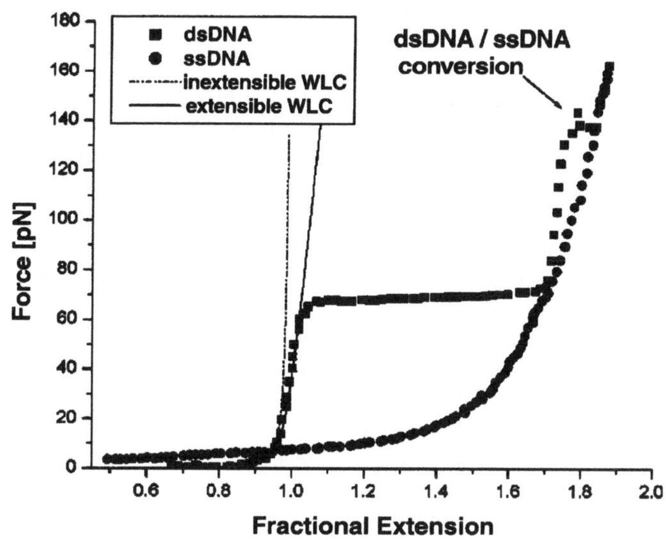

Fig. 2. Conversion of dsDNA into ssDNA (OT experiment, pTYB1 plasmid). For dsDNA single molecules (squares), the force vs. fractional extension (x/L) curve shows a typical overstretching plateau at 68 pN (150 mM NaCl). At low forces (smaller than 5–10 pN, entropic regime), the mechanics of dsDNA can be well described in terms of an inextensible WLC model (Equation 2, dots). At higher forces (smaller than 68 pN), enthalpic contributions have to be taken into account to correctly describe the observed mechanics [extensible WLC model (Equation 3, solid line)]. dsDNA can be converted into ssDNA if a high force (about 140 pN) is applied (circles).

enzymes (Husale *et al.*, 2002). Covalent coupling to polystyrene microspheres was performed in a similar procedure as described in the previous section (Hegner, 2000). At low forces [smaller than 10 pN (Odijk, 1995)], the mechanics of the dsDNA (squares) follows a WLC behavior [Equation (2), Figure 2 (dots)] with a persistence length A of 50 nm at 150 mM NaCl and pH 7.2 (Bustamante *et al.*, 2000a; Strick *et al.*, 2000b). Note that for thick polymer where the sub-units are held together by several bonds in parallel (e.g. dsDNA), the FJC model fails to describe the observed mechanics (Merkel, 2001). Although OT are widely used to investigate stretching of single molecules at low force (entropic regime), we would like to point out that magnetic tweezers is always a technique of choice due to its high force-resolution (see Introduction).

As described previously, an extensible WLC model [Figure 2 (solid line)] provides an excellent agreement for forces in the range of 10–68 pN. From a linear fit, enthalpic parameters such as the stretch modulus can be determined. Reported values at standard buffer conditions (150 mM NaCl, pH 7.2) are in the order of 1500 pN for bare dsDNA (Smith *et al.*, 1996; Bustamante *et al.*, 2000a).

Overstretching transition of bare dsDNA

At forces of about 68 pN (Cluzel *et al.*, 1996; Smith *et al.*, 1996; Rief *et al.*, 1999), a highly cooperative transition can be seen: over a range of 2–3 pN, the dsDNA is stretched to about 1.7 times its natural contour length (B-DNA form). The occurrence of this plateau, first seen with OT (Smith *et al.*, 1996), is still a subject of debate. Different explanations, either force-induced melting (Rouzina and Bloomfield, 2001a, b) or structural transition (S-DNA) (Lebrun and Lavery, 1996; Haijun *et al.*, 1999) have been proposed to explain the experimental observations. Due to (i) the high cooperativity and (ii) the observed hysteresis while relaxing the molecule, the overstretching plateau could be caused through a force-induced melting (i.e. conversion of dsDNA into ssDNA). When subsequently stretching the dsDNA apart its opposite 5–5 (Smith *et al.*, 1996) or its 5–3 ends (Hegner *et al.*, 1999; Husale *et al.*, 2002), the molecular link should either fall apart (5–5 pulling) or the force vs. extension curve should show a behavior typical of ssDNA (5–3 pulling). However, neither of these phenomena are observed in experiments. Moreover, it was shown that partially cross-linked DNA (5% intercalation of psoralen) only slightly modified the overstretching transition (Smith *et al.*, 1996). For this reason, many groups have suggested that B-DNA undergoes a structural transition into a new form called S-DNA. To date, calculations based on this assumption (i.e. DNA remains in a dsDNA form during the overstretching transition) were not able to predict the experimental trends (e.g. width of the transition, or value of the overstretching plateau). In contrast, models based only on purely thermodynamical arguments found that melting of B-DNA should occur at 60–80 pN of external applied force and correctly estimated the temperature-, salt- and pH-dependence of the overstretching plateau (Williams *et al.*, 2001a, b; Wenner *et al.*, 2002).

Mechanical properties of bare ssDNA under physiological conditions

Also shown in Figure 2 (circles) is the elastic property of bare ssDNA (150 mM NaCl, pH 7.2). Note that ssDNA was directly obtained from dsDNA by applying a high tension (see arrow in Figure 2) on the single molecule (Hegner *et al.*, 1999). For single-chain polymer, it is believed that a FJC model is applicable. This is especially true at 150 mM salt, where ions of different charges are perfectly counterbalanced (theta solvent, no excluded-volume effects). Although the force vs. fractional extension curve can be fitted with a modified FJC model [where the Kuhn segments are allowed to be stretched (Smith *et al.*, 1996)], it is necessary in this case to shrink the expected contour length by ~15%. In contrast, a non-extensible WLC model (with appropriate contour length) qualitatively describes the mechanics up to about 30 pN. Typical values reported for the persistence length are in the order of 0.75 nm (Smith *et al.*, 1996; Rivetti *et al.*, 1998).

AFM vs. *OT and magnetic tweezers*

Although AFM based techniques have been applied in the past to investigate mechanical properties of single polymers, intrinsic relevant parameters such as the persistence length are not accessible to this technique, mainly because of the large thermal noise of commercial AFM cantilever (10 pN, in liquid and at room temperature). However, AFM is a technique of choice to reveal overall mechanical trends, as shown for example by Rief *et al.* (1999). For instance, these studies have carefully studied DNA mechanics in the high-force regime and have shown that (i) the overstretching transition represents thermodynamical equilibrium (independent of pulling rate) and (ii) the melting of dsDNA into ssDNA (observed at higher forces, see Figure 2) is a non-equilibrium process (depends on pulling rate).

Mechanical properties of DNA under non-physiological conditions

When DNA is stretched under non-physiological conditions [change in solvent (Smith *et al.*, 1996), salt (Smith *et al.*, 1992, 1996; Baumann *et al.*, 1997; Wenner *et al.*, 2002), pH (Williams *et al.*, 2001a) or temperature (Williams *et al.*, 2001b)], the mechanics is strongly modified. Varying such external parameters affects both the value of the overstretching force as well as the persistence length. For instance, Smith *et al.* (1992) found an increase of the persistence length A as well as a decrease of the overstretching force when decreasing the salt concentration.

Mechanical properties of coated DNA: binding agents, proteins

We now investigate how mechanical properties are affected when some agents (Ethidium Bromide (EtBr) or SYBR® Green I) intercalate/bind to dsDNA (Figure 3). Although such measurements can be of fundamental interest, such studies can also be of great use for screening purposes without the need of competitive assays (Husale *et al.*, 2002). As expected, an extensible WLC model still provides an excellent description of the mechanical properties when dsDNA is exposed to such binding ligands. For EtBr [1 μg/ml] (Figure 3, stars), we find a decrease in persistence length by factor about two ($A = 25$ nm), a reduction in the stretch modulus of about six ($S = 250$ pN), and a change in contour length of ~25%, respectively. In addition, dsDNA does not show in this case any cooperativity at high force. In contrast (data not shown), SYBR® Green I only slightly

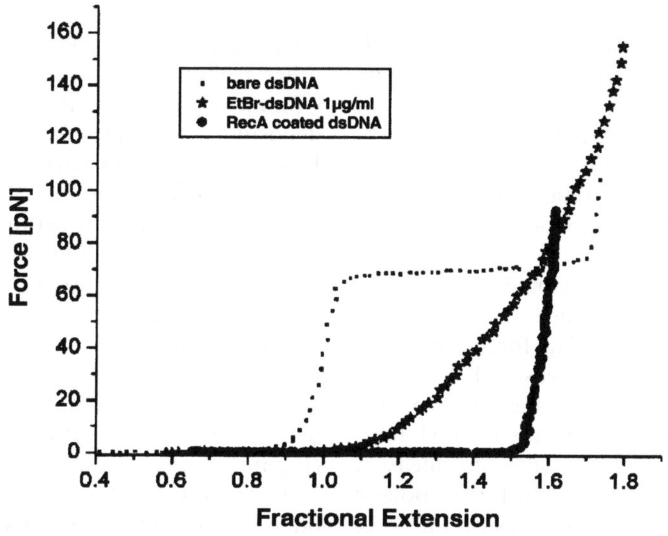

Fig. 3. Comparison of the mechanical properties of (i) naked dsDNA (dots) (ii) dsDNA coated with EtBr (stars) (Husale *et al.*, 2002) (ii) dsDNA coated with RecA (circles) (150 mM NaCl). For this experiment, a pTYB1 dsDNA plasmid (7477 bp) was used. The change in persistence length is well evidenced by the change in slope steepness when approaching the contour length. Coated dsDNA with EtBr (RecA) has a smaller (higher) persistence length than naked dsDNA and therefore the force vs. extension curve shows a less steep (steeper) slope upon approaching the contour length. Note the drastic increase of the contour length of bare dsDNA when coated with RecA.

affects the entropic elastic behavior of bare dsDNA mechanics ($A = 40$ nm). This would explain why – among all fluorescence dyes – SYBR® Green is best suited when enzymatic reactions have to be performed (Schäfer *et al.*, 2000). Note, however, that the mechanics are altered at higher forces since we measure a stretch modulus of about 500 pN. Still, a small overstretching plateau can be seen at ~80 pN.

We emphasize that the differences in mechanics obtained for these two latter compounds can be understood easily by the fact that – unlike EtBr – SYBR® Green I does not intercalate between stacked base pairs. Rather, SYBR® Green I is thought to bind in the minor groove of DNA.

Although most Protein-DNA complexes affect the mechanical properties on a local scale only [DNA-enzymes (Bustamante and Rivetti, 1996; Bustamante *et al.*, 2000b), initiation protein complexes (e.g. tbp)], some of them act on a larger scale (from a few tens of bps to a few thousand). For instance, RecA is known to lengthen and stiffen DNA chains (Figure 3, circles) to facilitate the pairing of homologous sequences (Rocca and Cox, 1997). Hegner *et al.* (1999) investigated the polymerization of RecA on dsDNA and the mechanics of both RecA-dsDNA and – ssDNA complexes. In agreement with biochemical ensemble experiments, the study of mechanical properties of RecA-dsDNA filaments showed that RecA binds to one DNA strand, while the second strand does not adhere tightly to the RecA helix and that this strand is able to slide past the protein component.

Finally, studies have been performed on single chromatin fibers to reveal how assembly and disassembly of chromatin takes place on dsDNA upon pulling (Cuy and Bustamante, 2000; Bennick *et al.*, 2001). At low force (smaller than ~10 pN), nucleosomes are able to assemble and the fiber can be reversibly stretched without disassemble the chromatin structure. Interestingly, the force vs. extension curves show discrete, sudden drops at higher forces. This suggests an irreversible dissociation of the nucleosomal core particles from the DNA.

Twisting dsDNA

Magnetic tweezers have the capability to easily and accurately rotate magnetic handles (see Strick *et al.*, 2000b for a recent review). It makes it therefore possible to study (i) torsionally constrained DNA single molecules [for instance the B to P-DNA transition where the helical pitch is strongly reduced (Allemand *et al.*, 1998)] and (ii) the relaxation of supercoiled DNA by individual enzymes (e.g. topoisomerase) (Strick *et al.*, 2000c).

Unzipping and unbinding dsDNA

In the preceeding sections we have seen that stretching or twisting a single DNA molecule yields important information on either the mechanics (persistence length, stretch modulus) or the stress-induced transitions (overstretching transition, P-DNA form, etc.). Additional information can be obtained when unzipping or unbinding dsDNA molecules, as briefly described below.

Despite inherent experimental and theoretical limitations, unzipping experiments might have potential applications for large-scale DNA sequencing (Essevaz-Roulet *et al.*, 1997; Rief *et al.*, 1999). An interesting phenomenon observed in such experiments was the occurrence of stick-slip events, similar to what can be found on a macroscopic scale (Bockelmann *et al.*, 1997). In other words, strain energy accumulates while stretching the dsDNA single molecule (the force increases) and at a certain force threshold (different for A–T and C–G base pairs) the DNA unzipps in a cooperative manner. These experimental findings have stimulated a number of theoretical investigations (Lubensky and Nelson, 2000; Cocco *et al.*, 2001; Bockelmann *et al.*, 2002).

Closely related experiments to unzipping are unbinding experiments, where the force needed to fully separate both strands is recorded. Unbinding forces of weak, non-covalent bonds have been measured by SFM (Dammer *et al.*, 1996) or biomembrane force probes (Evans and Ritchie, 1997). Initially, these SFM measurements focused on feasibility studies to measure single biomolecular interactions. For instance, Boland and Ratner (1995) have first tried to measure the force necessary to break a single DNA base pair. But recently a few groups showed that these single-molecule experiments give a direct link to ensemble experiments where thermodynamic data are measured. Moreover, such studies give insight into the geometry of the energy landscape of a biomolecular bond (Evans, 1998; Strunz *et al.*, 2000; Evans, 2001; Merkel, 2001). An inherent feature of these experiments is that unbinding forces depend on the rate of loading and on the details of the functional relationship between bond lifetime and the applied force. These points have been addressed in details for dsDNA: (i) loading rate and length (bp) dependence (Strunz *et al.*, 1999) and (ii) entropic contributions to the energy landscape (Schumakovitch *et al.*, 2002). An important result (in agreement with thermodynamic data) was the finding – on the single-molecule level – of cooperative unbinding of base pairs in the DNA duplex.

DNA imaged by SFM

SFM is capable of generating images within ranges of resolution that are of particular interest in biology. Although true atomic resolution may not be possible with biological samples at the moment, a great deal of information can still be obtained from images that show details at a slightly lower level of resolution (Bustamante and Rivetti, 1996; Fotiadis *et al.*, 2002). To date, only a few scanning tunneling microscopy (STM) experiments have been reported on DNA (Guckenberger *et al.*, 1994; Tanaka *et al.*, 1999). Since STM ope-ration in native conditions (i.e. liquid) is difficult to achieve, we believe that SFM is the technique of choice. For this reason, STM experiments will not be discussed in this review.

Imaging techniques

A decade ago, imaging of dsDNA under ambient conditions was performed using contact mode imaging were a cantilever tip is scanned across the surface while applying a constant force. The advantage of SFM over other techniques is that (i) direct imaging is possible without staining of the molecule and (ii) a simple deposition techniques onto flat surfaces can be applied. Erie and colleagues (Erie *et al.*, 1994) imaged lambda Cro protein when bound as a single dimer or multiple dimers to its three operator sites on dsDNA. They observed that bending of the non-specific sites is advantageous for a protein such as Cro that bends its specific site, because it increases the binding specificity of the protein. First images of dsDNA in liquids using contact mode showed that technical improvements were needed to allow higher resolution for imaging dsDNA interacting with enzymes (Hegner *et al.*, 1993) (Figure 4).

Nowadays, the dynamic-mode operation is the method of choice to image DNA molecules (in air or in liquids). Rivetti *et al.* (1999) analyzed the structure of *E. coli* RNA polymerase-sigma open promoter complex. The high-resolution capabilities of SFM allowed a detailed analysis of a large number of molecules and showed that the dsDNA contour length of open promoter is reduced by approximately 30 nm (90 bp) relative to the free dsDNA. The dsDNA bend angle measured with different methods varied from 55° to 88°. This strongly supported the notion that during transcription initiation, the promoter DNA wraps nearly 300° around the polymerase.

Mechanics of DNA deposited on flat surfaces

An important point to be considered is the fact that for imaging, the DNA is required to be deposited on a flat surface and that this deposition can affect parameters involving DNA mechanics. If proteins have to interact with the DNA, the conditions for binding the DNA to a surface might not be suitable for an optimal protein-DNA interaction.

Rivetti *et al.* (1996, 1998) developed suitable procedures to deposit DNA molecules onto freshly cleaved mica. This method allows the DNA to equilibrate on the surface as in an ideal two-dimensional solution. Under equilibration conditions, this study indicated that the SFM can be used to determine the persistence length of DNA molecules to a high degree of precision. They reexamined the DNA length measurement recently and revealed a discrete, size-dependent, shortening of DNA molecules deposited onto mica under low salt conditions (Rivetti and Codeluppi, 2001). Awareness of this structural alteration, which can be attributed to a partial transition from B- to A-form DNA, may lead to a more correct interpretation of DNA molecules or protein-DNA complexes imaged by SFM in the future.

We have shown previously that small agents can considerably affect the mechanics. In a recent study, Coury *et al.* (1996) presented a procedure to detect these properties by measuring the contour length of the

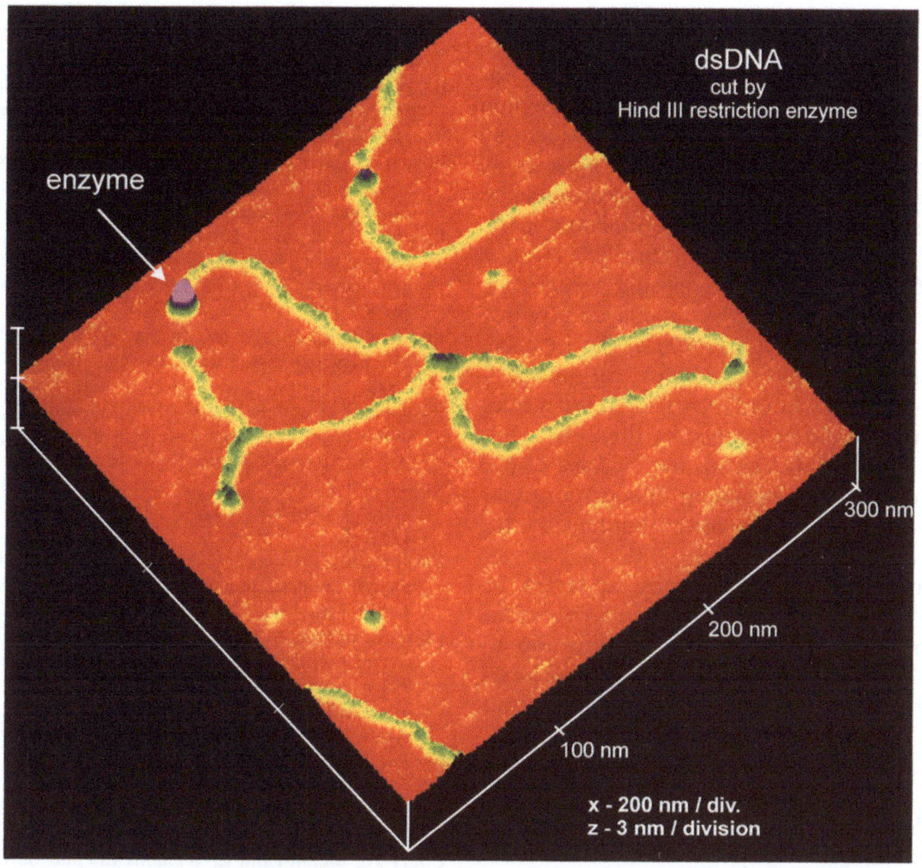

Fig. 4. Tapping-mode SFM image of an enzyme (HindIII) interacting with a plasmid DNA (pNEB193). As suggested by the 5 nm gap visible next to the enzyme, the deposition of the protein DNA complex took place immediately after the digestion of the closed plasmid. Measurements were made under ambient conditions [deposition buffer 4 mM Hepes pH 7.2, 10 mM NaCl, 2 mM $MgCl_2$ (Rivetti *et al.*, 1996)].

dsDNA molecule using SFM. They incubated the bare DNA molecules with the specific agent and subsequently deposited the modified molecules onto a mica surface. They investigated the amount of extension relative to the contour length. It was shown that the fraction of bound molecules could be estimated and an affinity could be determined by subjecting the DNA to various amounts of ligands. Such an approach reveals some of these parameters, but has the drawback that molecules have to be placed onto a surface in order to be accessible to the SFM, which affects the binding of the ligand to the DNA.

Protein-dsDNA interaction measured in physiological buffer solutions

The group of Bustamante provided a series of images with macromolecular resolution, which showed details on the mechanisms by which RNA polymerase non-specifically translocates along DNA (Guthold *et al.*, 1999). The dynamics of nonspecific and specific *Escherichia coli* RNA polymerase (RNAP)-DNA complexes were directly observed using SFM operating in buffer solution. Imaging conditions were found in which DNA molecules were adsorbed onto mica strongly enough to be imaged. Moreover, the same molecule is bound loosely enough to be able to diffuse on the surface. In sequential images of non-specific complexes, RNAP was seen to slide along DNA, performing a one-dimensional random walk. Note that in these studies the enzyme remained fixed on the surface whereas the DNA was sliding through (along) the polymerase enzyme. It was observed that mica-bound transcription complexes showed a transcription rate, which was about three times smaller than the rate of complexes in solution. This assay confirmed that the biological function of the enzyme is considerably affected by the surface but that 'native' imaging is possible using SFM in liquids.

Mapping of dsDNA with SFM and time-lapse imaging of protein-DNA interactions

Some other specific highlights in the literature include the mapping of DNA using restriction enzymes (Allison *et al.*, 1996, 1997). These experiments showed that SFM can be used to perform restriction mapping on individual cosmid clones. A mutant EcoRI endonuclease was site-specifically bound to DNA. Distances between endonuclease molecules bound to lambda DNA were measured and compared to known values. These experiments demonstrate the accuracy of SFM mapping to better than 1 kbp. These results may be extended to identify other important site-specific protein-DNA interactions, such as transcription factor and mismatch repair enzyme binding, which are difficult to resolve by other techniques. In addition, time-lapsed microscopy was used to observe dynamic interactions of proteins on dsDNA. In these series of experiments, dynamic interactions of the tumor suppressor protein p53 with a DNA fragment containing a p53-specific recognition sequence were directly observed by time-lapse tapping-mode SFM in liquid (Jiao *et al.*, 2001). As in other studies (Guthold *et al.*, 1999) the divalent cation Mg^{2+} was used to loosely attach both DNA and p53 to a mica surface so that they could be imaged by the SFM while interacting with each other. Various interactions of p53 with DNA were observed, including dissociation/reassociation, sliding and possibly direct binding to the specific sequence.

These authors visualized two modes of target recognition of p53: (a) direct binding, and (b) initial non-specific binding with subsequent translocation by one-dimensional diffusion of the protein along the DNA to the specific site. It is to be foreseen that in the future, i.e. when both higher resolution (Li *et al.*, 1999) and faster imaging in liquids (time to require one image <1 min) will be routine, time lapsed-imaging of SFM in liquids will allow to get insights of unprecedented quality and dynamics.

Acknowledgements

Financial support of the NCCR 'Nanoscale Science' the Swiss National Science Foundation and the ELTEM Regio Project Nanotechnology is gratefully acknowledged.

References

Allemand F, Bensimon D, Lavery R and Croquette V (1998) Stretched and overwound DNA forms a Pauling-like structure with exposed bases. *Proc Natl Acad Sci USA* **95:** 14,152–14,157.

Allison DP, Kerper PS, Doktycz MJ, Spain JA, Modrich P, Larimer FW, Thundat T and Warmack RJ (1996) Direct atomic force microscope imaging of EcoRI endonuclease site specifically bound to plasmid DNA molecules. *Proc Natl Acad Sci USA* **93:** 8826–8829.

Allison DP, Kerper PS, Doktycz MJ, Thundat T, Modrich P, Larimer FW, Johnson DK, Hoyt PR, Mucenski ML and Warmack RJ (1997) Mapping individual cosmid DNAs by direct AFM imaging. *Genomics* **41:** 379–384.

Baumann CG, Smith SB, Bloomfield VA and Bustamante C (1997) Ionic effects on the elasticity of single DNA molecules. *Proc Natl Acad Sci USA* **94:** 6185–6190.

Bennink ML, Leuba SH, Leno, GH, Zlatanova, J, DE Grooth BJ and Greve J (2001) Unfolding individual nucleosomes by stretching single chromatin fibers with optical tweezers. *Nat Struct Biol* **8:** 606–610.

Bensimon D, Simon AJ, Croquette V and Bensimon A (1995) Stretching DNA with receding meniscus: experiments and models, *Phys Rev Lett* **74:** 4754–4757.

Bockelmann U, Essevaz-Roulet B and Heslot M (1997) Molecular stick-slip motion revealed by opening DNA with piconewton forces. *Phys Rev Lett* **79:** 4489–4492.

Bockelmann U, Thomen PH, Essevaz-Roulet B, Viasnoff V and Heslot M (2002) Unzipping DNA with optical tweezers: high sensitivity and force flips. *Biophys J* **82:** 1537–1553.

Boland T and Ratner BD (1995) Direct measurement of hydrogen bonding in DNA nucleotide bases by atomic force microscopy. *Proc Natl Acad Sci USA* **92:** 5297–5301.

Bouchiat C, Wang MD, Allemand JF, Strick T, Block SM and Croquette V (1999) Estimating the persistence length of a worm-like chain molecule from force-extension measurements. *Biophys J* **76:** 409–413.

Bustamante C, Marko JF, Siggia ED and Smith S (1994) Entropic Elasticity Of Lambda-Phage DNA. *Science* **265:** 1599–1600.

Bustamante C and Rivetti C (1996) Visualizing protein-nucleic acid interactions on a large scale with the scanning force microscope. *Ann Rev Biophys Biomol Struct* **25:** 395–429.

Bustamante C, Smith SB, Liphardt J and Smith D (2000a) Single-molecules studies of DNA mechanics. *Curr Opi Struct Biol* **10:** 279–285.

Bustamante C, Macosko JC and Wuite JL (2000b) Grabbing the cat by the tail: manipulating molecules one by one. *Nat Rev Mol Cell Biol* **1:** 130–136.

Cluzel P, Lebrun A, Heller C, Lavery R, Viovy JL, Chatenay D and Caron F (1996) DNA: an extensible molecule. *Science* **271:** 792–794.

Chrisey LA, Lee GU and O'Ferrall CE (1996) Covalent attachment of synthetic DNA to self-assembled monolayer films. *Nucleic Acids Res* **24:** 3031–3039.

Cocco S, Monasson R and Marko JF (2001) Force and kinetic barriers to unzipping of the DNA double helix. *Proc Natl Acad Sci USA* **98:** 8608–8613.

Coury JE, Mcfail-Isom L, Williams LD and Bottomley LA (1996) A novel assay for drug-DNA binding mode, affinity, and exclusion number: scanning force microscopy. *Proc Natl Acad Sci USA* **93:** 12,283–12,286.

Cuy Y and Bustamante C (2000) Pulling a single chromatin fiber reveals the forces that maintained its higher order structure. *Proc Natl Acad Sci USA* **97:** 127–132.

Davenport RJ, Wuite GJL, Landick R and Bustamante C (2000) Single-molecule study of transcriptional pausing and arrest by *E. coli* RNA Polymerase. *Science* **287:** 2497–2500.

Dammer U, Hegner M, Anselmetti D, Wagner P, Dreier M, Huber W and Güntherodt HJ (1996) Specific antigen/antibody interactions measured by force microscopy. *Biophys J* **70:** 2437–2443.

Erie DA, Yang G, Schultz HC and Bustamante C (1994) DNA bending by Cro protein in specific and nonspecific complexes: implications for protein site recognition and specificity. *Science* **266:** 1562–1566.

Essevaz-Roulet B, Bockelmann U and Heslot F (1997) Mechanical separation of the complementary strands of DNA. *Proc Natl Acad Sci USA* **94:** 11,935–11,940.

Evans E and Ritchie K (1997) Dynamic strength of molecular adhesion bonds. *Biophys J* **72:** 1541–1555.

Evans E (1998) Energy landscapes of biomolecular adhesion and receptor anchoring at interfaces explored with dynamic force spectroscopy. *Faraday Discuss* **111:** 1–16.

Evans E (2001) Probing the relation between force-lifetime-and chemistry in single molecular bonds. *Ann Rev Biophys Biomol Struct* **30:** 105–128.

Flory PJ (1969) In: *Statistical Mechanics of Chain Molecules*, Interscience Publishers, New York.

Fotiadis D, Scheuring S, Muller SA, Engel A and Muller DJ (2002) Imaging and manipulation of biological structures with the AFM. *Micron* **33:** 385–397.

Grange W, Husale S, Guntherodt HJ and Hegner M (2002) Optical tweezers system measuring the change in light momentum flux. *Rev Sci Instrum* **73:** 2308–2316.

Guckenberger R, Heim M, Cevc G, Knapp HF, Wiegrabe W and Hillebrand A (1994) Scanning-tunneling-microscopy of insulators and biological specimens based on lateral conductivity of ultrathin water films. *Science* **226:** 1536–1538.

Guthold M, Zhu XS, Rivetti C, Yang GL, Thomson NH, Kasas S, Hansma HG, Smith B, Hansma PK and Bustamante C (1999) Direct observation of one-dimensional diffusion and transcription by *Escherichia coli* RNA polymerase. *Biophys J* **77:** 2284–2294.

Haber C and Wirtz D (2000) Magnetic tweezers for DNA micromanipulation. *Rev Sci Instrum* **71:** 4561–4570.

Haijun Z, Yang Z and Zhong-Can O-Y (1999) Bending and base-stacking interactions in double-stranded DNA. *Phys Rev Lett* **82:** 4560–4563.

Hegner M, Wagner P and Semenza G (1993) Immobilizing DNA on gold via thiol modification for atomic force microscopy. Imaging in buffer solutions. *FEBS Lett* **336:** 452–456.

Hegner M, Smith SB and Bustamante C (1999) Polymerization and mechanical properties of single RecA-DNA filaments. *Proc Natl Acad Sci USA* **96:** 10,109–10,114.

Hegner M (2000) DNA handles for single molecule experiments. *Single Mol* **1:** 139–144.

Husale S, Grange W and Hegner M (2002) DNA mechanics affected by small DNA interacting ligands. *Single Mol* **3:** 91–96.

Jiao YK, Cherny DI, Heim G, Jovin TM and Schaffer TE (2001) Dynamic interactions of p53 with DNA in solution by time-lapse atomic force microscopy. *J Mol Biol* **314:** 233–243.

Lebrun A and Lavery R (1996) Modelling extreme stretching of DNA. *Nucleic Acids Res* **24:** 2260–2267.

Leger JF, Robert J, Bourdieu L, Chatenay D and Marko JF (1998) RecA binding to a single double-stranded DNA molecule: a possible role of DNA conformational fluctuations. *Proc Natl Acad Sci USA* **95:** 12,295–12,299.

Leger JF, Romano G, Sakar A, Robert J, Bourdieu L, Chatenay D and Marko JF (1999) Structural transitions of a twisted and stretched DNA molecule. *Phys Rev Lett* **83:** 1066–1069.

Li J, Cassell AM and Dai HJ (1999) Carbon nanotubes as AFM tips: measuring DNA molecules at the liquid/solid interface. *Surf Int Anal* **28:** 8–11.

Lubensky DK and Nelson DR (2000) Pulling pinned polymers and unzipping DNA. *Phys Rev Lett* **85:** 1572–1575.

Marko JF and Siggia ED (1995) Stretching DNA. *Macromolecules* **28:** 7016–7018.

Merkel R (2001) Force spectroscopy on single passive biomolecules and single biomolecular bonds. *Phys Rep* **346:** 343–385.

Odijk T (1995) Stiff chains and filaments under tension. *Macromolecules* **28:** 7016–7018.

Otobe K and Ohani T (2001) Behavior of DNA fibers stretched by precise meniscus motion control. *Nucl Acid Res* **29:** U15–U20.

Perkins TT, Smith DE and Chu S (1997) Single polymer dynamics in an elongational flow. *Science* **276:** 2016–2021.

Rief M, Clausen-Schaumann H and Gaub HE (1999) Sequence dependent mechanics of single DNA molecules. *Nat Struct Biol* **6:** 346–349.

Rivetti C, Guthold M and Bustamante C (1996) Scanning force microscopy of DNA deposited onto mica: equilibration versus kinetic trapping studied by statistical polymer chain analysis. *J Mol Biol* **264:** 919–932.

Rivetti C, Guthold M and Bustamante C (1999) Wrapping of DNA around the *E. coli* RNA polymerase open promoter complex. *EMBO J* **18:** 4464–4475.

Rivetti C, Walker C and Bustamante C (1998) Polymer chain statistics and conformational analysis of DNA molecules with bends or sections of different flexibility. *J Mol Biol* **280:** 41–59.

Rivetti C and Codeluppi S (2001) Accurate length determination of DNA molecules visualized by atomic force microscopy: evidence for a partial B- to A-form transition on mica. *Ultramicroscopy* **87:** 55–66.

Rocca AI and Cox MM (1997) RecA protein: structure, function, and role in recombinational DNA repair. *Prog Nucleic Acid Res Mol Biol* **56:** 129–223.

Rouzina I and Bloomfield VA (2001a) Force-Induced Melting of the DNA Double Helix. 1. Thermodynamic Analysis. *Biophys J* **80:** 882–893.

Rouzina I and Bloomfield VA (2001b) Force-Induced Melting of the DNA Double Helix. 2. Effect of Solution Conditions. *Biophys J* **80:** 894–900.

Schäfer B, Gemeinhardt H, Uhl V and Greulich KO (2000) Single molecule DNA restriction analysis in the light microscope. *Single Mol* **1:** 33–40.

Schumakovitch I, Grange W, Strunz T, Bertoncini P, Güntherodt H-J and Hegner M (2002) Temperature dependence of unbinding forces between complementary DNA strands. *Biophys J* **82:** 517–521.

Smith SB, Finzi L and Bustamante C (1992) Direct mechanical measurements of the elasticity of single DNA molecules by using magnetic beads. *Science* **258:** 1122–1126.

Smith SB, Cui Y and Bustamante C (1996) Overstretching B-DNA: the elastic response of individual double stranded and single stranded DNA molecules. *Science* **271:** 795–799.

Strick T, Allemand JF, Croquette V and Bensimon D (2000a) Twisting and stretching single DNA molecules. *Prog Biophys Mol Biol* **74:** 115–140.

Strick T, Allemand JF, Bensimon D and Croquette V (2000b) Stress-induced structural transitions in DNA and proteins. *Ann Rev Biophys Biomol Struct* **29:** 523–543.

Strick T, Croquette V and Bensimon D (2000c) Single-molecule analysis of DNA uncoiling by type II topoisomerase. *Nature* **404:** 901–904.

Strunz T, Oroszlan K, Schäfer R and Güntherodt HJ (1999) Dynamic force spectroscopy of single DNA molecules. *Proc Natl Acad Sci USA* **96:** 11,277–11,284.

Strunz T, Oroszlan K, Schumakovitch I, Güntherodt H-J and Hegner M (2000) Model energy landscapes and the force-induced dissociation of ligand-receptor bonds. *Biophys J* **79:** 1206–1212.

Tanaka H, Hamai C and Kawai T (1999) High-resolution scanning tunneling microscopy imaging of DNA molecules on Cu(111) surfaces. *Surf Sci* **432:** L611–L616.

Wenner JR, Williams MC, Rouzina I and Bloomfield VA (2002) Salt dependence of the elasticity and overstretching transition of single DNA molecules. *Biophys J* **82:** 3160–3169.

Williams MC, Wenner JR, Rouzina I and Bloomfield VA (2001a) Entropy and heat capacity of DNA melting from temperature dependence of single molecule stretching. *Biophys J* **80:** 1932–1939.

Williams MC, Wenner JR, Rouzina I and Bloomfield VA (2001b) Effect of pH on the overstretching Transition of double-stranded DNA: evidence of force-induced DNA melting. *Biophys J* **80:** 874–881.

Wuite GJL, Davenport RJ, Rappaport A and Bustamante C (2000) An integrated laser trap/flow control video microscope for the study of single biomolecules. *Biophys J* **79:** 1155–1167.

Journal of Muscle Research and Cell Motility **23**: 377–395, 2002.

Stretching and imaging single DNA molecules and chromatin

JORDANKA ZLATANOVA[1,*] and SANFORD H. LEUBA[2]
[1]*Department of Chemistry and Chemical Engineering, Polytechnic University, Brooklyn, NY 11201;* [2]*Department of Cell Biology and Physiology, University of Pittsburgh School of Medicine, Hillman Cancer Center, UPCI Research Pavilion, Pittsburgh, PA 15213, USA*

Abstract

The advent of single-molecule biology has allowed unprecedented insight into the dynamic behavior of biological macromolecules and their complexes. Unexpected properties, masked by the asynchronous behavior of myriads of molecules in bulk experiments, can be revealed; equally importantly, individual members of a molecular population often exhibit distinct features in their properties. Finally, the single-molecule approaches allow us to study the behavior of biological macromolecules under applied tension or torsion: understanding the mechanical properties of these molecules helps us understand how they function in the cell. The aim of this chapter is to summarize and critically evaluate the properties of single DNA molecules and of single chromatin fibers. The use of the high-resolution imaging capabilities of the atomic force microscopy has been covered, together with manipulating techniques such as optical fibers, optical and magnetic tweezers, and flow fields. We have learned a lot about DNA and how it responds to applied forces. It is also clear that even though the study of the properties of individual chromatin fibers has just begun, the single-molecule approaches are expected to provide a wealth of information concerning the mechanical properties of chromatin and the way its structure changes during processes like transcription and replication.

Forces in biology and their range

The need to assess interactive forces in biology has come with the realization of the importance of forces in the functioning of cells and organisms. Forces are exerted at both the extracellular and intracellular levels. Some examples of interactive extracellular forces include cellular interactions in epithelium that control turgor and the interactions between leukocytes and endothelium cells during leukocytes rolling along blood vessels in response to inflammatory signals. Examples of intracellular forces include the forces exerted during the reversible structural transformations of chromatin and chromosomes during the cell-division cycle, chromosome movements during mitosis and meiosis and the action of molecular motors involved in mechanical movements (myosin moving along actin filaments during muscle contraction, kinesin moving along microtubules during vesicle trafficking, RNA- and DNA polymerases threading template DNA through their active sites, enzymes moving between extended and contracted forms during function). Another class of intracellular forces governs protein–protein and protein–DNA/RNA interactions. Intramolecular forces, on the other hand, create and maintain proper structuring and folding of macromolecules, properties that are of crucial importance for their functioning. Such forces hold the two strands of double-helical DNA together, fold RNA molecules into functional entities, and govern protein folding and denaturation.

The magnitude of forces acting at the molecular level varies within several orders of magnitude (Figure 1). The smallest forces acting on a molecule are those due to thermal agitation (Langevin forces), causing Brownian motion. For objects a couple of microns in size (beads, cells) in water at room temperature, the Langevin force is ~10 fN (this force, albeit so small in absolute terms, is huge in the micro-world: every second a cell experiences a thermal knock equal to its weight (Strick *et al.*, 2000b). Because of its random and ubiquitous nature, the force that causes Brownian motion of the sensors used for force measurements (see below) represents the instrumental force noise.

Somewhat larger are the entropic forces that are connected to creating and maintaining order in macromolecules (reducing the number of possible configurations of the macromolecule/solvent system). Both the random-coil configuration of DNA in solution and the denatured state of a polypeptide chain have maximum entropy, which is reduced upon mechanically stretching the DNA molecule or as a result of the intramolecular folding of the protein. Introducing order into biological macromolecules requires work against entropy to be done, hence, application of force. The magnitude of entropic forces is on the order of a few pN (a thousand times higher than the thermal agitation forces); indeed, such forces have been experimentally measured for DNA stretching in the low force regime (see below).

* To whom correspondence should be addressed: Tel.: + 1-718-260-3176; Fax: + 1-786-524-5899; E-mail: jzlatano@duke.poly.edu

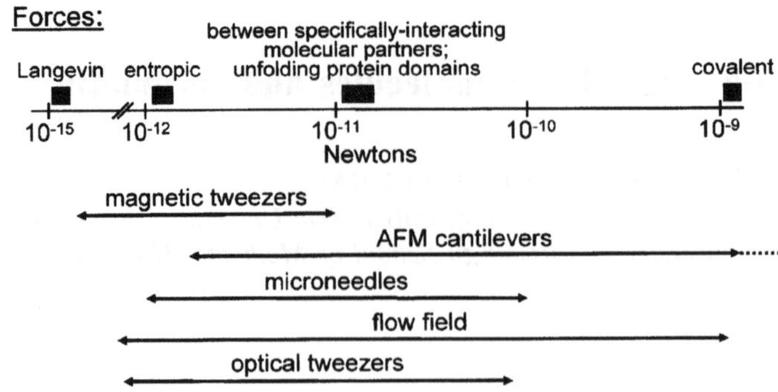

Fig. 1. Schematic of the ranges of forces in biology. The line represents the range of forces encountered in biology, and the filled squares above the line denote the specific force ranges characteristic of the interactions of biological molecules with molecules from the environment and/or the forces involved in intra- or inter-molecular interactions. The arrows below the line illustrate the ranges of forces that can be sensed or applied to biological macromolecules using the different techniques covered in this chapter.

The next group within the force magnitude range consists of forces encountered in non-covalent interactions between molecular partners in specific pairs (ligand/receptor, antigen–antibody, etc.). The interactions between such partners involve the creation of a whole range of new non-covalent bonds (van der Waals, hydrogen, electrostatic) and breaking of such preexisting interactions within each of the partners. These modifications of the molecular structure of the partners require significant energies, and hence the input of significant forces, usually ~200–300 pN (Zlatanova *et al.*, 2000). The forces needed to deform internal molecular structures, like unfolding of individually folded domains in polypeptide chains, are of the same order of magnitude.

The strongest molecular forces are those involved in covalent bonds. A receding water meniscus breaks dsDNA molecules at ~1 nN (Bensimon *et al.*, 1994, 1995) (the published scission force of 480 pN has been corrected for Young's modulus, see Bustamante *et al.*, 2000), whereas AFM-stretched short DNA fragments preserve their integrity to at least 1.7 nN (Lee *et al.*, 1994). AFM-stretching of polysaccharide chains at force-loading rates of 10 nN/s cause rupture of silicon–carbon bonds at ~2.0 nN, while sulfur–gold bonds break at ~1.4 nN (Grandbois *et al.*, 1999).

Single-molecule imaging and manipulation techniques

The atomic force microscope (AFM) and microneedles

AFM and microneedles can be classified as mechanical force transducers, in which forces are applied or sensed through a bendable beam.

The AFM can be used for imaging, force measurements, and manipulations. In the AFM, a sharp tip mounted on a flexible cantilever is allowed to interact with the sample deposited on an atomically flat surface: atoms at the tip apex experience attractive or repulsive forces of interaction with atoms on the sample. Depending on the nature of the interactions and their magnitude, the cantilever bearing the tip deflects towards (attraction or adhesion) or away (repulsion) from the surface to a different degree. These deflections are registered by a laser beam reflected off the backside of the cantilever onto a photodiode position detector (Figure 2a); the signal from the position detector is transformed into a topographic image (the probe is raster-scanned in the x–y direction), or is recorded as a force curve (the probe is moved in the z-direction only, upwards and downwards). In the AFM, the bendable cantilever is used as a mechanical force transducer that can both generate or detect forces and displacements.

Other single-molecule techniques that are based on the use of mechanical force transducer are microneedles and optical fibers (Figure 2b). Glass microneedles are softer than AFM cantilevers, and can thus measure smaller forces (Bustamante *et al.*, 2000). The displacement detection in such instruments relies on imaging the microneedle itself (Kishino and Yanagida, 1988); alternatively, an optical fiber that projects light from its tip onto a photodiode can be used in lieu of a needle (Cluzel *et al.*, 1996; Leger *et al.*, 1998). Such instruments are not yet commercially available which limits their application.

Optical tweezers (OT), magnetic tweezers (MT), and flow fields

Optical and magnetic tweezers are representatives of another class of single-molecule instruments, the so-called external field manipulators. In these instruments, the molecule is acted upon from a distance, by application of external fields (magnetic, photonic, or hydrodynamic) to the molecule itself or to an appropriate handle to which the molecule is attached. In optical tweezers the handle is a transparent polystyrene bead, in magnetic tweezers it is typically a Dynabead®, composed of highly cross-linked polystyrene with magnetic material (Fe_2O_3 and Fe_3O_4) evenly distributed throughout the pores of the bead. Dynabeads are superparamagnetic, i.e. they exhibit magnetic properties only when placed within a magnetic field. In order to manipulate single molecules with external fields, it is

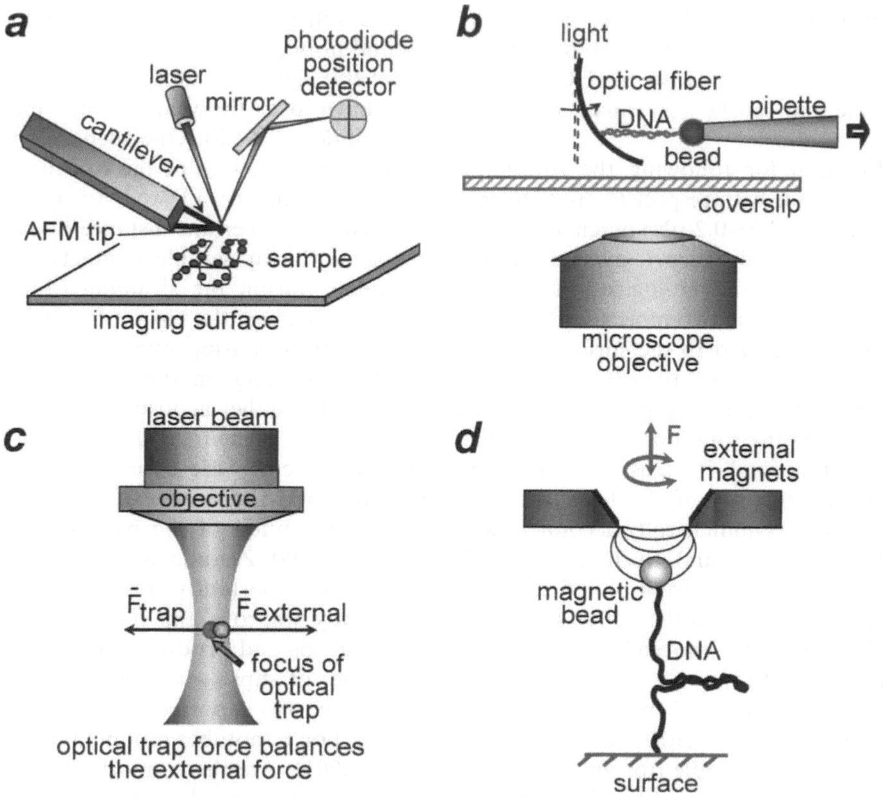

Fig. 2. Principle of operation of AFM, optical fibers, optical tweezers, and magnetic tweezers. (a) AFM. The tip/cantilever raster scans the biological sample on an atomically flat surface, and a topographic image is created from the changes in the laser signal caused by the deflections of the cantilever; these, in turn, are caused by tip/sample interactions. (b) Stretching of a single DNA with an optical fiber. Force is determined by Hooke's law, $F = k\Delta x$, where k is the spring constant (stiffness) of the optical fiber and Δx is its deflection. (c) Optical trap. Force is determined by $F = k\Delta x$, where k is the trap stiffness and Δx is the displacement of the bead in the trap. (d) Magnetic tweezers. A single DNA molecule is tethered between a superparamagnetic bead and a surface. Based on the equipartition theorem, force is determined by $F = lk_B T/\langle \Delta x^2 \rangle$, where l is the distance between the DNA-tethered bead and the surface, k_B is Boltzmann's constant, T is the temperature, and $\langle \Delta x^2 \rangle$ is the Brownian fluctuations.

necessary to attach the other end of the molecule to a surface or to an additional bead. A micropipette can be used to hold this second bead with suction. Such an approach to suspend a single DNA molecule between 2 μm-sized beads – one held in a force-measuring optical trap and the other one held in a pipette – has been introduced by Smith *et al.* (1996), and successfully used by other investigators in the field (e.g. Bennink *et al.*, 2001a).

Optical tweezers (Figure 2c) were first developed in the 1980s by Ashkin to levitate atoms (Ashkin, 1997), but their application to biology has by far exceeded their use in physics. As a photon hits a dielectric bead, its momentum changes as a result of the difference in refractive indexes of the medium and the bead; by conservation of momentum, the bead experiences force from light, and is pushed into a direction opposite to that of the refracted photon. In a laser beam focused through an objective, the bead will experience forces from multiple photons hitting it; in addition, the bead will scatter light. The resultant of these forces will create a potential well just below the waist of the focused laser beam that will hold the particle suspended. A bead moved out of this equilibrium position (by external force) will experience forces that will bring it back to this position. If a macromolecule is attached to the bead and is subjected to pulling and/or twisting at its other end (e.g. by holding it in a pipette by suction, see above), it will displace the bead out of its equilibrium position. Since the external force applied to the macromolecule (and hence to the trapped bead) is counterbalanced by the optical trap, the displacement of the bead from its zero-force equilibrium position can be used to calculate the force applied to the molecule (see legend to Figure 2).

Magnetic tweezers make use of external magnetic fields to apply and measure forces to biological polymers that had been tethered between a surface and a small superparamagnetic bead (Figure 2d). Each particle of ferric oxide within the bead acquires a magnetic moment when placed in a magnetic field, and the net magnetic moment aligns the bead with respect to the external field; thus, controlled rotation of the field will induce a synchronous rotation of the bead. If the macromolecule is attached to the surface and the bead in a topologically constrained manner (e.g. in the case of DNA, the attachment to the surfaces is through multiple contacts on all four ends of the molecule), then the molecule will experience a controlled torque. The forces created by permanent magnets (Smith *et al.*, 1992) or by electromagnets (Ziemann *et al.*, 1994; Gosse and Croquette, 2002) are very stable and can be very small, in the fN range (Figure 1).

MT can be also used for following the rotational movement of one molecule with respect to another one. In such cases a low-strength (~0.2 pN) magnetic field is applied to the bead to merely keep it away from the surface of the cuvette. If the biological interaction involves relative rotation of the two molecular partners, and one of them is immobilized to the surface and the other one is attached to the bead, then the rotation of this second molecule with respect to the immobilized can be followed by following the rotation of the bead. The rotation of the bead around its axis can be easily monitored by video-microscopy, if the bead is asymmetrical, or alternatively, asymmetrically decorated with smaller beads (Mickey Mouse ears). Such an approach has been used to visualize the rotation of a DNA template molecule with respect to the active site of an immobilized RNA polymerase molecule during transcription (Harada *et al.*, 2001).

Finally, flow fields can be applied to exert controlled forces on macromolecules, again either directly on the molecule, or through a bead handle (for an example of such an approach, see Figure 8a, and text below). This experimental set-up allows for easy replacement of buffers and introduction of different soluble biofactors into the flow cell to follow biochemical reactions; forces up to 10 nN can be readily applied and measured using Stokes' law (for further details on the technique, see Bustamante *et al.*, 2000).

AFM imaging of DNA and chromatin fibers

DNA was one of the first biological macromolecules to be imaged with the AFM. The structure of the DNA double helix varies with the environmental conditions and has been well characterized by conventional methods, including electron microscopy (EM). This detailed knowledge of the structure and its dependence on the milieu, together with its nanometer dimensions, made it a favorite imaging substrate in the early days of the AFM: DNA was used as a gold standard to verify the usefulness of AFM imaging and to optimize conditions for deposition, treatment of substrate and sample for optimal surface attachment, and imaging (Bustamante *et al.*, 1993; Hansma, 2001). It became clear that optimal lateral resolution can be achieved if the sample is attached to the surface strongly enough to preclude movement of the imaged material by the scanning tip so that stable imaging is allowed, but weakly enough to avoid sample deformation. Such intermittent attachment is difficult to predict *a priori* and has to be experimentally determined for different samples and different imaging substrates. Another rule of thumb for successful imaging is to use the lowest possible force that would still give stable imaging and reduce the instrumental noise.

Several significant improvements of the instrumentation have, over the years, helped improve the imaging capabilities of AFM. One such important improvement was the introduction of the so-called *dynamic* mode, in which the tip is allowed to oscillate at a certain frequency at a certain distance above the sample, thus reducing the tip-sample contacts to only a few brief moments during the scanning (during the alternative *contact* imaging, the tip is dragged across the sample surface, thus causing much stronger sample deformation). Depending on the way the tip is brought into oscillation – either acoustically or through the use of a magnetically coated tip acted upon by an alternating electric current – the dynamic imaging mode is called tapping- or MAC-mode, respectively (for a more detailed description of the modes of AFM imaging, see Lindsay, 2000; Zlatanova *et al.*, 2000).

Equally importantly, researchers and manufacturers constantly improve the quality and sharpness of the imaging tips and produce softer and softer cantilevers. The use of tips with electronically deposited thin extensions was reported, as well as the use of single-wall carbon nanotubes mounted on regular tips (Hafner *et al.*, 2001). Another important instrumental development that significantly improved spatial resolution was introduced in the laboratory of Shao (Han *et al.*, 1995). These authors constructed a cryo-AFM that allowed imaging under adjustable temperature from 77 to 220 K in liquid nitrogen vapor. The cryo conditions improve the mechanical strength of the biological specimens and reduce the instrumental noise due to thermal fluctuations of the cantilever/tip assembly.

Once high-resolution DNA imaging became routine, a broad range of structures, including linear DNA molecules, relaxed and supercoiled plasmids, small DNA rings, cruciform DNA, bend DNA, etc., has been imaged. A list of some important contributions has been compiled in Table 1, and Figure 3 presents example images of some of these structures. It is conceivable that future technical improvements will increase the imaging capabilities of the AFM still further, and more in-depth studies on DNA molecules and their biologically relevant structural transitions will be forthcoming.

The recent resurrection of interest in chromatin structure and its dynamic changes during DNA replication, recombination and repair, as well as in gene expression has been reflected in attempts to visualize individual chromatin fibers with high resolution imaging techniques, such as EM and AFM. Some of the more important studies are summarized in Table 2 (for a more detailed recent discussion, see Zlatanova and Leuba, in press), and some representative images are shown in Figure 4. Despite the obvious contributions of AFM imaging to understanding chromatin fiber structure and dynamics, it is clear that much more effort and the active participation of many more laboratories are needed.

Table 1. Some major results and conclusions from AFM imaging of DNA

Imaged substrate	Imaging conditions	Major results/conclusions	Reference
Plasmid DNA	Mica Contact mode Air	Reliable AFM images of DNA	Vesenka *et al.* (1992a) Bustamante *et al.* (1992)
Plasmid DNA or DNA restriction fragments	Mica Contact mode Liquid	In the work from H. Hansma's laboratory, DNA was visualized under propanol In the work from Z. Shao's laboratory, DNA was visualized complexed within a cationic lipid bilayers (1,2-dipalmitoyl-3-trimethylammonium-propane) In both laboratories, coiling of a size comparable to the turns of the double helix was seen	Hansma *et al.* (1995) Mou *et al.* (1995)
Plasmid DNA, λ-DNA	Mica Contact and tapping mode Air	Some DNA molecules were found to be stretched by up to 80%, consistent with an overstretching transition from B-form to S-form DNA (see text)	Thundat *et al.* (1994)
Single-, double-, and triple-stranded DNA fragments, plasmids, λ-DNA, and RNA homopolymer poly(A)	Mica or oxidized silicon Tapping mode Air	It was possible to observe a range of lengths of DNA from 25 bases to the entire 48,502 bp of a single λ-DNA	Hansma *et al.* (1996)
Linear DNA restriction fragments from 79 to 1057 bp	Mica Tapping mode In aqueous buffer containing $NiCl_2$, $CoCl_2$, $ZnCl_2$, $MnCl_2$, $CdCl_2$, or $HgCl_2$	The smaller the ionic radius of the divalent ion, the more reproducible imaging of DNA was obtained.	Hansma and Laney (1996)
Supercoiled plasmid DNA	APTES-treated mica Tapping mode Air or in 20 mM Tris–HCl, pH 7.6, 1 mM EDTA (TE) with NaCl up to 160 mM	Clear images of supercoiled plasmid DNA in various salts: the plasmids compacted upon salt addition; movement of DNA between successive images while imaging in buffer Note that while the authors were able to get images with a particular ionic condition, the conditions were not changed *in situ*: 25 μl droplets were imaged at a time without fluid exchange	Lyubchenko and Shlyakhtenko (1997)
Cruciform DNA	Same conditions as above with or without 200 mM NaCl	Supercoiled DNA extruded cruciforms in a plasmid with a 106 bp inverted repeat Two alternative conformations of the cruciform were observed	Shlyakhtenko *et al.* (1998)
Synthetic oligonucleotides designed to self-assemble during annealing into two-dimensional lattices	Mica Contact mode Air dried Imaged under isopropanol	Sheets of self-assembled four-way junction DNA were observed The two dimensional lattices, appearing like woven fabrics, had close to the expected periodicities	Winfree *et al.* (1998) Mao *et al.* (1999)
Restriction fragments mixed with various concentrations of spermidine	Mica	Imaged in 1.5–3 μm spermidine, DNA molecules were unremarkably linear; in 7.5–15 μm spermidine, most DNA molecules had at least one intramolecular loop; in 30 μm spermidine, the DNA molecules often appeared as flowers; and in 150 μm spermidine, DNA molecules appeared as largely planar extremely large multimolecular aggregates	Fang and Hoh (1998)

Table 1. (*Continued*)

Imaged substrate	Imaging conditions	Major results/conclusions	Reference
	Tapping mode Air		
Linearized plasmid DNA	Mica	Visualization of DNA in A- or B-form by manipulating the percentage of ethanol that the sample was flushed with on the mica. Treatment with 30% ethanol produced images of DNA in A-form	Fang *et al.* (1999)
	Tapping mode Air		

DNA manipulation

DNA stretching

Attempts to study the elastic properties of double- and single-stranded DNA molecules stem from the realization that DNA is subjected to different types of mechanical stresses during its functioning as a template for replication and transcription. Other processes such as genetic recombination and repair also impose significant strains and torsions to the DNA molecules. The single-molecule techniques allow for unprecedented assessment of the mechanical properties of macromolecules, including DNA.

A variety of methods have been used to study the response of single DNA molecules to applied tension: flow, receding meniscus, magnetic tweezers (with or without concurrent flow), glass microneedles, and optical traps (for references and further discussion, see Bustamante *et al.*, 2000). There is good agreement among published results from different groups, although the interpretation of the structural transitions at different forces is still in debate. Here we will summarize the major findings.

When the DNA double helix experiences tensile forces of up to ~10 pN, it behaves as an elastic rod that can be accurately described by an inextensible worm-like chain (WLC) (Figure 5a). The stiffness (bending rigidity) of such a rod is described by the persistence length *P*: the length of the chain over which is preserves its local direction. For a polyelectrolyte, the persistence length *P* has two components: intrinsic $P(P_i)$, which in the case of DNA is determined by base stacking interactions, and electrostatic $P(P_e)$ which reflects the intrachain repulsive interactions. P_e is a function of the ionic environment and tends to decrease as the ionic strength of the medium increases. Wenner *et al.* (2002) have reported the use of OT to stretch DNA molecules over a broad range of monovalent salt concentrations and have estimated a change in the persistence length from 59 to 46 nm upon changing the sodium ion concentration from 2.57 to 1000 mM. The persistence length for a double helical DNA molecule at physiological ionic strength is ~50 nm; stretching of a randomly coiled polymer chain with such persistence length would require low forces on the order of only 0.1 pN (Bustamante *et al.*, 2000). This stretching regime is entirely entropic in nature.

Above 10 pN, the behavior of DNA deviates from the inextensible WLC model, stretching beyond its B-from length (Figure 5a). This additional lengthening of the molecule can occur if the molecule changes its chemical structure, i.e. behaves as a stretchable solid with a certain elastic stretch modulus S. The stretch modulus has been estimated to be ~1 nN at physiological ionic strength (Smith *et al.*, 1996).

If the force applied to the DNA molecule exceeds 65 pN, there is a very abrupt change in the force curve: the molecule suddenly yields and overstretches to ~1.7 times its B-form contour length (Figure 5a and b). This structural transition (known as *overstretching transition*) occurs over a very narrow force range (only ~2 pN for λ-DNA stretched in 150 mM NaCl). Interestingly, the overstretching transition force decreases rather significantly as the salt concentration is decreased, being 68 pN at 1000 mM NaCl, and dropping down to 52 pN at 2.57 mM salt (Wenner *et al.*, 2002). This reduction in overstretching force is attributed to a decrease in the stability of the DNA double helix with decreasing salt concentration (this observation was instrumental in the interpretation of the structural changes occurring during the overstretching transition, see below). The overstretching transition force is also sequence-dependent, being ~65 pN in poly(dG-dC), and only ~35 pN in poly(dA-dT) (Rief *et al.*, 1999). Also, while the transition force measured by a glass needle has been reported as ~65 pN for nicked DNA, for intact double-stranded molecule the force exceeded ~110 pN (Leger *et al.*, 1999).

Finally, the force rises rapidly again following the overstretching transition, to reach a new smaller plateau at ~150 pN (Figure 5b). The portion of the force curve following this second plateau overlays the force curve of ss-DNA, both curves being asymptotic to the same length at high forces. This convergence of the ds- and ss-DNA curves at higher forces has been interpreted as an indication of total untwisting of the two strands from around each other (Smith *et al.*, 1996), or, alternatively, as complete melting of the double helix (Rief *et al.*, 1999; Wenner *et al.*, 2002, see below).

During relaxation, the ds-DNA curves exhibit *hysteresis* (Figure 5a). A coincidence of stretching and relax-

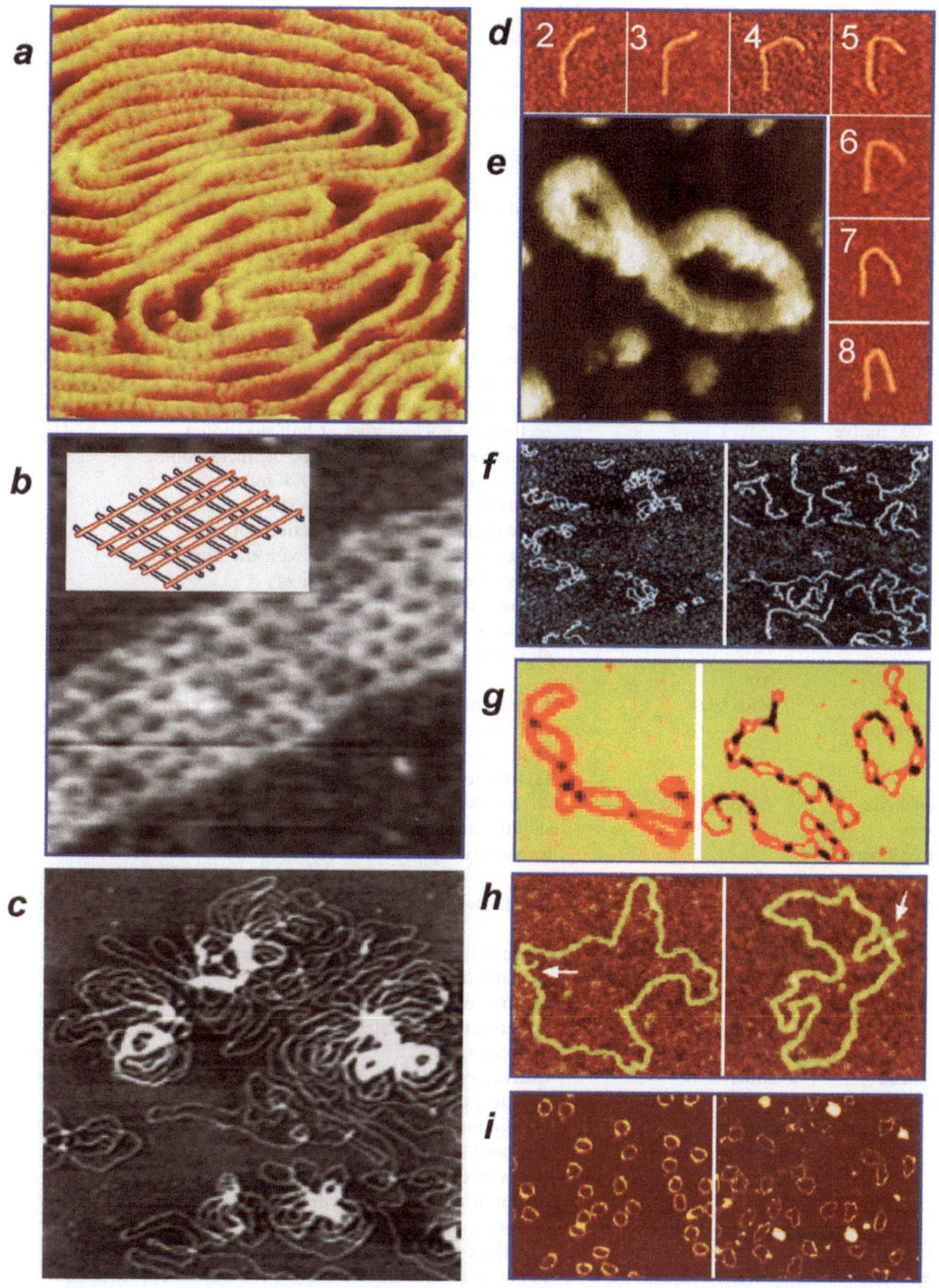

Fig. 3. Some examples of AFM images of naked DNA. (a) DNA in a cationic lipid bilayer (Mou *et al.*, 1995). Striations visualized in the naked DNA correspond to the helical repeat of DNA. (Courtesy of Z. Shao). (b) DNA tapestry constructed from annealing of specific four-way junction DNA (Mao *et al.*, 1999). Inset is model of the expected DNA construction. (Courtesy of N. Seeman). (c) Plasmid DNA complexed with 500 mM spermine (Hansma *et al.*, 1998). (Courtesy of H. Hansma). (d) Curved DNA fragments (~400 bp) with a different number of phased A-tracks near the center. The number in each image indicates the number of A-tracks in the DNA (Rivetti *et al.*, 1998). (Courtesy of C. Rivetti). (e) Supercoiled plasmid with recognizable handedness (it is possible to determine which DNA strand is on top) (Samori *et al.*, 1993). (Courtesy of B. Samori). (f) Plasmid DNA deposited from buffer with 0 mM NaCl (left panel) and 160 mM NaCl (right panel) (Lyubchenko and Shlyakhtenko, 1997). Increasing the salt concentration leads to a transition from a loosely interwound superhelices to much more compacted structures with regions of close helix contacts. (Courtesy of Y. Lyubchenko). (g) Supercoiled DNA imaged in low ionic strength buffer (left panel) and in same buffer with 200 mM NaCl (right panel) (conditions similar to those described in Lyubchenko and Shlyakhtenko, 1997). (Courtesy of Y. Lyubchenko). (h) Relaxed plasmids with cruciform extrusions (Shlyakhtenko *et al.*, 1998). Two alternative conformations of the cruciforms can be visualized: extended and compact (X-type) conformations in the left and right panels, respectively. (Courtesy of Y. Lyubchenko). (i) DNA minicircles (168 bp created by ligation of a 42 bp sequence) in 1 mM $MgCl_2$ (left panel) and 1 mM $ZnCl_2$ (right panel) (Han *et al.*, 1997). The DNA minicircles are uniformly bent in the magnesium solution, whereas they are kinked in the zinc solution. (Courtesy of S. Lindsay). Figures reprinted with permission of the publishers.

ation curves is a characteristic feature of processes at equilibrium [the molecule passes through a succession of time-independent equilibrium states (Hill, 1963, cited in Bustamante *et al.*, 2000)]. Thus, the presence of

Table 2. Major results and conclusions from AFM imaging of chromatin

Imaged substrate	Imaging conditions	Major results	Reference
Chicken erythrocyte (CE)[a] chromatin	Mica	First beads-on-a-string AFM images	Vesenka *et al.* (1992b)
	Contact mode Air/propanol		
Nucleosomal arrays reconstituted from histone octamers and 208-18[b] DNA	Glass	Beads-on-a-string morphology; center-to-center distances of ~37 nm	Allen *et al.* (1993)
	Contact mode Air		
Hypotonically spread CE nuclei	Glass	Beads-on-a-string morphology in hypotonic spreads; the detergent spreads are supranucleosomal chains	Fritzsche *et al.* (1994), (1995)
Detergent-treated nuclei from human B lymphocytes; native, dry, or rehydrated samples	Contact mode	Image processing (extraction of cross-sections of nucleosomes at half-maximum height) reveals ellipsoid shape of nucleosomes with an aspect ratio of 1.2–1.4 and a relatively smooth perimeter; The orientation of the virtual ellipsoid cross-sections of nucleosomes was correlated with the direction of the fiber axis, with >50% of nucleosomes aligned with the axis (could be partly due to interaction with glass and/or drying)	Fritzsche and Henderson (1996), (1997)
	Air		
CE chromatin fibers at different salt concentrations	Mica/glass	Loose, three-dimensional, 30 nm irregular structures even in the absence of salt; beads-on-a-string fibers seen only in H1/H5-depleted fibers	Zlatanova *et al.* (1994) Leuba *et al.* (1994) Yang *et al.* (1994)
Unfixed or glutaraldehyde-fixed chromatin fibers; native or LH-depleted fibers	Tapping mode	At 10 mM NaCl the fiber condenses slightly; at 80 mM NaCl highly compacted, irregularly segmented fibers	
rDNA minichromosomes from *Tetrahymena thermophila*	Air Mica	Condensed 30 nm fibers near center of mica	Martin *et al.* (1995)
	Tapping mode	Extended fibers at the mica periphery with partially dissociated nucleosomes	
	Air at less than 10% relative humidity	Clusters of smaller particles within these nucleosomes suggested to be individual histone molecules	
Progressively trypsinized CE chromatin fibers	Mica	Cleavage of LH tails is associated with fiber lengthening whereas cleavage of H3 N-tails results in fiber flattening; Zig-zag fiber morphology persists at later stages of digestion and is attributed to retention of the globular domain of LH in fiber	Leuba *et al.* (1998a, b)
Reconstitution of CE chromatin fibers depleted of LH or of LH and the N-tails of H3 with either intact H5 or its isolated globular domain	Tapping mode	The three-dimensional organization of nucleosomes in extended (low ionic strength) chromatin fibers requires the globular domain of LHs and either the tails of LH or the N-terminal tails of H3	
	Air		
LH-stripped mono-, di-, and oligonucleosomes from CE	Mica	Occasional visualization of the DNA wrapped around the histone octamer and of linker DNA	Zhao *et al.* (1999)
	Tapping mode	Occasional superbeads observed	
	Air		
Hypotonically spread CE nuclei	Mica	AFM can visualize nucleosome positioning	Sato *et al.* (1999)
	Tapping mode	Addition of H1 reportedly compacts the dinucleosome; suggested stem-structure formation by H1	
	Air		
Chromatin fibers from control or poly(ADP-ribosyl)ated CE nuclei	Mica	Poly(ADP-ribosyl)ation induces decondensation of chromatin structure which remains significantly decondensed even in the presence of Mg ions	d'Erme *et al.* (2001)
In vitro poly(ADP-ribosyl)ated fibers	Tapping mode	Mg ions cannot substitute for linker histones to induce compaction	

Table 2. (Continued)

Imaged substrate	Imaging conditions	Major results	Reference
	Air		
HeLa mononucleosomes Nucleosome arrays reconstituted from modified 208-12 and core histones	Mica and carbon nanotube tips	Dimers form from mononucleosomes with ~60 bp more weakly bound by histones than in control mononucleosomes	Schnitzler *et al.* (2001)
The arrays remodeled with hSWI/SNF	Tapping mode	Control arrays with evenly spaced nucleosomes are disorganized by SWI/SNF; compact dimers within these arrays could not be positively identified	
	Air		
Chromatin fibers isolated from cells with normal or elevated levels of m^{5}C Nucleosome arrays reconstituted from either unmethylated or *in vitro* methylated 208-12 and core histones; additional reconstitution of LH	Mica	DNA methylation induced chromatin fiber compaction only in the presence of bound LH; AFM results substantiated by MNase digestion patterns and sucrose gradient centrifugation	Karymov *et al.* (2001)
	Tapping mode	AFM imaging can visualize alternative nucleosome positioning on adjacent 208-bp repeats (the distribution of center-to-center distances on 208-12 is bimodal)	
	Air		
Nucleosome arrays reconstituted from 208-18 and either histone octamers, H3/H4 tetramers or the histone-fold protein HMf from *Archaea*	Mica	The HMf-nucleoprotein complexes can be considered *bona fide* chromatin structures	Tomschik *et al.* (2001)
	Tapping mode	The HMf-containing mononucleosomes are less stable than the canonical octasomes	
	Air		
Nucleosomal arrays reconstituted from a 5.4 kbp circular template and control or totally-tailless recombinant histones	Mica	Beads-on-a-string	An *et al.* (2002)
	Tapping mode	Center-to-center distance frequency distributions indistinguishable for the control and tailless reconstitutes	
	Air		
CE chromatin	CryoAFM	CryoAFM gives higher resolution of chromatin fibers	Shao (1999)
	Mica		
	Tapping mode Air/liquid nitrogen	At DNA entry/exit point, added mass suggests visualization of linker histone	
Control and hyperacetylated mononucleosomes isolated from HeLa cells	Mica	Low force images of control and hyperacetylated mononucleosomes appear to be the same	Dunker *et al.* (2001)
	Tapping mode	Large imaging force causes flattening of mononucleosomes	
	Air	Reduction to normal force allows control mononucleosomes to regain original heights whereas hyperacetylated mononucleosomes fail to do so	

[a] CE – chicken erythrocyte.

[b] 208-18, a tandemly repeated DNA sequence that has a nucleosome positioning 208-bp-sequence repeated 18 times (Simpson *et al.*, 1985).

hysteresis during relaxation is indicative of kinetic constraints on the reformation of the initial structure: the rate of relaxing the force is higher than the rate of the slowest relaxation process in the molecule, in other words, the force relaxation is too fast to allow reversibility of the structural transitions that had occurred during stretching. A plausible structural mechanism for the occurrence of hysteresis is the partial melting of the double-helix: the separated base pairs may not pair immediately when the molecule is relaxed (Wenner *et al.*, 2002).

What happens during the overstretching transition? This unusual feature of the force–extension curve immediately attracted the attention of theoreticians that

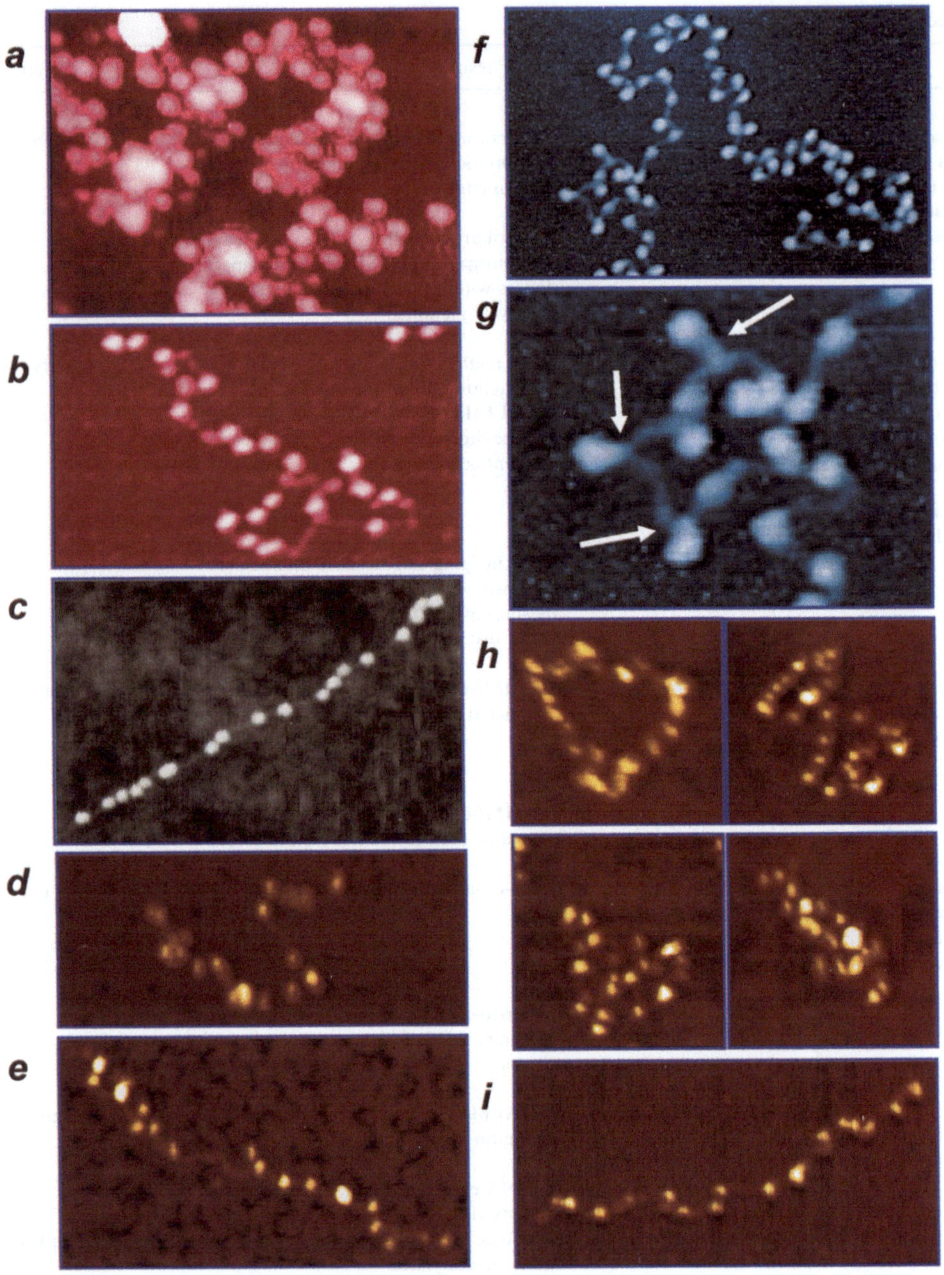

Fig. 4. Some example AFM images of chromatin fibers. (a) Unfixed chicken erythrocyte chromatin fiber imaged on glass in air (S. Leuba and G. Yang, unpublished). (b) Linker histone-depleted chicken erythrocyte chromatin fiber on mica in air (S. Leuba, unpublished). (c) Nucleosomal array, reconstituted from 208-18 DNA and core histones by salt dialysis, on mica in air (Allen *et al.*, 1993). (Courtesy of M. Allen). (d) Nucleosomal array, reconstituted from 208-18 DNA and core histones by salt dialysis, on mica in air (Tomschik *et al.*, 2001). (e) H3/H4 tetrasome array, reconstituted from 208-18 DNA and H3/H4 tetramers by salt dialysis, on mica in air (Tomschik *et al.*, 2001). (f) Cryo-AFM image of chicken erythrocyte chromatin fiber on mica (Shao, 1999). Nucleosomes are well resolved along with linker DNA. (Courtesy of Z. Shao). (g) Cryo-AFM image (zoom of a portion of (f)) suggesting visualization of linker histone. Arrows point to increased mass at DNA entry/exit points of nucleosome. (h) Plasmids reconstituted with recombinant assembly factors and intact recombinant core histones (upper left-hand panel), or core histones missing their N-terminal extensions (other three panels) (An *et al.*, 2002). (i) Nucleosomal array, reconstituted from 208-18 DNA and archaeal-histone HMf (Histone from *Methanothermus fervidus*) by salt dialysis, imaged on mica in air (Tomschik *et al.*, 2001). Figures reprinted with permission of the publishers.

came up with different models for the structure of the overstretched DNA (Konrad and Bolonick, 1996; Lebrun and Lavery, 1996; Kosikov *et al.*, 1999). None of these earlier models is, however, universally accepted. Smith *et al.* (1996), who were the first to report the overstretching transition, interpreted it in terms of unwinding of the double helix, with preservation of the base pairing (Figure 5c, left-hand side of drawing). Recently, in a series of experimental and theoretical papers, Bloomfield and coworkers have suggested that

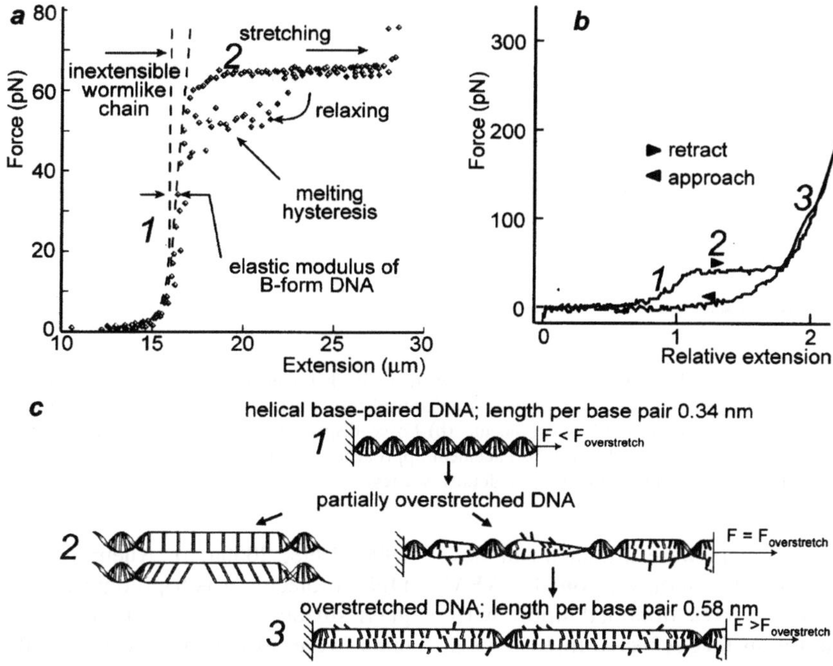

Fig. 5. DNA-stretching curves with possible explanations of the overstretching transition. (a) Force–extension curve for a single λ-DNA molecule in 150 mM NaCl, 10 mM Tris–HCl, pH 8.0, 1 mM EDTA as determined by optical tweezers (redrawn from Smith *et al.*, 1996). (b) Force-extension curve for a 1.5 µm long DNA fragment as determined by AFM (redrawn from Rief *et al.*, 1999). (c) Schematic explanation of the overstretching transitions. Right-hand side of schematic redrawn from Wenner *et al.* (2002) and left-hand side redrawn from Smith *et al.* (1996). The numbers 1, 2, and 3 in the force–extension curves in (a), (b) correspond to their respective numbers in the schematic in (c). For further explanation, see text.

the overstretching transition represents an equilibrium form of DNA melting, and that the rise in force observed at 1.7 times the contour length represent non-equilibrium rate-dependent melting of the double helix (Wenner *et al.*, 2002, and references cited therein) (Figure 5c, right-hand side of drawing). The transition that occurs at high forces after overstretching has been first reported and interpreted as complete melting of the double helix by Rief *et al.* (1999). These authors also found that both the overstretching and the melting transitions are sequence-dependent, occurring at significantly lower forces in poly(dA-dT) than in poly(dG-dC).

Unzipping of double-stranded DNA

Single-molecule approaches have been also used to study DNA melting by mechanically unzipping of the double helix. Lee *et al.* (1994) measured the interaction forces between two complementary oligonucleotides, 20 bases in length, covalently immobilized to self-assembled monolayers on a silica AFM tip and a silica surface. The sequence of the oligonucleotides was designed to restrict base pairing so that stable double stranded complexes of only 20, 16, or 12 bp could be formed. The forces needed to rupture those were centered about 1.52, 1.11, and 0.83 nN, respectively. In the experiments of Essevaz-Roulet *et al.* (1997), a single λ-DNA molecule was unzipped with a glass microneedle. The force-extension curve exhibited a saw-tooth pattern in the interval of 12–13 pN, and this variation in the force along the sequence roughly corresponded to its GC content. A recent paper from the same laboratory (Bockelmann *et al.*, 2002) reported higher-resolution unzipping data obtained with OT on λ-DNA. The experimental set-up is diagrammatically presented in Figure 6a, and the force–displacement curve, in Figure 6b. The measurements revealed sequence-specific features at a scale of 10 bp. In addition, the unzipping force exhibited characteristic flips between different values at specific positions of the sequence; these flips were attributed to bistabilities in the position of the opening fork, due to the existence of two local energy minima of similar magnitudes, separated by a low potential barrier.

Higher resolution force data were also obtained by unzipping of a short hairpin consisting of a 12 bp poly(GC) DNA stem and a short single-stranded loop; the hairpin was flanked by 600 bp dsDNA handles (unpublished work from Bustamante's lab, cited in Bustamante *et al.*, 2000). Stretching of this DNA construct by OT produced a force curve in which the steep increase of force in the stretchable solid regime of the dsDNA handles (see above) was interrupted by a force plateau region at 16 pN over 10 nm, the length of the unwound hairpin region. The forces measured were similar to those reported by Rief *et al.* (1999) for AFM-unzipped short poly(dG-dC) and poly(dA-dT) stretches.

Finally, DNA unzipping experiments were performed using AFM by C. Lieber's group (Noy *et al.*, 1997).

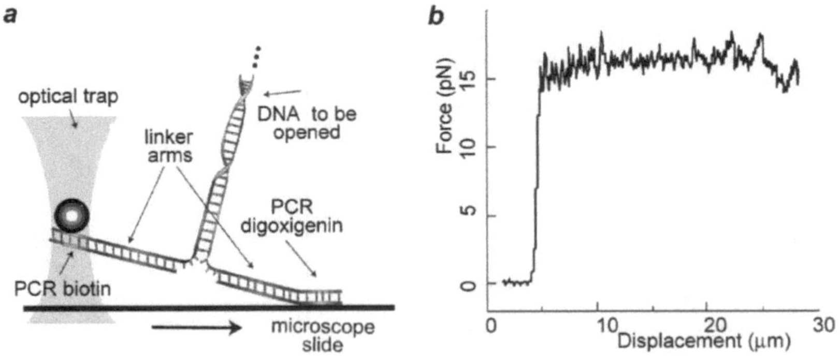

Fig. 6. Schematic of DNA unzipping and experimental data redrawn from Bockelmann *et al.* (2002). (a) A single λ-DNA molecule is held between a glass surface and a polystyrene bead in a force-measuring optical trap via two double-stranded DNA handles, each of ~7 kbp in length. The microscope slide is moved to unzip the DNA molecule. (b) Force–displacement curves of the mechanical unzipping of the λ-DNA double helix. The force starts to rise significantly when the displacement approaches the sum of the contour lengths of the double-stranded linker arms, after which the double helix starts to unzip. For further details, see text.

Complementary DNA 14-mers were attached via flexible linkers to self-associated monolayers on the AFM tip and the surface. The curves obtained upon withdrawal of the tip from the surface reflected three steps in the separation of the complementary strands: stretching of B-form DNA, structural transition to an overstretched conformation, and finally, separation into single-stranded oligonucleotides. Despite these overall similarities with DNA stretching with OT, the forces involved were higher: ~120 pN for the overstretching transition, and ~460 pN for the unzipping reaction. The reason for these discrepancies in numbers remains unclear.

DNA twisting

The elastic behavior of single supercoiled DNA molecules has been extensively studied by the group of David Bensimon and Vincent Croquette in Paris (Strick *et al.*, 1996, 1998, 2000a, b; Allemand *et al.*, 1998). A single DNA molecule was topologically constrained between a glass surface and a magnetic bead, i.e. all four DNA ends were attached through multiple points to the respective surfaces so that DNA was prevented from swiveling about its anchoring points. A magnetic field created by a pair of external magnets was used to both pull the DNA and coil it in a controlled and reversible manner (see Figure 2 for illustration of the experimental set-up and the formula to determine the stretching force). Force–extension curves were recorded at fixed values of superhelical density σ and constant forces, ranging from 6 fN to 20 pN. Extension versus σ curves revealed intriguing differences in the behavior of positively versus negatively supercoiled molecule, and these differences were dependent on the degree of superhelicity (Figure 7a).

Below 0.4 pN, DNA responded in a symmetrical manner to positive and negative supercoiling, forming

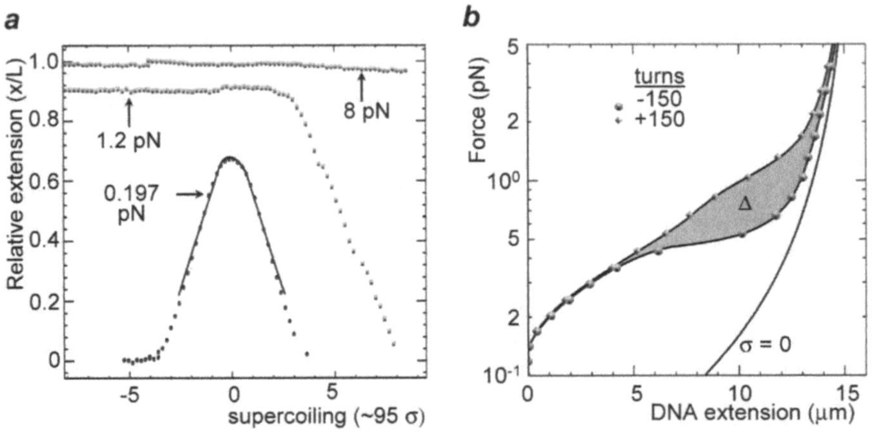

Fig. 7. DNA supercoiling experimental data redrawn from Strick *et al.* (2000a, b). (a) Relative extension of a DNA molecule vs. the degree of supercoiling. At the low force of 0.197 pN, the relative extension of the DNA decreases symmetrically for either positive or negative supercoiling. At the force of 1.2 pN, the relative extension only decreases during positive supercoiling, and a force of 8 pN prevents changes in relative extension for either positive or negative supercoiling. (b) Force–extension curves for DNA with either 150 negative (circles) or positive (diamonds) turns. The difference in work of stretching the two kinds of DNA is indicated in the shaded region by Δ. Righthand curve is for DNA without supercoiling ($\sigma = 0$). The value of Δ can be used to estimated the elastic torsional persistence length of the molecule (see text).

plectonemes (reducing extension). At intermediate forces (e.g. 1.2 pN), the extension of the negative supercoiled DNA was insensitive to changes in the molecule's linking number; supercoils still formed for positive coiling while local denaturation absorbed the torsional stress for negative σ. Finally, in the high force regime (>3 pN), no plectonemes were observed for either negatively or positively supercoiled DNA. The positively supercoiled DNA underwent a transition to a new phase called P-DNA (Allemand *et al.*, 1998). Chemical reactivity studies and numerical simulations suggest that in this conformation the phosphate-sugar backbone is winding inside the structure, while the bases are exposed in solution. A structure similar to P-DNA (Pauling-DNA) presumably exists in nature for the DNA of a specific virus, Pf1, where the unusual conformation is stabilized by viral coat proteins.

As detailed in Strick *et al.* (2000a), measuring the work done while stretching negatively or positively supercoiled DNA of equal superhelical density, allows to evaluate the elastic torsional persistence length of the molecule (Figure 7b). The shaded area between the two curves represents the work difference Δ: plotting the square root for the work difference vs. the number of turns the molecule is over- or under-wound gives a straight line whose slope can be used for extracting the torsional constant. The estimate gave 86 $\pm$ 10 nm, in reasonable agreement with the current and very imprecise estimate of 75 $\pm$ 25 nm.

Chromatin fiber manipulation

Chromatin fiber assembly under applied force

Chromatin assembly *in vivo* takes place massively during DNA replication; nucleosomes have to assemble in the wake of the transcriptional machinery, too, since the transcribing RNA polymerase has to remove the nucleosomes out of its way to be able to read the genetic message. The naked DNA stretches are expected to quickly reform chromatin, so that the roles of chromatin in both compacting the DNA and regulating its functions are restored. This re-formation of nucleosomes in the wake of RNA polymerase (and other DNA-tracking enzymes as well) takes place while the DNA molecule is still under tension as a result of the pulling exerted by the stationary RNA polymerase (Cook, 1999) on the transcribed DNA. Both RNA and DNA polymerases have been shown to be amongst the strongest molecular motors, developing forces of up to 30–40 pN (Wang *et al.*, 1998; Wuite *et al.*, 2000). If the forces measured *in vitro* are also developed *in vivo*, then the question arises whether the DNA under tension can be assembled into nucleosomes and what the force dependence of the assembly process is. Three groups have approached this issue at the single fiber level, using three different approaches.

Viovy and coworkers used video microscopy to follow chromatin assembly on a single λ-DNA molecule (48.5 kbp, 16.4 μm contour length) attached at one end to a glass surface, the other end being free so that it could be stretched by flow (Figure 8a) (Ladoux *et al.*, 2000). Chromatin assembly was achieved by flowing in of *Xenopus* or *Drosophila* cell-free extracts and observed by real-time fluorescence microscopy (DNA was fluorescently labeled by intercalation of YOYO-1). There was a clear dependence of the rate of chromatin assembly on the shear rate, and hence on the force applied to the DNA. Assembly could proceed, albeit at a much-reduced rate, up to forces of 12 pN.

Bennink *et al.* (2001b) used a similar biological system (λ-DNA assembled by *Xenopus* egg extract) in an OT set-up. Initially, a single DNA molecule was suspended between two micron-sized beads, one held by a pipette, and the other one in an optical trap (Figure 8b). Since the presence of cell debris in the extract precluded the use of the optical trap for force measurements, the trap was switched off during the assembly and forces were estimated using either the parameters of the laminar flow (Stokes law) or by measuring the Brownian motion of the freely suspended bead. The addition of the extract led to visible shortening of the distance between the two beads, reflecting chromatin formation. The kinetics was strongly dependent on the applied force, with complete inhibition of assembly at forces exceeding 10 pN.

More recently, MT have been used to approach the same issue (Leuba/Zlatanova, unpublished). A simple horizontal set-up was constructed, in which a single permanent magnet is placed at an angle with respect to the path of the light beam; this placement allows easy observation of the movement of the magnetic bead across the video screen. A λ-DNA molecule, attached to the inner surface of the cuvette and the bead, was assembled with purified histone octamers and nucleosome assembly factor 1 (NAP-1) (Figure 8c). The instrument allows recording assembly curves under constant force; in addition, it is possible to change the force several times during a single round of assembly in a stepwise manner. The overall dependence of the assembly rate on the tension applied to the DNA molecule agrees with the results of Ladoux *et al.* (2000) and Bennink *et al.* (2001b). The experiments in which the force was changed during individual assembly reactions allowed the first-ever assessment of the speed with which the system responded to changing forces: the response was instantaneous. This observation provides an unprecedented insight of how changes in the *in vivo* rate of transcription (and hence in the tension experienced by the DNA being threaded through the RNA polymerase) may regulate the rate of reformation of nucleosomes in the wake of the polymerase. Finally, the data provided the first real-time demonstration of the dynamic equilibrium between an assembled and a disassembled state of individual nucleosomes in the fiber context; such equilibrium has been previously suggested on the basis of a few biochemical experiments

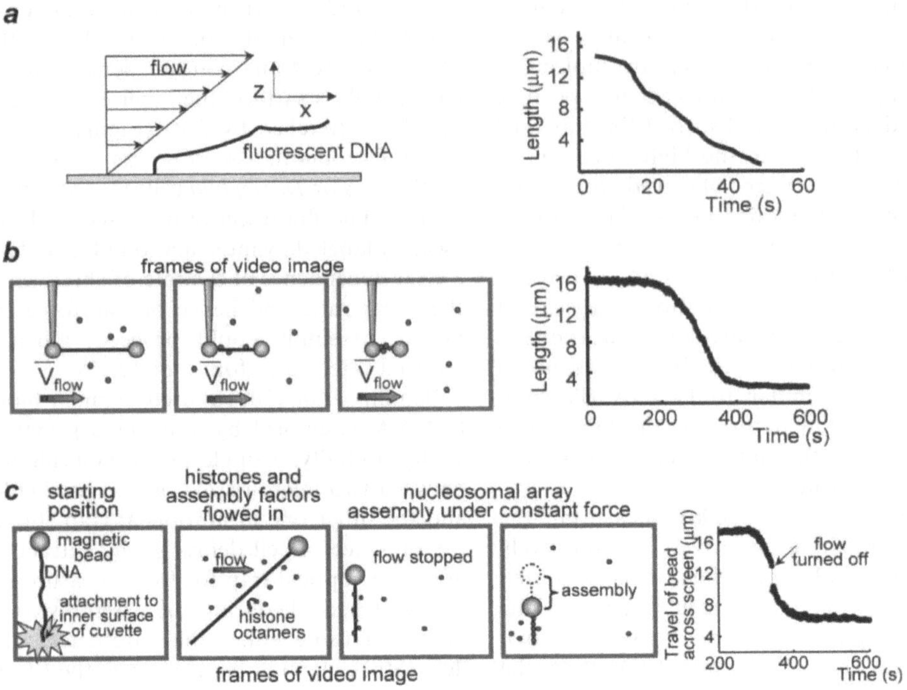

Fig. 8. Schematics of methods applied to assemble single chromatin fibers using (a) flow, (b) optical tweezer/flow, and (c) magnetic tweezer/flow setups and the resulting graphs of change in length over time. (a) The experiments of Ladoux *et al.* (2000) (see text). (b) The experiments of Bennink *et al.* (2001b) (see text). (c) Magnetic tweezers experiments (Leuba/Zlatanova, unpublished; see text).

performed at the mononucleosomal level, and theory (van Holde, 1988; Widom, 1999).

Chromatin fiber disassembly under applied force: unraveling of individual nucleosomes

Chromatin fiber disassembly under applied force has been successfully studied so far only with optical tweezers.

Attempts to use the AFM for mechanically stretching chromatin fibers (Figure 9a) were discontinued because of an unexpected artifact. The stretching curves of native or reconstituted chromatin fibers exhibited a multi-peak, saw-tooth pattern, similar to the patterns obtained upon stretching of multi-domain proteins like titin (Rief *et al.*, 1997) or tenascin (Oberhauser *et al.*, 1998). A closer look at the distances between individual peaks, however, made it clear that the sought-after unraveling of individual nucleosomes as a result of mechanical stretching of the DNA did not occur, despite the relatively high forces applied (300–600 pN): the jumps in the force curves corresponded instead to removal of successive intact nucleosomes from the glass surface, followed by stretching of the naked DNA between the nucleosomes attached to the tip and the surface.

Three laboratories have reported results on stretching individual chromatin fibers with OT, with consistent overall results (see Figure 9b–d for schematics of the corresponding approaches). Cui and Bustamante (2000) stretched isolated chicken erythrocyte fibers, Bennink *et al.* (2001a) pulled on fibers directly reconstituted in the flow cell from λ-DNA and histones with the help of *Xenopus* extracts, and Michelle Wang and colleagues (Brower-Toland *et al.*, 2002) stretched 208-17 nucleosomal arrays preassembled in bulk by the salt-dialysis method and then suspended between a glass surface and a polystyrene microbead. The first two groups presented data showing that up to ~20 pN the fibers underwent reversible stretching, but applying stretching forces above 20 pN led to irreversible alterations, interpreted in terms of removal of histone octamers from the fibers with recovery of the mechanical properties of naked DNA (Figure 10a–c).

The high speed of data acquisition in Bennink *et al.* (2001a) allowed recording of force curves in which discrete, sudden drops in force could be observed upon fiber stretching, reflecting discrete opening events in fiber structure (Figure 10b and c). These opening events were quantized as increments in fiber lengths of ~65 nm and were attributed to unwrapping of individual nucleosomal particles. The forces to unspool individual nucleosome particles ranged between 20 and 40 pN.

In the more recent paper from Wang's laboratory (Brower-Toland *et al.*, 2002), the force curve recorded at constant pulling rate exhibited 17 peaks, corresponding to successive unraveling of 17 particles in the nucleosomal array. The low rate of stretching [typical rate ~28 nm/s, compare with typical rate ~1 µm/s in Bennink *et al.* (2001a)] presumably allowed a step-wise

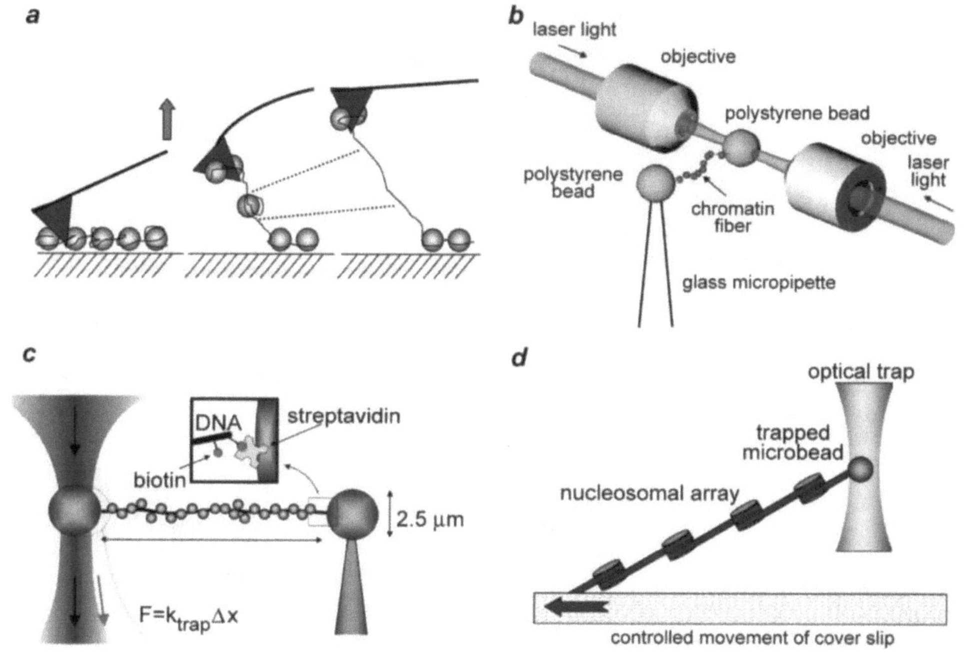

Fig. 9. Schematics of AFM and optical tweezers approaches to unravel single chromatin fibers by force. (a) AFM-mediated unraveling of nucleosomes (conceptual ideas leading to the experiments of Leuba *et al.*, 1999, 2000). An AFM tip is pushed into a chromatin fiber adhering to a glass surface in buffer. When the cantilever is raised, it deflects because of the chromatin tether between the tip and the surface. If a nucleosome unravels, the chromatin tether will lengthen abruptly leading to restoring the normal geometry of the cantilever. (b) Schematic of a chromatin fiber pulled between two beads by dual-beam optical tweezers, redrawn from Cui and Bustamante (2000). One bead is held by suction with a glass micropipette, and the other bead is held in a force-measuring optical trap. Pipette was moved to stretch the chromatin fiber. (c) Schematic of a chromatin fiber pulled between two beads in the single-beam optical tweezers (Bennink, 2000) as used for the λ-chromatin-stretching experiments of Bennink *et al.* (2001a) (see text). (d) Schematic of a single nucleosomal array suspended between a glass coverslip and a polystyrene bead held in an optical trap. Mechanical stretching was achieved by controlled movement of the coverslip (redrawn from Brower-Toland *et al.*, 2002).

release of the DNA from its interactions with the histone core, each step corresponding to breaking of DNA-histone interactions of different chemical stability, as revealed by the crystal structure of the core particle. Brower-Toland *et al.* (2002) suggested that a portion of each nucleosome (approximately one half) unraveled at low forces, while the remainder unraveled at forces exceeding 20 pN. Thus, the opening events in Bennink *et al.* (2001a) correspond to unraveling of the entire particle at once (~65 nm steps), while the discernable opening events of Brower-Toland *et al.* (2002) correspond only to the second phase of the DNA unwrapping from around the histone core (~27 nm). [Please note that the length of the nucleosomal DNA in Bennink *et al.* (2001a) corresponds to more than two full turns of the DNA superhelix around the histone core, probably due to the presence of HMG1/2 in the *Xenopus* extract, whereas the core particles in Brower-Toland *et al.* (2002) contain DNA that is constrained by the histone octamer only, i.e. is shorter. The DNA that unravels during the low-force regime was estimated to be 76 bp, and that which unravels during the high force regime is 80 bp. Thus, the total length of the nucleosomal DNA is ~156 bp, which, although higher that the 146 bp in the crystal structure of the particle, is not unreasonably high, having in mind the relatively high concentrations of ions (van Holde and Zlatanova, 1999)]. Although Brower-Toland's model suggests that both ends of the nucleosomal DNA unravel simultaneously from the particle to stop at the relatively strong contacts at positions +4 and −4 of the DNA superhelix, an alternative stepwise model is conceivable, in which one half of the nucleosome unravels unilaterally, with the strong protein/DNA contacts at the dyad axis serving as a roadblock to a total quick release of the entire DNA from the surface (J. Widom, personal communication).

The authors also provide data on repeated stretching and relaxation of the same fiber: reformation of *some* (by no means all) nucleosomes during the relaxation portion of the cycle was reported (Brower-Toland *et al.*, 2002). This nucleosome re-formation was a function of the force applied during the initial stretching: if the force was below that of the overstretching DNA transition, some nucleosomes reformed. If, however, the initial stretching force approached the B–S transition force (~60 pN, the reversibility was impeded. The authors interpret these results as 'contrasting significantly' with those of Bennink *et al.* (2001a) (Figure 11a). It should be pointed out, however, that the two types of experiments are not directly comparable: the maximum force exerted in the reversibility experiments of Bennink *et al.* (where the second and all further stretchings were taken all the way to the overstretched DNA length) was such that it distorted the double helical DNA structure. This would, of course, lead to complete dissociation of the octamer from the DNA, as actually stated by

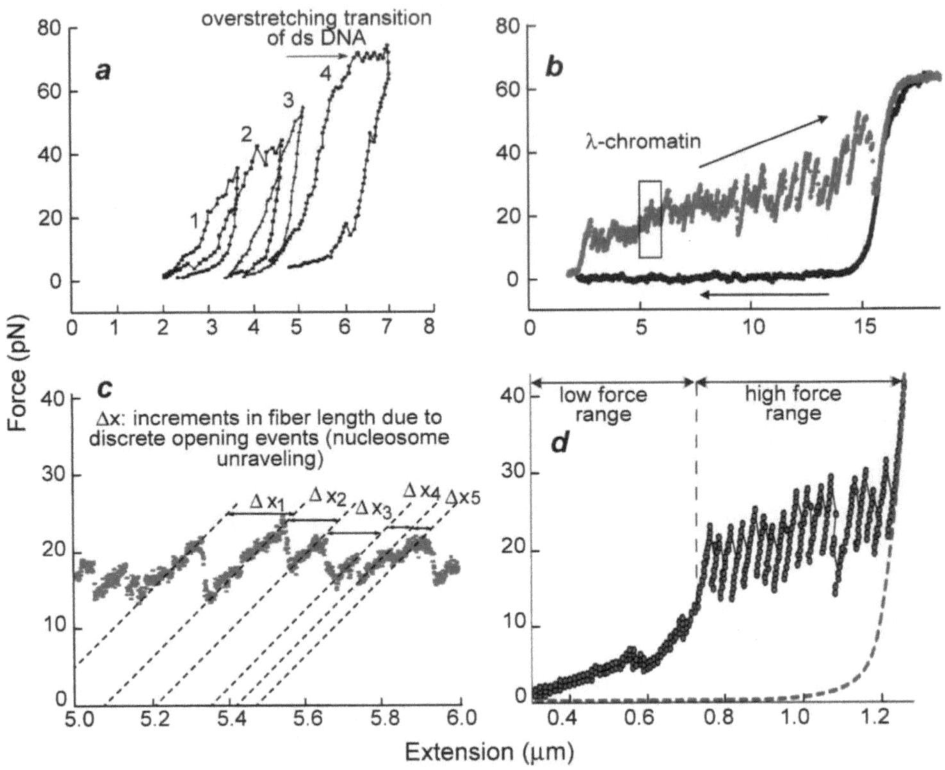

Fig. 10. Force–extension curves of unraveling individual chromatin fibers from three different groups. (a) Force-induced irreversible changes in a chicken erythrocyte chromatin fiber in low ionic strength solution (redrawn from Cui and Bustamante, 2000). The number next to each stretch and release cycle indicates the order of stretching. The fiber gets longer with each cycle. When the fiber is stretched to ~65 pN, a plateau corresponding to the overstretching transition of dsDNA can be observed (4th curve and horizontal arrow). (b) Force–extension curve of a chromatin fiber assembled on a single λ-DNA molecule using a *Xenopus laevis* egg extract (Bennink *et al.*, 2001a). After assembly of the chromatin fiber with the *Xenopus* extract, the extract was replaced with a physiological ionic strength solution. (c) More detailed view of the force signal from 5 to 6 μm of the stretch curve depicted in (b) (see boxed portion of curve). Dashed lines through linear portions of the force curve are used to estimate the lengths of the stretched intermediates before the abrupt drops in the force, which represent the unraveling of nucleosomes. Length increments are indicated as Δx_n. (d) Force–extension curve of a 3684 bp nucleosomal array containing 17 positioned nucleosomes (redrawn from Brower-Toland *et al.*, 2002). At a force >15 pN, a sawtooth pattern with 17 peaks was observed with a uniform distance of ~27 nm, reflecting partial unraveling of each nucleosome (see text).

Brower-Toland *et al.* (2002) themselves. As far as the lack of reversibility following the very first stretch in Bennink *et al.* (2001a) is concerned, it may be attributed to the high rate of pulling (~1 μm/s, as contrasted to the much lower pulling rate of ~28 nm/s in the experiments of Brower-Toland *et al.*, 2002). Thus the explanation for the contrasting results of Bennink *et al.* (2001a) lies in the way the two sets of experiments were performed: they cannot be directly compared.

Thus, all three papers on stretching chromatin fibers with OT estimate that nucleosomes are unraveled at forces exceeding 20 pN. Interestingly, these forces are in the same range as the stall forces developed by RNA and DNA polymerases (Wang *et al.*, 1998; Wuite *et al.*, 2000), the enzymes that encounter nucleosomes while reading the information in the DNA. This may mean that the polymerases are by themselves capable of removing the nucleosomes out of their way.

Finally, Figure 12 presents a direct comparison of the stretching curves in the low-force regime for naked DNA (data from two laboratories) and for chromatin. Undoubtedly, the force–extension curves, and hence the elastic properties for the two structures are quite different, as expected. The figure also presents some mechanical parameters for DNA and chromatin, as derived from these curves.

Concluding remarks

The brief overview of the DNA and chromatin single-molecule work presented above reveals the power of the new methodology in not only exploring the mechanical properties of these molecules in real time, but in revealing new features in their behavior that remained masked in population experiments. It has also become clear that there is a large variability in the behavior of individual members of the respective populations of molecules. The set of methods described will be undoubtedly enriched by the addition of other single-molecule approaches that have already been applied in other research areas, like single-molecule fluorescence and single-pair fluorescence resonance energy transfer. New approaches are also expected to be developed. The future seems to be

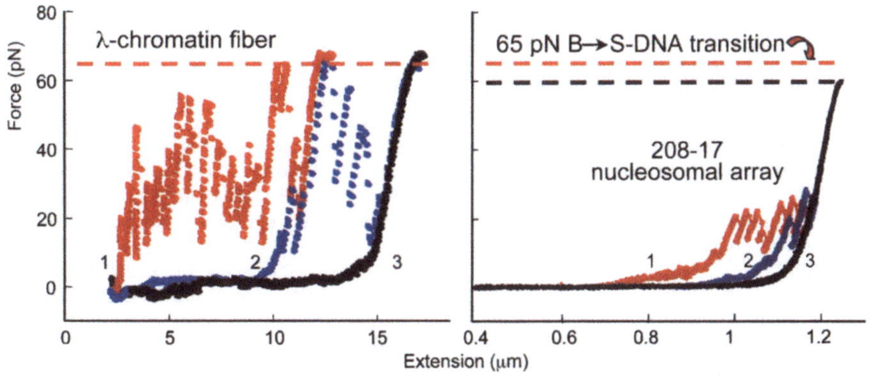

Fig. 11. Comparison of reversibility in chromatin stretch curves from two different groups. (a) Successive stretches on the same λ-chromatin fiber (data from Bennink *et al.*, 2001a). Numbers next to each stretch curve indicate the order of stretching. The first stretch was taken to an extension of ~11 μm and subsequent stretches were to the fully stretched contour length of the DNA molecule (λ-DNA, 48,502 bp, 16.4 μm contour length). The second stretch removed nucleosomes remaining after the first partial stretch. Forces over 60 pN were reached in each stretch. (b) Successive stretches on the same nucleosomal array (redrawn from Brower-Toland *et al.*, 2002). Numbers next to each stretch curve indicate the order of stretching. The first curve was stretched to a maximum force of 50 pN and in subsequent stretches the force was increased to 60 pN.

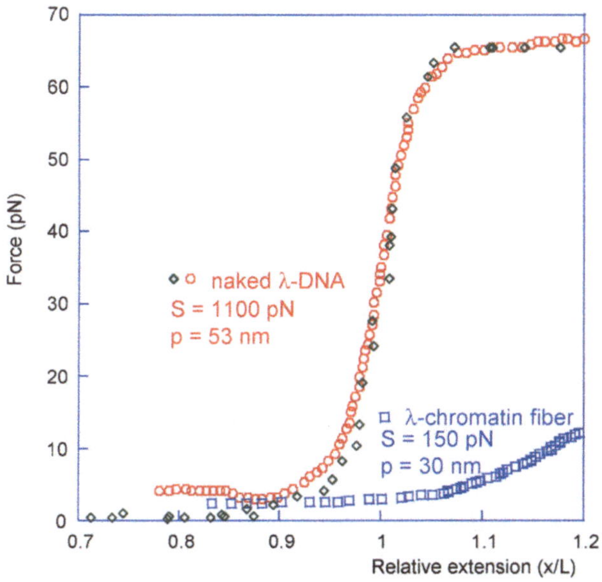

Fig. 12. Comparison of naked λ-DNA and λ-chromatin fiber stretch curves. (a) Naked DNA force–relative extension curve redrawn from Bustamante *et al.* (2000) (diamonds) or plotted from the data of Bennink *et al.* (2001a) (circles) using a contour length of 16 μm. The force–relative extension curve of the λ-chromatin fiber (squares) was calculated from the data of the completely reversible portion of the stretch curve in the low force regime (see 2–3 μm portion in Figure 10b) using a contour length of 2.3 μm (determined as the *x*-intercept of a linear fit to the data). The numerical values for the stretch modulus (*S*) and the persistence length (*P*) for both naked λ-DNA and λ-chromatin fiber are listed in the graph.

challenging and bright for the new generation of scientists interested in understanding the intricate behavior of individual biological macromolecules.

Acknowledgements

We would like to thank Drs M. Allen, H. Hansma, S. Lindsay, Y. Lyubchenko, C. Rivetti, B. Samori, Z. Shao, and N. Seeman for AFM images, and M. Karymov and M. Tomschik for assistance with figures. Our research has been supported by a NCI K22 grant (SHL), University of Pittsburgh School of Medicine startup funds (SHL), and Polytechnic University startup funds (JZ).

References

Allemand JF, Bensimon D, Lavery R and Croquette V (1998) Stretched and overwound DNA forms a Pauling-like structure with exposed bases. *Proc Natl Acad Sci USA* **95:** 14,152–14,157.

Allen MJ, Dong XF, O'Neill TE, Yau P, Kowalczykowski SC, Gatewood J, Balhorn R and Bradbury EM (1993) Atomic force microscope measurements of nucleosome cores assembled along defined DNA sequences. *Biochemistry* **32:** 8390–8396.

An W, Palhan VB, Karymov MA, Leuba SH and Roeder RG (2002) Selective requirements for histone H3 and H4 N termini in p300-dependent transcriptional activation from chromatin. *Mol Cell* **9:** 811–821.

Ashkin A (1997) Optical trapping and manipulation of neutral particles using lasers. *Proc Natl Acad Sci USA* **94:** 4853–4860.

Bennink ML (2000) Force spectroscopy of single DNA–protein complexes. PhD thesis (p. 24) University of Twente, Enschede, The Netherlands.

Bennink ML, Leuba SH, Leno GH, Zlatanova J, de Grooth BG and Greve J (2001a) Unfolding individual nucleosomes by stretching single chromatin fibers with optical tweezers. *Nat Struct Biol* **8:** 606–610.

Bennink ML, Pope LH, Leuba SH, de Grooth BG and Greve J (2001b) Single chromatin fibre assembly using optical tweezers. *Single Mol* **2:** 91–97.

Bensimon D (1996) Force: a new structural control parameter? *Structure* **4:** 885–889.

Bensimon A, Simon A, Chiffaudel A, Croquette V, Heslot F and Bensimon D (1994) Alignment and sensitive detection of DNA by a moving interface. *Science* **265:** 2096–2098.

Bensimon D, Simon AJ, Croquette V and Bensimon A (1995) Stretching DNA with a receding meniscus – experiments and models. *Phys Rev Lett* **74:** 4754–4757.

Bockelmann U, Thomen P, Essevaz-Roulet B, Viasnoff V and Heslot F (2002) Unzipping DNA with optical tweezers: high sequence sensitivity and force flips. *Biophys J* **82:** 1537–1553.

Brower-Toland BD, Smith CL, Yeh RC, Lis JT, Peterson CL and Wang MD (2002) Mechanical disruption of individual

nucleosomes reveals a reversible multistage release of DNA. *Proc Natl Acad Sci USA* **99:** 1960–1965.
Bustamante C, Keller D and Yang G (1993) Scanning force microscopy of nucleic acids and nucleoprotein assemblies. *Curr Opin Struct Biol* **3:** 363–372.
Bustamante C, Smith SB, Liphardt J and Smith D (2000) Single-molecule studies of DNA mechanics. *Curr Opin Struct Biol* **10:** 279–285.
Bustamante C, Vesenka J, Tang CL, Rees W, Guthold M and Keller R (1992) Circular DNA molecules imaged in air by scanning force microscopy. *Biochemistry* **31:** 22–26.
Cluzel P, Lebrun A, Heller C, Lavery R, Viovy JL, Chatenay D and Caron F (1996) DNA: an extensible molecule. *Science* **271:** 792–794.
Cook PR (1999) The organization of replication and transcription. *Science* **284:** 1790–1795.
Cui Y and Bustamante C (2000) Pulling a single chromatin fiber reveals the forces that maintain its higher-order structure. *Proc Natl Acad Sci USA* **97:** 127–132.
d'Erme M, Yang G, Sheagly E, Palitti F and Bustamante C (2001) Effect of poly(ADP-ribosyl)ation and Mg^{2+} ions on chromatin structure revealed by scanning force microscopy. *Biochemistry* **40:** 10,947–10,955.
Dunker AK, Lawson JD, Brown CJ, Williams RM, Romero P, Oh JS, Oldfield CJ, Campen, AM, Ratliff CM, Hipps KW, Ausio J, Nissen MS, Reeves R, Kang C, Kissinger CR, Bailey RW, Griswold MD, Chiu W, Garner EC and Obradovic Z (2001) Intrinsically disordered protein. *J Mol Graph Model* **19:** 26–59.
Essevaz-Roulet B, Bockelmann U and Heslot F (1997) Mechanical separation of the complementary strands of DNA. *Proc Natl Acad Sci USA* **94:** 11,935–11,940.
Fang Y and Hoh JH (1998) Early intermediates in spermidine-induced DNA condensation on the surface of mica. *J Am Chem Soc* **120:** 8903–8909.
Fang Y, Spisz TS and Hoh JH (1999) Ethanol-induced structural transitions of DNA on mica. *Nucleic Acids Res* **27:** 1943–1949.
Fritzsche W and Henderson E (1996) Scanning force microscopy reveals ellipsoid shape of chicken erythrocyte nucleosomes. *Biophys J* **71:** 2222–2226.
Fritzsche W and Henderson E (1997) Chicken erythrocyte nucleosomes have a defined orientation along the linker DNA–a scanning force microscopy study. *Scanning* **19:** 42–47.
Fritzsche W, Schaper A and Jovin TM (1994) Probing chromatin with the scanning force microscope. *Chromosoma* **103:** 231–236.
Fritzsche W, Schaper A and Jovin TM (1995) Scanning force microscopy of chromatin fibers in air and in liquid. *Scanning* **17:** 148–155.
Gosse C and Croquette V (2002) Magnetic tweezers: micromanipulation and force measurement at the molecular level. *Biophys J* **82:** 3314–3329.
Grandbois M, Beyer M, Rief M, Clausen-Schaumann H and Gaub HE (1999) How strong is a covalent bond? *Science* **283:** 1727–1730.
Hafner JH, Cheung CL, Woolley AT and Lieber CM (2001) Structural and functional imaging with carbon nanotube AFM probes. *Prog Biophys Mol Biol* **77:** 73–110.
Han W, Dlakic M, Zhu YJ, Lindsay SM and Harrington RE (1997) Strained DNA is kinked by low concentrations of Zn^{2+}. *Proc Natl Acad Sci USA* **94:** 10,565–10,570.
Han W, Mou J, Sheng J, Yang J and Shao Z (1995) Cryo atomic force microscopy: a new approach for biological imaging at high resolution. *Biochemistry* **34:** 8215–8220.
Hansma HG (2001) Surface biology of DNA by atomic force microscopy. *Ann Rev Phys Chem* **52:** 71–92.
Hansma HG and Laney DE (1996) DNA binding to mica correlates with cationic radius: assay by atomic force microscopy. *Biophys J* **70:** 1933–1939.
Hansma HG, Golan R, Hsieh W, Lollo CP, Mullen-Ley P and Kwoh D (1998) DNA condensation for gene therapy as monitored by atomic force microscopy. *Nucleic Acids Res* **26:** 2481–2487.
Hansma HG, Laney DE, Bezanilla M, Sinsheimer RL and Hansma PK (1995) Applications for atomic force microscopy of DNA. *Biophys J* **68:** 1672–1677.
Hansma HG, Revenko I, Kim K and Laney DE (1996) Atomic force microscopy of long and short double-stranded, single-stranded and triple-stranded nucleic acids. *Nucleic Acids Res* **24:** 713–720.
Harada Y, Ohara O, Takatsuki A, Itoh H, Shimamoto N and Kinosita K Jr (2001) Direct observation of DNA rotation during transcription by *Escherichia coli* RNA polymerase. *Nature* **409:** 113–115.
Karymov MA, Tomschik M, Leuba SH, Caiafa P and Zlatanova J (2001) DNA methylation-dependent chromatin fiber compaction *in vivo* and *in vitro*: requirement for linker histone. *FASEB J* **15:** 2631–2641.
Kishino A and Yanagida T (1988) Force measurements by micromanipulation of a single actin filament by glass needles. *Nature* **334:** 74–76.
Konrad MW and Bolonick JI (1996) Molecular dynamics simulation of DNA stretching is consistent with the tension observed for extension and strand separation and predicts a novel ladder structure. *J Am Chem Soc* **118:** 10,989–10,994.
Kosikov KM, Gorin AA, Zhurkin VB and Olson WK (1999) DNA stretching and compression: large-scale simulations of double helical structures. *J Mol Biol* **289:** 1301–1326.
Ladoux B, Quivy JP, Doyle P, du Roure O, Almouzni G and Viovy JL (2000) Fast kinetics of chromatin assembly revealed by single-molecule videomicroscopy and scanning force microscopy. *Proc Natl Acad Sci USA* **97:** 14,251–14,256.
Lebrun A and Lavery R (1996) Modelling extreme stretching of DNA. *Nucleic Acids Res* **24:** 2260–2267.
Lee GU, Chrisey LA and Colton RJ (1994) Direct measurement of the forces between complementary strands of DNA. *Science* **266:** 771–773.
Leger JF, Robert J, Bourdieu L, Chatenay D and Marko JF (1998) RecA binding to a single double-stranded DNA molecule: a possible role of DNA conformational fluctuations. *Proc Natl Acad Sci USA* **95:** 12,295–12,299.
Leger JF, Romano G, Sarkar A, Robert J, Bourdieu L, Chatenay D and Marko JF (1999) Structural transitions on a twisted and stretched DNA molecule. *Phys Rev Lett* **83:** 1066–1069.
Leuba SH, Bustamante C, van Holde K and Zlatanova J (1998a) Linker histone tails and N-tails of histone H3 are redundant: scanning force microscopy studies of reconstituted fibers. *Biophys J* **74:** 2830–2839.
Leuba SH, Bustamante C, Zlatanova J and van Holde K (1998b) Contributions of linker histones and histone H3 to chromatin structure: scanning force microscopy studies on trypsinized fibers. *Biophys J* **74:** 2823–2829.
Leuba SH, Karymov MA, Liu Y, Lindsay SM and Zlatanova J (1999) Mechanically stretching single chromatin fibers. *Gene Ther Mol Biol* **4:** 297–301.
Leuba SH, Yang G, Robert C, Samori B, van Holde K, Zlatanova J and Bustamante C (1994) Three-dimensional structure of extended chromatin fibers as revealed by tapping-mode scanning force microscopy. *Proc Natl Acad Sci USA* **91:** 11,621–11,625.
Leuba SH, Zlatanova J, Karymov MA, Bash R, Liu Y.-Z, Lohr D, Harrington RE and Lindsay SM (2000) The mechanical properties of single chromatin fibers under tension. *Single Mol* **1:** 185–192.
Lindsay SM (2000) The scanning probe microscope in biology. In: Bonnell D (ed.) *Scanning Probe Microscopy and Spectroscopy: Theory, Techniques, and Applications*, 2nd edn. Wiley, New York.
Lyubchenko YL and Shlyakhtenko LS (1997) Visualization of supercoiled DNA with atomic force microscopy *in situ*. *Proc Natl Acad Sci USA* **94:** 496–501.
Mao C, Sun W and Seeman NC (1999) Designed two-dimensional DNA Holliday junction arrays visualized by atomic force microscopy. *J Am Chem Soc* **121:** 5437–5443.
Martin LD, Vesenka JP, Henderson E and Dobbs DL (1995) Visualization of nucleosomal substructure in native chromatin by atomic force microscopy. *Biochemistry* **34:** 4610–4616.

Mou J, Czajkowsky DM, Zhang Y and Shao Z (1995) High-resolution atomic-force microscopy of DNA: the pitch of the double helix. *FEBS Lett* **371:** 279–282.

Noy A, Vezenov DV, Kayyem JF, Meade TJ and Lieber CM (1997) Stretching and breaking duplex DNA by chemical force microscopy. *Chem Biol* **4:** 519–527.

Oberhauser AF, Marszalek PE, Erickson HP and Fernandez JM (1998) The molecular elasticity of the extracellular matrix protein tenascin. *Nature* **393:** 181–185.

Rief M, Clausen-Schaumann H and Gaub HE (1999) Sequence-dependent mechanics of single DNA molecules. *Nat Struct Biol* **6:** 346–349.

Rief M, Gautel M, Oesterhelt F, Fernandez JM and Gaub HE (1997) Reversible unfolding of individual titin immunoglobulin domains by AFM. *Science* **276:** 1109–1112.

Rivetti C, Walker C and Bustamante C (1998) Polymer chain statistics and conformational analysis of DNA molecules with bends or sections of different flexibility. *J Mol Biol* **280:** 41–59.

Samori B, Nigro C, Armentano V, Cimieri S, Zuccheri G and Quagliariello C (1993) DNA supercoiling imaged in 3 dimensions by scanning force microscopy. *Angew Chem Int Edit* **32:** 1461–1463.

Sato MH, Ura K, Hohmura KI, Tokumasu F, Yoshimura SH, Hanaoka F and Takeyasu K (1999) Atomic force microscopy sees nucleosome positioning and histone H1-induced compaction in reconstituted chromatin. *FEBS Lett* **452:** 267–271.

Schnitzler GR, Cheung CL, Hafner JH, Saurin AJ, Kingston RE and Lieber CM (2001) Direct imaging of human SWI/SNF-remodeled mono- and polynucleosomes by atomic force microscopy employing carbon nanotube tips. *Mol Cell Biol* **21:** 8504–8511.

Shao Z (1999) Probing Nanometer Structures with Atomic Force Microscopy. *News Physiol Sci* **14:** 142–149.

Shlyakhtenko LS, Potaman VN, Sinden RR and Lyubchenko YL (1998) Structure and dynamics of supercoil-stabilized DNA cruciforms. *J Mol Biol* **280:** 61–72.

Simpson RT, Thoma F and Brubaker JM (1985) Chromatin reconstituted from tandemly repeated cloned DNA fragments and core histones: a model system for study of higher order structure. *Cell* **42:** 799–808.

Smith SB, Cui Y and Bustamante C (1996) Overstretching B-DNA: the elastic response of individual double- stranded and single-stranded DNA molecules. *Science* **271:** 795–799.

Smith SB, Finzi L and Bustamante C (1992) Direct mechanical measurements of the elasticity of single DNA molecules by using magnetic beads. *Science* **258:** 1122–1126.

Strick TR, Allemand JF, Bensimon D, Bensimon A and Croquette V (1996) The elasticity of a single supercoiled DNA molecule. *Science* **271:** 1835–1837.

Strick TR, Allemand JF, Bensimon D and Croquette V (2000b) Stress-induced structural transitions in DNA and proteins. *Annu Rev Biophys Biomol Struct* **29:** 523–543.

Strick TR, Allemand J-F, Croquette V and Bensimon D (1998) Physical approaches to the study of DNA. *J Stat Phys* **93:** 647–672.

Strick T, Allemand J, Croquette V and Bensimon D (2000a) Twisting and stretching single DNA molecules. *Prog Biophys Mol Biol* **74:** 115–140.

Thundat T, Allison DP and Warmack RJ (1994) Stretched DNA structures observed with atomic force microscopy. *Nucleic Acids Res* **22:** 4224–4228.

Tomschik M, Karymov MA, Zlatanova J and Leuba SH (2001) The archaeal histone-fold protein HMf organizes DNA into bona fide chromatin fibers. *Structure* **9:** 1201–1211.

van Holde KE (1988) *Chromatin.* Springer-Verlag, New York.

van Holde K and Zlatanova J (1999) The nucleosome core particle: does it have structural and physiologic relevance? *Bioessays* **21:** 776–780.

Vesenka J, Guthold M, Tang CL, Keller D, Delaine E and Bustamante C (1992a) Substrate preparation for reliable imaging of DNA molecules with the scanning force microscope. *Ultramicroscopy* **42–44:** 1243–1249.

Vesenka J, Hansma H, Siegerist C, Siligardi G, Schabtach E and Bustamante C (1992b) Scanning force microscopy of circular DNA and chromatin in air and propanol. *SPIE* **1639:** 127–137.

Wang MD, Schnitzer MJ, Yin H, Landick R, Gelles J and Block SM (1998) Force and velocity measured for single molecules of RNA polymerase. *Science* **282:** 902–907.

Wenner JR, Williams MC, Rouzina I and Bloomfield VA (2002) Salt dependence of the elasticity and overstretching transition of single DNA molecules. *Biophys J* **82:** 3160–3169.

Widom J (1999) Equilibrium and dynamic nucleosome stability. In: Becker PB (ed.) *Methods Mol Biol* (pp. 61–77) Humana Press, Totowa, New Jersey.

Winfree E, Liu F, Wenzler LA and Seeman NC (1998) Design and self-assembly of two-dimensional DNA crystals. *Nature* **394:** 539–544.

Wuite GJ, Smith SB, Young M, Keller D and Bustamante C (2000) Single-molecule studies of the effect of template tension on T7 DNA polymerase activity. *Nature* **404:** 103–106.

Yang G, Leuba SH, Bustamante C, Zlatanova J and van Holde K (1994) Role of linker histones in extended chromatin fibre structure. *Nat Struct Biol* **1:** 761–763.

Zhao H, Zhang Y, Zhang SB, Jiang C, He QY, Li MQ and Qian RL (1999) The structure of the nucleosome core particle of chromatin in chicken erythrocytes visualized by using atomic force microscopy. *Cell Res* **9:** 255–260.

Ziemann F, Radler J and Sackmann E (1994) Local measurements of viscoelastic moduli of entangled actin networks using an oscillating magnetic bead micro-rheometer. *Biophys J* **66:** 2210–2216.

Zlatanova J and Leuba SH (in press) Chromatin structure and dynamics: lessons from single molecule approaches. In: Zlatanova J and Leuba SH (eds.) *Chromatin Structure and Dynamics: State-of-the-Art.* Elsevier, Amsterdam.

Zlatanova J, Leuba SH, Yang G, Bustamante C and van Holde K (1994) Linker DNA accessibility in chromatin fibers of different conformations: a reevaluation. *Proc Natl Acad Sci USA* **91:** 5277–5280.

Zlatanova J, Lindsay SM and Leuba SH (2000) Single molecule force spectroscopy in biology using the atomic force microscope. *Prog Biophys Mol Biol* **74:** 37–61.

Journal of Muscle Research and Cell Motility **23**: 397–407, 2002.

Optical tweezers stretching of chromatin

LISA H. POPE*, MARTIN L. BENNINK and JAN GREVE
Biophysical Techniques, Department of Applied Physics and MESA$^+$ Research Institute, University of Twente, PO Box 217, 7500 AE Enschede, The Netherlands

Abstract

Recently significant success has emerged from exciting research involving chromatin stretching using optical tweezers. These experiments, in which a single chromatin fibre is attached by one end to a micron-sized bead held in an optical trap and to a solid surface or second bead via the other end, allows manipulation and force detection at a single-molecule level. Through force-induced stretching of chromatin, mechanical properties, specific inter-molecular bond strengths and DNA–protein association and dissociation kinetics have been determined. These studies will be extremely fruitful in terms of understanding the function of chromatin structure and its dynamics within the cell.

Introduction

Chromatin structure

The eukaryotic genome is packaged inside the cell nucleus in the form of a compact DNA–protein structure – chromatin. Within a human cell, ~2 m of the genetic code, has to undergo an enormous compaction in length. In its most condensed state, this compaction is of the order 10^5. So how is this formidable task actually accomplished? The first compaction level is through the formation of the nucleosome core particle (NCP) (Figure 1a, b). Each NCP occurs on average every 200 base pairs (bp) along the DNA template, and along with a variable length of linker DNA, is referred to as a nucleosome. Higher levels of compaction are thought to involve binding of linker histones and inter-nucleosomal interactions that are mediated to some extent via histone tails that protrude from the NCP; however much less is known about this compaction level. The ability of linker histone to bind the entry and exit DNA segments of the NCP obviously has some influence over the relative positions of adjacent nucleosomes. Despite earlier proposed models for the more compact '30 nm chromatin fibre' (Finch and Klug, 1976; Widom and Klug, 1985), recent cryoelectron microscopy (cryo-EM) (Horowitz *et al.*, 1994; Bednar *et al.*, 1995, 1998) and AFM studies (Leuba *et al.*, 1994, 1998a; Bustamante *et al.*, 1997) have revealed an irregular zig–zag configuration (Figure 1d) in which the extent of compaction is sensitive to salt conditions (Zlatanova *et al.*, 1994, 1998; Horowitz *et al.*, 1997). Interactions with the histone core probably impose limitations on the motions of nucleosome-bound DNA, hence the free linker along with the linker histone interactions are likely to govern the torsional positioning of successive nucleosomes. Our current picture of chromatin structure has emerged following an enormous amount of research undertaken over the past few decades, however remains incomplete. In particular if we consider cellular requirements we can envisage that this structure must be highly dynamic (Workman and Kingston, 1998; Widom, 1998). Hence current research is now focusing on the role of chromatin dynamics in the regulation of cellular processes.

Structural dynamics

In order to maintain DNA in both an organized and compact state the nucleosome must be a relatively stable unit. So why is chromatin structure dynamic? It is evident that through the course of the cell cycle there is variation in the extent to which access to the genetic code is required. In particular dynamics are crucial for processes such as gene expression, replication, recombination and repair. These processes all rely on the ability of enzymes, regulatory proteins and molecular motors to gain access to the DNA template. So how is chromatin structure regulated in order to maintain cellular function? This is a question that is far from being fully understood; slowly many new mechanisms are being identified that are involved in chromatin remodelling. It is known that DNA–histone interactions and changes in the strength of these interactions play an important role. Post-translational histone modifications such as acetylation, phosphorylation, methylation and ubiquitination have all been identified (van Holde, 1989) and some of which have been correlated to different chromatin states such as transcriptionally active, inactive or condensed (Kornberg and Lorch, 1999). In particular, histone tails that protrude from the NCP appear to be a main target for these modifications and have been shown to be important for the stability of

*To whom correspondence should be addressed: Tel.: +31-53-4893157; Fax: +31-53-4891105; E-mail: l.h.pope@tn.utwente.nl

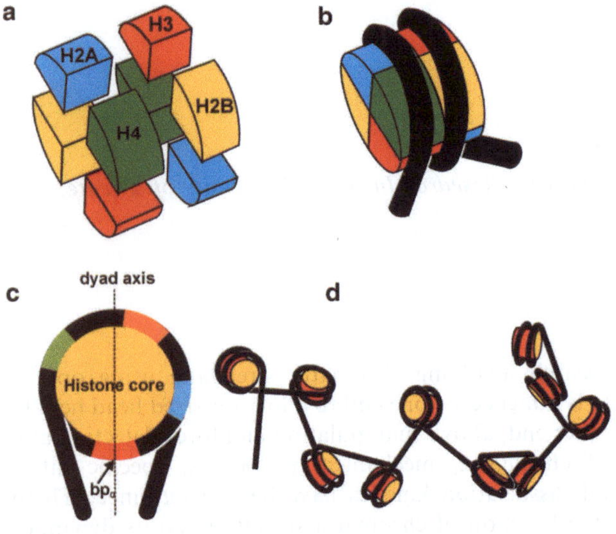

Fig. 1. (a) The NCP comprises of two of each of the four histone proteins, H2A, H2B, H3 and H4, which come together to form a core histone octamer. (b) Around which 146 bp of DNA are wrapped 1.65 times in a left-handed superhelix. (c) For clarity interactions of the DNA with the core histone octamer (Luger and Richmond, 1998) are illustrated only for the first superhelical turn (four of the seven regions), other interactions are symmetrical about the central bp_0. The order in interaction strengths is red > orange > blue > green (each colour represents a region of two consecutive interaction sites, each individual site being at a location where the minor groove of the DNA turns inwards to face the core octamer). In particular at the entry–exit regions the DNA is most weakly bound (green) and the strongest interaction point (red) is at the central DNA region, at the same position as the symmetry dyad. (d) Each NCP along with linker DNA is known as a nucleosome and is repeated along the entire length of the DNA molecule providing the first level of chromatin compaction.

higher order structure (Polach *et al.*, 2000). This has been demonstrated using AFM by the removal of the N-terminal histone tails from both H3 and linker histones, which consequently resulted in the loss of higher order structure (Leuba *et al.*, 1998a, b). Intra-nucleosomal stability, dictated by interactions between the DNA and histone core octamer, is also known to be of importance. For example, Cryo-EM has revealed that the NCP does not necessarily have a fixed 146 bp of DNA wrapped around it, but is in fact dynamic in nature (Furrer *et al.*, 1995). In particular the DNA ends appear to easily peel away from the core, giving variability in the entry–exit angles of the DNA. In fact many experiments have demonstrated that changes in the nature of DNA-histone interactions are central to controlling the amount of DNA that wraps around the histone octamer and that this value can range anywhere between 100 and 170 bp (van Holde and Zlatanova, 1999). This variation is of particular significance since it suggests a mechanism through which access to the DNA template may be easily regulated simply through DNA peeling or sticking. Also of importance is the status of the DNA. For example, it is known that DNA methylation is central to gene silencing (Hsieh and Fire, 2000; Dobosy and Selker, 2001). Besides DNA and histone modifying enzymes, the cell has many more chromatin remodelling systems and DNA binding proteins that regulate chromatin structure (Ramsperger and Stahl, 1995; Workman and Kingston, 1998; Havas *et al.*, 2000).

Although structural changes are obviously important, little is known about the magnitude and origins of the interaction forces that maintain and stabilize these conformations. One way to investigate the nature of these interactions is by studying the response of a single chromatin fibre to an external force. This type of study is of particular relevance since molecular motors, such as polymerases, that copy or transcribe the DNA template, are capable of force-induced remodelling of chromatin structure. Through single-molecule studies it has been shown that these motors can produce forces up to tens of piconewtons (Yin *et al.*, 1995; Wang *et al.*, 1998). It remains to be determined whether these forces alone disrupt nucleosomal structure and provide access to the DNA template, or whether they act in combination with any of the aforementioned interaction-modifying mechanisms that alter chromatin structure. In addition to force, torque is also an important parameter to consider. For example, it has been shown that molecular motors are capable of inducing supercoiling within the DNA template (Liu and Wang, 1987; Harada *et al.*, 2001). It is becoming increasingly evident that single-molecule studies will be central to addressing important questions about chromatin structure and its dynamic behaviour within the living cell, in particular on the mechanisms underlying complex biological processes.

Single-molecule studies of chromatin

One of the first single-molecule chromatin stretching studies involved the micromanipulation of a whole chromosome suspended between two glass micropipettes (Houchmandzadeh *et al.*, 1997). Fascinating results have emerged using this technique; including the force–extension behaviour for mitotic chromosomes when stretched up to 80 times their native length (Poirier *et al.*, 2000) along with the dynamics of force relaxation (Poirier *et al.*, 2001). Studies of chromatin stretching through other single-molecule techniques are few. An attempt was made using AFM force spectroscopy, where a synthetic nucleosomal array was stretched between a cantilever and surface (Leuba *et al.*, 2000). Unfortunately these molecules proved to be stuck to the underlying support, which in turn introduced non-specific interactions into the data. Obviously surface effects such as these need to be resolved if AFM is to fulfil its unique potential of imaging the molecules before and after stretching. This review will focus on insights into chromatin through the stretching of single fibres using optical tweezers. Figure 2 illustrates the optical tweezers technique and shows the experimental configuration used by our own group.

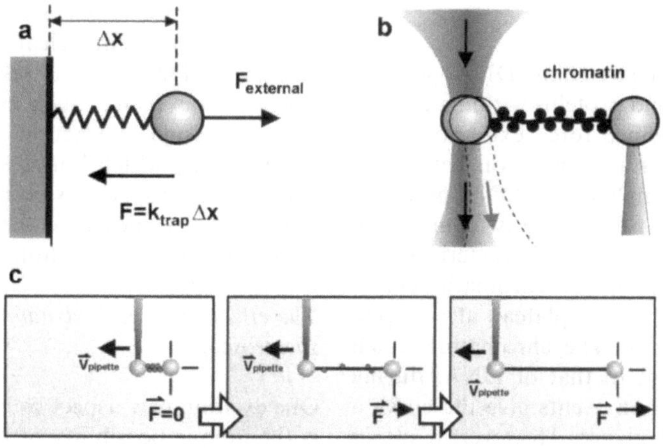

Fig. 2. Stretching single chromatin fibres using optical tweezers. (a) An illustration of the force transducing properties of the optical tweezers. Identical to a Hooke's spring with zero rest length, the force exerted on the trapped bead increases linearly with the displacement of the bead from the centre of the trap. Application of an external force to the trapped bead, results in a bead displacement, Δx, such that this force is exactly balanced with the optical force. (b) When an attachment of the bead is made to a single molecule, using streptavidin–biotin chemistry or similar, the external force is equal to the force within the molecule. The deflection of the laser is used to accurately measure the displacement of the bead with respect to the trap centre, and therefore the tension within the DNA or chromatin fibre. (c) The stretching of a chromatin fibre involves piezo-controlled movement of the non-trapped end of the fibre away from the end held by the trapped bead. This end could be attached to either a second bead held by a micropipette, as shown here, or another solid surface such as a glass coverslip. The force–extension characteristics of this stretching process can then be followed at high temporal resolution.

Force spectroscopy

Molecular stretching is a technique that provides information on the mechanical properties of biomolecules. Experimentally these investigations involve the application of a force to one end of the molecule while the other end remains fixed. The extension of the molecule as a function of stretching force can then be recorded. Already many studies have looked at the stretching of polymer chains, which consist of hundreds to thousands of monomer units, each unit typically having many degrees of freedom. Along with experimental results, theoretical descriptions of polymer stretching based on statistical physics have been developed. These models consider two types of molecular restoring forces under an external load. At small extensions entropic forces dominate. Measurement of these forces provides information about molecular flexibility. Upon further extension the conformational space explored by a molecule is quickly reduced and enthalpic elasticity starts to dominate the mechanical behaviour. At a structural level the intrinsic elasticity is due to an increasing tension within the backbone of the polymer. Under high-external loads, bond deformation, disruption of salt bridges, hydration networks and intra-molecular hydrogen bonds can ultimately induce gross conformational changes. Single-molecule techniques that have been used to study polymer stretching include AFM, optical tweezers, magnetic tweezers and glass microneedles. Before attempting to understand optical tweezers stretching of chromatin fibres, it is helpful to consider the stretching of a DNA molecule.

Stretching single DNA molecules

Since the advent of single-molecule techniques, DNA was an obvious biological polymer to study. Through a number of techniques it was quickly shown that when stretched up to forces of ~10 pN the worm like chain (WLC) model (Bustamante *et al.*, 1994; Marko and Siggia, 1995) could accurately describe its mechanical behaviour. However, when put under tensions greater than 10 pN the molecule extended beyond its contour length as a simple spring, hence a linear increase in length with force was observed (intrinsic elasticity). Interestingly, under quasi-physiological conditions at forces of between 65 and 70 pN, a 'force plateau' was observed, during which the DNA extended to ~1.7 times the length of B-DNA (Cluzel *et al.*, 1996; Smith *et al.*, 1996). The structure adopted at the end of this plateau was termed S-DNA and an experiment in which DNA was untwisted and stretched was used to predict the helical parameters characteristic of this conformation (Léger *et al.*, 1999). Although it was initially thought that this plateau reflects a transition from B-DNA to a stretched double helical structure, more recent interpretations suggest Watson–Crick bp melt into single stranded DNA regions held in close proximity by short oligomeric double stranded stretches (Williams *et al.*, 2002). Through DNA stretching it is rapidly becoming evident that we can measure molecular parameters that could previously only be theoretically estimated. More recently elegant studies using torsionally constrained DNA have revealed insights into the mechanical response of supercoiled DNA to force (Strick *et al.*, 2000).

Stretching chromatin

From current knowledge of stretching DNA, we can examine how additional bound proteins within a chromatin fibre alter these well-defined force–extension characteristics. Figure 3(a) shows a comparison between a force–extension curve of a single naked DNA molecule (inset) and a single chromatin fibre. The two curves are distinctly different, however some characteristics of DNA can still be identified in the chromatin curve. For example, there remains a force plateau at ~65 pN, and the relax curve is similar. The chromatin stretch curve is clearly very different to that of DNA; during stretching a series of disruption events give the curve a noisy appearance at a first glance. However, a closer look at a small portion of the stretch curve reveals very distinct steps (Figure 3b). Furthermore we can consider the generally accepted interpretations of such data. Each step, identified by a gradual rise followed by a sudden sharp drop in force, can be attributed to a DNA–protein unbinding event resulting in partial disruption of the chromatin structure. The disruption force, *F*, for each event can be related to the strength of the interaction that is being broken and the dissociation kinetics of the interaction. The sharp drop in force reflects a release of tension within the fibre due to an increase in contour length, ΔL_0. However, interpretation of this data is not necessarily straightforward. For example, since ΔL_0 for a single disruption event may be very small, the force–extension data describing some of these structural intermediates may be limited to a very small range, which in turn makes an accurate fit with a mechanical model almost impossible. However, if we consider that for adjacent small steps the mechanical properties of the fibre are very similar, we can assume that for a given force the change in extension, Δx, will essentially be ΔL_0. These measurements clearly give an estimate of the amount of DNA that is released per unbinding event. Forces however are much more difficult to interpret; in order to assign interaction strengths, it is necessary to know exactly which bonds are being broken and under what conditions. At this point it is important to consider exactly what is measured through force-induced rupture of biomolecular interactions.

The effect of force – 'dynamic' and 'static' force spectroscopy

One exciting new aspect of measuring interaction forces is the unique possibility of mapping energy landscapes. Considering experiments in which two interacting biomolecules, held together by non-covalent bonds, are forced to rupture; recent theoretical descriptions have attempted to correlate the measured rupture forces to thermodynamic properties. These forces are not a direct measure of the strength of the specific interaction under investigation, but are a measure of the interaction when influenced by an external force. In the unperturbed system, i.e. under zero force conditions, an interaction will rupture given a sufficient amount of time, simply through thermal fluctuations. If we consider that the load applied to the interaction either increases relatively slowly or is relatively small when applied instantaneously, then there is still time for thermal fluctuations to drive the system over an energy barrier separating the bound from the unbound state. Hence a thermally activated process can be studied under the influence of an external force. However, it must be taken into account that the probability of bond rupture increases with both force and time. For a single energy barrier

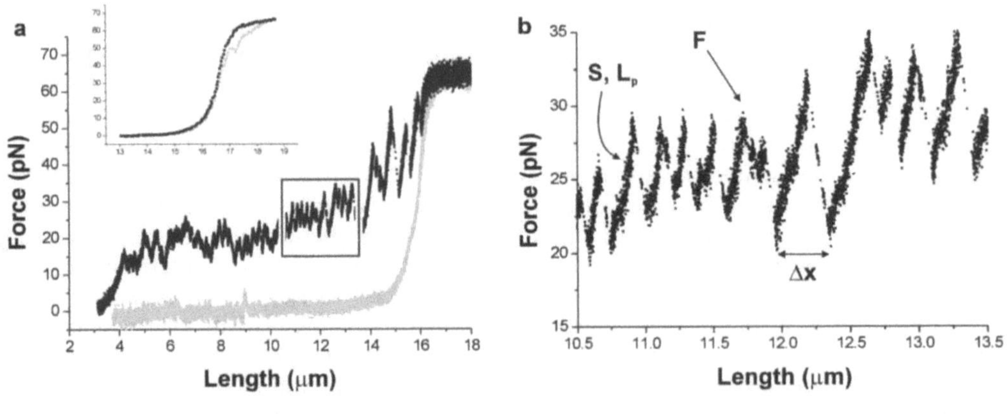

Fig. 3. (a) A comparison between a force–extension curve collected from the same DNA molecule before (inset) and after incubating with *Xenopus laevis* egg extract as described earlier (Bennink *et al.*, 2001a). This can be used to illustrate the general features that are observed as a result of disrupting interactions that hold together a chromatin fibre. The stretch curve is shown in black and the relax curve in grey. (b) An enlargement of a small portion (box in (a)) of the chromatin stretch curve reveals discrete unbinding events, observed here as steps in the data. From each individual step, information can be extracted about the mechanical properties of the fibre, such as contour length (L_0), persistence length (L_p) and stretch modulus (S) (provided that the data covers a sufficient range in both force and extension to allow proper fitting of a polymer-stretching model). Additionally a study of the rupture forces (F), over a large range of loading rates can be related to the energetics and kinetics of the unbinding process. The change in length Δx, measured at a single force value, is essentially the change in contour length, ΔL_0, assuming that the mechanical properties of the fibre only change by a small amount upon a single unbinding event and that the fibre is being stretched in the enthalpic force regime.

along the unbinding pathway, theory predicts a scaling of the measured unbinding force with the logarithm of the loading rate (the rate at which force is increased per unit time, dF/dt) (Evans and Ritchie, 1997). In the case of biomolecular interactions, it can be expected that the unbinding energy landscape is significantly more complex than a single energy barrier, and that there may be many different pathways to dissociation. The instantaneous application of an external force on an interaction can be thought of as a means by which energy is added to the system. This additional energy reduces the height of the energy barrier(s) through tilting the energy landscape and hence governs both the location and size of the barrier(s). Tilting the landscape by a set amount can be achieved using a 'force-clamp' method in single-molecule experiments. This technique ideally instantaneously applies a set force to the interaction under investigation and the lifetime of the interaction is measured. Alternatively, the same information can be obtained if the applied force is increased over time, hence the force is not clamped but rather ramped, and again the lifetime is measured. In this case the rate at which the force increases dictates the rate at which the energy landscape is tilted. This method in which rupture forces are studied over a range of loading rates has been termed dynamic force spectroscopy and has already been used in AFM (Strunz *et al.*, 1999; Pope *et al.*, 2001), optical tweezers (Brower-Toland *et al.*, 2002) and biomembrane force probe studies (Evans and Ritchie, 1999; Merkel *et al.*, 1999) of biomolecular interactions. Furthermore, loading rates spanning many orders of magnitude allow controlled tilting of the energy landscape to study multiple barriers in turn. This technique has revealed the locations and heights of three individual barriers for the avidin–biotin system and two barriers for the streptavidin–biotin system (Merkel *et al.*, 1999). In addition to the heights and locations of energy barriers that are crossed during an unbinding event, kinetic information is also obtained. For an in depth review on this subject see the work of Evans (1998, 2001).

If we consider the optical tweezers experiments described here, where various types of chromatin fibres are disrupted through force, we realize that this data is difficult to interpret. It is known that for a single NCP there are 14 interaction sites, each one at a location where the minor groove of the DNA double helix turns inwards to face the histone core. Following publication of a high-resolution crystal structure of the NCP (Luger *et al.*, 1997) these interaction sites were qualitatively assessed and have been assigned various strengths (Figure 1c) (Luger and Richmond, 1998). It was suggested that interactions are symmetric about the central bp (located at the dyad position), the entry–exit regions of the DNA are weakly bound and the central DNA region is tightly bound. If in addition to DNA-core histone interactions there are also linker histone, non-histone protein and inter-nucleosomal interactions, then for a fibre containing hundreds of NCPs it will be very difficult to assign specific interactions to each unbinding event. Finally it should also be realized that force-induced rupture of DNA–protein interactions that hold together a complex structure such as chromatin do not necessarily reflect the processes that are actually occurring within the cell. It is more likely that force, torque and histone-modifying enzymes, along with other remodelling systems, all act in combination during cellular processes that involve dynamic structural changes within chromatin.

So how can force spectroscopy be used to obtain thermodynamic data regarding interactions within a chromatin fibre? If we consider previous dynamic force spectroscopy experiments it is clear that the systems investigated have largely focused on the study of a single interacting pair. Even for a very simple system, rupture forces have to be collected over exponential scales of loading rate in order to map prominent barriers in the energy landscape. Achieving such a large range in loading rate is by no means trivial. If the stiffness of the system (the force transducer of the technique itself plus any polymer tethers used to apply force) is assumed to be constant then speed can be used to control loading rate (simply the product of the speed, v, and the spring constant, k). However, we know that polymer tethers have a non-linear stiffness, hence true loading rates are often more difficult to determine. Using speed to alter loading rate also has limitations; one can imagine that movement of relatively large objects such as an AFM cantilever or glass micropipette through an aqueous medium will introduce flow and disturb the environment directly surrounding the molecules under investigation. Furthermore, this problem is likely to be more pronounced the faster the pulling speed. Alternatively, static force spectroscopy could be used to study bond rupture. This technique involves the instantaneous application of a given force to a single interaction or series of interactions. Here a linear increase in force would result in the measurement of exponentially shortening lifetimes. This also is difficult to achieve, since lifetimes will range from years to microseconds under linear force in the mapping of energy barriers. In terms of chromatin, the simplest system to start with would be a single NCP. However, with 14 interaction sites where close contacts between the DNA and core histone octamer are made, the way in which a single NCP will dissociate will be by no means easy to measure, or the results easy to interpret. In addition to there being 14 interaction sites, their rupture may also be uncorrelated in time. Hence this presents difficulties in assigning events to specific interactions. It is likely that to study these interactions and accurately assign strengths will require site-directed mutagenesis studies of each of the 14 NCP binding sites. This kind of approach has previously been used to study a single titin immunoglobulin domain (Marszalek *et al.*, 1999). The NCP itself must be understood fully before one can move on to study the effect of linker histone and inter-nucleosomal interactions. However difficult, this area of research is extremely exciting, challenging and no doubt will be the focus of a number of leading groups in the single-molecule field.

Optical tweezers studies of chromatin – a review of published results and discussion

Currently there are three published studies of chromatin stretching using optical tweezers. The first study used various sized fragments of native chromatin, extracted and purified from the nuclei of chicken erythrocytes (Cui and Bustamante, 2000). The experimental configuration was very similar to that shown in Figure 2. Of particular interest is that the fibres were stretched in various salt conditions, 5 mM (low) and 40 and 150 mM (high) NaCl concentrations. The rationale for these experimental conditions stems from the observation that chromatin fibres have a compact structure in high-salt and an extended structure in low-salt (Zlatanova, 1994, 1998). Hence modulation of ionic conditions was used to investigate the mechanical properties of extended vs. compact forms of chromatin. In this study all data was collected at one single loading rate (rate not specified). In contrast, the second study focused on the structural information that could be gained through chromatin disruption (Bennink *et al.*, 2001a). The experimental configuration used is shown in Figure 2. Following the capture of a single DNA molecule between the two beads, a nucleosomal array was reconstituted by replacing the buffer within the flow cell with *Xenopus laevis* egg extract. This extract, known to assemble nucleosomal arrays on naked DNA molecules (Laskey *et al.*, 1977), contained all core histones, but did not contain somatic linker histone. However, other extract proteins such as the high-mobility group proteins (HMG1 and HMG2) and embryonic linker histone B4 were present and known to be capable of locking DNA in a two-turn wrap around the NCP (~60 nm of NCP bound DNA). This methodology also allowed the rate of compaction due to NCP assembly to be followed in real-time using videomicroscopy (Bennink *et al.*, 2001b). Stretching of these reconstituted chromatin fibres was carried out using a single apparent loading rate of 100 pN/s in 150 mM NaCl. The most recent optical tweezers study, investigated well-defined short nucleosomal arrays (Brower-Toland *et al.*, 2002). A DNA sequence was chosen that contained 17 tandem repeats of the sea urchin 5S nucleosome positioning element. Prior to the experiment a salt dialysis method was used to assemble the nucleosomal arrays. Instead of using a glass micropipette, the DNA was attached between the trapped bead and a piezo-controlled glass microscope coverslip. In contrast to the other studies, two different types of disruption experiments were undertaken, whereby either a force clamp or a velocity clamp mode of operation was used. The optical tweezers set-up in this study differed from the previous experimental configurations since it additionally incorporated a feedback enhanced optical trap (Wang *et al.*, 1997). During the velocity clamp mode, modulation of the light intensity allowed continuous change of the trap stiffness in order to maintain a constant position of the bead in the trap. All of these experiments were undertaken in conditions of 100 mM NaCl and 1.5 mM $MgCl_2$. Additionally, in contrast to the other two studies, a dynamic force spectroscopy study was carried out in which loading rate conditions spanned three orders of magnitude. Notably all three studies have used histones that have unknown post-translational modifications.

Insights into chromatin mechanics

From these first studies significant insight has been gained into the mechanical properties of chromatin. In particular a stretch modulus, *S*, described as the force required to extend a polymer to twice its length (Smith *et al.*, 1996), of 5 pN, and a persistence length, L_P, of 30 nm were determined for a single native chromatin fibre in low-salt (Cui and Bustamante, 2000). These values were obtained from the force–extension data in the non-destructive 0–20 pN force regime, where a continuous deformation process was observed that could be described by an extensible WLC model (Wang *et al.*, 1997). The chromatin stretch modulus is notably two orders of magnitude less than that of DNA in similar conditions (~500 pN (Baumann *et al.*, 1997) and ~1000 pN (Wang *et al.*, 1997) suggesting that chromatin has a much weaker intrinsic elasticity). The 30 nm persistence length suggested that chromatin is more flexible than DNA, which was reported as ~75 nm (Baumann *et al.*, 1997) and ~50 nm (Wang *et al.*, 1997) under similar low monovalent salt conditions. In contrast, a stretch modulus for chromatin reconstituted with egg extract was reported as ~150 pN (Bennink *et al.*, 2001a). Although the fibres investigated are known to be different in terms of composition, these values differ by an order of magnitude. This difference clearly requires further investigation and the study of well-defined fibres of different composition may help to shed light on this anomaly. From these reported values it is clear that chromatin stretching cannot be accounted for simply by the stretching of linker DNA, but is likely to reflect the stretching of weak inter-nucleosomal interactions.

Chromatin structure

It is known that chromatin undergoes many structural changes that regulate accessibility to the DNA template. Since changes in ionic conditions are known to alter chromatin structure through changes in compaction, it is interesting to consider the force plateau at ~5 pN observed only for native chromatin in high-salt conditions (Cui and Bustamante, 2000). From both Cryo-EM and AFM there is evidence that chromatin in high-salt conditions adopts a rod-like structure, with small inter-nucleosomal distances and DNA–NCP entry–exit angles, as compared with the low-salt extended irregular zig–zag configuration (Horowitz *et al.*, 1997; Zlatanova *et al.*, 1998). It was suggested that this compaction was possibly due to neutralization of the negatively charged linker DNA and increased inter-nucleosomal interactions. Hence the 5 pN plateau may reflect a transition from the condensed to extended conformational state through the breaking of weak salt-mediated inter-nucleosomal interactions.

Notably, our own work has revealed the potential of studying chromatin disruption in order to investigate structure and interactions central to its dynamic behaviour in greater detail (Bennink *et al.*, 2001a). In contrast to the first study (Cui and Bustamante, 2000), a much higher data acquisition rate, in the order of kHz rather than Hz, was used which allowed similar data to be reproduced at much higher resolution. For the single experimental loading rate studied, the tension within the fibre steadily rose to a value of ~20 pN. This was followed by a 'destructive regime' in which a series of unevenly spaced disruption events could be identified as steady rises in force followed by discrete, sharp drops. This series of force peaks rose from ~20–40 pN, and ended when the fibre reached a contour length corresponding to that of the original naked DNA molecule. Prior to each disruption event the rise in force was found to be essentially linear. In this regime the extensible WLC model could be approximated by a Hooke's spring. The contour length of each intermediate structure during the dissociation process was then obtained from the intercept of the linear fit with the length axis (Bennink *et al.*, 2001a). Analysis of ΔL_0 revealed a strong tendency for the quantized release of DNA lengths that corresponded to the disruption of individual (65 nm) or multiple nucleosomes. However, a closer examination of this data reveals a significant spread in these quantized values. If we consider the nature of the reconstituted fibre, it is possible that the unbinding of non-histone proteins and the disruption of inter-nucleosomal interactions gives uncertainty in the assignment of these rupture events solely to DNA-core histone interactions. Interestingly although the reconstituted chromatin fibre was investigated under 'high-salt' conditions of 150 mM NaCl (Bennink *et al.*, 2001a), force–extension characteristics were found to be more similar to the low-salt native fibre (Cui and Bustamante, 2000). In particular there was no evidence of a plateau at ~5 pN. This observation suggests that the reconstituted fibre has weak inter-nucleosomal interactions and is therefore possibly an extended chromatin conformation. Moreover if the egg extract reconstituted fibre does have a structure with few inter-nucleosomal interactions, this along with quantized length increments suggests that the majority of interactions disrupted are DNA-core histone interactions. It is clear however that assignment of individual events to specific interactions is too ambitious for a chromatin fibre that consists of ~240 nucleosomes, along with other protein molecules, and an unknown higher order structure.

To be noted is that the first two optical tweezers studies (Cui and Bustamante, 2000; Bennink *et al.*, 2001a) unfortunately do not provide insight into the effect of loading rate on the measured disruption forces and possible unbinding mechanisms for a chromatin fibre. More recent data from our group (Figure 4) has revealed that *Xenopus laevis* extract reconstituted chromatin fibres start to dissociate under forces of just 10 pN, under lower loading rate conditions (Pope *et al.*, 2002). These forces are considerably different to the 20–40 pN force range previously reported (Bennink *et al.*, 2001a). Analysis of the average step size revealed an increase with increasing loading rate, suggesting that different loading rates promote the unravelling of chromatin via different unbinding pathways. Significantly, this new data reveals that nucleosomes do not necessarily unravel as an 'all or none' process, as shown at relatively higher loading rates (Bennink *et al.*, 2001a); but under lower loading rates can involve much smaller disruption events, consistent with minimal energy costs for intra-nucleosomal disruption events (Polach and Widom, 1995) within a reconstituted chromatin fibre (L. H. Pope *et al.*, unpub.) and in agreement with the recent low-force regime data published on unravelling a short nucleosomal array (Brower-Toland *et al.*, 2002).

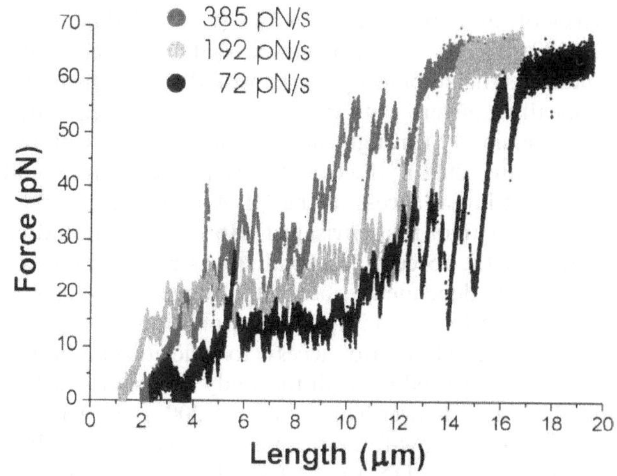

Fig. 4. Dynamic force spectroscopy stretch data for chromatin reconstituted using *Xenopus laevis* egg extract. Exactly the same procedure was used as described in the earlier paper by Bennink *et al.* (2001a), the only difference is that these fibres were disrupted over a range of apparent loading rates, from 72 to 385 pN/s, rather than for just a single loading rate. Notably a dramatic effect is observed where forces are greatly reduced under a relatively small decrease in loading rate. Also it can be seen from the data that it is possible to reach the 65 pN plateau, particularly at higher loading rates, before the fibre has reached a contour length that would correspond to naked DNA (in this case 16.4 μm).

Mechanisms of access to nucleosome-bound DNA

Force-induced rupture of chromatin has provided insights into the mechanisms through which access to the genetic code is attained. For example, for native chromatin in high-salt conditions, the proposal that the ~5 pN force plateau was due to inter-nucleosomal interaction disruption, allowed a calculation that the corresponding energy of unbinding was $\sim$3.4 $k_B T$ (Cui and Bustamante, 2000). Hence it was proposed that in physiological conditions this unbinding energy would allow chromatin to locally convert between a compact and extended structure simply through thermal fluctuations (Cui and Bustamante, 2000). If correct this proposition would imply that initial access to the DNA template could be much easier than first envisaged. However, it should be noted that the force plateau value was not checked for loading rate dependency, hence the suggested $\sim$3.4 $k_B T$ may be even lower and hence need revision.

Further insights into access to nucleosome-bound DNA were revealed through the study of a short nucleosomal array (Brower-Toland *et al.*, 2002). It was deduced that in a low-force regime there was the gradual release of 76 bp of DNA per NCP. Measurements of ΔL_0 from the saw-tooth data in the higher force regime revealed that individual disruption events corresponded to the release of 82 bp of DNA per NCP. These observations along with evidence that nucleosome disruption was reversible if tension within the fibre was kept below a certain force threshold, allowed the authors to suggest a three-stage process for nucleosome unbinding. Taking into account an assessment of NCP interaction strengths (Luger and Richmond, 1998), the model proposed was as follows: under low force, the entry and exit DNA was peeled symmetrically from the NCP releasing 76 bp of DNA (stage 1) (disruption of green interaction site, Figure 1c); 80 bp of DNA was then released, first through disruptions of strong interactions located $\pm$40 bp from the dyad (orange interaction site, Figure 1c), giving rise to the high-force saw-tooth observations, at which point the process remained reversible (stage 2), and then at higher forces the remaining 11 bp of DNA attached at the dyad position (red interaction site, Figure 1c) were disrupted and the histones completely dissociated from the DNA (stage 3).

Such a model, derived from experimental measurements, is the first of its type and clearly demonstrates the power of this technique in providing insight into the relative strengths of interactions that must be overcome in gaining access to the DNA template. Other studies such as restriction enzyme accessibility assays (Polach and Widom, 1995) and Cryo-EM (Furrer *et al.*, 1995) have given insight into the accessibility of NCP-bound DNA. These studies have shown that the amount of DNA associated with the NCP is not fixed at 146 bp but is dynamic in nature. In particular the ends of the DNA could easily be peeled away from the core octamer. This along with the optical tweezers data provides an indication that these are weak interactions that can be broken under a relatively low force, or through thermal fluctuations that induce spontaneous peeling. All of these findings are consistent with the assessment made by Luger and Richmond (1998) into the relative strengths of intra-NCP interactions. Indeed thermally activated spontaneous peeling of DNA from the NCP would provide a convenient mechanism through which proteins have enough access to the template to actually bind whilst the structural integrity of chromatin is maintained through the stronger interactions. If this mechanism provided initial access, then motor proteins, such as RNA and DNA polymerases, could use force to completely remove the octamer from the DNA template as transcription or replication progresses. Undoubtedly the ability of this peeling to occur will rely on the modified state of the histones, and also interactions with other proteins such as linker histone.

Energetics and kinetics

Considering chromatin assembly kinetics, it should be highlighted that the formation of a chromatin fibre directly within the flow cell, is an extremely attractive technique (Bennink *et al.*, 2001b). Through these studies it was shown that assembly under forces of just a few pN proceeds at a rate of approximately 2–3 nucleosomes per second (Bennink *et al.*, 2001b), much faster than measured in previous bulk measurement experiments (Stein, 1979; Daban and Cantor, 1982) and in excellent agreement with an earlier single-molecule study in which assembly was followed in real-time using fluorescently labelled DNA, along with the same chromatin assembly extract (Ladoux *et al.*, 2000). Clearly continuation of these optical tweezers studies will be extremely valuable in terms of understanding chromatin structure and formation. A striking advantage of this technique is that both assembly and disruption of exactly the same fibre can be directly studied.

From dynamic force spectroscopy studies it is clear that a wealth of information can be gained through the study of rupture forces. Unfortunately the first two optical tweezers studies were both undertaken using a single loading rate to induce chromatin fibre disruption (Cui and Bustamante, 2000; Bennink *et al.*, 2001a). In both cases irreversible disruption events began at forces of $\sim$20 pN, however we now have evidence that this is very different for other loading rates (Figure 4). Hence it is not at all surprising that this original 20 pN value is ten times larger than that predicted by an equilibrium model, where it is calculated that forces of $\sim$2 pN should be sufficient to peel the DNA from the core octamer (Marko and Siggia, 1997). Also worth considering is that optical tweezers-induced dissociation is a non-equilibrium process since rebinding of the histone core is unfavourable in histone free buffer and at fast pulling speeds (Cui and Bustamante, 2000). Additionally the measured rupture forces could be higher if dissociation requires twisting the NCPs in such a way that force acts perpendicular to the plane in which the DNA wraps around the histones (Cui and Bustamante, 2000). The theoretical 2 pN clearly does not reflect geometrical constraints and other limitations of the optical tweezers experiment. Moreover it is clear that before any interpretation can be given to the measured forces, the exact interactions that are disrupted must be known. The idea that force is required to rotate the NCP in a geometrically constrained system is interesting; hence torque-induced nucleosome rupture will undoubtedly be an interesting future study (Sarkar and Marko, 2001).

A first attempt to use dynamic force spectroscopy in the study of a short nucleosomal array was a significant step forward towards characterization of specific chromatin interactions (Brower-Toland *et al.*, 2002). The advantages of such a synthetically designed system are clear; the maximum possible number of NCPs is exactly known, and the only interactions that will be disrupted will be between the DNA and core histones, since other DNA binding proteins are not present. Measurement of loading-rate dependent forces corresponding to the rupture of specific interactions believed to be at positions ±40 bp from the dyad (orange interaction site, Figure 1c), i.e. the saw-tooth forces, allowed a first attempt at experimentally measuring the width and height of a prominent energy barrier crossed during nucleosome dissociation. If we consider that this experimental system consisted of a nucleosomal array containing a maximum number of 17 NCPs, and that each individual NCP possibly had 14 interaction sites; then it is still difficult to appreciate that these events can be unambiguously assigned to the breaking of exactly the same interactions per event. However, the constant 80 bp disruption signature does give significant grounds to this assignment. This study was undertaken over three orders in magnitude in loading rate, and rupture forces ranged from ~16 to ~24 pN; still significantly higher than theoretically predicted. A plot of rupture force vs. loading rate on a logarithmic scale, as expected, gave a linear relationship. From this data it was deduced that the two interaction sites, at ±40 bp, disrupted cooperatively via a single energy barrier, the height of which was estimated at 36–38 k_BT and its position was 3.2 nm from an energy minimum. Additionally the rate constant for simultaneous disruption of these two symmetrical interaction points was determined as 3×10^{-7} s^{-1}. It is evident that it would be attractive to study this system in further detail over many more orders in magnitude in loading rate. It must be stressed however that to obtain statistically relevant numbers from these experiments requires hundreds of measurements of rupturing the same interaction for just one single loading rate. Unfortunately the authors in this paper do not state the number of measurements that were actually made per loading rate. In contrast to the other studies, this model system cannot reflect the true nature of intact native chromatin. However, its simplicity does provide a good starting point to slowly unravel the characteristics and nature of specific interactions that determine chromatin structure and dynamics. Whether this system is simple enough to give an in depth description of the interactions central to the maintenance of the NCP structure remains to be seen.

Modelling chromatin stretching

Central to understanding the experimental force–extension behaviour of chromatin stretching and disruption, will be modelling studies. As yet only a few studies are published in the literature, the first of which derived a low-resolution chromatin model from the mechanical properties of naked DNA along with some known structural parameters (Katritch *et al.*, 2000). This model was mainly developed to provide further insight into the low-force regime data, < 20 pN, of native chromatin where nucleosomal structure remains intact (Cui and Bustamante, 2000). Force–extension data were derived from Monte Carlo sampling of various nucleosomal array models. The computed force–extension curves were found to be most sensitive to changes in the equilibrium entry–exit angle of the linker DNA, which set a limit of 50 ± 10°, and the effective linker length of DNA, which was limited to a value of 40 ± 5 bp. Importantly, these values gave a good fit to the experimental relax trace at 5 mM NaCl (Cui and Bustamante, 2000). Furthermore, the simulations suggested that these values remained almost constant over the 0–20 pN force range. Hence it was concluded that the chromatosome (NCP plus DNA associated with bound linker histone) is a fairly rigid unit that is stable to mechanical stress. The low-salt fibre behaviour could be modelled simply by elastic linker deformations, however the transition from a compact to an extended state at high-salt required the incorporation of attractive interactions between close nucleosomes. From both the experimental and simulated data, the depth of the attractive inter-nucleosomal potential was estimated to be ~2 k_BT. Furthermore, this small attractive potential was shown to be sufficient to induce chromatin condensation, since a greater than twofold shrinkage of the fibre model was observed for forces close to zero. The final chromatin model was an irregular zig–zag structure with intact chromatosomes and deformable linkers. Of particular relevance to the dynamic behaviour of chromatin in response to cellular requirements is the observation that this structure could undergo large-scale structural rearrangements with minimal change to the chromatosomes, and that a highly condensed conformation could be adopted in response to very small (~2 k_BT) inter-nucleosomal attractions.

Conclusions

From these first optical tweezers studies of chromatin stretching, significant insight into structure, mechanics, energetics, kinetics and dynamics has been attained. Studies undertaken in both high- and low-salt conditions provided insight into the compact and extended forms of chromatin. Data acquisition at high resolution has revealed that this technique is capable of studying disruption events that correspond to the dissociation of single nucleosomes. Furthermore it has been shown that nucleosomes do not necessarily unravel as a single event but can be induced to dissociate as a multistage process that involves possibly three or more consecutive events.

The experimental pathway to dissociation is probably affected by the rate at which load is applied to the system, and hence should be studied using techniques such as dynamic or static force spectroscopy in order to determine what happens in an unperturbed system. Already dynamic force spectroscopy using optical tweezers has allowed the measurement of an energy barrier crossed upon unbinding a specific intra-nucleosomal interaction. In addition to chromatin disruption, the assembly of chromatin within the optical tweezers flow cell has allowed measurements of assembly kinetics as a function of force applied to the template. In terms of dynamics, all these insights are particularly relevant for understanding the structural response of chromatin to both thermal fluctuations and applied force, for example by molecular motors that interact with the DNA template. Importantly these first results have shown that optical tweezers will be invaluable in future studies to further our knowledge of how processes are regulated through structural remodelling of chromatin. It is evident that we have only yet scratched the surface and many more experiments remain to be carried out.

So where does the future of these studies lie? Experimentally much is to be learnt about the energetics and kinetics involved in the structural dynamics of chromatin that are central to cellular function. It is evident that an in depth study of the unbinding characteristics of a single nucleosome will be extremely interesting and no doubt experimentally challenging. Such studies will require a 'pure system', for example a positioning sequence could be used to assemble a single nucleosome, ideally using recombinant histones that can subsequently be enzymatically modified. The study of a single nucleosome will no doubt lead to the study of nucleosomal arrays. Along with histone modifications other effects such as linker DNA sequence and length, and the effect of ionic conditions on weak inter-nucleosomal interactions will also be of interest. One of the most exciting challenges will be the study of transcription and replication along a chromatin template. Native chromatin clearly has more biological relevance, hence studies should of course also continue in this direction and hopefully, through insights gained from the simple synthetic system, this data will become easier to interpret.

Acknowledgements

The authors gratefully acknowledge G. Leno for the kind gift of *Xenopus laevis* egg extract used for the chromatin data presented here (prepared by Z.H. Lu). We would like to thank P. M. Williams, University of Nottingham, and J. Langowski, DKFZ Heidelberg, for many useful discussions. Presented research is supported by the Dutch Foundation for Fundamental Research on Matter. Technical support was kindly provided by K. van Leijenhorst.

References

Baumann CG, Smith SB, Bloomfield VA and Bustamante C (1997) Ionic effects on the elasticity of single DNA molecules. *Proc Natl Acad Sci USA* **94:** 6185–6190.

Bednar J, Horowitz RA, Dubochet J and Woodcock CL (1995) Chromatin conformation and salt-induced compaction: three-dimensional structural information from cryoelectron microscopy. *J Cell Biol* **131:** 1365–1376.

Bednar J, Horowitz RA, Grigoryev SA, Carruthers LM, Hansen JC, Koster AJ and Woodcock CL (1998) Nucleosomes, linker DNA, and linker histone form a unique structural motif that directs the higher order folding and compaction of chromatin. *Proc Natl Acad Sci USA* **95(24):** 14,173–14,178.

Bennink ML, Leuba SH, Leno GH, Zlatanova J, de Grooth BG and Greve J (2001a) Unfolding individual nucleosomes by stretching single chromatin fibers with optical tweezers. *Nature Str Biol* **8:** 606–610.

Bennink ML, Pope LH, Leuba SH, de Grooth BG and Greve J (2001b) Single chromatin fibre assembly using optical tweezers. *Single Mol* **2:** 91–97.

Brower-Toland BD, Smith CL, Yeh RC, Lis JT, Peterson CL and Wang MD (2002) Mechanical disruption of individual nucleosomes reveals a reversible multistage release of DNA. *Proc Natl Acad Sci USA* **99:** 1960–1965.

Bustamante C, Marko JF, Siggia ED and Smith SB (1994) Entropic elasticity of λ-phage DNA. *Science* **265:** 1599–1600.

Bustamante C, Zuccheri G, Leuba SH, Yang G and Samori B (1997) Visualization and analysis of chromatin by scanning force microscopy. *Methods: Comp Methods Enzymol* **12:** 73–83.

Cluzel P, Lebrun A, Heller C, Lavery R, Viovy J-L, Chatenay D and Caron F (1996) DNA: an extensible molecule. *Science* **271:** 792–794.

Cui Y and Bustamante C (2000) Pulling a single chromatin fiber reveals the forces that maintain its higher-order structure. *Proc Natl Acad Sci USA* **97:** 127–132.

Daban JR and Cantor CR (1982) Role of histone pairs H2A, H2B and H3, H4 in the self-assembly of nucleosome core particles. *J Mol Biol* **156:** 771–789.

Dobosy JR and Selker EU (2001) Emerging connections between DNA methylation and histone acetylation. *Cell Mol Life Sci* **58:** 721–727.

Evans E (1998) Energy landscapes of biomolecular adhesion and receptor anchoring at interfaces explored with dynamic force spectrocopy. *Faraday discuss* **111:** 1–16.

Evans E (2001) Probing the relation between force – lifetime – and chemistry in single molecular bonds. *Annu Rev Biophys Biomol Str* **30:** 105–128.

Evans E and Ritchie K (1997) Dynamic strength of molecular adhesion bonds. *Biophys J* **72:** 1541–1555.

Evans E and Ritchie K (1999) Strength of a weak bond connecting flexible polymer chains. *Biophys J* **76:** 2439–2447.

Finch JT and Klug A (1976) Solenoidal model for superstructure in chromatin. *Proc Natl Acad Sci USA* **73:** 1897–1901.

Furrer P, Bednar J, Dubochet J, Hamiche A and Prunell A (1995) DNA at the entry–exit of the nucleosome observed by cryoelectron microscopy. *J Struct Biol* **114:** 177–183.

Harada Y, Ohara O, Takatsuki A, Itoh H, Shimamoto N and Kinosita K (2001) Direct observation of DNA rotation during transcription by *Escherichia coli* RNA polymerase. *Nature* **409:** 113–115.

Havas K, Flaus A, Phelan M, Kingston R, Wade PA, Lilley DMJ and Owen-Hughes T (2000) Generation of superhelical torsion by ATP-dependent chromatin remodelling activities. *Cell* **103:** 1133–1142.

Horowitz RA, Agard DA, Sedat JW and Woodcock CL (1994) The three-dimensional architecture of chromatin in situ: electron tomography reveals fibers composed of a continuously variable zig–zag nucleosomal ribbon. *J Cell Biol* **125:** 1–10.

Horowitz RA, Koster AJ, Walz J and Woodcock CL (1997) Automated electron microscope tomography of frozen-hydrated chromatin: the irregular three-dimensional zigzag architecture persists in compact, isolated fibers. *J Struct Biol* **120:** 353–362.

Houchmandzadeh B, Marko JF, Chatenay D and Libchaber A (1997) Elasticity and structure of eukaryote chromosomes studied by micromanipulation and micropipette aspiration. *J Cell Biol* **139:** 1–12.

Hsieh J and Fire A (2000) Recognition and silencing of repeated DNA. *Annu Rev Genet* **34:** 187–204.

Katritch V, Bustamante C and Olson WK (2000) Pulling chromatin fibers: computer simulations of direct physical micromanipulations. *J Mol Biol* **295:** 29–40.

Kornberg RD and Lorch Y (1999) Twenty-five years of the nucleosome, fundamental particle of the eukaryote chromosome. *Cell* **98:** 285–294.

Ladoux B, Quivy J-P, Doyle P, du Roure O, Almouzni G and Viovy J-L (2000) Fast kinetics of chromatin assembly revealed by single-molecule videomicroscopy and scanning force microscopy. *Proc Natl Acad Sci USA* **97:** 14,251–14,256.

Laskey RA, Mills AD and Morris NR (1977) Assembly of SV40 chromatin in a cell-free system from Xenopus eggs. *Cell* **10:** 237–243.

Léger JF, Romano G, Sarkar A, Robert J, Bourdieu L, Chatenay D and Marko JF (1999) Structural transitions of a twisted and stretched DNA molecule. *Phys Rev Lett* **83:** 1066–1069.

Leuba SH, Bustamante C, van Holde K and Zlatanova J (1998b) Contributions of linker histones and histone H3 to chromatin structure: scanning force microscopy studies on trypsinized fibers. *Biophys J* **74:** 2823–2829.

Leuba SH, Bustamante C, Zlatanova J and van Holde K (1998a) Linker histone talis and N-tails of histone H3 are redundant: scanning force microscopy studies of reconstituted fibers. *Biophys J* **74:** 2830–2839.

Leuba SH, Yang G, Robert C, Samori B, van Holde K, Zlatanova J and Bustamante C (1994) Three-dimesional structure of extended chromatin fibers as revealed by tapping-mode scanning force microscopy. *Proc Natl Acad Sci USA* **91:** 11,621–11,625.

Leuba SH, Zlatanova J, Karymov MA, Bash R, Liu Y-Z, Lohr D, Harrington RE and Lindsay SM (2000) The mechanical properties of single chromatin fibres under tension. *Single Mol* **1:** 185–192.

Liu LF and Wang JC (1987) Supercoiling of the DNA template during transcription. *Proc Natl Acad Sci USA* **84:** 7024–7027.

Luger K, Meader AW, Richmond RK, Sargent DF and Richmond TJ (1997) X-ray structure of the nucleosome core particle at 2.8 Å resolution. *Nature* **389:** 251–259.

Luger K and Richmond TJ (1998) DNA binding within the nucleosome core. *Curr Op Str Biol* **8:** 33–40.

Marko JF and Siggia ED (1995) Stretching DNA. *Macromolecules* **28:** 8759–8770.

Marko JF and Siggia ED (1997) Driving proteins off DNA using applied tension. *Biophys J* **73:** 2173–2178.

Marszalek PE, Lu H, Li H, Carrion-Vazquez M, Oberhauser AF, Schulten K and Fernandez JM (1999) Mechanical unfolding intermediates in titin modules. *Nature* **402:** 100–103.

Merkel R, Nassoy P, Leung A, Ritchie K and Evans E (1999) Energy landscapes of receptor–ligand bonds explored with dynamic force spectroscopy. *Nature* **397:** 50–53.

Poirier M, Eroglu S, Chatenay D and Marko JF (2000) Reversible and irreversible unfolding of mitotic newt chromosomes by applied force. *Mol Biol Cell* **11:** 269–276.

Poirier MG, Nemani A, Gupta P, Eroglu S and Marko JF (2001) Probing chromosome structure with dynamic force relaxation. *Phys Rev Lett* **86:** 360–363.

Polach KJ, Lowary PT and Widom J (2000) Effects of core histone tail domains on the equilibrium constants for dynamic DNA site accessibilty in nucleosomes. *J Mol Biol* **298:** 211–223.

Polach KJ and Widom J (1995) Mechanism of protein access to specific DNA sequences in chromatin: a dynamic equilibrium model for gene regulation. *J Mol Biol* **254:** 130–149.

Pope LH, Bennink ML, Arends MP and Greve J (2002) Optical tweezers for force-induced rupture of chromatin. *Biophys J* **82:** 2476.

Pope LH, Davies MC, Laughton CA, Roberts CJ, Tendler SJB and Williams PM (2001) Force-induced melting of a short DNA double helix. *Eur Biophys J* **30:** 53–62.

Ramsperger U and Stahl H (1995) Unwinding of chromatin by the SV40 large T-antigen DNA helicase. *EMBO J* **14:** 3215–3225.

Sarkar A and Marko JF (2001) Removal of DNA-bound proteins by DNA twisting. *Phys Rev E* **64:** 1–9.

Smith SB, Cui Y and Bustamante C (1996) Overstretching B-DNA: the elastic response of individual double-stranded and single-stranded DNA molecules. *Science* **271:** 795–799.

Stein A (1979) DNA folding by histones: the kinetics of chromatin core particle reassembly and the interaction of nucleosomes with histones. *J Mol Biol* **130:** 103–134.

Strick T, Allemand J-F, Croquette V and Bensimon D (2000) Twisting and stretching single DNA molecules. *Prog Biophys Mol Biol* **74:** 115–140.

Strunz T, Oroszlan K, Schäfer R and Guntherödt H (1999) Dynamic force spectroscopy of single DNA molecules. *Proc Natl Acad Sci USA* **96:** 11,277–11,282.

Van Holde KE (1989) In: Rich A (ed.) Chromatin, pp. 111–148, Springer-Verlag, New York.

Van Holde K and Zlatanova J (1999) The nucleosome core particle: does it have structural and physiologic relevance? *BioEssays* **21:** 776–780.

Wang M, Schnitzer MJ, Yin H, Landick R, Gelles J and Block SM (1998) Force and velocity measured for single molecules of RNA polymerase. *Science* **282:** 902–907.

Wang MD, Yin H, Landick R, Gelles J and Block SM (1997) Stretching DNA with optical tweezers. *Biophys J* **72:** 1335–1346.

Widom J (1998) Structure, dynamics, and function of chromatin in vitro. *Ann Rev Biophys Biomol Struct* **27:** 285–327.

Widom J and Klug A (1985) Structure of the 300 Å chromatin filament: X-ray diffraction from oriented samples. *Cell* **43:** 207–213.

Williams MC, Rouzina I and Bloomfield VA (2002) Thermodynamics of DNA interactions from single molecule stretching experiments. *Acc Chem Res* **35:** 159–166.

Workman JL and Kingston RE (1998) Alteration of nucleosome structure as a mechanism of transcriptional regulation. *Annu Rev Biochem* **67:** 545–579.

Yin H, Wang MD, Svoboda K, Landick R, Block SM and Gelles J (1995) Transcription against an applied force. *Science* **270:** 1653–1657.

Zlatanova J, Leuba SH and van Holde K (1998) Chromatin fiber structure: morphology, molecular determinants, structural transitions. *Biophys J* **74:** 2554–2566.

Zlatanova J, Leuba SH, Yang G, Bustamante C and van Holde K (1994) Linker DNA accessibility in chromatin fibers of different conformations: a reevaluation. *Proc Natl Acad Sci USA* **91:** 5277–5280.

Journal of Muscle Research and Cell Motility **23**: 409–431, 2002.

Micromechanical studies of mitotic chromosomes

M.G. POIRIER[1] and J.F. MARKO[1,2,*]
[1]*Department of Physics, University of Illinois at Chicago, 845 West Taylor Street, Chicago IL 60607-7059, USA;* [2]*Department of Bioengineering, University of Illinois at Chicago, 851 South Morgan Street, Chicago IL 60607-7052, USA*

Abstract

We review micromechanical experiments on mitotic chromosomes. We focus on work where chromosomes were extracted from prometaphase amphibian cells, and then studied by micromanipulation and microfluidic biochemical techniques. These experiments reveal that chromosomes have well-behaved elastic response over a fivefold range of stretching, with an elastic modulus similar to that of a loosely tethered polymer network. Perturbation by microfluidic 'spraying' of various ions reveals that the mitotic chromosome can be rapidly and reversibly decondensed or overcondensed, i.e. that the native state is not maximally compacted. Finally, we discuss microspraying experiments of DNA-cutting enzymes which reveal that the element which gives mitotic chromosomes their mechanical integrity is DNA itself. These experiments indicate that chromatin-condensing proteins are not organized into a mechanically contiguous 'scaffold', but instead that the mitotic chromosome is best thought of as a cross-linked network of chromatin. Preliminary results from restriction-enzyme digestion experiments indicate a spacing between chromatin 'cross-links' of roughly 15 kb, a size similar to that inferred from classical chromatin-loop-isolation studies. We compare our results to similar experiments done by Houchmandzadeh and Dimitrov (J Cell Biol **145:** 215–213 (1999)) on chromatids reconstituted using *Xenopus* egg extracts. Remarkably, while the stretching elastic response of the reconstituted chromosomes is similar to that observed for chromosomes from cells, the reconstituted chromosomes are far more easily bent. This result suggests that reconstituted chromatids have a large-scale structure which is quite different from chromosomes in somatic cells. More generally our results suggest a strategy for the use of micromanipulation methods for the study of chromosome structure.

Nomenclature

Symbol	*Name*	*Value*
A	persistence length (mitotic newt chromosome)	0.1 m
B	bending modulus (mitotic newt chromosome)	10^{-22} N m^2
f_0	force constant (mitotic newt chromosome)	10^{-9} N
η	fluid viscosity (water and cell culture media)	10^{-3} Pa s
η'	internal viscosity (mitotic newt chromosome)	100 Pa s
k	Boltzmann constant (equal to R/N_A)	1.381×10^{-23} J/K
kT	unit of thermal energy (at 300 K = 27°C)	4.1×10^{-21} J
k_p	pipette stiffness (typical force-measuring pipette)	10^{-3} N/m
ℓ	distance to pipette (for bending experiment)	10^{-6} m
L	length (mitotic newt chromosome)	20×10^{-6} m
N_A	Avagadro's number	6.022×10^{23}
r	cross-section radius (mitotic newt chromosome)	1×10^{-6} m
R	gas constant (equal to $N_A k$)	8.316 J/K
u	transverse fluctuation (mitotic newt chromosome)	1×10^{-7} m
Y	Young modulus	500 Pa

Distance:	1 m = 10^6 μm = 10^9 nm = 10^{10} Å
Force:	1 newton (N) = 1 kg m/sec^2 = 10^5 dyne 1 nN = 10^{-9} N 1 pN = 10^{-12} N 1 kT/nm = 4.1 pN (at 300 K = 27°C)
Energy:	1 Joule (J) = 1 kg m^2/s^2 = 10^7 erg = 0.239 cal 1 kT = 0.59 kcal/mol (at 300 K = 27°C)
Pressure:	1 Pascal (Pa) = 1 N/m^2
Bending modulus:	1 N m^2 = 1 J m
dsDNA:	1 Gbp (10^9 bp) of dsDNA = 1.013 picograms (pg = 10^{-12} g) = 0.34 m contour length

*To whom correspondence should be addressed: E-mail: jmarko@uic.edu

Introduction

The question of how the millimeter to centimeter-long double-stranded DNAs that encode the genomes of cells are physically organized, or 'folded' is a fundamental yet unresolved problem of cell biology. This is remarkable, given the large amount of effort that has been devoted to traditional microscopy of higher-order chromatin structure. The fact that new models for large-scale chromosome structure (Kimura *et al.*, 1999; Machado and Andrew, 2000; Dietzel and Belmont, 2001; Losada and Hirano, 2001; Stack and Anderson, 2001) continue to be proposed in the literature indicates that this question remains open.

There are many reasons why determination of chromosome structure in any cell is challenging. However, one of the main problems is certainly *that chromosomes have a dynamic structure* which changes drastically during the cell cycle (Figure 1). In this article, the focus is on the folding of the chromosome in vertebrate cells during mitosis, specifically at the stage between prophase and metaphase when chromosomes are completely condensed and the nuclear envelope has been disassembled, but where the chromosomes are not yet attached to the mitotic spindle. We will be mainly considering chromosomes from newt (*Notophthalmus viridescens*) and frog (*Xenopus laevis*). These are model organisms for study of mitotic chromosome structure in part simply because their chromosomes are large (Figure 2).

A second problem that chromosome researchers must confront is that *chromosomes are soft physical objects*, with elastic stiffness far less than that of the DNAs and proteins from which they are composed. This means that the structures of chromosomes can be destroyed – or changed – by preps which leave protein and DNA secondary structures intact. This paper is concerned with reviewing recent studies of mechanical properties of mitotic chromosomes that quantify their softness. Emphasis will be placed on the idea that mechanical measurements can be used to assay for structural changes introduced biochemically.

'Architecture and components of Eukaryote chromosomes' provides a brief review of previous biophysical studies of chromosome structure, and DNA and chromatin physical properties. 'Elasticity of mitotic chromosomes' reviews experiments studying stretching, bending, and dynamic elastic responses of mitotic chromosomes (Table 1). 'Combined biochemical-micromechanical study of mitotic chromosomes' then discusses experiments that modify chromosome structure chemically and biochemically, while monitoring the changes in chromosome mechanical properties. This includes discussion of the effects of shifts in salt concentration, and DNA-cutting enzymes. Experiments discussed in this section have clear implications for mitotic chromosome structure, and in particular rule out the contiguous protein scaffold model which posits that chromatin fibers are organized as loop domains tethered to an internal and

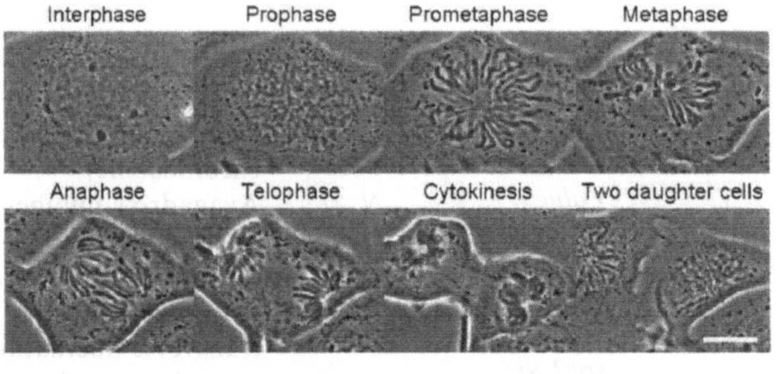

Fig. 1. Cell cycle in a newt cell. During mitosis, chromosomes condense inside the nucleus during prophase, the nuclear envelope disassembles and chromosomes float loose in the cytoplasm during prometaphase, they are captured and aligned by the spindle at metaphase, and the two duplicate chromatids of each chromosome are pulled apart at anaphase. Bar is 20 μm, image is phase contrast, 60x oil objective.

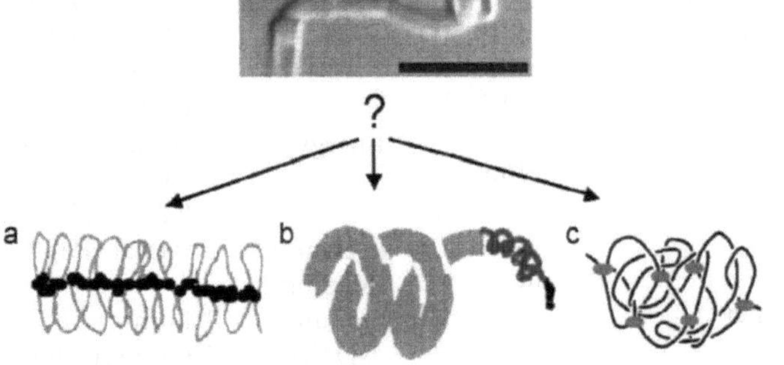

Fig. 2. Prometaphase chromosome, attached at its ends to pipettes outside a cell. Bar is 10 μm, image is DIC, 60x oil objective. Below are three possible models of how chromatin is arranged within a mitotic chromosome.

Table 1. Physical properties of mitotic chromosomes

Chromosome type	Experimental conditions	Young's modulus, Y (Pa)	Bending rigidity, B (J m)	Internal viscosity (Pa s)	References
Drosophila metaphase chromosome	*In vivo*	ND	$\sim 6 \times 10^{-24}$	ND	Marshall *et al.* (2001)
Grasshopper metaphase I & anaphase I chromosome	*In vivo*	200–1000 (avg = 430)	ND	~ 100	Nicklas and Staehly (1967), Nicklas (1983)
Newt (*N. viridescens*) prometaphase chromosome	Cell culture medium	100 to 1000	1 to 3×10^{-22}	100	Houchmandzadeh *et al.* (1997), Poirier *et al.* (2000, 2001a, b)
Newt prometaphase chromosome	*In vivo*	ND	$2\text{–}5 \times 10^{-23}$	ND	Poirier *et al.* (2002b)
Xenopus prometaphase chromosome	Cell culture medium	200–800	$0.5\text{–}2 \times 10^{-23}$	ND	Poirier *et al.* (2002b)
Xenopus prometaphase chromatid	Cell culture medium	~ 300	$\sim 5 \times 10^{-24}$	ND	Poirier *et al.* (2002b)
Xenopus reconstituted chromatid	*Xenopus* egg extract	1000	1.2×10^{-26}	ND	Houchmandzadeh and Dimitrov (1999)

ND indicates quantity not directly measured.
Ranges for values indicate the width of distribution of measured values, and not measurement errors.

physically connected protein skeleton. Finally, 'Conclusion' presents a preliminary model of mitotic chromosome structure based on these results, and then discusses some of the many open questions, including the controversial topic of DNA connections between mitotic chromosomes.

Work of Poirier *et al.* (2000, 2001a, 2002a, b, c) is described in more detail in the PhD thesis of Poirier (2001b). These documents plus images and movies of experiments are available at http://www.uic.edu/~jmarko.

Architecture and components of eukaryote chromosomes

In this section, we review current understanding of the components of chromosomes and overall chromosome structure (see Koshland and Strunnikov, 1996; Hirano, 2000 for more detail on this topic). We also discuss physical properties of DNA and chromatin fiber, with emphasis on recent micromanipulation experiments.

Eukaryote chromosomes are made of chromatin fiber

Chromosomes of animals contain on the order of 10 Mb to 1 Gb of dsDNA (for dsDNA 1 Gbp = 1 pg). At all stages of the cell cycle, the DNA is organized into nucleosomes (Kornberg, 1974), octamers of histone proteins around which dsDNA is wrapped. Each nucleosome is about 10 nm in diameter, and involves about 200 bp of dsDNA (146 bp wrapped, with the balance as internucleosomal 'linker' DNA). The structure of the nucleosome has been precisely determined using X-ray crystallography (Richmond *et al.*, 1984; Arents *et al.*, 1991; Luger *et al.*, 1997). Discovery of the remodeling of nucleosome structure and chemical modification of histones themselves during gene expression (Wolffe and Guschin, 2000) indicates that there are most likely many chemical-structural states of chromatin to understand.

The molecular weight of 200 bp of dsDNA is about 120 kD, and the molecular weight of the histone octamer plus one 'linker' histone (which sits on the linker DNA) is about 125 kD. Thus the relative weight of dsDNA and histones in chromosomes is roughly equal; histones are a major protein component of chromosomes.

It is known that the DNA bound to nucleosomes is able to transiently unbind. Precise experiments (Polach and Widom, 1995; Widom, 1997; Anderson and Widom, 2000) show that restriction-enzyme access to DNA is exponentially attenuated as one moves into nucleosome-bound DNA. This raises the interesting question of on what timescale, and for what factors, transient access to DNA may occur, via conformational fluctuation of the nucleosome itself.

Electron microscope (Thoma *et al.*, 1979) and X-ray diffraction (Widom and Klug, 1985) studies suggest that the nucleosomes fold into a chromatin fiber of ~30 nm diameter, possibly with a helical structure. However, little else about supranucleosomal organization ('higher-order chromatin structure') is solidly understood. This is a result of the relative softness of chromatin fiber, which leads to chromatin's apparent flexible-polymer properties (Cui and Bustamante, 2000; Marko and Siggia, 1997a, see below), plus the inhomogeneity inherent to chromatin. Polymer-like flexibility may account for observations of non-helical chromatin fiber structures (Horowitz *et al.*, 1994; Woodcock and Horowitz, 1995).

Chromatin is sensitive to ionic conditions. When chromatin fibers are extracted into solution at sub-physiological 10 mM univalent ionic strength, they are observed in the electron microscope as an 10 nm-thick 'beads-on-a-string'. At the more physiological ionic strength of 100–150 mM univalent ions, nucleosomes stack into a more condensed, and thicker, 30 nm-thick fiber (Figure 3). At physiological ionic strength, lateral internucleosomal attractions tend to lead to aggregation of isolated fibers (Van Holde, 1989).

The sensitivity of chromatin fiber to ionic strength indicates that nucleosome–nucleosome interactions have

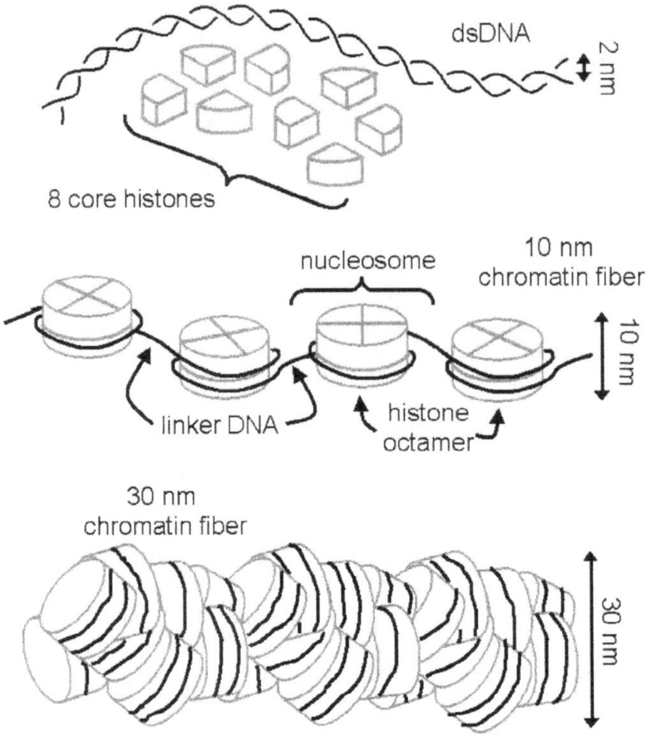

Fig. 3. dsDNA, histones, nucleosome, 10 nm chromatin fiber, 30 nm chromatin fiber. Structural-biological studies of chromatin have focused on the ultrastructure of isolated nucleosomes, and on studying the conformation of nucleosomes in the 10 and 30 nm fiber.

a strong electrostatic component. At low ionic strength, electrostatic interactions have a long range, and the like-charged nucleosomes (chromatin fiber has a net negative charge, similar to dsDNA) repel sufficiently to open chromatin fiber up. At higher ionic strength, this repulsion is overcome by attractive nucleosome–nucleosome interactions, and the fiber folds up.

The drastic structural effect of the change in ionic strength from 10 to 100 mM shows that chromatin fiber is relatively soft, or equivalently that internucleosomal interactions are relatively weak. During the unfolding of chromatin fiber by changing ionic conditions, the nucleosomes themselves do not undergo major conformational changes; the strong electrostatic histone-dsDNA interactions are relatively unperturbed until much higher ionic strengths ($\sim$0.8 M Na^+) are reached. Similarly, dsDNA structure is essentially insensitive to this change in ionic strength.

The 30 nm chromatin fiber is thought to be anywhere from 10- to 50-fold shorter in contour length than the underlying dsDNA. A widely used estimate results from the compaction of the 1200 bp associated with six nucleosomes, into one 10-nm-thick turn of helical chromatin fiber: the resulting 120 bp/nm for chromatin is about 40 times less than the 3 bp/nm for dsDNA. In fact, this 40-fold compaction factor has not been convincingly shown to apply *in vivo*. Given that it is known that some nucleosomes are positioned, some are mobile, and that there are a wide range of histone modifications and variants, it seems unlikely that there is a universal chromatin fiber structure or length compaction factor.

Micromechanics of dsDNA

A new approach to biophysical characterization of DNA is mechanical manipulation of single molecules, with molecular tension as an experimentally controllable and measurable quantity. Methods used to study single dsDNAs are all based on attaching the ends of the molecule to large objects which act as 'handles' (Bustamante *et al.*, 2000). The handles are used to apply controllable forces and to provide an optical marker for the molecule ends and therefore end-to-end extension. Although these techniques usually are restricted in application to molecules of at least a few kilobases in length, ingenious techniques (Bustamante *et al.*, 2000) have been developed to measure relative positions to as little as a few nanometers (Liphardt *et al.*, 2001). We use the example of dsDNA to introduce some basic ideas of polymer elasticity that will later be used to discuss chromosome extensibility.

The double helix has a persistence length of about $A = 50$ nm (150 bp for B-DNA, Hagerman, 1988). The persistence length is the contour length over which thermal (Brownian) fluctuations typically bend the double helix through a 60° bend. Over dsDNA lengths of less than 150 bp, the contour is of fixed shape (the double helix is nominally roughly straight, but some sequences are intrinsically rather severly bent). Over

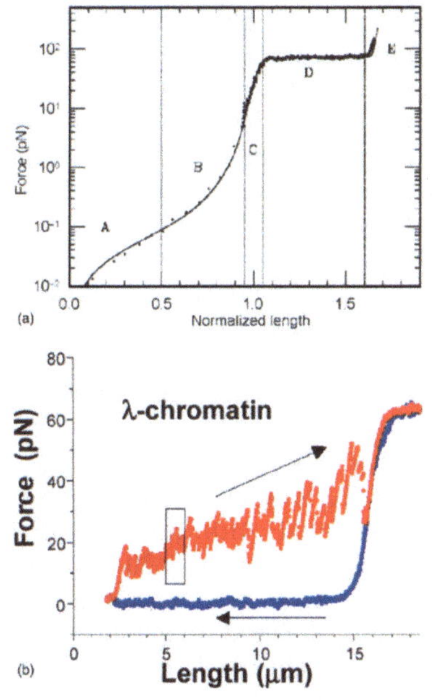

Fig. 4. Comparison of elastic response of (a) single dsDNAs, and (b) chromatin fiber. dsDNA and chromatin fiber both display an initial low-force (sub-pN) elastic regime, followed by a higher-force (few pN) regime. However, dsDNA (a) shows a very stiff and non-linear response, while chromatin fiber (b) shows a more gradual elastic response (Bennink *et al.*, 2001, figure reproduced with permission) as it is extended. This is believed to be due to driving the chromatin fiber opening transition (10–30 nm fiber transition of Figure 3) by force. At ~20 pN, force jumps corresponding to nucleosome removal events are observed.

distances longer than 150 bp, a dsDNA undergoes appreciable dynamic bending.

Thermally excited bends along a long DNA are straightened out by tensions (Figure 4a) of greater than kT/A ($kT = 4.1 \times 10^{-21}$ J, the energy of a single thermal fluctuation at room temperature $T \sim 300$ K; recall $k = R/N_A$ where R is the familiar gas constant and N_A is Avogadro's number), or about 0.1 piconewtons (pN). At 0.1 pN a dsDNA is extended to slightly greater than half its total contour length. At higher forces (0.1–10 pN) dsDNA elasticity is highly non-linear, with tension increasing quickly as the length approaches that of B-form (3 bp/nm) (Smith *et al.*, 1992; Bustamante *et al.*, 1994).

The characteristic tension to begin to extend a dsDNA (0.1 pN) is a small force, even by single-molecule standards. Cellular motor proteins generate forces ranging from a few pN (myosin: 5 pN, kinesin: 8 pN) to tens of pN (RNA polymerase: 40 pN, Yin *et al.*, 1995), roughly because they convert chemical energy at the rate of a few kT per nanometer of motion (note that 1 kT/nm = 4 pN). Another source of tension on dsDNA *in vivo* are DNA-protein interactions; for example it has been demonstrated that polymerization of RecA onto dsDNA generates forces in excess of 50 pN (Leger *et al.*, 1998). In the cell dsDNA thus can be stretched out and structurally modified by forces generated by the machinery which transcribes (Yin *et al.*, 1995), replicates (Wuite *et al.*, 2000), and repairs it.

From forces of 0.1–10 pN, dsDNA elastic response is well expressed by the empirical force law (Bustamante *et al.*, 1994)

$$f = \frac{kT}{A}\left[\frac{x}{L} + \frac{1}{4(1 - x/L)^2} - \frac{1}{4}\right] \quad (1)$$

where A is the persistence length of 50 nm, and where x is the molecule end-to-end extension and L its total B-form contour length. Equation (1) captures the weak initial elastic response where force increases from 0 to about kT/A as x/L increases from 0 to about 0.5, and the strong non-linear force increase as x approaches L. These two features are generic for all flexible polymers which undergo random-walk-like bending fluctuations when unstretched.

For even larger forces (10–100 pN), dsDNA secondary structure starts first to stretch (10–50 pN), and then the double helix is disrupted, and stretches to an extended form at ~65 pN (Cluzel *et al.*, 1996; Smith *et al.*, 1996); this disruption has a strong DNA twisting-dependence (Allemand *et al.*, 1998; Leger *et al.*, 1999). For forces of 10–50 pN, dsDNA can be thought of as an elastic rod, with elastic Young modulus $Y \sim 300$ MPa. The meaning of Y comes from the force needed to stretch an elastic rod of uniform and circular cross-section and equilibrium (unstretched) length L, so as to increase its length by ΔL (Landau and Lifshitz, 1986):

$$f = \pi r^2 Y \frac{\Delta L}{L} \quad (2)$$

Here, r is the cross-sectional radius of the rod (for dsDNA, $r = 1$ nm; note that for general cross section shape, πr^2 can be replaced by the rod cross-sectional area).

The Young modulus is thus the stress (force per cross-sectional area) at which an elastic rod would be doubled in length, if its initial linear elasticity could be extrapolated: Y characterizes the stretching elasticity of a material in a shape-independent way. Similarly, $f_0 = \pi r^2 Y$ is the force at which a rod would double in length, based on extrapolation of its linear elasticity. For dsDNA, $f_0 \sim 1000$ pN, and like most solid materials, Equation (2) applies only for $\Delta L/L$ much less than unity (for dsDNA, the regime where (2) applies is from $\Delta L/L = 1.0$–1.05, where 1.0 refers to the B-DNA length).

The bending flexibility of an elastic rod is also related to Y. An elastic rod's bending modulus, again assuming linear elasticity and circular uniform cross-section, is

$$B = \frac{\pi}{4} r^4 Y \quad (3)$$

This quantity has dimensions of energy times length. If our elastic rod is bent into a circular arc of bending radius R, the torque that must be applied is B/R, and the force that must be applied is B/R^2. For dsDNA, $Y = 300$ MPa gives $B = 2 \times 10^{-28}$ Jm.

For rods which are thin enough to be bent by thermal fluctuation (e.g. the double helix), it is useful to relate B to the bending persistence length A:

$$A = \frac{B}{kT} = \frac{\pi r^4 Y}{4\,kT} \tag{4}$$

For dsDNA, we therefore see that $Y = 300$ MPa gives rise to an estimate of $A = 50$ nm, essentially the observed value. The connection between the value of Y obtained from stretching the double helix, with that obtained from separate measurement of the persistence length A, show that elementary concepts of elasticity apply at the nanometer scale of the interior of the double helix.

Micromechanics of chromatin fibers

Recently, the force-extension properties of chromatin fiber extracted from chicken erythrocytes (Cui and Bustamante, 2000) were measured. Three different force regimes were observed. First, very low-force 'entropic elasticity' regime is observed, similar to that seen for dsDNA. This initial low-force (below 0.1 pN) force response is thought to be due to the polymer flexibility of chromatin, and allows an estimate of chromatin persistence length of about 30 nm, slightly shorter than dsDNA itself. This low persistence length is possible due to the zig–zag path of the linker DNA: a spring (a 'Slinky' toy is a good example) can be far more easily bent than the wire from which it is formed. However, to date completely convincing data for chromatin low-force (< 0.1 pN) 'polymer' elasticity under physiological conditions have not yet been published.

At higher forces (0.1–5 pN), what is observed depends strongly on ionic conditions, as one would expect based on the 10–30 nm fiber transition observed with increasing ionic strength. At relatively low (10 mM Na^+) ionic strength, a strongly non-linear elastic response similar to that of dsDNA is observed. However, at closer to physiological ionic strength (40 mM Na^+), a more gradual, nearly linear elastic response is observed for forces between 0.1 and 5 pN (barely visible in Figure 4b, data from Bennink *et al.*, 2001). This can be explained in terms of the unstacking of adjacent nucleosomes, i.e. by the idea that force can be used to drive the 30–10 nm fiber transition. This transition is observed to be reversible, and is characterized by a force constant $f_0 \sim 5$ pN and a high degree of smooth extensibility (compare with 'bare' dsDNA which has a stretching force constant of 1000 pN, and can be stretched by only about 5% before transforming to a new stretched form). The doubling in length of the chromatin fiber over a 5 pN increase in force observed by Cui and Bustamante (2000) can be combined with the native fiber 30 nm diameter to estimate an effective Young modulus, $Y \sim 10^5$ Pa, far below the effective modulus of straight DNA ~300 MPa. As DNA is folded up, its effective modulus is reduced.

At higher forces (20 pN), irreversible extension of chromatin fiber occurs (Cui and Bustamante, 2000). Recent experiments observe this to be in the form of a series of jumps of quantized length (Figure 4b). These jumps are thought to be associated with removal of single nucleosomes, and possibly individual ~80 bp winds of DNA (Brower-Toland *et al.*, 2002). It is likely that this threshold for nucleosome removal is highly extension-rate dependent, since the known binding free energy ~20–30 kT/nucleosome indicates that one should expect equilibrium between bound and free nucleosomes for forces near 2–3 pN (Marko and Siggia, 1997b).

Observation of this equilibrium for pure chromatin fiber would require long experimental time scales since the barrier associated with nucleosome removal or rebinding is likely close to the 20 kT binding energy. However, use of nucleosome-assembly factors such as NAP-1 which act in thermal equilibrium may make it practical to observe chemical equilibrium between octamer on- vs. off-states (S. Leuba, private communication). Recent experiments have moved in this direction, assembling chromatin fibers *in vitro*, onto initially bare molecules of dsDNA, using cell-extract-derived chromatin-assembly systems (Ladoux *et al.*, 2000; Bennink *et al.*, 2001), which have allowed measurement of the ~10 pN forces applied by chromatin-assembly enzymes.

Chromosome structure at scales larger than the chromatin fiber

Beyond the chromatin fiber, it is thought that many species of protein act to define chromosome structure. During interphase, this includes the machinery of gene regulation and expression, centers of DNA replication (Cook, 1991), and the nuclear matrix (Wolffe, 1995, Sec. 2.4.2). We focus on large-scale chromosome structure as observed by traditional microscopy, in recent three-dimensional studies of chromosome structure and dynamics.

Structural-biological studies of mitotic chromosome structure

Much of our understanding of mitotic chromosome structure at larger scales is based on relatively invasive electron microscopy (EM) studies, and on optical microscopy. Based on EM visualization of DNA loops extending from an apparent protein-rich chromosome body after histone depletion (Paulson and Laemmli, 1977; Paulson, 1988), and to some extent on direct visualization of these chromatin loops in fixed cells, one commonly discussed model for mitotic chromosome structure is based on labile chromatin loops (Figure 2) interconnected by a protein-rich 'scaffold' (Marsden and Laemmli, 1979). Other studies suggest that the scaffolding is coiled (Boy de la Tour and Laemmli, 1988).

Although these experiments are often taken to imply the existence of a connected protein 'skeleton' inside the mitotic chromosome (see the textbooks Lewin, 2000, pp. 551–552, Lodish *et al.*, 1995, pp. 349–353, Wolffe, 1995, pp. 52–55), Laemmli has recently emphasized to one of us that the conclusion that the internal protein skeleton is mechanically contiguous does not follow from his group's experiments and is not meant to be implied by any of his publications (see discussion and figures in Laemmli, 2002). The issue of the connectivity and mechanical integrity of the DNA and non-DNA components of the mitotic chromosome will be a primary focus of the section 'Combined biochemical-microchemical study of mitotic chromosomes'.

Other microscopy studies suggest a hierarchical structure (Figure 2) formed from a succession of coils at larger length scales (Belmont *et al.*, 1987, 1989). Proposals have since been made for mitotic chromosome structure which combine loop and helix folding motifs (Saitoh and Laemmli, 1993). Existing microscopy studies do not give a consistent picture of mitotic chromosome structure, in part because of the invasive preparations necessary for EM visualization, and the inability of light microscopy to resolve detail smaller than ~200 nm.

The folding scheme of interphase chromatin inside the nucleus pre-1990 was highly unsettled. With no techniques to differentiate different chromosomes or chromosomal regions, light microscopy by itself reveals little, and EM again leads to conflicting views of chromatin structure at length scales from 10 to 100 nm. Biochemical analysis of chromatin domains (Jackson *et al.*, 1990) suggest that interphase chromatin is organized into ~50 kb domains.

Three-dimensional microscopy study of chromosome structure and dynamics

Increasing use of fluorescent labeling and optical sectioning microscopy techniques in the 1990s has allowed many features of chromosome structure to be determined by mapping physical position of specific DNA sequences with ~300 nm precision. Fluorescent *in situ* hybridization and other techniques applied to whole chromosomes shows that different chromosomes occupy different regions or territories of the interphase nucleus (Cremer *et al.*, 1993; Zink *et al.*, 1998), and has also shown the existence of interchromosomal regions.

Similar studies where specific chromosome loci were tagged have been used to measure the real-space distance between genetic markers as a function of the chromatin length between the markers. Remarkably, these studies show interphase chromosomes to have a random-walk-like organization at < 1 Mb scales, and a 'loop' organization at 1–100 Mb scales (Yokota *et al.*, 1995). Similar studies have been used to study attachments of chromosomes to the nuclear envelope (Marshall *et al.*, 1996). The structure of the bulk of the interphase nucleus remains uncertain, with the role of a nucleoskeleton ('nuclear matrix') in chromosome organization still unclear (Pederson, 2000).

FISH study of loci along metaphase chromosomes has also been done to verify that genes are in linear order at > 1 Mb scales. However, markers spaced by less than 1 Mb are often seen in random order, indicating that at the corresponding < 1 μm scale, metaphase chromatin is not rigidly ordered (Trask *et al.*, 1993). This lack of determined structure is consistent with the flexible-loop-domain picture of metaphase chromosome structure (Figure 2), although one might argue that the fixation used somehow distorted structures at these scales.

Structural studies have also been done *in vivo*, by the use of live-cell dyes for specific structures, by incorporation of fluorescent nucleotides (Manders *et al.*, 1999), and by expression of fusions of chromosome-specific proteins with green fluorescent protein (GFP) (Tsukamoto *et al.*, 2000; Belmont, 2001). One study used both techniques to show that there are ~1 μm position fluctuations of interphase chromosome loci from a range of species (Marshall *et al.*, 1997). These fluctuations persisted even in poisoned cells, suggesting that ~Mb chromosome segments are free to undergo thermal fluctuation, in the manner of flexible polymers. This result is at odds with the idea of a dense, rigid nucleoskeleton and suggests instead that chromosomes have intermittent attachments, with ~Mb regions of chromatin free to move on micron length-scales.

A recent study of yeast (*S. cerevisiae*) interphase chromosome structure by Dekker *et al.*, (2002) is unique in its methodology and results. This study used cross-linking of isolated nuclei, followed by restriction-enzyme digestion. The fragments were self-ligated, and the resulting fragments were PCR-amplified and analyzed. The result was a statistical 'map' of *in vivo* chromatin contacts, giving a statistical three-dimensional chromosome model.

Chromosome-folding proteins identified using cell-free chromosome assembly systems

It is possible to study chromosomes assembled *in vitro*. *Xenopus* egg extracts provide an excellent system for doing this, converting *Xenopus* sperm chromatin into either interphase nuclei, or metaphase-like chromatids (Smythe and Newport, 1991). This system has permitted identification of proteins thought to be critical to organizing mitotic chromosomes, most notably the SMC protein family (Hirano and Mitchinson, 1994; Strunnikov *et al.*, 1993, 1995; Strunnikov, 1998).

Hirano and Mitchison (1994) showed that if the XCAP-C/E proteins (two of the SMC proteins in *Xenopus*) were removed from *Xenopus* egg extracts, then only a cloud of tangled chromatin fibers would result, instead of mitotic chromatids. Furthermore, anti-XCAPs were found to destabilize assembled mitotic chromatids, indicating that XCAP-C/Es were needed both for assembly, and for maintenance of mitotic chromosome structure. Hirano and Mitchison also found that the XCAPs were localized inside the mitotic chromatids. Further work established that XCAPs in condensin complexes (Hirano, 1997) show an ATP-dependent DNA-coiling capability (Kimura *et al.*, 1997, 1999). Other SMC-type proteins have other roles in modulating chromosome structure (Strunnikov and Jessberger, 1999), notably holding sister mitotic chromatids together during prophase ('cohesins', see Michaelis *et al.*, 1997; Guacci *et al.*, 1997; Losada *et al.*, 1998). Losada and Hirano (2001) have suggested that the balance between condensin and cohesin SMCs determines large-scale metaphase chromosome morphology.

Many questions remain about the SMC proteins, which have a remarkable structure of $\sim$100 nm coiled-coils with a central hinge (Melby *et al.*, 1998), and ATP-binding and hydrolysing end domains. Their distribution inside mitotic chromatids, clear revelation of their function in chromosome condensation, and the question of whether or not they are the major proteins of the 'mitotic protein scaffold' remain unanswered. Thanks to the biochemical characterizations described above, these questions may be answerable through gradual 'biochemical dissection' (Hirano, 1995, 1998, 1999).

Topoisomerase II

One of the most common proteins found in mitotic chromosomes is topo II (Gasser *et al.*, 1986), the enzyme which passes dsDNA through dsDNA, and which is assumed to be the enzyme primarily responsible for removing entanglements of chromatin fiber during chromosome condensation and segregation. This idea is strongly supported by experiments using mitotic Xenopus egg extracts: when topo II is depleted, sperm chromatin just forms a cloud of apparently entangled chromatin fibers, which never form condensed and segregated chromatids (Adachi *et al.*, 1991).

However, a second hypothesis that topo II also plays a structural role in mitotic chromosomes is contentious (Warburton and Earnshaw, 1997). Immunofluorescence studies show that topo II is localized into helical tracks inside chromatids (Boy de la Tour and Lamelli, 1988; Sumner, 1996). Combined with the fact that topo II interacts with two strands of dsDNA, this result suggests that topo II might be part of an internal protein structure in the mitotic chromosome. However, other experiments use salt treatment to deplete topo II from mitotic chromosomes, with no apparent deleterious effect on their structure (Hirano and Mitchison, 1993). Recently it was reported that the axial distribution of topo II may be triggered by cell lysis (Christensen *et al.*, 2002). At present, these experiments can be reconciled by supposing that topo II is critical for establishment of mitotic chromosome structure by allowing dsDNA disentanglement, that it is present in high copy number on the assembled mitotic chromosome, but that it does not play a crucial role in holding the mitotic chromosome together.

Chromosomal titin

Titin is a huge protein of filamentous structure, and is the elastic restoring element of sarcomeres (Trinick, 1996). The mechanical response of isolated titin molecules has been precisely measured using single-molecule manipulation (Kellermayer *et al.*, 1997; Reif *et al.*, 1997; Tskhovrebova *et al.*, 1997). Because of its structure, a long series of independently folded globular domains, titin displays initial linear elasticity followed by a series of irreversible force jumps associated with successive domain unfolding events. Remarkably, it was found that muscle titin antibodies localize onto mitotic chromosomes (Machado *et al.*, 1998; Machado and Andrew, 2000a, b). It has been therefore speculated that a chromosomal titin might play a role in chromosome condensation, and might be a contributor to chromosome elastic response (Houchmandzdeh and Dimitrov, 1999).

Why study mitotic chromosomes micromechanically?

The structure of chromosomes, beyond the nucleosomal scale, is poorly understood, in spite of a huge focus of effort. This is because chromosome structure is dynamic, and because chromatin is inhomogeneous and soft. Mitotic chromosomes are a logical starting point for study of chromosome structure, since they are packaged (condensed), segregated from one another, and gene expression is halted, all of which are simplifying factors. Study of mitotic chromosome structure will presumably shed light into the mechanism of chromosome disentanglement and condensation (Hirano, 2000).

Basic questions about the mitotic chromosome of interest to us include the following: what is the physical arrangement of chromatin fiber (randomly or regularly coiled or folded?). What are the molecules (proteins?) which accomplish this folding? What molecules are necessary to keep the mitotic chromosome folded up? How are the processes of chromosome condensation, and disentanglement coordinated? All of these questions have a mechanistic as well as structural character, and might be attacked using a combination of biochemical and *micromechanical* experimental methods.

In addition to studying chromosome structure, biophysical chromosome experiments provide information relevant to understanding a range of *in vivo* chromosome biology questions. For example, stresses applied to chromosomes are known to play a role in chromosome alignment and segregation during mitosis (Alut and Nicklas, 1989; Nicklas and Ward, 1994; Li and Nicklas, 1995, 1997; Nicklas *et al.*, 1995, 1998, 2001; Nicklas, 1997; King *et al.*, 2000). Kinetochore chromatin elasticity is central to in a recent model for capture of mitotic chromosomes on the mitotic spindle (Joglekar and Hunt, 2002), and chromosome stretching has been used to study the roles of specific proteins in chromatin compaction (Thrower and Bloom, 2001). Chromosome stiffness has also been proposed to play a role in the mechanism of meiotic synapsis (Kleckner, 1996; Zickler and Kleckner, 1999).

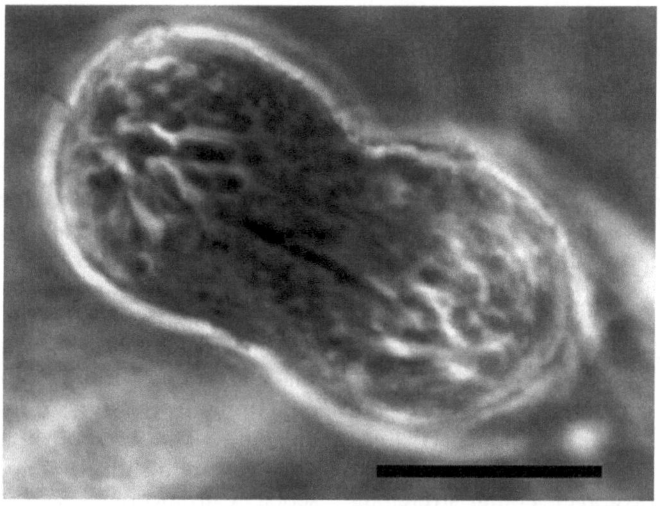

Fig. 5. Newt (*N. viridescens*) tissue culture cell, showing a chromosome being stretched to about twice its native length, by the mitotic spindle during anaphase. The spindle forces are known to be on the order of 1 nN, indicating that the force constant of a whole chromosome is on a similar scale. Bar is 20 μm. Photo courtesy of Prof J. Tang.

Elasticity of mitotic chromosomes

Mitotic chromosomes can be observed to occasionally be stretched out by the ~nN (nanonewton) spindle forces *in vivo* (Figure 5). This has led to a number of studies of chromosome stretching (Callan, 1955; Bak *et al.*, 1977, 1979; Nicklas, 1983, 1988; Claussen, 1994).

Lampbrush chromosomes

One of the earliest discussions of chromosome extensibility was by Callan (1955) who carried out manipulation experiments on amphibian lampbrush chromosomes using glass microneedles. These observations plus DNA-ase experiments of Gall (1963) were used to support the hypothesis that each chromatid contains a single linear DNA.

The lampbrush phase occurs during female meiotic prophase in birds and amphibians, and has played a special role in cell biology for three reasons. First, lampbrush chromosomes are huge, even by amphibian standards, up to ~1 mm long. Second, they display large, clearly flexible loop domains, tethered to a central axis (Gall, 1956). The basic idea of chromatin loops tethered to a central chromosome axis, clearly the case for lampbrushes, has been used as a basic model for chromosome structure at other cell stages, notably mitosis. Third, the large lampbrush loops are 'puffed up' by huge numbers of RNA transcripts coming off tandem polymerases. Electron-microscope observation of the tandem transcription units along lampbrush loops provided early and convincing evidence of the processive nature of transcription (Miller and Beatty, 1969; Miller and Hamkalo, 1972; Morgan, 2002).

Marvelous pictures of lampbrush chromosomes can be found in the monograph of Callan (1986); it should be noted that the large loops are apparently not in sharp focus, despite the use of flash photography. This is because the loops are in *motion*, i.e. undergoing thermal conformational fluctuation (Callan, 1986, p. 28–29). This feature of lampbrush chromosomes is an example of flexible-polymer behavior of chromatin, on a huge and directly observable scale (Marko and Siggia, 1997b).

Mitotic chromosome extensibility and elasticity

Observation of stretching of chromatids by the mitotic spindle, plus the huge length of DNA per chromatid, leads naturally to the notion that mitotic chromosomes should be extensible. This expectation was verified by Nicklas and Staehly (1967), who used microneedles to hook chromosomes inside grasshopper spermatocytes, and demonstrated that meitoic chromosomes (metaphase I through anaphase I) were extensible and elastic, i.e. would return to native length after being stretched by up to eight times.

Nicklas' fundamental study of chromosome elasticity in grasshopper spermatocytes

The first experiment to quantify the elastic response of a chromosome *in vivo* was carried out by Nicklas (1983), using microneedles to carry out experiments inside living cells. The cells used were grasshopper (*Melanoplus sanguinipes*) spermatocytes which have a soft cell cortex which allows needles to grab chromosomes without breaking the cell membrane. Forces were measured by observing microneedle bending. Microneedles were used which required between 0.076 and 0.25 nN/μm per micron of deflection (1 nN $= 10^{-9}$ N; recall 1 N $=$ 1 kg m/s^2). Using a film analysis technique, the resolvable deflection of about 0.25 μm gave force resolution of roughly 0.05 nN.

Nicklas (1983) noted that during anaphase I it was possible to measure the elastic response of one and two chromatids independently, by carrying out experiments on chromosomes either before, or after, their chromatid separation (during anaphase I, the chromatids 'unpeel' except for the kinetochore). Using a statistical analysis of data on a number of chromosomes, he showed that attached pairs of chromatids required twice as much force to be doubled in length as did single chromatids. The elasticity observed was linear (force proportional to change in length, and to cross-sectional area, see Equation 2). The force needed to double a grasshopper meiotic anaphase I chromosome (two chromatids) was determined to be $f_0 = 0.75$ nN; single chromatids were found to have $f_0 = 0.32$ nN (when reading Nicklas' paper, keep in mind 1 nN $= 10^{-4}$ dyne). This result was used to infer that the (average) Young stretching modulus of an anaphase I chromosome was 430 Pa (again, note 1 Pa $= 1$ N/m^2 $= 10$ dyne/cm^2). The range of linear elastic response was reported to be at least up to $\Delta L/L = 2$ (threefold extension).

The experiments of Nicklas (1983) are superb in being *in vivo* measurements, which are sufficiently quantitative that it is completely convincing that the elastic response of the chromosomes, and not some aspects of the cell membrane or cytoskeleton, are being measured. However, this depended on the very fluid cell surface of insect spermatocytes (Nicklas, 1983; Zhang and Nicklas, 1995, 1999) a feature not shared by mammalian somatic cells. This is emphasized by Skibbens and Salmon (1997) who were able to do elegant chromosome manipulations inside cultured newt lung cells during mitosis only using very stiff microneedles, with consequently no possibility to use their bending to measure forces.

Stretching mitotic chromosomes after their removal from cells

Given that stretching chromosomes inside mitotic vertebrate cells is not possible, the next best approach to study of chromosome stretching is to remove chromosomes from cells into the cell buffer. This approach will always be subject to the criticism that chemical conditions outside the cell will alter chromosome structure, but using comparisons with available *in vivo* information, the relation between *in vivo* and *ex vivo* chromosome structure can be understood. As we will describe below, our own experiments combined with those of others convince us that there is little or no change in chromosome structure at least initially after removal from a mitotic cell.

Classen *et al.* (1994) noted that metaphase chromosomes could be highly extended, and have used chromosome stretching to develop high-resolution chromosome banding techniques (Hliscs *et al.*, 1997a, b). The first measurement of the elastic response of a mitotic chromosome extracted from a cell was carried out by Houchmandzadeh *et al.* (1997), using mitotic cells in primary cultures of newt lung epithelia (*Notophthalmus viridescens*). This organism is attractive for chromosome research since it is a vertebrate with relatively few (haploid $n = 11$), large (haploid genome ~35 pg of dsDNA) chromosomes (Gregory, 2001). Each *N. viri* chromatid therefore contains about 3 pg = 3 Gbp, or about one meter, of DNA. At metaphase, the chromosomes are between 10 and 20 μm long, and have a diameter of about 2 μm. Newt epitheilia cells are easily cultured as a monolayer on dishes built on cover glass which are open to room atmosphere, making them well-suited for micromanipulation experiments (Reider and Hard, 1990).

Houchmandzadeh *et al.*, used glass micropipettes (inside diameter ~2 μm, Brown and Flaming, 1986) to puncture mitotic cells, and then to grab onto the chromosomes. The micropipettes were introduced into the open culture dish from above, using an inverted microscope. Chromosomes were grabbed by aspirating the chromosome end into the pipette opening, with the other chromosome end anchored in the cell. The main method used by Houchmandzadeh *et al.*, to apply controlled stretching forces to chromosomes was to use aspiration into a pipette that had been treated with BSA so that the chromosome could slide freely while in contact with the bore of the pipette. The chromosome acted as a piston, and by controlling the aspiration pressure, it could be stretched. This technique allows sensitive measurements, but has the defect that the chromosome-pipette seal is not perfect, and the 'piston' will be leaky. This will result in an overestimation of the modulus, since part of the pressure applied to the pipette drives flow.

The results were essentially that mitotic chromosomes are elastic, with a Young modulus estimated to be approximately 1000 Pa at prometaphase (i.e. chromosomes condensed, but not yet attached to spindle), compatible with the results of Nicklas after taking account of the flow effect mentioned above. Over a range of two-fold extension, the elasticity was remarkably linear (see Figure 8 of Houchmandzadeh *et al.*, 1997). Experiments were also carried out just after nuclear envelope breakdown (end of prophase), and it was found that chromosome had a higher elastic modulus $Y = 5000$ Pa. In addition, Houchmandzadeh *et al.* (1997) discuss the result of severe deformation of chromosomes, using untreated pipettes, to which chromosomes adhere permanently. It was found that prometaphase chromosomes could be extended to as large as 100 times their native length without breaking. For extensions beyond 10 times length, the chromosomes did not return to their native length.

Poirier *et al.* (2000) then used the micropipette-based manipulation technique to more quantitatively measure newt chromosome mechanical properties. Calibrated micropipette bending was used as the force-measurement scheme, for chromosomes removed from cells and suspended between two pipettes (Figure 6). This allows both ends of the chromosome to be monitored, and therefore chromosome extension can be precisely controlled. Digital image acquisition and analysis were used to measure pipette bending. Measurement of the corre-

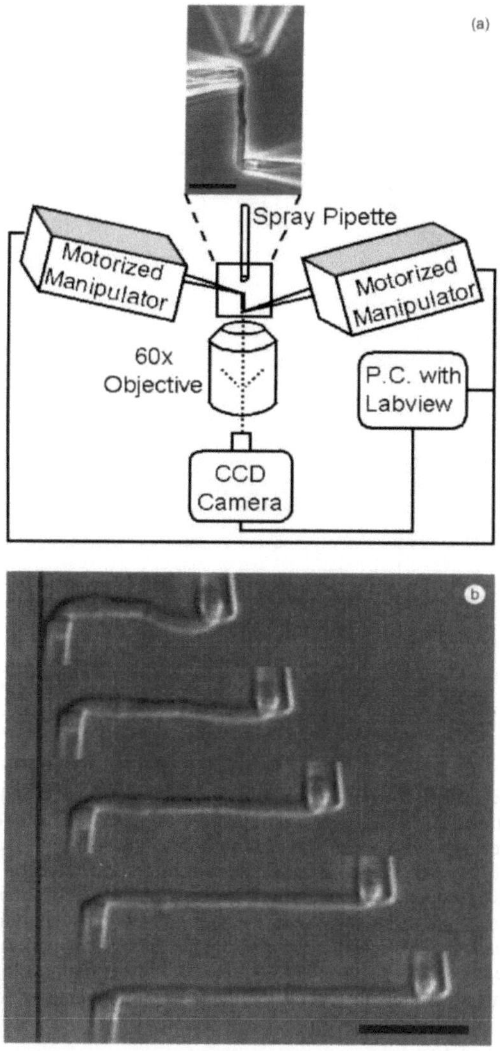

Fig. 6. Two-pipette chromosome experiment, as carried out in Poirier (2000). (a) Schematic diagram of experimental setup. Two pipettes are used to hold a mitotic chromosome, with one pipette fabricated with a deflection force constant $\sim$1 nN/μm to allow chromosome tension to be measured. A third pipette can be moved near to the chromosome to microspray reagents for combined chemical-micromechanical experiments. (b) Example images collected during force-extension experiment. As the left pipette is moved, the right pipette is observed to deflect. Digital image analysis allows pipette deflections to be measured to about 10 nm accuracy. Bar is 10 μm.

lation between pipette images allows pipette shifts (and therefore deflections) to be determined to about 10 nm accuracy. Pipettes were used with bending moments $\sim$1 nN/μm micron of deflection, thus setting a theoretical limit on force resolution of 0.01 nN = 10 pN. In practice, force resolution is usually limited by slow mechanical drifts of the pipettes.

The force-extension response of single mitotic (prometaphase) chromosomes are shown in Figure 7. Completely reversible elastic force response was observed for extensions up to about five times native extension, with a force constant $f_0 \sim 1$ nN. Given the 1.6 μm diameter of the chromosomes, this corresponds to a Young modulus near 500 Pa, near to the value obtained by Nicklas (1983). [The 300 Pa quoted in Poirier *et al.*, 2000 is based on a slight overestimate of the chromosome thickness; our current best estimate is a prometaphase chromosome diameter of about 1.6 μm.] Although on the same order of magnitude as the modulus measured by Houchmandzadeh *et al.* (1997), the lower modulus observed by Poirier *et al.* (2000) indicates that the aspiration technique overestimates chromosome elasticity. To date we have carried out about 100 chromosome stretching experiments on newt mitotic chromosomes, and in accord with Nicklas (1983), we find appreciable variation in the force constant, roughly from $f_0 = 0.5$–2 nN (see histograms of Figure 7c). Unfortunately there are no obvious cytological markers on newt chromosomes (for a karyotype see Hutchison and Pardue, 1975) so we are unable at present to determine whether particular newt chromosomes have consistently higher or lower force constants.

A feature of chromosome stretching which is quite obvious in all the above studies is that mitotic chromosomes do not narrow as they are stretched, in the reversible elastic regime. Our measurements (Figure 7a inset) indicate that the fractional decrease in chromosome width is less than 0.1 times the fractional chromosome length increase. For a solidly bonded elastic medium, this ratio is usually close to 0.5, corresponding to volume conservation. By contrast, the volume of a mitotic chromosome actually increases as it is being stretched. This can only occur if the fluid medium surrounding the chromosome flows into it as it is stretched, and in turn this indicates that the chromatin fibers inside a mitotic chromosome do not adhere to one another.

We have recently improved a number of aspects of this experiment. First, we obtained a newt eye lens epithelial tissue culture line (TVI line, Reese, 1976) which provides many more metaphase cells per experiment dish. Second, we developed a technique of using a micropipette loaded with a 0.05% solution of Triton X-100 in 60% PBS, which we spray onto the surface of a mitotic cell to produce a hole through which the mitotic chromosomes are disgorged. Finally, we now generally anchor the force-measuring pipette to the sample to reduce its mechanical drift. The updated method leads to results consistent with those of Poirier (2000). A chromosome force measurement using a chromosome Triton-extracted from a TVI cell is shown in Figure 7b; its initial elastic response is reversible and linear with a force constant near to 1 nN.

We have also carried out experiments on *Xenopus* A6 tissue culture cells. These amphibian cells are very similar to newt cells, but have smaller chromosomes ($n = 18$, haploid DNA content $\sim$3 pg = 3 Gb, or about 160 Mb/chromosome). These chromosomes can be isolated and manipulated at prometaphase; they show the same general elastic properties as newt chromosomes, with a force constant of about 1 nN, and a Young modulus $Y \sim 1000$ Pa (Poirier *et al.*,

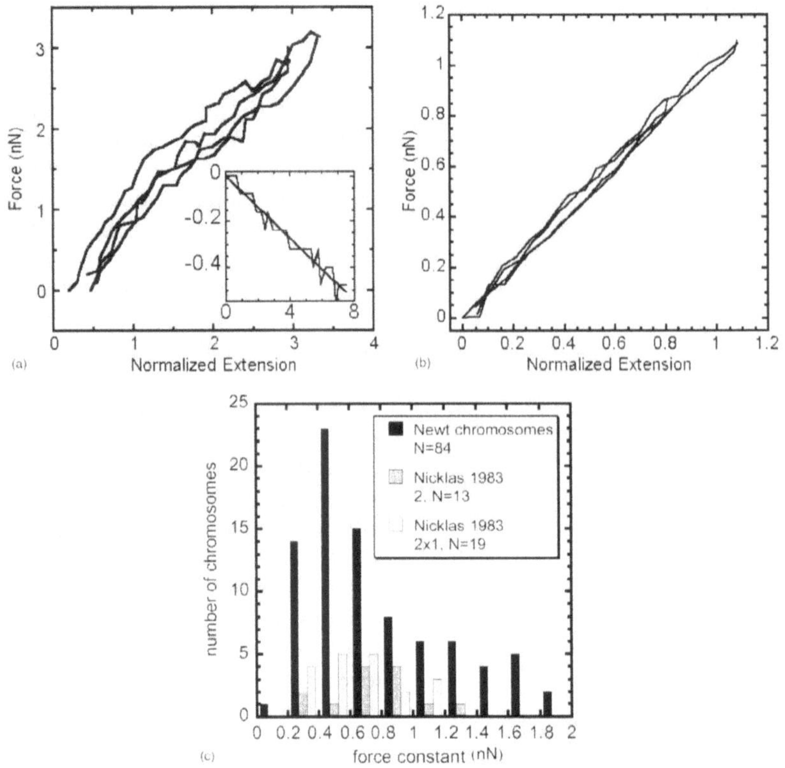

Fig. 7. Force-extension data for newt chromosomes. (a) Data from Poirier *et al.* (2000) for primary cultures of newt lung cells. The different curves show successive extension-retraction cycles; their coincidence indicates that the chromosome has reversible elasticity over the fourfold range of extension shown. The elastic response is nearly linear, and the initial force increase shows that the chromosome force constant is about 1 nN. Inset of (a) shows the fractional change in chromosome width as a function of extension, and indicates that the chromosome Poisson ratio is less than 0.1. (b) Data for newt TVI cell line for small extensions (up to two times native length), after chromosome extraction using dilute Triton X-100 (see text). In this range, the chromosome elastic response is strikingly linear, again with a force constant near 1 nN. (c) Histogram of force constants of 84 extracted newt prometaphase chromosomes, plus histograms of *in vivo* force constants of grasshopper spermatocyte metaphase I chromosomes (2 chromatids), and single chromatids (Nicklas, 1983). The single-chromatid grasshopper data has had forces doubled for direct comparison with the two-chromatid data sets. The distributions of force constants are essentially identical in the newt and grasshopper systems.

2002b). Thus, grasshopper, newt and frog mitotic chromosomes all require roughly 1 nN of force to be doubled in length; this level of force constant corresponds to Young moduli of roughly 500 Pa.

In vitro *assembled chromosomes*

Houchmandzadeh and Dimitrov (1999) carried out an important study of the mechanical properties of mito tic chromatids assembled *in vitro*, using *Xenopus* egg extracts. It is important to note that the system studied is assembled from sperm DNA, and as a result isolated chromatids are assembled. Micropipettes were used to grab, manipulate, and stretch the chromatids; force measurement was done via observation and calibration of micropipette bending (stiffnesses ~1 nN/μm) using the same general scheme as shown in Figure 6. Stretching experiments were carried out in buffer, following chromosome assembly.

The *in vitro*-assembled chromatids display stretching elasticity similar to that of chromosomes isolated from cells. For small extensions, linear elasticity was observed, with a force constant ~1 nN, and Young modulus $Y \sim 1000$ Pa. However, for extensions beyond about two times native length, the force observed during retraction is significantly less than the force observed during extension, indicating that irreversible changes have occurred. Finally, for extensions about 10 times native length, the *in vitro*-assembled chromatids show a force 'plateau'. Houchmandzadeh and Dimitrov (1999) also present an explanation for mitotic chromatid elastic response in terms of a titin-like elastic 'core'.

Roughly, the *in vitro* chromatids have stretching elasticity rather similar to chromosomes from cells, but are somewhat more fragile at higher extensions. It would be of great interest to have stretching data on *replicated in vitro*-assembled chromosomes, which would have two duplicate chromatids; replicated chromosomes can be assembled using 'cycling' egg extracts (Smythe and Newport, 1991).

Relation to chromatin fiber elasticity

The initial elastic response of mitotic chromosomes is not due to gross alteration of chromatin fiber structure, as can be seen from comparison of chromosome (Figure 7) and chromatin (Figure 4b) elastic responses. Chromatin fibers display reversible elasticity with a force constant of roughly 5 pN (pN $= 10^{-12}$ N). Since there will be on the order of a few thousand chromatin fibers in a chromosome cross-section (the chromosomes discussed above are roughly a micron in cross-section, and each fiber is roughly 30 nm thick), the 1 nN force at which chromosome length is doubled corresponds to a maximum force per chromatin fiber of a fraction of a pN. Therefore chromatin fiber structure is not being appreciably altered when chromosomes are being stretched by a factor of two; the initial elastic response of chromosomes must be due to modification of larger-scale condensed chromatin structure. The relatively low modulus of the chromosome indicates that large-scale chromatin structure is remarkably soft, yet elastic.

Bending elasticity of chromosomes

The Young modulus definition (2) strictly applies only to homogeneous materials. The degree to which this assumption is correct can be checked by measuring chromosome bending, and then comparing the bending stiffness with (3), the result expected for a rod made of a homogeneous elastic material.

The main result of experiments that compare elongational and bending stiffness of chromosomes is that *in vivo* (and for chromosomes extracted from cells), bending and stretching properties are related in the way that we expect for uniform elastic media (Poirier *et al.*, 2002b). This indicates that chromosome elasticity is due to the bulk of the cross-section of the chromosome, and is not mainly due to a thin, stiff, central structure. Remarkably, the *in vitro* assembled *Xenopus* chromatids studied by Houchmandzadeh and Dimitrov (1999) are far more flexible than one would expect from their Young modulus of about 1000 Pa. This is a strong indication that *in vitro* assembled mitotic chromatids have an internal structure which is distinct from that of *in vivo* mitotic chromosomes.

Bending moduli of chromosomes have been measured using observation of spontaneous thermal bending fluctuations. This approach has been used to measure the bending elasticity of a number of filamentous cell structures; elegant experiments of this type by Gittes *et al.* (1993) were used to measure the bending rigidity of actin filaments and microtubules. The bending modulus B is inversely proportional to the amplitude-squared of bending fluctuations. The simplest experiment to envision is one where one end of the filament being studied is anchored to a solid object (e.g. a very stiff micropipette, Figure 8a). As one moves down the rod from the anchor point, the amplitude of fluctuation perpendicular to the rod will increase. In the case where the fluctuations are small (rod length small compared to rod persistence length), we expect:

$$\overline{u^2} = \frac{32kT\ell^3}{\pi^4 B} \tag{5}$$

where the bar indicates the average of the fluctuation-amplitude-squared (Poirier *et al.*, 2002b).

Bending flexibility of extracted mitotic chromosomes

In the previous section we saw that the Young modulus of a mitotic newt chromosome inferred from stretching was

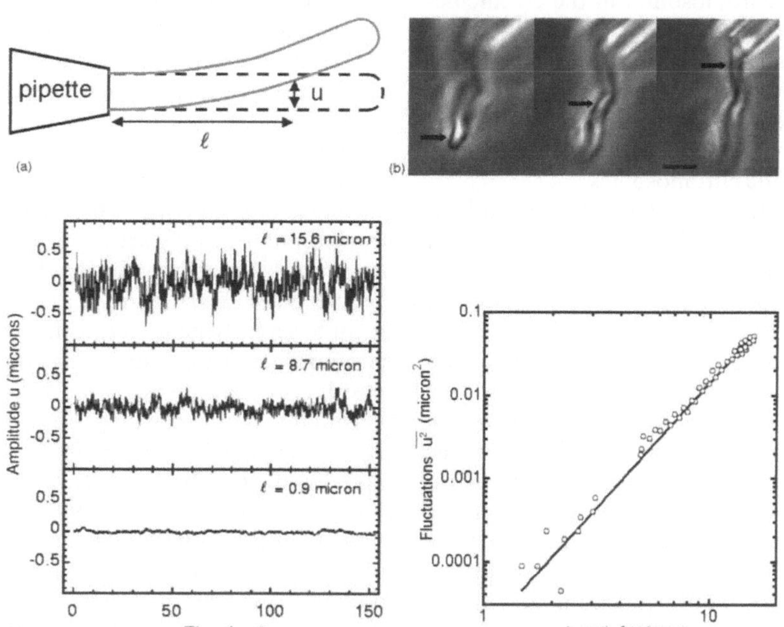

Fig. 8. Measurement of bending fluctuations for newt (*N. viridescens*) prometaphase chromosomes. (a) Rod geometry showing transverse fluctuation-amplitude u as a function of distance ℓ from the pipette along the rod. (b) Chromosome anchored at one end in a pipette. Bar is 4 μm. (c) Amplitude time series as a function of time for the three positions indicated by arrows in part b; the distance ℓ from the anchored end is indicated in each panel. With increasing distance from the anchor, the fluctuations increase. (d) Mean square fluctuations vs. distance from anchor point; on a log–log plot the data fall on the cubic power law given by Equation (3). The fit shown allows the bending modulus to be extracted from the data.

roughly $Y = 500$ Pa. Using Equation 3, a chromosome cross-section radius of $r = 0.8$ μm, we obtain an expected $B = 1.6 \times 10^{-22}$ N m^2. Plugging this expected value of B into Equation 5, and the maximum chromosome length $L = 20$ μm, we find the root-mean-square fluctuation $\sqrt{\overline{u^2}} = 0.3$ μm. Observed fluctuations of newt chromosomes (Figure 8c) have about this amplitude, and lead to an estimate of B between 1 and 3×10^{-22} N (Poirier *et al.*, 2002b). The newt chromosomes have bending stiffness consistent with their stretching elasticity via (3). Furthermore, they show no sign of hinges or other easily bent regions along their length.

Xenopus A6 chromosomes were found to be somewhat more flexible, with bending moduli between 5 and 20×10^{-24} N m^2. This flexibility is due to the smaller cross-section of the frog chromosomes ($r \sim 0.5$ μm) and is consistent with measured A6 Young moduli (200–800 Pa, Poirier *et al.*, 2002b).

Bending fluctuations of chromosomes in vivo

To check the relation between bending moduli of newt chromosomes extracted from cells and *in vivo*, it would be useful to have data for mitotic chromosomes in live newt cells. Marshall *et al.* (2001) first did this, using observations of bending fluctuations to measure the bending rigidity of mitotic chromosomes in *Drosophila* embryo cells. During mitosis, the mitotic spindle induces large bending fluctuations. Marshall *et al.* therefore compared native cells (large non-thermal bending fluctuations) with colchicine-treated cells (no microtubules, and therefore much smaller bending fluctuations). The small fluctuations of the *Drosophila* chromosomes in the colchicine-treated cells led to an estimate of $B = 6 \times 10^{-24}$ N m^2, and a Young modulus estimate of 40 Pa. The much larger fluctuations in the native cells were then used to quantify the forces being applied to the chromosomes by the mitotic apparatus. No stretching data are available for *Drosophila* embryo mitotic chromosomes.

We used the basic method of Marshall *et al.* colchicine-treating newt cells, to obtain *in vivo* thermal bending data for mitotic chromosomes (Poirier *et al.*, 2002b). Using a variation on the approach described above an *in vivo* estimate of B of 0.2–0.5 $\times 10^{-22}$ N m^2 was obtained, about a factor of four smaller than obtained for isolated chromosomes.

The somewhat smaller values of B obtained *in vivo* may reflect a change in physical properties due to the chemical differences between cytoplasm and the extracellular medium. Alternately, there may be sources of non-thermal fluctuation which are weak and which are not disrupted by colchicine treatment. SMC 'condensin' proteins have a possible motor function, and could result in forces on top of thermal forces which tend to move chromosomes around. Also, the live-cells continue to crawl on their substrate, and it may be that cytoplasmic flows driven by cell crawling cause non-thermal fluctuations. Since non-thermal forces will generally increase bending fluctuations, we can expect the *in vivo* measurements to provide lower bounds on B. We conclude that newt chromosomes have B *in vivo* comparable to that measured in the extracellular medium, roughly 10^{-22} N m^2.

Bending flexibility of in vitro *assembled Xenopus chromatids*

Houchmandzadeh and Dimitrov (1999) measured the bending stiffness of mitotic chromatids assembled in *Xenopus* egg extracts. They observed that the roughly 20 μm-long chromatids were very flexible, and in precise experiments measured $B = 1.2 \times 10^{-26}$ N m^2. This is about 1000 times smaller than the value of B that we have obtained for chromosomes from *Xenopus* A6 cells. The *in vitro*-assembled chromatids are so flexible that they undergo polymer-like bending fluctuations. The thermal persistence length of the *in vitro*-assembled chromatids is $A = B/(kT) = 2.5$ μm, and movies of *in vitro*-assembled chromatids display observable dynamical bending on a few-micron length scale (S. Dimitrov, private communication).

This extreme flexibility led Houchmandzadeh and Dimitrov (1997) to suggest that the *in vitro*-assembled chromosomes should be organized around a thin core, which would provide stretching elasticity, but with very little bending rigidity. They propose that a few molecules of titin, suspected to be a chromosomal component (Machado *et al.*, 1998; Machado and Andrew, 2000a, b) could produce the observed elastic response.

The 1000-fold difference in B indicates that *in vitro*-assembled chromosomes must have a different internal structure from chromosomes in somatic cells. An interesting question is whether *in vitro*-assembled chromosomes which are run through a round of DNA replication so that they are chromatid pairs, have a larger bending rigidity consistent with their Young modulus.

Bending of chromosomes during mitosis

If one observes cells in culture going through mitosis, chromosomes can be observed to be bent during prometaphase as they are being aligned, and then during anaphase as the chromatids are being pulled towards the spindle poles. During anaphase, the chromosomes can be quite severely bent, and to the eye it appears that the chromosome arms are being pulled back by some retarding force.

Roughly, the retarding force needed to bend a chromosome into an anaphase 'U' shape is the bending modulus divided by the square of the width of the 'U' (Houchmandzadeh *et al.*, 1997). For a newt chromosome with $B = 10^{-22}$ N m^2 and a U-width of a few microns this retarding force is roughly 10^{-11} N. A basic question is whether this force is plausibly due to viscous drag. The drag force on the chromosome will be roughly its length times viscosity times its velocity; for newt chromosomes ($L = 10$ μm, cytoplasm viscosity = 0.01 Pa s, velocity = 0.01 μm/s) we obtain a drag force of about 10^{-15} N. Drag cannot generate the relatively large force needed to bend an anaphase chromosome (Nicklas, 1983).

Viscoelasticity of mitotic chromosomes

If one stretches a chromosome rapidly enough, a stretching force in excess of the equilibrium force will be required, since the stress in the chromosome will be partly due to the intrinsic elasticity, plus additional *viscous* stress associated with the fact that the chromosome internal structure is not able to reach its equilibrium at each moment in time. This effect was observed (Poirier *et al.*, 2000). The internal viscosity of a chromosome will also limit the rate at which it retracts following release of stress, as observed by Nicklas and Staehly (1967). This relaxation time can be related to a viscosity using the Young modulus via $t_0 Y = \eta'$ (Poirier *et al.*, 2001a). Plugging in $t_0 = 1$ s and $Y = 500$ Pa estimates an internal viscosity of η' of 500 Pa s, more than 10^5 times that of water.

Experiments specifically aimed at determining internal viscosity of a prometaphase newt chromosome (Poirier *et al.*, 2001a) were done using chromosomes attached to pipettes, by rapidly moving one pipette while acquiring visual data for the deflection of the other, force-measuring pipette. Generally, one observes an initial force pulse just shortly after the pipette is moved, followed by a decay to some final equilibrium force. In the linear regime (stretching to less than three times native length) gives $\eta' = 100$ Pa s, about 10^5 times the viscosity of water. The relaxation of observed width is faster than that of force, ruling out the possibility that the slow dynamics are due to gel-filling hydrodynamics (Tanaka and Fillmore, 1979).

A remaining possibility to explain the large η' is that when the chromosome is rapidly stretched, the chromatin inside it must rearrange, and the time needed for this rearrangement is on the order of 1 s. Entanglement dynamics of the long, tethered chromatin domains can easily be on this time scale (de Gennes, 1979, pp. 230–233, Poirier *et al.*, 2001a). A second measurement of chromosome internal viscosity is obtained from analysis of bending fluctuation dynamics (Poirier and Marko, 2002c). The characteristic lifetime of the fluctuations ($t_0 \sim 0.7$ s, see Figure 8c) again indicates $\eta' \sim 100$ Pa s, providing further evidence of chromatin domains undergoing conformational fluctuations on the second timescale.

Combined biochemical-micromechanical study of mitotic chromosomes

A direct method to analyze mitotic chromosome structure is to use changes in chromosome elasticity as an indicator of *changes* in chromosome structure introduced chemically. The strategy of real-time observation of chemical reactions on whole chromosomes is rather old. For example, actinomycin-D was used to release the RNA transcripts from the large 'puffed up' loops on amphibian lampbrush chromosomes; the subsequent collapse of the loops showed that their open morphology was due to active transcription (Izawa, 1963; Callan, 1982, 1986, p. 109). The new feature discussed below is mechanical measurement during such experiments.

Whole-genome-extraction experiments

Maniotis *et al.* (1997) developed a technique for extracting whole genomes from human and bovine tissue culture cells, during interphase (i.e. from the nucleus) and during mitosis. Microneedles were used to 'harpoon' either interphase nucleoli, or mitotic chromosomes. Chemical experiments were then done on the extracted genomes while observing on the inverted microscope. Remarkably, when these extractions are done, the whole genome (i.e. essentially all the chromatin) is obtained, due to mechanical connections between the chromosomes. These interchromosome connections are invisible fibers (evidenced by their mechanical effects), which are RNAase and protease sensitive, but which are cut by DNAase and micrococcal nuclease. They conclude that the chromosomes of mammalian cells are *connected together* at the chromatin level, i.e. that the molecule which holds the genome together is DNA.

Two experiments of Maniotis *et al.* (1997) on metaphase chromosomes are highly relevant here. First, it was observed that mitotic chromosomes can be rapidly decondensed by introduction of drops of high concentrations of ions (500 mM $MgCl_2$, 500 mM $CaCl_2$, 1 M $CuCl_2$, 1 M NaCl), and that this decondensation was reversible, unless very high concentrations of ions were used. These experiments indicate that mitotic chromatin is compacted by interactions of primarily ionic character, and suggest that the condensation of the mitotic chromosome is not a precise folding, since it can be cycled chemically on a short timescale.

Second, Maniotis *et al.* (1997) use drops of proteases (trypsin 5 mg/ml, proteinase K 50 mg/ml) to examine the role of proteins in mitotic chromosome organization. It is found that these enzymes cause rapid decondensation of chromosomes into 'swollen clouds'. Remarkably, the decondensed chromosomes could be recondensed by adding linker histone H1 at 1 mg/ml. Core histones and other non-histone proteins could not produce this effect. Apparently, the main effect of protein digestion is to disrupt nucleosome stacking interactions, since mitotic chromosome morphology can be 'rescued' using H1.

It is striking that H1 is sufficient for this rescue, since one might imagine that other, rarer proteins which define higher-order chromatin structure (i.e. above the level of the 30 nm fiber) would be cut by the proteases, and that this would limit the degree of recondensation that H1 could effect. Perhaps the large concentration of H1 and its accessibility (H1 is chemically exchanging on short timescales, Lever *et al.* (2000), Misteli *et al.* (2000)) make it a main target in this experiment, while the rare and perhaps other well-buried proteins which stabilize higher-level chromosome structure remain undamaged.

Combined micromechanical-chemical experiments

Our recent experiments focus on combining the *in situ* biochemical treatments with single-chromosome elasticity measurements, the aim being to do real-time quantitative monitoring of chromosome structure changes. Our focus is on study of mitotic chromosome structure. Mitotic chromosomes are extracted and set up for two-pipette micromanipulation, and their initial, native stretching elastic response is measured. Then, a third pipette of ID ~4 μm, larger than the chromosome-grabbing pipettes, which has been loaded with some reagent in suitable solution (typically 60% PBS or Tris buffer, pH 7.6; see Figure 6) is brought within 10 μm of the chromosome. Pressure is then used to spray the reagent at the chromosome. Calibration experiments using fluorescent dyes show that this results in a jet of reagent exiting the pipette, with concentrations near to those in the pipette up to 20 μm away. Beyond this distance, the reagent rapidly diffuses into the large (~1 ml) volume of the sample dish. In a typical experiment, volumes of a few thousand cubic microns are typically sprayed (1000 cubic microns is 10^{-12} l = 1 picolitre). Any reagent can be used, with reaction kinetics micromechanically observed via the force-measuring pipette. When reagent flow is stopped, the chromosome is rapidly (<1 s) returned to the initial (extracellular) buffer condition, in which the reaction's effect on elastic properties can be measured.

Shifts in ionic conditions can decondense or hypercondense mitotic chromosomes

Using our microspraying techniques we quantified the effect of shifts in salt concentration, and we have reproduced the abrupt decondensation effects reported by Maniotis *et al.* (1997) with >200 mM univalent and divalent salt concentrations (Poirier *et al.*, 2002a). In experiments where force was monitored, we found that applied tension could be entirely eliminated using high concentrations of Na^+ and Mg^{2+}; however, after the ~10 s sprays end, the chromosome folds back up into a native-like structure. Following sufficiently long (>10 min) exposures to high salt concentrations, mitotic chromosomes do not fully recondense, presumably as a result of protein loss.

We have also found that in the range 20–100 mM Mg^{2+} and Ca^{2+} concentration, mitotic chromosomes go through a range of rather strong *condensation*, generating contractile forces of up to 0.2 nN. As divalent cation concentration is ramped up from zero, we observe condensation near 20 mM, followed by an abrupt return to the native degree of compaction near 50 mM, and then finally strong decondensation at >200 mM. A similar condensation–decondenation effect is observed with increasing concentration of trivalent ions. No compaction occurs with any concentration of Na^+ or K^+. All these decondensation and condensation effects occur *isotropically*; under zero tension, the fractional length and fractional width changes are nearly equal. This behavior is easily reconciled with a chromosome model, which is an isotropic network of chromatin fibers, and is difficult to square with an anisotropic chromatin-loop-attached-to-scaffold model. The condensation effects show that mitotic chromatids are not near their maximum possible compaction (we observe up to a 30% volume decrease using trivalent ions, Poirier *et al.* (2002a)), and that charge interactions are important to controlling chromatin compaction.

Effects of divalent ions are in line with similar reentrant bundling (i.e. a bundling followed by an dissolution as divalent ion concentration is raised) of stiff polyelectrolytes recently observed, in DNA (Pelta *et al.*, 1996; Saminathan *et al.*, 1999) and actin solutions (Tang *et al.*, 1996). The condensation may be due to bridging interactions, i.e. net attractive interactions induced by localization of the multivalent ions (Ha and Liu, 1997). An alternative explanation is that the condensation occurs when the charge-neutral point is reached, eliminating coulomb repulsion and allowing other, attractive interactions to dominate (Nguyen *et al.*, 2000; Nguyen and Shklovskii, 2001).

Micrococcal nuclease completely disintegrates mitotic chromosomes

Micrococcal nuclease (MNase) non-specifically cuts dsDNA, and is widely used to cut chromatin between nucleosomes. We were motivated to use MNase to determine whether or not the internal protein 'scaffold' (Earnshaw and Laemmli, 1983; Boy de la Tour and Laemmli, 1988; Saitoh and Laemmli, 1994) was mechanically contiguous. A second aim of the experiments was to determine just how much of the chromosome elastic response was due to chromatin (i.e. dsDNA) itself.

We sprayed isolated newt TVI mitotic chromosomes with 1 nM MNase in suitable reaction buffer (60% PBS plus 1 mM $CaCl_2$), with a small tension (0.1 nN) initially applied. When the spray starts, chromosome tension (Figure 9) briefly jumps due to the slight compaction induced by the divalent Ca^{2+}, but then the tension drops below our force resolution (~0.01 nN = 10 pN) after 30 s. During this initial period, the morphology of the chromosome is unaffected. Then, over the time interval 100–200 s, the chromosome disintegrates, and is eventually severed. This experiment indicates that the force-bearing and structural element of the mitotic chromosome is DNA-based, i.e. chromatin itself, and indicates that the chromosome is not held together by a mechanically contiguous internal protein 'scaffold' (Poirier and Marko, 2002d).

A second type of experiment where digestion was done with zero applied force was stopped before the chromosome was morphologically altered (at 30 s of 1 nM exposure). The chromosome then could be extended into a string of blobs, connected by thin chromatin strands. The strands could be severed by a

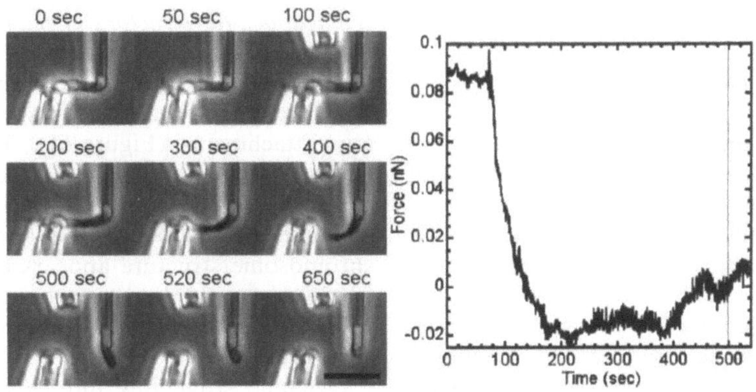

Fig. 9. Time course of tension in a chromosome, and chromosome morphology, during digestion by 1 nM MNase, with initial tension 0.1 nN. Spraying starts at 80 s; force decays after 30 s; chromosome is cut after 450 s. The spray pipette can be seen in the upper center of the $t > 120$ s frames. Bar is 10 μm.

brief spray of MNase, where the peak tension applied was < 100 pN. This latter experiment makes clear that the disassembly effect observed using MNase is not tension-dependent. The forces applied in this experiment are below those required to break single protein or nucleic acid chains.

These experiments imply that the mitotic chromosome is essentially a cross-linked network of chromatin, i.e. that higher-order chromosome structure is stabilized by non-DNA molecules (most likely proteins), which are isolated from one another. It is difficult to reconcile the MNase results with chromatin loops hanging from an internal mechanically contiguous protein scaffold.

Restriction enzymes with 4-base specificity can disintegrate mitotic chromosomes

We also carried out experiments with blunt-cutting restriction enzymes, which cut dsDNA at specific base-pair sequences (Poirier and Marko, 2002d). Enzymes were selected that were active in physiological-like buffers (i.e. pH near 7, ionic conditions near 100 mM univalent plus ~10 Mm divalents). Two enzymes with 4-base recognition sequences, Alu I (AG^CT) and Hae III (GG^CC) (which occur every 256 bases on random-sequence DNA) were used, and they cut up mitotic chromosomes in the fashion of MNase. Alu I severs the chromosome completely after < 100 s (Figure 10; force increase at spraying onset is condensing effect of the ~10 mM Mg^{2+} in the enzyme buffer, easily understood given our salt experiments). After factoring in the 10-fold reduction in sequence accessibility in chromatin vs. bare DNA, this experiment shows that mitotic chromosomes are not cross-linked more often than once every few kb.

Experiments with 6-base-recognition sequence enzymes Stu I (AGG^CCT) and Dra I (TTT^AAA) show essentially zero force effect (Figure 10), indicating that the accessible 6-base sites are rarer than chromatin cross-links. To test further the accessibility of 6-base-wide sites, we also used Cac8I where 4 bases are

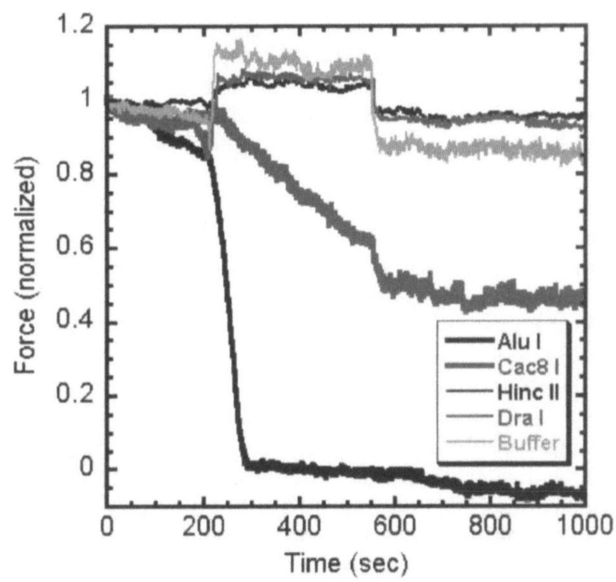

Fig. 10. Time course of tension in a chromosome during digestion by blunt-cutting restriction enzymes. Initial force in all experiments was 0.6 nN; each force curve is normalized to this initial value. Enzyme exposure is from 200 to 550 s. Bottom curve shows Alu I completely reducing force to zero (cutting chromosome completely). Middle curve shows Cac 8I only partially relaxing applied tension (partially cutting chromosome) Top curves show only small effects of Hinc II, Dra I and restriction-enzyme activity buffer (no enzyme). The step increases in force for the top curves reflect the slight condensation of the chromosome reversibly driven by the divalent ions of the activity buffer.

recognized out of a 6-base region (GCN^NGC). This enzyme shows an intermediate effect (Figure 10), partially reducing applied force, but not totally severing the chromosome. Thus, the 6-base site size is partially available to the restriction enzymes. As for MNase, these results are consistent with a chromatin network model with a cross-link every few tens of kb, and are inconsistent with an internal-protein-scaffold model, unless the 'scaffold' has the form of many small localized protein structures which is of course again a cross-linked network of chromatin.

Conclusion

Mechanical properties of mitotic chromosomes

Mitotic chromosomes stretch and bend as if they are classical elastic media, but with an enormous range of extensibility. Mitotic newt chromosomes can be reversibly stretched fivefold, and over this range their elastic response is nearly linear with a Young (stretch) modulus of about 500 Pa. The mitotic chromosomes of newt and *Xenopus* are therefore doubled in length by forces $\sim$1 nN, similar to the elastic response of grasshopper spermatocyte metaphase I chromosomes (Nicklas, 1983), and also similar to the maximum forces applied by the mitotic spindle to chromosomes during anaphase. Houchmandzadeh and Dimitrov (1999) found similar stretching behavior in their study of *in vitro* assembled *Xenopus* chromatids.

Our observation of newt and *Xenopus* chromosome bending stiffnesses in accord with their stretching properties is distinct from the extreme bending flexibility observed for *in vitro* assembled *Xenopus* chromatids (Houchmandzadeh and Dimitrov, 1999). *In vitro* assembled chromatids must have an internal structure quite distinct from metaphase chromosomes *in vivo*. An important experiment is therefore measurement of the bending flexibility of *in vitro chromosomes* assembled in egg extracts and cycled through one round of DNA replication.

DNA-cutting experiments

Cutting dsDNA inside the mitotic newt chromosome with sufficient frequency completely disconnects the chromosome. MNase and 4-base blunt-cutting restriction-enzymes dissolve the chromosome into optically invisible fragments. By far the simplest interpretation of this experiment is that the elastic response and mechanical continuity of the mitotic chromosome is due to chromatin fiber, i.e. DNA itself. A rough estimate of the genomic distance between cuts required to disconnect the chromosome is 15 kb, based on the gradual reduction in effect of more rarely cutting restriction enzymes. 6-base blunt-cutting restriction enzymes have no effect on the mechanical properties of whole mitotic newt chromosomes.

Implications for structure of the mitotic chromosome

We suggest that the mitotic chromosome has a *network* structure, i.e. is organized by isolated chromatin-chromatin attachments (Figure 11). The purely mechanical measurements (stretching and bending) indicate that chromosome stretching is supported by stress spread across its whole cross-section, and therefore that mitotic chromosome structure appears to be, at the scale of a whole chromatid, homogeneous. This hypothesis is also supported by the homogeneous way whole chromosomes elongate.

Dynamic stretching and bending experiments both show that mitotic chromatin relaxes extremely slowly, on a roughly 1 s time. We hypothesize that this long timescale is due to chromatin conformational fluctuation, and that the long timescale has its origin in entanglements. This implies that mitotic chromatin is not heavily constrained by chromatin-folding proteins, i.e. that there are long stretches of chromatin between 'cross-links'. These stretches of chromatin are apparently free to undergo slow conformational motions.

Shifts in ionic conditions can rapidly decondense and overcondense a mitotic chromosome. These morphological changes are reversible for short (10 s) salt treatments, and at zero stress are isotropic, again suggesting a homogeneous and disordered mitotic chromatin organization. At least 1/3 of chromosome volume is mobile cytoplasm or buffer based on condensation experiments. Finally, the DNA-cutting experiments indicate that the mechanically contiguous structural element of the mitotic chromosome is DNA (i.e. chromatin) itself. The non-chromatin fiber content of the mitotic chromosome must be disconnected. We rule out models for mitotic chromosome structure based on mechanically contiguous non-DNA skeletons or scaffolds, in favor of a network model. At present, the identity of the chromatin cross-linkers (the network 'nodes') is unknown; at present the most likely suspects are the condensin-type SMC protein complexes.

Future experiments

Combined chemical-micromechanical experiments provide complementary information to usual biochemical

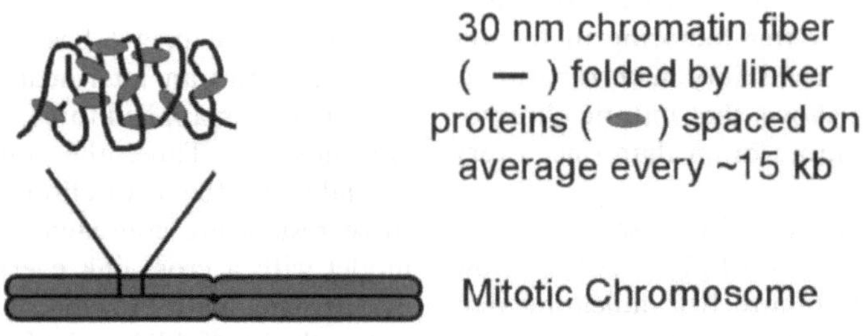

Fig. 11. Network model of a mitotic chromatid. Black curve indicates the single linear chromatin fiber, gray blobs show isolated non-DNA cross-linking elements. If the chromatin is cut sufficiently often, the chromosome will be severed; the non-DNA cross-linkers are not mechanically contiguous through the chromosome.

assay and microscopy approaches. Traditional biochemical approaches give information about local interactions and the products of chemical reactions. Traditional microscopy gives information about morphology and structure in a given cell state, or in a given preparation of molecules. Our approach allows the direct study of elasto-mechanical properties of chromosomes, and to observe how those properties are modified dynamically by chemical reactions. Flexibility and connectivity of chromatin fiber in the mitotic chromosome are difficult to study by traditional biochemical and microscopy approaches, but are rather obvious results of a combined chemical-micromechanical approach. The question of mitotic chromosome organization therefore can be profitably attacked by integrating information from all these approaches.

A basic variation on the stretching experiments would be study of relative elasticity of different regions of the mitotic chromosome. Use of labels for centromere, telomere and euchromatin regions of the chromosome would allow the elasticity of different types of chromatin to be studied. For example, elasticity of the kinetochore is relevant to modeling of chromosome capture by the mitotic spindle (Joglekar and Hunt, 2002).

To further analyze our network model of the mitotic chromosome it is important to analyze the sizes of chromatin fragments produced by enzyme microdigestion. This could be done via aspiration of the fragments followed by fluorescence quantification of them after dispersal onto a slide. Also, further digestion experiments using other DNA cutters, RNAases, and proteases need to be done. Effects of other chemical modifications of chromatin (e.g. acetylation, phosphorylation) on mitotic chromosome condensation, monitored precisely via elasticity, would also be interesting. These kinds of experiments in general give information on the poorly understood question of enzyme access in dense chromatin.

Development of techniques to study the structural roles of specific proteins might be possible. We have already demonstrated antibody labeling using microspraying, for antihistone (Poirier *et al.*, 2000) and for anti-XCAPs (unpublished). The simplest types of experiments would be visualization of antibody binding patterns as a function of chromosome stretching. This general technique might also be useful for chromosome mapping (Clausen, 1994; Hliscs, 1997a, b), especially if different parts of a chromosome could be exposed to different reagents using microchannel arrays.

More ambitious experiments would use fluorophores which can generate large amounts of hydroxyls, lysing the antibody targets (Beerman and Jay, 1994). This technique has been used successfully to study disruption of cytoskeletal proteins, and might be used in conjunction with chromosome elasticity measurement to study the effect of condensin or cohesin disruption on mitotic chromosome structure. This type of experiment could directly test models of SMC function such as that of Losada and Hirano (2001).

Study of the orientational ordering of chromatin using polarization microscopy could be informative. Purified and concentrated nucleosomes have been demonstrated to form chiral liquid crystal phases (Leforestier *et al.*, 1999; Livolant and Leforestier, 2000); optical activity has also been observed for certain chromosomes (Livolant, 1978; Livolant and Maestre, 1988). A major question is whether animal chromosomes have similar liquid crystal organization, either in native or stretched forms. Preliminary experiments on newt chromosomes in our showed no detectable birefringence for chromosomes stretched up to four times native length, suggesting that ordered domains of mitotic chromatin are smaller than the wavelength of light, i.e. <100 nm, and that appreciable stretching of chromosomes does not induce strong orientational ordering of chromatin.

We have repeatedly observed interchromosome fibers between mitotic chromosomes as discussed by Maniotis *et al.* (1997), and these objects require further study. In initial experiments we have verified the result of Maniotis *et al.* (1997) that these fibers are cut by MNase, and therefore contain nucleic acid (most likely DNA). Rough stretching experiments show that these fibers are highly and reversibly extensible, with an estimated force constant in the nanonewton range. These are therefore a more folded structure than the 30 nm fiber, but because they are barely visible in the light microscope, we estimate their thickness to be less than 200 nm. DNA staining and quantification are an objective of our current studies. We also hypothesize that these fibers are telomeric structures (the interchromosome fibers at metaphase almost always come from chromosome ends), and therefore probes for telomere DNA should be tested. An interesting question is whether these fibers are intrinsic to transformed cells (most of our work is in tumor cell lines) and therefore parallel studies in primary cell cultures are of strong interest.

Other chromosome structures could be studied by combined chemical-micromechanical techniques. We are interested in comparing mitotic chromosomes to meiotic chromosomes. The range of physical structures occurring during meiosis provides a motivation for micromechanical experiments. Mechanical properties of meiotic chromosomes may play a crucial role in general recombination (Kleckner, 1996; Zickler and Kleckner, 1999), and may be related to polymer physics of the chromatin loops (Marko and Siggia, 1997a). Interphase chromosomes would be extremely interesting to study as isolated objects. Maniotis *et al.* (1997) have used purely mechanical techniques to extract whole interphase genomes, an important first step. We are searching for a biochemical method to open the nuclear envelope to allow gentler interphase genome extractions.

Finally, we note that Hinnebusch and Bendich (1997) have demonstrated that bacterial chromosomes can be extracted and physically studied. Cunha *et al.* (2001a, b) have succeeded in isolating and chemically manipulating *E. coli* nucleoids, which might also be studied using micromechanical techniques. The wide range of genetic and biochemical tools developed for *E. coli*, plus the many very basic and open questions regarding bacterial chromosome structure, make it a highly attractive system for micromanipulation study.

Acknowledgements

It is a pleasure to acknowledge the help and advice of Prateek Gupta, Tamar Monhait, Chee Xiong, Eric Siggia, Herbert Macgregor, Peter Moens, Chris Woodcock, Susan Gasser, Nancy Kleckner, Lynn Zecheidrich, Nick Cozzarelli, Tatsuya Hirano, Carlos Bustamante, Didier Chatenay, Bahram Houchmandzadeh, Albert Libchaber, Peter Moens, Joe Gall, Andrew Maniotis, Paul Janmey, Josef Kas, Wallace Marshall, John Sedat, Rebecca Heald, Abby Dernburg, Stefan Dimitrov, Ulrich Laemmli, Jon Widom, Ted Salmon, Lon Kaufman, and Arnold Kaplan. This research would not have been possible without the kind gift of the TVI cell line from David Reese. Experiments at UIC were supported by grants from the Whitaker Foundation, the NSF (DMR-9734178), Research Corporation, the Johnson and Johnson Focused Giving Program, the Petroleum Research Foundation of the American Chemical Society, and the University of Illinois Foundation.

References

Adachi Y, Luke M and Laemmi UK (1991) Chromosome assembly *in vitro*: topoisomerase II is required for condensation. *Cell* **64:** 137–148.

Allemand JF, Bensimon D, Lavery R and Croquette V (1998) Stretched and overwound DNA forms a Pauling-like structure with exposed bases. *Proc Natl Acad Sci* USA **95:** 14152–14157.

Alut JG and Nicklas RB (1989) Tension, microtubule rearrangements, and the proper distribution of chromosomes in mitosis. *Chromosoma* **98:** 33–39.

Anderson JD and Widom J (2000) Sequence and position-dependence of the equilibrium accessibility of nucleosomal DNA target sites. *J Mol Biol* **296:** 979–987.

Arents G, Burlingame RW, Wang BC, Love WE and Moudrianakis EN (1991) The nucleosomal core histone octamer at 3.1 A resolution: a tripartite protein assembly and a left-handed. *Proc Natl Acad Sci* USA **22:** 10148–10152.

Bak AL, Zeuthen J and Crick FH (1977) Higher-order structure of human mitotic chromosomes. *Proc Nat Acad Sci* USA **74:** 1595–1599.

Bak P, Bak AL and Zeuthen J (1979) Characterization of human chromosomal unit fibers. *Chromosoma* **73:** 301–315.

Beerman AEL and Jay DG (1994) Chromophore-assisted laser inactivation of cellular proteins. *Meth Cell Biol* **44:** 715–731.

Belmont AS, Sedat JW and Agard DA (1987) A three-dimensional approach to mitotic chromosome structure: evidence for a complex heirarchical organization. *J Cell Biol* **105:** 77–92.

Belmont AS, Braunfeld MB, Sedat JW and Agard DA (1989) Large-scale chromatin structural domains within mitotic and interphase chromosomes *in vivo* and *in vitro*. *Chromosoma* **98:** 129–143.

Belmont AS (2001), Visualizing chromosome dynamics with GFP. *Trends Cell Biol* **11:** 250–257.

Bennink ML, Leuba SH, Leno GH, Zlatanova J, de Grooth BG, Greve J (2001) Unfolding individual nucleosomes by stretching single chromatin fibres with optical tweezers, *Nat. Struct. Biol.* **8:** 606–610.

Boy de la Tour E and Laemmli UK (1988) The metaphase scaffold is helically folded: sister chromatids have predominantly opposite helical handedness. *Cell* **55:** 937–944.

Brown KT and Flaming D-C (1986) *Advanced Micropipette Techniques for Cell Physiology*, John Wiley & Sons, New York, pp. 139–141.

Brower-Toland BD, Smith CL, Yeh RC, Lis JT, Peterson CL and Wang MD (2002) Mechanical disruption of individual nucleosomes reveals a reversible multistage release of DNA. *Proc Nat Acad Sci USA* **99:** 1752–1754.

Bustamante C, Marko JF, Siggia ED and Smith S (1994) Entropic elasticity of lambda-phage DNA. *Science* **265:** 1599–1600.

Bustamante C, Smith SB, Liphardt J and Smith D (2000) *Curr Opin Struct Biol* **10:** 279–285.

Callan HG (1955) Recent work on the structure of cell nuclei. In: *Fine Structure of Cells. Symposium of the VIIIth Congress in Cell Biology, Leiden 1954*, Noordhof, Groningen, pp. 89–109.

Callan HG (1982) The Croonian Lecture, 1981. Lampbrush chromosomes. *Proc Roy Soc London B* **214:** 417–448.

Callan HG (1986) *Lampbrush Chromosomes*, Springer, New York.

Christensen MO, Larsen MK, Barthelmes HU, Hock R, Andersen CL, Kjeldsen E, Knudsen BR, Westergaard O, Boege F and Mielke C (2002) *J Cell Biol* **157:** 31–44.

Claussen U, Mazur A and Rubstov N (1994) Chromosomes are highly elastic and can be stretched. *Cytogenet Cell Gen* **66:** 120–125.

Cluzel P, Lebrun A, Heller C, Lavery R, Viovy JL, Chatenay D and Caron F (1996) DNA: an extensible molecule. *Science* **271:** 792–794.

Cook PR (1991) The nucleoskeleton and the topology of replication. *Cell* **66:** 627–637.

Cremer T, Kurz A, Zirbel R, Dietzel S, Rinke B, Schrock E, Speicher MR, Mathieu U, Jauch A, Emmerich P, *et al.* (1993) Role of chromosome territories in the functional compartmentalization of the cell nucleus. *Cold Spring Harb Symp Quant Biol* **58:** 777–792.

Cui Y, Bustamante C (2000) Pulling a single chromatin fiber reveals the forces that maintain its higher-order structure. *Proc Natl Acad Sci USA* **97:** 127–132.

Cunha S, Odijk T, Suleymanoglu E and Woldringh CL (2001a) Isolation of the *Escherichia coli* nucleoid. *Biochimie* **83:** 149–154.

Cunha S, Woldringh CL and Odijk T (2001b) Polymer-mediated compaction and internal dynamics of isolated *Escherichia coli* nucleoids. *J Struct Biol* **136:** 53–166.

de Gennes P-G (1979) *Scaling Concepts in Polymer Physics*, Cornell University Press, Ithaca NY.

Dekker J, Rippe K, Dekker M and Kleckner N (2002) Capturing chromosome conformation. *Science* **295:** 1306–1311.

Dietzel S and Belmont AS (2001) Reproducible but dynamic positioning of DNA in chromosomes during mitosis. *Nat Cell Biol* **3:** 767–770.

Earnshaw WC and Laemmli UK (1983) Architecture of metaphase chromosomes and chromosome scaffolds. *J Cell Biol* **96:** 84–93.

Gall JG (1956) On the submicroscopic structure of chromosomes. *Brookhaven Symp Biol* **8:** 17–32.

Gall JG (1963) Kinetics of deoxyribonuclease action on chromosomes. *Nature* **198:** 36–38.

Gasser SM, Laroche T, Falquet J, Boy de la Tour E and Laemmli UK (1986) Metaphase chromosome structure. Involvement of topoisomerase II. *J Mol Biol* **188:** 613–629.

Gittes F, Mickey B, Nettleson J and Howard J (1993) Flexural rigidity of microtubules and actin filaments measured from thermal fluctuations in shape. *J Cell Biol* **120:** 923–934.

Gregory TR (2001) Animal genome size database, http://www.genomesize.com

Guacci V, Koshland D and Strunnikov A (1997) A direct link between sister chromatid cohesion and chromosome condensation revealed through the analysis of MCD1 in *S. cerevisiae*. *Cell* **91:** 47–57.

Ha B-Y and Liu AJ (1997) Counterion-mediated attraction between two like-charged rods. *Phys Rev Lett* **79:** 1289–1292.

Hagerman PJ (1988) Flexibility of DNA. *Ann Rev Biophys Biochem* **17:** 265–286.

Harnau L and Reineker P (1999) *Phys Rev E* **60:** 4671–4676.

Hinnebusch BJ and Bendich AJ (1997) The bacterial nucleoid visualized by fluorescence microscopy of cells lysed within agarose: comparison of *Escherichia coli* and spirochetes of the genus Borrelia. *J Bacteriol* **179:** 2228–2237.

Hirano T and Mitchison J (1993) Topoisomerase II does not play a scaffolding role in the organization of mitotic chromosomes assembled in *Xenopus* egg extracts. *J Cell Biol* **120:** 601–612.

Hirano T and Mitchison J (1994) A heterodimeric coiled-coil protein required for mitotic chromosome condensation *in vitro*. *Cell* **79:** 449–458.

Hirano T (1995) Biochemical and genetic dissection of mitotic chromosome condensation. *TIBS* **20:** 357–361.

Hirano T, Kobayashi R and Hirano M (1997) Condensins, chromosome condensation protein complexes containing XCAP-C, XCAP-E and a Xenopus homolog of the Drosophila barren protein. *Cell* **89:** 511–521.

Hirano T (1998) SMC protein complexes and higher-order chromosome dynamics. *Curr Opin Cell Biol* **10:** 317–322.

Hirano T (1999) SMC-mediated chromosome mechanics: a conserved scheme from bacteria to vertebrates? *Genes and Dev* **13:** 11–19.

Hirano T (2000) Chromosome cohesion, condensation, and separation. *Ann Rev Biochem* **69:** 115–144.

Hliscs R, Muhlig P and Claussen U (1997a) The nature of G-bands analyzed by chromosome stretching. *Cytogenet Cell Genet* **79:** 162–166.

Hliscs R, Muhlig P and Claussen U (1997b) The spreading of metaphases is a slow process which leads to a stretching of chromosomes, *Cytogenet Cell Genet* **76:** 167–171.

Horowitz RA, Agard DA, Sedat JW and Woodcock CL (1994) The three dimensional architecture of chromatin *in situ*: electron tomograph reveals fibers composed of a continuously variable zig-zag nucleosomal ribbon. *J Cell Biol* **125:** 1–10.

Houchmandzadeh B, Marko JF, Chatenay D and Libchaber A (1997) Elasticity and structure of eukaryote chromosomes studied by micromanipulation and micropipette aspiration. *J Cell Biol* **139:** 1–12.

Houchmandzadeh B and Dimitrov S (1999) Elasticity measurements show the existence of thin rigid cores inside mitotic chromosomes. *J Cell Biol* **145:** 215–223.

Hutchison N and Pardue ML (1975) The mitotic chromosomes of Nothophthalamus (= Triturus) viridescens: localization of C banding regions and DNA sequences complementary to 18S, 28S, and 5S ribosomal DNA. *Chromosoma* **53:** 51–69.

Izawa M, Allfrey VG and Mirsky AL (1963) The relationship between RNA synthesis and loop structure in lampbrush chromosomes. *Proc Nat Acad Sci. USA* **49:** 544–551.

Jackson DA, Dickinson P and Cook PR (1990) The size of chromatin loops in HeLa cells. *EMBO J* **9:** 567–571.

Joglekar A and Hunt AJ (2002) A simple mechanistic model for directional instability during mitotic chromosome movement. *Biophys J* **83:** 42–58.

Kellermayer MSZ, Smith SB, Granzier HL and Bustamante C (1997) Folding-unfolding transitions in single titin molecules characterized with laser tweezers. *Science* **276:** 1112–1116.

Kimura K and Hirano T (1997) ATP-dependent positive supercoiling of DNA by 13S condensin: a biochemical implication for chromosome condensation. *Cell* **90:** 625–634.

Kimura K, Rybenkov VV, Crisona NJ, Hirano T and Cozzarelli NR (1999) 13S condensin actively reconfigures DNA by introducing global positive writhe: implications for chromosome condensation. *Cell* **98:** 239–248.

King JM, Hays TS and Nicklas RB (2000) Tension on chromosomes increases the number of kinetochore microtubules, but only within limits. *J Cell Sci* **113:** 3815–3823.

Kleckner N. Meiosis: how could it work? (1996) *Proc Natl Acad Sci USA* **93:** 8167–8174.

Kornberg RD (1974) Chromatin structure: a repeating unit of histones and DNA. Chromatin structure is based on a repeating unit of eight histone molecules and about 200 base pairs of DNA. *Science* **184:** 868–871.

Koshland D and Strunnikov A (1996) Mitotic chromosome condensation. *Annu Rev Cell Dev Biol* **12:** 305–333.

Laemmli UK (2002) Packaging genes into chromosomes, http://www.molbio.unige.ch/PACKGENE/PAGE1.html.

Ladoux B, Quivy JP, Doyle P, du Roure O, Almouzni G and Viovy JL (2000) Fast kinetics of chromatin assembly revealed by single-molecule videomicroscopy and scanning force microscopy. *Proc Nat Acad Sci USA* **97:** 14,251–14,256.

Landau LD and Lifshitz IM (1986) *Theory of Elasticity*. Pergamon, New York.

Leforestier A, Fudaley S and Livolant F (1999) Spermidine-induced aggregation of nucleosome core particles: evidence for multiple liquid crystalline phases. *J Mol Biol* **290:** 481–494.

Leger JF, Robert J, Bourdieu L, Chatenay D and Marko JF (1998) RecA binding to a single double-stranded DNA molecule: A possible role of DNA conformational fluctuations. *Proc Natl Acad Sci USA* **95:** 12,295–12,299.

Leger JF, Romano G, Sarkar A, Robert J, Bourdieu L, Chatenay D and Marko JF (1999) Structural transitions of a twisted and stretched DNA molecule. *Phys Rev Lett* **83:** 1066–1069.

Lever MA, Th'ng JP, Sun X and Hendzel MJ (2000) Rapid exchange of histone H1.1 on chromatin in living human cells. *Nature* **408:** 873–876.

Lewin B (2000) *Genes VII*. Oxford University Press, New York.

Li X and Nicklas RB (1995) Mitotic forces control a cell-cycle checkpoint. *Nature* **373:** 630–632.

Li X and Nicklas RB (1997) Tension-sensitive kinetochore phosphorylation and the chromosome distribution checkpoint in praying mantis spermatocytes. *J Cell Sci* **110:** 537–545.

Liphardt J, Onoa B, Smith SB, Tinoco I Jr. and Bustamante C (2001) *Science* **292:** 733–737.

Livolant F (1978) Positive and negative birefringence in chromosomes. *Chromosoma* **21:** 45–58.

Livolant F and Maestre MF (1988) Circular dichroism microscopy of compact forms of DNA and chromatin *in vivo* and *in vitro*: cholesteric liquid-crystalline phases of DNA and single dinoflagellate nuclei. *Biochemistry* **27:** 3056–3068.

Livolant F and Leforestier A (2000) Chiral discotic columnar germs of nucleosome core particles. *Biophys J* **78:** 2716–2729.

Lodish H, Baltimore D, Berk A, Zipursky SL, Matsudaria P and Darnell J (1995) *Molecular Cell Biology*, Scientific American, New York.

Losada A, Hirano M and Hirano T (1998) Identification of Xenopus SMC protein complexes required for sister chromatid cohesion. *Genes Dev* **12:** 1986–1997.

Losada A and Hirano T (2001) Shaping the metaphase chromosome: coordination of cohesion and condensation. *Bioessays* **23:** 924–935.

Luger K, Mader AW, Richmond RK, Sargent DF and Richmond TJ (1997) Crystal structure of the nucleosome core particle at 2.8 A resolution. *Nature* **389:** 251–260.

Machado C, Sunke CE and Andrew DJ (1998) Human autoantibodies reveal titin as a chromosomal protein. *J Cell Biol* **141:** 321–333.

Machado C and Andrew DJ (2000a) D-titin: a giant protein with dual roles in chromosomes and muscles. *J Cell Biol* **151:** 639–652.

Machado C and Andrew DJ (2000b) Titin as a chromosomal protein. *Adv Exp Med Biol* **481:** 221–236.

Manders EMM, Kimura H and Cook PR, Direct imaging of DNA in living cells reveals the dynamics of chromosome formation (1999) *J Cell Biol* **144:** 813–821.

Maniotis AJ, Bojanowski K and Ingber DE (1997) Mechanical continuity and reversible chromosome disassembly within intact genomes removed from living cells. *J Cell Biochem* **65:** 114–130.

Marko JF and Siggia ED (1997a) Polymer models of meiotic and mitotic chromosomes. *Mol Biol Cell* **8:** 2217–2231.

Marko JF and Siggia ED (1997b) Driving proteins off DNA with applied tension. *Biophys J* **73:** 2173–2178.

Marsden MP and Laemmli UK (1979) Metaphase chromosome structure: evidence for a radial loop model. *Cell* **17:** 849–858.

Marshall WF, Dernburg AF, Harmon B, Agard DA and Sedat JW (1996) Specific interactions of chromatin with the nuclear envelope: positional determination within the nucleus in Drosophila melanogaster. *Mol Biol Cell* **7:** 825–842.

Marshall WF, Straight A, Marko JF, Swedlow J, Dernburg A, Belmont A, Murray AW, Agard DA and Sedat JW (1997) Interphase chromosomes undergo constrained diffusional motion in living cells. *Curr Biol* **1:** 930–939.

Marshall WF, Marko JF, Agard DA and Sedat JW (2001) Chromosomal elasticity and mitotic polar ejection force measured in living Drosophila embryos by four-dimensional microscopy-based motion analysis. *Curr Biol* **11:** 1–20.

Melby T, Ciampaglio CN, Briscoe G and Erickson HP (1998) The symmetrical structure of structural maintenance of chromosomes (SMC) and MukB proteins: long, antiparallel coiled coils, folded at a flexible hinge. *J Cell Biol* **142:** 1595–1604.

Michaelis C, Ciock R and Nasmyth K (1997) Cohesins: chromosomal proteins that prevent premature separation of sister chromatids. *Cell* **91:** 35–45.

Miller OL and Beatty BR (1969) Visualization of nucleolar genes. *Science* **164:** 955–957.

Miller OL and Hamkalo BA (1972) Visualization of RNA synthesis on chromosomes. *Int Rev Cytol* **33:** 1–25.

Misteli T, Gunjan A, Hock R, Bustin M and Brown DT (2000) Dynamic binding of histone H1 to chromatin in living cells. *Nature* **408:** 877–881.

Morgan GT (2002) Lampbrush chromosomes and associated bodies: new insights into principles of nuclear structure and function. *Chromosome Res* **10:** 177–200.

Nguyen TT, Rouzina I and Shklovskii BI (2000) Reentrant Condensation of DNA induced by multivalent counterions. *J Chem Phys* **112:** 2562–2568.

Nguyen TT and Shklovskii BI (2001) Complexation of DNA with positive spheres: phase diagram of charge inversion and reentrant condensation. *J Chem Phys* **115:** 7298–7308.

Nicklas RB and Staehly CA (1967) Chromosome micromanipulation. I The mechanics of chromosome attachment to the spindle. *Chromosoma* **21:** 1–16.

Nicklas RB (1983) Measurements of the force produced by the mitotic spindle in anaphase. *J Cell Biol* **97:** 542–548.

Nicklas RB (1988) The forces that move chromosomes in mitosis. *Annu Rev Biophys Biophys Chem* **17:** 431–449.

Nicklas RB and Ward SC (1994) Elements of error correction in mitosis: microtubule capture, release, and tension. *J Cell Biol* **126:** 1241–1253.

Nicklas RB, Ward SC and Gorbsky GJ (1995) Kinetochore chemistry is sensitive to tension and may link mitotic forces to a cell cycle checkpoint. *J Cell Biol* **130:** 929–939.

Nicklas RB (1997) How cells get the right chromosomes. *Science* **275:** 632–637.

Nicklas RB, Campbell MS, Ward SC and Gorbsky GJ (1998) Tension-sensitive kinetochore phosphorylation *in vivo*. *J Cell Sci* **111:** 3189–3196.

Nicklas RB, Waters JC, Salmon ED and Ward SC (2001) Checkpoint signals in grasshopper meiosis are sensitive to microtubule attachment, but tension is still essential. *J Cell Sci* **114:** 4173–4183.

Paulson JR and Laemmli UK (1977) The structure of histone-depleted metaphase chromosomes. *Cell* **12:** 817–828.

Paulson JR (1988) Scaffolding and radial loops: the structural organization of metaphase chromosomes. In: *Chromosomes and Chromatin, Vol. III*. Adolph KW, (ed.) (pp. 3–30) CRC Press, Boca Raton, FL.

Pederson T (2000) Half a century of "the nuclear matrix". *Mol Biol Cell* **11:** 799–805.

Pelta J, Livolant F and Sikorav JL (1996) DNA aggregation induced by polyamines and cobalthexamine. *J Biol Chem* **27:** 5656–5662.

Poirier M, Eroglu S, Chatenay D and Marko JF (2000) Reversible and Irreversible Unfolding of Mitotic Chromosomes by Applied Force. *Mol Biol Cell* **11:** 269–276.

Poirier MG, Nemani A, Gupta P, Eroglu S and Marko JF (2001a) Probing Chromosome Structure Using Dynamics of Force Relaxation. *Phys Rev Lett* **86:** 360–363.

Poirier MG, Combined biochemical-micromechanical study of mitotic chromosomes (2001b) Ph.D. thesis (University of Illinois at Chicago), available from http://www.uic.edu/~jmarko.

Poirier MG, Monhait T and Marko JF (2002a) Condensation and decondensation of mitotic chromosomes driven by shifts in ionic conditions. *J Cell Biochem* **85:** 422–434.

Poirier MG and Marko JF (2002b) Bending Rigidity of Mitotic Chromosomes. *Mol Biol Cell* **13:** 2170–2179.

Poirier MG and Marko JF (2002c) Effect of internal viscosity on biofilament dynamics. *Phys Rev Lett* **88:** 228103.

Poirier MG, Marko JF (2002d) Mitotic chromosomes are chromatin networks without a contiguous protein scaffold. *Proc. Natl. Acad. Sci. USA* **99:** 15393–15397.

Polach KJ and Widom J (1995) Mechanism of protein access to specific DNA sequences in chromatin: a dynamic equilibrium model for gene regulation. *J Mol Biol* **254:** 130–149.

Reese DH, Yamada T and Moret R (1976) An established cell line from the newt Notophthalmus viridescens. *Differentiation* **6:** 75–81.

Reif M, Guatel M, Oesterhelt F, Fernandez JM and Gaub HE (1997) Reversible unfolding of individual titin immunoglobulin domains by AFM. *Science* **276:** 1109–1112.

Richmond TJ, Finch JT, Rushton B, Rhodes D and Klug A (1984) Structure of the nucleosome core particle at 7 A resolution. *Nature* **311:** 532–537.

Rieder CL and Hard R (1990) Newt lung epithelial cells: cultivation, use, and advantages for biomedical research. *Int Rev Cytol* **122:** 153–220.

Saminathan M, Antony T, Shirahata A, Sigal LH and Thomas TJ (1999) Ionic and Structural Specificity Effects of Natural and Synthetic Polyamines on the Aggregation and Resolubilization of Single-, Double-, and Triple-Stranded DNA. *Biochemistry* **38:** 3821–3830.

Saitoh Y and Laemmli UK (1993) From the chromosomal loops and the scaffold to the classic bands of metaphase chromosomes. *Cold Spring Harb Symp Quant Biol* **58:** 755–765.

Saitoh Y and Laemmli UK (1994) Metaphase chromosome structure: bands arise from a differential folding path of the highly AT-rich scaffold. *Cell* **76:** 609–622.

Skibbens RV and Salmon ED (1997) Micromanipulation of chromosomes in mitotic vertebrate tissue cells: tension controls the state of kinetochore movement. *Exp. Cell Res.* **15:** 235, 314–324.

Smith SB, Finzi L and Bustamante C (1992) Direct mechanical measurements of the elasticity of single DNA molecules by using magnetic beads. *Science* **258:** 1122–1126.

Smith SB, Cui Y and Bustamante C (1996) Overstretching B-DNA: the elastic response of individual double-stranded and single-stranded DNA molecules. *Science* **271:** 795–799.

Smythe C and Newport JW (1991) Systems for the study of nuclear assembly, DNA replication, and nuclear breakdown in *Xenopus laevis* egg extracts. *Methods Cell Biol* **35:** 449–468.

Stack SM and Anderson LK (2001) A model for chromosome structure during the mitotic and meiotic cell cycles. *Chromosome Res* **9:** 175–198.

Strunnikov AV, Larionov VL and Koshland D (1993) SMC1: an essential yeast gene encoding a putative head-rod-tail protein is required for nuclear division and defines a new ubitquitous family. *J Cell Biol* **123:** 1635–1648.

Strunnikov AV, Hogan E and Koshland D (1995) SMC2, a *Saccharomyces cerivisiae* gene essential for chromosome segregation and condensation, defines a subgroup within the SMC family. *Genes Dev* **9:** 587–599.

Strunnikov AV (1998) SMC proteins and chromosome structure. *Trends Cell Biol* **8:** 454–459.

Strunnikov AV and Jessberger R (1999) Structural maintenance of chromosomes (SMC) proteins: conserved molecular properties for multiple biological functions. *Eur J Biochem* **263:** 6–13.

Sumner AT (1996) The distribution of topoisomerase II on mammalian chromosomes. *Chromosome Res* **4:** 5–14.

Tanaka T and Fillmore DJ (1979) Kinetics of swelling of gels. *J Chem Phys* **70:** 1214–1218.

Tang JX and Janmey PA (1996) The polyelectrolyte nature of F-actin and the mechanism of actin bundle formation. *J Biol Chem* **271:** 8556–8563.

Thoma F, Koller T and Klug A (1979) Involvement of histone H1 in the organization of the nucleosome and of the salt-dependent superstructures of chromatin. *J Cell Biol* **83:** 403–427.

Thrower DA and Bloom K (2001) Dicentric chromosome stretching during anaphase reveals roles of Sir2/Ku in chromatin compaction in budding yeast. *Mol Biol Cell* **12:** 2800–2812.

Trask BJ, Allen S, Massa H, Fertitta A, Sachs R, van den Engh G and Wu M (1993) Studies of metaphase and interphase chromosomes using fluorescence *in situ* hybridization. *Cold Spring Harb Symp Quant Biol* **58:** 767–775.

Trinick J (1996) Titin as scaffold and spring. *Curr Biol* **6:** 258–260.

Tskhovrebova L, Trinick J, Sleep J-A and Simmons R-M (1997) Elasticity and unfolding of single molecules of the giant muscle protein titin. *Nature* **387:** 308–312.

Tsukamoto T, Hashiguchi N, Janicki SM, Tumbar T, Belmont AS and Spector DL (2000) Visualization of gene activity in living cells. *Nature Cell Biol* **2:** 871–878.

Van Holde K (1989) *Chromatin.* Springer, New York.

Warburton PE and Earnshaw WC (1997) Untangling the role of DNA topoisomerase II in mitotic chromosome structure and function. *Bioassays* **19:** 97–99.

Widom J and Klug A (1985) Structure of the 300A chromatin filament: X-ray diffraction from oriented samples. *Cell* **43:** 207–213.

Widom J (1997) Chromosome structure and gene regulation. *Physica A* **244:** 497–509.

Wolffe A (1995) *Chromatin.* Academic, San Diego.

Wolffe AP and Guschin D (2000) Chromatin Structural Features and Targets That Regulate Transcription. *J Struct Biol* **129:** 102–122.

Woodcock CL and Horowitz RA (1995) Chromatin organization reviewed. *Trends Cell Biol* **5:** 272–277.

Wuite GJ, Smith SB, Young M, Keller D and Bustamante C (2000) Single-molecule studies of the effect of template tension on T7 DNA polymerase activity. *Nature* **404:** 103–106.

Yin H, Wang MD, Svoboda K, Landick R, Block SM and Gelles J (1995) Transcription against an applied force. *Science* **270:** 1653–1657.

Yokota H, van den Engh G, Hearst JE, Sachs R and Trask RJ (1995) Evidence for the organization of chromatin in megabase pair-sized loops arranged along a random walk path in the human G0/G1 interphase nucleus. *J Cell Biol* **130:** 1239–1249.

Zhang D and Nicklas RB (1995) The impact of chromosomes and centrosomes on spindle assembly as observed in living cells. *J Cell Biol* **129:** 1287–1300.

Zhang D and Nicklas RB (1999) Micromanipulation of chromosomes and spindles in insect spermatocytes. *Meth Cell Biol* **61:** 209–218.

Zickler D and Kleckner N (1999) Meiotic chromosomes: integrating structure and function. *Ann Rev Genet* **33:** 603–754.

Zink D, Cremer T, Saffrich R, Fischer R, Trendelenburg MF, Ansorge W and Stelzer EH (1998) Structure and dynamics of human interphase chromosome territories *in vivo*. *Hum Genet* **102:** 241–251.

Elastic Invertebrate Muscle Proteins

Journal of Muscle Research and Cell Motility **23**: 435–447, 2002.

Varieties of elastic protein in invertebrate muscles

BELINDA BULLARD[1,*], WOLFGANG A. LINKE[2] and KEVIN LEONARD[1]
[1]*European Molecular Biology Laboratory, D-69012 Heidelberg;* [2]*Institute of Physiology and Pathophysiology, University of Heidelberg, D-69120 Heidelberg, Germany*

Abstract

Elastic proteins in the muscles of a nematode (*Caenorhabditis elegans*), three insects (*Drosophila melanogaster, Anopheles gambiae, Bombyx mori*) and a crustacean (*Procambus clarkii*) were compared. The sequences of thick filament proteins, twitchin in the worm and projectin in the insects, have repeating modules with fibronectin-like (Fn) and immunoglobulin-like (Ig) domains conserved between species. Projectin has additional tandem Igs and an elastic PEVK domain near the N-terminus. All the species have a second elastic protein we have called SLS protein after the *Drosophila* gene, *sallimus*. SLS protein is in the I-band. The N-terminal region has the sequence of kettin which is a spliced product of the gene composed of Ig-linker modules binding to actin. Downstream of kettin, SLS protein has two PEVK domains, unique sequence, tandem Igs, and Fn domains at the end. PEVK domains have repeating sequences: some are long and highly conserved and would have varying elasticity appropriate to different muscles. Insect indirect flight muscle (IFM) has short I-bands and electron micrographs of *Lethocerus* IFM show fine filaments branching from the end of thick filaments to join thin filaments before they enter the Z-disc. Projectin and kettin are in this region and the contribution of these to the high passive stiffness of *Drosophila* IFM myofibrils was measured from the force response to length oscillations. Kettin is attached both to actin near the Z-disc and to the end of thick filaments, and extraction of actin or digestion of kettin leads to rapid decrease in stiffness; residual tension is attributable to projectin. The wormlike chain model for polymer elasticity fitted the force-extension curve of IFM myofibrils and the number of predicted Igs in the chain is consistent with the tandem Igs in *Drosophila* SLS protein. We conclude that passive tension is due to kettin and projectin, either separate or linked in series.

Introduction

But, as the world, harmoniously confused:
Where order in variety we see,
And where, though all things differ, all agree.

Alexander Pope
'Windsor Forest' (1711)

Invertebrate animals encompass a wide range of species with muscles of differing function. So far, the elastic proteins in the muscles have been studied in only two invertebrate phyla: arthropods (insects and crustacea) and nematodes. But even this limited selection reveals the diversity in the architecture of proteins with an elastic function. The arthropods are the most varied group in the animal kingdom and are the subject of much current work on the relationship between the evolution of genes and diversity in morphology (Averof, 1997; Akam, 2000). Among arthropods, insects represent the greatest number of species, a success that is largely due to the development of flight. Here we will compare the elastic proteins of the invertebrates for which the sequence is known, and give a more detailed account of the function of these proteins in insect flight muscle.

*To whom the correspondence should be addressed: Tel.: +49-6221-387313; Fax: +49-6221-387306; E-mail. bullard@embl-heidelberg.de

The narrow focus of recent work on vertebrate titin, especially the human cardiac isoform, has been at the expense of a broader picture of possible ways in which modular proteins may control muscle elasticity. The invertebrate elastic proteins have some similarity in structure to vertebrate titin, and the large body of information on titin makes this a useful model from which to predict the function of the invertebrate proteins. However, the muscles in many invertebrates have functions not found in vertebrates, and the elastic properties of the proteins will be peculiar to these muscles.

Recent re-appraisal of evolutionary relationships, based on the fossil record and ribosomal RNA sequences, places nematodes and arthropods in the same superclade of Ecdysozoa or moulting animals, and annelids and molluscs in the Lophotrochozoa (Aguinaldo *et al.*, 1997; de Rosa *et al.*, 1999; Conway Morris, 2000). In previous classifications, arthropods and annelids were in the same clade based on their segmentation, and nematodes were considered more primitive because simpler. Insects and crustaceans are also now thought to be more closely related than they were previously (Averof and Cohen, 1997). However, the new clade of Ecdysozoa is controversial (Blair *et al.*, 2002) and the evolutionary relationship of invertebrates is by no means settled. New classifications may be tested by an analysis of protein families such as the elastic proteins in muscle, though there are few examples at present.

The genome sequences of *Caenorhabditis elegans* and *Drosophila melanogaster* have been completed, and more recently that for a mosquito, *Anopheles gambiae*. In addition there is partial genomic DNA sequence for other invertebrates such as the silkworm, *Bombyx mori*. The four species for which the complete sequence of some modular elastic proteins is known: *C. elegans*, *D. melanogaster*, *A. gambiae* and *Procambus clarkii* (crayfish) belong to groups that are thought to be genetically similar, despite the obvious differences in phenotype between worm, insects and crustacean. As we shall see, there are clear homologies in the elastic muscle proteins of the four species, but parts of the sequences diverge to accommodate widely differing functions of the muscles. The modular design of the proteins is particularly suited to stepwise modification not requiring evolution of new genes to change function.

Mechanical properties of the muscles

A current challenge is to determine the relationship between the passive stiffness of different muscle types and the properties of elastic proteins in the sarcomere. Invertebrates have more than one type of modular elastic protein which compounds the problem. Whereas the elasticity of vertebrate striated muscle can be considered in terms of titin, often in different isoforms, predictions of the elasticity of invertebrate muscles must take account of both projectin and a second modular protein known variously as kettin, D-titin and I-connectin.

In the case of *C. elegans*, nothing is known about the elastic properties of the body wall muscle, beyond what can be deduced from the movements of the worm and the structure of the muscle. Fibres are obliquely striated and have dense bodies instead of Z-discs; the thick filaments are 10 μm long, so the sarcomere length (the distance between dense bodies) is at least this length. Thin filaments are ~6 μm which suggests the muscle is capable of developing tension over a wide range of sacromere lengths (Waterston, 1988). The dense bodies are linked to the cuticle so that contraction of sarcomeres causes the cuticle to shorten; coordination of local contractions produces waves along the worm (Waterston, 1988). Crayfish claw fibres are striated and have long sarcomeres (8–13 μm) with long I-bands. Fibres are easily extensible when relaxed and do not develop appreciable passive tension until stretched to about 140% rest length (Fukuzawa *et al.*, 2001). The wide range of sarcomere lengths over which *C. elegans* body wall fibres and crayfish claw fibres are thought to operate is in contrast to the nearly isometric contraction of *Drosophila* flight muscle.

Insect striated muscles may be synchronous or asynchronous. In synchronous muscles, there is a direct correspondence between nervous impulses to the muscle and contractions. Non-flight muscles are synchronous and the structure of the sarcomere is typical of striated muscles with an I-band of varying width, depending on the function of the muscle. Insects with lower wingbeat frequencies, such as dragonflies, butterflies and locusts have synchronous flight muscle. Small insects with high wingbeat frequencies, such as flies, mosquitos, bees and wasps, and some larger insects, including the waterbug *Lethocerus*, have asynchronous flight muscle. In these muscles, oscillatory contractions are produced by delayed activation in response to stretch, combined with the resonant properties of the thorax (Pringle, 1978). The upstroke and downstroke of the wings result from distortions of the thorax rather than direct action of the muscles on the base of the wing, hence the muscles are known as indirect flight muscles (IFM). The sarcomeres in the IFMs are 3.4 μm long in *Drosophila*; there is almost no I-band, and the fibres change in length by only 3.3% during flight (Chan and Dickinson, 1996). Rapid oscillatory contraction requires that the sarcomeres are stiff so that strain is transmitted to the thick and thin filaments. The proteins responsible for the high stiffness will be described. The contrasting sarcomere structures of an IFM and a synchronous muscle in *Drosophila* thorax are shown in Figure 1.

Drosophila has become the natural choice for investigating insect elastic proteins, especially since the completion of the genome sequence (Adams *et al.*, 2000). The small size of the muscle makes mechanical measurements on fibres difficult but two groups have been successful in getting the muscle to do oscillatory work (Peckham *et al.*, 1990; Kreuz *et al.*, 1996). The disadvantage of whole fibre preparations for estimating the elasticity of the sarcomere is that other components have to be taken into account, such as collagen and surrounding mitochondria, particularly in the case of insect flight muscle where mitochondria are packed tightly between myofibrils. In addition, neighbouring myofibrils may be linked through connections between Z-discs. The measurement of tension in isolated myofibrils gives a direct estimate of elasticity without these complications and in the case of *Drosophila*, there is the added advantage that myofibrils are easier to manipulate than fibres (Kulke *et al.*, 2001).

Elastic proteins of invertebrates

Twitchin

Early work on the genes involved in muscle function in *C. elegans* led to the identification of the first large muscle protein with a modular structure made up of immunoglobulin-like (Ig) and fibronectin-like (Fn) domains. Worms with mutations in the *unc-22* gene showed characteristic twitching of the body wall muscles (Waterston *et al.*, 1980) and genetic evidence suggested an interaction between the gene product, twitchin, and

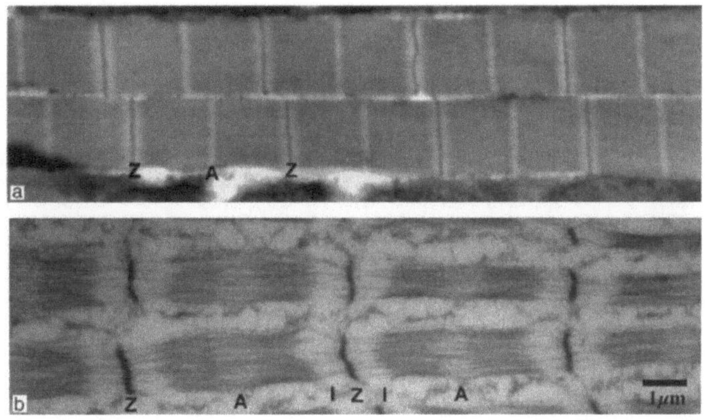

Fig. 1. Variation in the structure of *Drosophila* muscle fibres. Electron micrographs of cryo-sections: (a) IFM showing sarcomeres in which thick filaments extend nearly the whole length of the sarcomere. (b) A synchronous thoracic muscle with less well ordered, longer sarcomeres with longer I-bands. Z, Z-disc; A, A-band; I, I-band. Scale bar 1 μm for a and b.

myosin. Twitchin is associated with thick filaments in the obliquely striated body wall muscle (Moerman *et al.*, 1988). A large part of the 753 kDa molecule has a repeating structure consisting of two Fn and one Ig domain. The pattern of modules varies towards the C-terminus and is followed by a protein kinase domain with associated auto-inhibitory sequence (Benian *et al.*, 1989, 1993 and Figure 2a). At each end of the molecule there are Ig domains separated by unique sequence. Interestingly, a nearly normal phenotype results when up to six Fn–Ig modules are deleted from the protein, so in the case of *C. elegans*, the exact length of the molecule is not critical (Kiff *et al.*, 1988). Twitchin is unlikely to have an effect on the passive stiffness of nematode muscle because it is confined to the thick filaments. However, it is of interest here because of the similarity between twitchin and projectin. Twitchin is also associated with thick filaments in the catch muscle of the mollusc, *Mytilus edulis* (Vibert *et al.*, 1993), but although sequence is available for part of the molecule (Funabara *et al.*, 2001), there is not enough to be useful in a comparison of sequences.

Projectin

Projectin is a protein of 800–1000 kDa associated with thick filaments in many invertebrate muscles. The sequence of *Drosophila* genomic and cDNA codes for a core region with a repeating pattern of Fn and Ig domains like twitchin but with four extra modules (Ayme-Southgate *et al.*, 1991; Fyrberg *et al.*, 1992; Daley *et al.*, 1998 and Figure 2a). The variation in the pattern of Fn–Ig modules towards the C-terminus is identical to the one seen in the twitchin sequence. There is also a kinase domain near the C-terminus, which has homology to the kinase domain in twitchin. Recently, the projectin sequence has been extended towards the N-terminus using the completed *Drosophila* genome sequence, and here the protein diverges from nematode twitchin (Southgate and Ayme-Southgate, 2001 and Figure 2a). Alternative splicing in this region produces isoforms of differing size. Some isoforms have sequence predicted to be elastic because they are homologous to an extensible region in vertebrate titin which has a high proportion of proline (P), glutamic acid (E), valine (V), and lysine (K) (Labeit and Kolmerer, 1995). At the N-terminus, all isoforms have eight Ig domains upstream of a variable length PEVK region which is followed by five or six Ig domains. There are 13 exons in the PEVK region of the gene which show a rich pattern of alternative splicing reminiscent of the numerous splice forms in the PEVK sequence of vertebrate titin (Bang *et al.*, 2001). Thus the sequence of the longer isoforms of projectin has the same general pattern seen in vertebrate skeletal muscle titin: tandem Ig domains either side of a variable length PEVK region, followed by Fn–Ig modules with a kinase domain near the C-terminus. There is a predicted projectin sequence in the *Anopheles* genome which corresponds to a slightly shorter molecule than the full length *Drosophila* projectin: there is one less Fn–Ig module, shorter tandem Ig regions at the N-terminus, and shorter PEVK sequence (Figure 2a). The conservation in the repeating pattern of Fn–Ig modules and the kinase domain in nematode twitchin and the two insect projectins suggest an essential function for this part of the molecule, whereas the variability at the N-terminus is likely to be an adaptation to the mechanical requirements of the different insects. The crayfish claw muscle has a projectin-like protein associated with the thick filament but the sequence is not known (Hu *et al.*, 1990).

The position of projectin in the sarcomere differs in different types of insect muscle. In synchronous muscles of *Lethocerus* and *Drosophila*, there are projectin molecules across the whole A-band, with the exception of a gap in the centre of the sarcomere at the H-zone; while in IFM, projectin is confined to the end of the A-band and extends across the short I-band to the Z-disc (Lakey *et al.*, 1990; Saide *et al.*, 1990; Vigoreaux *et al.*, 1991). The distance spanned by projectin along the end of the thick filament up to the Z-disc is about 300 nm, which is approximately the length of the molecule (Nave and

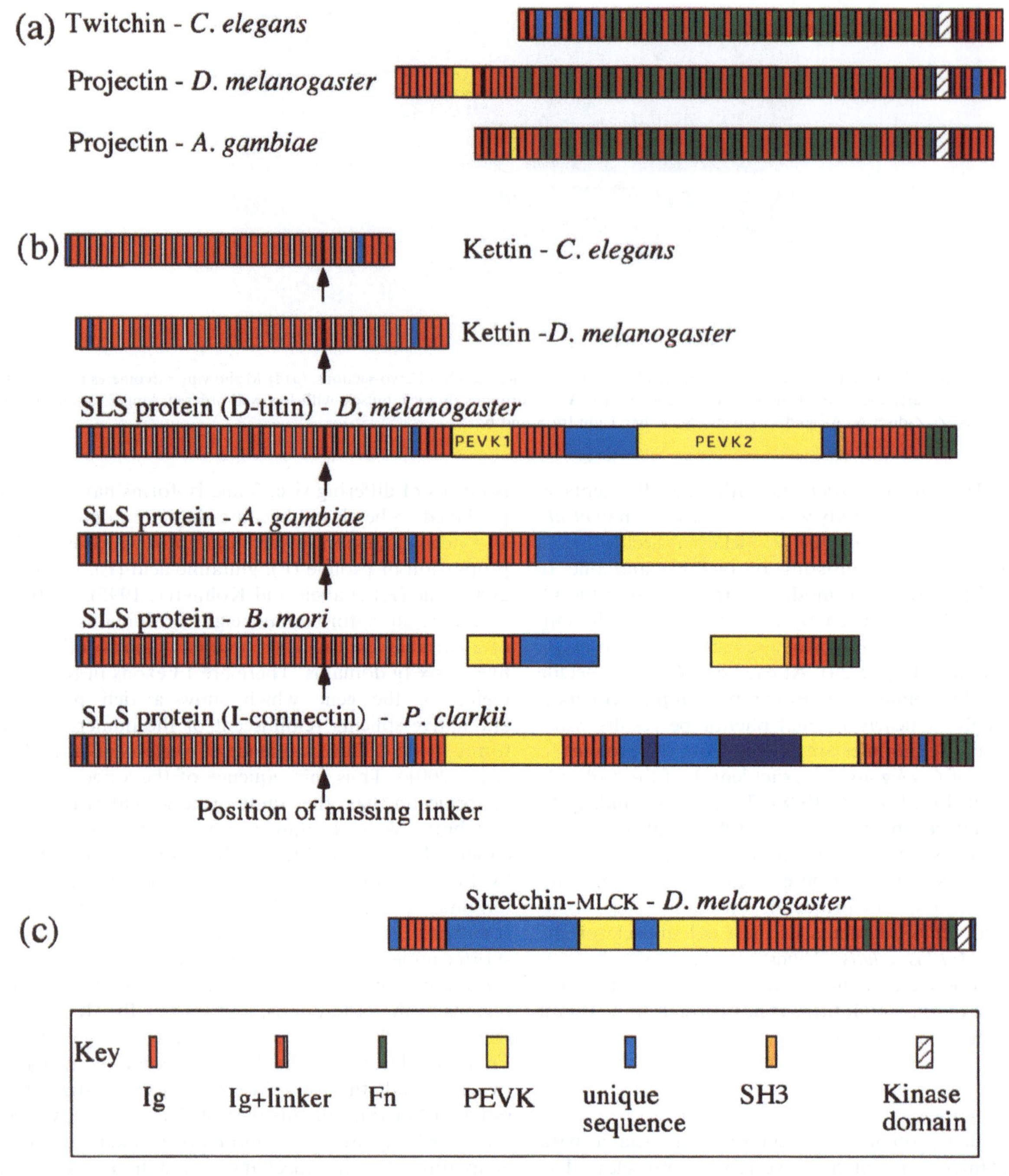

Fig. 2. Modular structure of elastic proteins in invertebrate muscles. (a) Twitchin and projectin. These proteins are associated with thick, myosin-containing filaments and have characteristic Fn–Ig modules and a kinase domain at the C-terminus; the pattern of Fn–Ig modules changes towards the C-terminus. Insect projectins have an N-terminal extension with tandem Igs and PEVK. (b) SLS protein in different species. The N-terminal region of the gene is kettin which is associated with thin filaments. The *C. elegans* genome has no further sequence downstream. In *D. melanogaster* (probably also in the other insects) and *P. clarkii*, kettin is expressed as an isoform of SLS protein. All kettins have a missing linker at the same position in the Ig-linker region. The complete sequence predicted from the genome is given, though there are several isoforms in *D. melanogaster* and probably all the species. Downstream of kettin, SLS protein has two PEVK domains, tandem Igs, unique sequence, and Fns at the C-terminus. Repeats within the unique region of the *P. clarkii* sequence are dark blue. All the insects and the crayfish have a conserved SH3 domain near the C-terminus. (c) Stretchin is a protein predicted from the *D. melanogaster* genome sequence. It has some similarity to the C-terminal region of SLS proteins, except for the kinase domain. The length of the domains in the figure is proportional to the number of amino acids, except for the Ig and Fn domains which are smaller than shown. Sequences are compiled from FlyBase, Wormpep and the EMBL and NCBI data bases. The *Bombyx mori* sequence was assembled from three EMBL data base entries (Y. Koike et al., unpublished data): AB079865 (Bmkettin) which is a homologue of kettin, AB079866 (Bmtitin) which corresponds to the central region of SLS protein and AB079867 which corresponds to the C-terminal region of SLS protein.

Weber, 1990; Bullard and Leonard, 1996). By analogy with vertebrate titin, it is usually assumed that the N-terminus of projectin is in the Z-disc and the C-terminal region, which has the Fn–Ig modules and the kinase domain, is in the A-band, though this has not been established by labelling with antibodies. The *Drosophila* projectin sequence does not have the Z-repeats near the N-terminus that are characteristic of this region of titin, and it is possible that projectin, unlike titin, does not enter the Z-disc. There are uncertainties about the position of different isoforms of projectin in the sarcomere of different muscle types. In *Drosophila*, the IFM isoform is smaller than the isoforms in leg and jump muscle (Vigoreaux *et al.*, 1991), probably because it has less PEVK sequence. There are many muscle types in the insect thorax and it is possible that IFMs are not the only muscles in which projectin spans the I-band; some projectin isoforms with longer PEVK sequence may link thick filaments to Z-disc in fibres with longer sarcomeres.

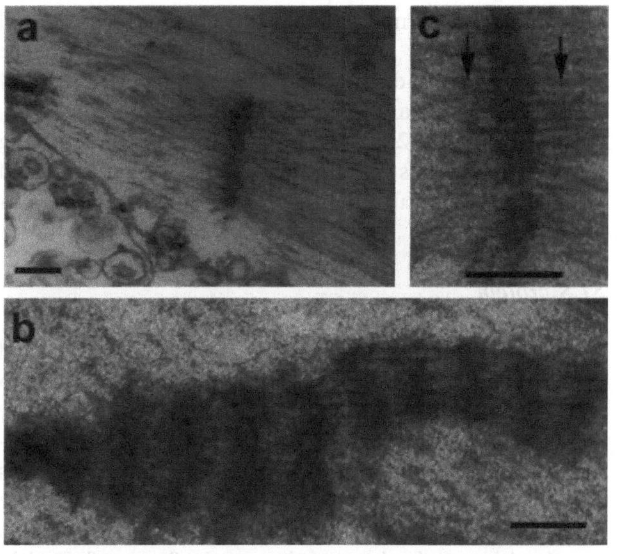

Fig. 3. Kettin in thin filaments. Cryo-sections of IFM in the *Drosophila* mutant Mhc^7 which has no thick filaments in the IFM. Thin filaments emerging from Z-discs vary in length. (a) Low magnification view of Z-disc and thin filaments. (b) Tiger tails formed by regular repeats of Z-discs are commonly seen in sections. The minimum distance from the edge of one Z-disc to the next corresponds to the maximum extent of kettin along an actin filament. (c) Either side of the Z-disc, the thin filaments appear straighter for a distance corresponding to the length of filament spanned by kettin (arrows). Scale bars 200 nm.

Kettin

Kettin is an unusual member of the Ig-containing modular proteins in that it is bound to the thin filament along most of the length of the molecule. The *Drosophila* protein is 540 kDa and the complete cDNA sequence codes for a protein of this size. The sequence consists predominately of Ig domains separated by linkers of 35 residues, with the exception of a module two-thirds of the way along the molecule, where the linker is missing. At the N- and C-termini there are consecutive Ig domains with regions of unique sequence. An important difference from projectin is that there are no Fn domains (Hakeda *et al.*, 2000; Kolmerer *et al.*, 2000 and Figure 2b). The *Anopheles* and *Bombyx* genomic sequences predict these insects have a kettin similar in size and sequence to the *Drosophila* protein (Figure 2b). Kettin has been isolated from the crayfish, *P. clarkii*, (Maki *et al.*, 1995) and the cDNA sequence codes for a kettin remarkably similar to the insect proteins (Fukuzawa *et al.*, 2001 and Figure 2b). The *C. elegans* genome also has kettin sequence but in this case, there are four fewer Ig-linker modules downstream of the missing linker (Hakeda *et al.*, 2000; Kolmerer *et al.*, 2000 and Figure 2b). The similarity in the pattern of modules in different kettin sequences is striking: the missing linker is common to all and, in the case of the arthropods, the number of Ig-linker modules is the same.

The uniformity in the primary structure of kettin and the invariance in different species are explained by its position in the sarcomere. The orientation and layout of the molecule have been determined in the IFM of *Lethocerus* and *Drosophila* by labelling fibres with antibodies to different parts of the sequence (van Straaten *et al.*, 1999). The N-terminus is in the Z-disc and the molecule extends along the thin filament for about 100 nm in *Drosophila* IFM (40 nm within the Z-disc and 60 nm outside). Kettin binds to actin with high affinity and the stoichiometry between kettin Ig-linker modules and actin monomers is close to one-to-one. The most likely model for kettin on the thin filament, based on the stoichiometry, is one in which the molecule follows the helix of actin and Ig domains, separated by linker sequences, are bound to actin monomers (van Straaten *et al.*, 1999).

In crayfish claw muscle, kettin is close to the borders of the Z-disc and stays in the same position when sarcomeres are stretched to 12 μm (Fukuzawa *et al.*, 2001). Since the length of the kettin sequence (Figure 2b) is independent of the length of the I-band, as illustrated by the extremes in insect IFM and crayfish claw muscle, the molecule probably acts to stabilize actin filaments in the region of the Z-disc rather than along the whole length of the I-band. This is supported by the appearance of thin filaments in electron micrographs of a *Drosophila* mutant, Mhc^7 (previously called *Ifm*(*2*)2), which has no thick filaments in the IFM. The length of thin filaments between Z-discs varies, but the minimum distance is about 60 nm, forming characteristic 'tiger-tails' (Figure 3). This suggests that actin is stabilised by kettin where the filaments emerge from the Z-disc. In other sarcomeres, filaments appear straighter and more ordered for a similar distance outside the Z-disc. Kettin promotes the anti-parallel association of actin filaments (van Straaten *et al.*, 1999) and this may be the driving force behind the formation of tiger-tails, and possibly also behind the initial formation of sarcomeres in normal muscle. In *C. elegans* body wall muscle, kettin is associated with dense bodies, the homologues of Z-discs, and here the precise length of the molecule appears to be less critical (Kolmerer *et al.*, 2000 and Figure 2b). The significance of the missing linker in all kettins remains a mystery, though it would be expected to disturb the regular binding of Ig domains to actin.

SLS protein

A gene containing Ig and PEVK domains was identified in *Drosophila* by Andrew and colleagues (Machado *et al.*, 1998) and different regions of the sequence were shown to be expressed in IFM. The gene was called *D-titin* because of the large size of the protein in embyos and the homology to the I-band region of vertebrate titin. It was suggested that, in addition to functioning as an elastic protein in muscle, an isoform of D-titin associated with chromosomes is responsible for their elasticity (Machado *et al.*, 1998; Machado and Andrew, 2000). The *Drosophila* genome sequence shows that D-titin has the whole kettin sequence near the N-terminus (Machado and Andrew, 2000; Zhang *et al.*, 2000 and Figure 2b) and kettin is now known to be an alternatively spliced product. The gene is called *sallimus* (*sls*) in FlyBase and is the locus of *sls* mutations (Kennison and Tamkun, 1988); sallimus is the Finnish word for 'fate' or 'destiny'. As we have seen, the module pattern of projectin is similar to the pattern in vertebrate titin. D-titin is rather different. The whole molecule is predicted to be 1.8–1.9 MDa but there are several alternatively spliced isoforms and it is not clear if the entire sequence is expressed as a single large molecule in muscle. Unlike vertebrate titin, the molecule has no Fn–Ig super repeats and the only Fn domains are the four at the C-terminus (Figure 2b). The lack of titin-like myosin binding modules immediately suggests that the molecule does not extend along the A-band region of the sarcomere (Machado and Andrew, 2000); D-titin is therefore unlikely to be a template regulating thick filament length, as is thought to be the case for vertebrate titin (Trinick, 1994). Surprisingly, nearly half the sequence downstream of kettin is PEVK-like and occurs in two domains. Tandem Igs and unique sequence make up the rest of the C-terminal half of the molecule. Close to the C-terminus there is an SH3 domain, likely to be the site of a proline-rich ligand.

A similar sequence can be identified in the genome of *Anopheles* and in the partial genomic sequence of *Bombyx* (Figure 2b); cDNA from crayfish claw muscle has sequence generally similar to those of the three insects (Fukuzawa *et al.*, 2001 and Figure 2b). The protein needs a name that can be used for all species; D-titin is not appropriate. We will use the general term SLS protein for all species in the tradition of calling proteins after the *Drosophila* gene, and in order not to confuse the protein with vertebrate titin. The crayfish protein has been called I-connectin (Fukuzawa *et al.*, 2001), but again this suggests it is titin (connectin)-like. All the *sls* genes have the sequence of kettin at the N-terminus and, as mentioned above, kettin is a spliced product of the gene which has been identified as a distinct protein in *Lethocerus*, *Drosophila* and crayfish. The *C. elegans* genome has kettin sequence, but unlike other *sls* genes, there is no further sequence downstream. The crayfish is the only species for which the entire cDNA from the *sls* gene has been sequenced and there are some unusual repeats in the region of unique sequence in the C-terminal half of the molecule which are not found in the insect sequences (Fukuzawa *et al.*, 2001 and Figure 2b). The SH3 domain near the C-terminus is conserved in all the SLS proteins, suggesting an essential function. In view of the common overall pattern in the sequence of the genes sequenced so far, we propose the name SLS protein for the gene product in all invertebrates; SLS protein would include the spliced product, kettin.

Several different isoforms of the *Drosophila* SLS protein produced by splicing together different ORFs of the gene, have been identified from the cDNA and by labelling thoracic muscles with antibodies to different regions of the sequence (Bullard *et al.*, 2002). The predominant isoforms in IFM are 540 kDa (kettin), and a less abundant 700 kDa isoform in which the extreme C-terminal region of the gene, including the tandem Igs (Figure 2b), is spliced to the end of kettin. These isoforms are present within the same myofibril (K. Leonard *et al.*, unpublished data). In *Lethocerus*, the molecular weight of the major IFM isoform is 700 kDa (Lakey *et al.*, 1993). Leg muscles, and other non-flight muscles in the *Drosophila* thorax that have sarcomeres with longer I-bands, have larger SLS isoforms in addition to kettin. Antibodies to different regions of the sequence downstream of kettin label the I-band or the edge of the A-band in many different muscle types. Therefore, even the longer isoforms do not extend far into the A-band, as was predicted from the lack of titin-like Fn–Ig repeats in the sequence. The Fn domains at the C-terminus probably bind to the end of the thick filament and may be spliced to isoforms of varying length.

In contrast to *Drosophila* IFM and many of the non-flight muscles, the crayfish claw muscle has only one isoform of SLS protein (I-connectin) in addition to kettin (Fukuzawa *et al.*, 2001). The 1.9 kDa molecule extends to as much as 3.5 µm in stretched sarcomeres of the claw muscle, which is equivalent to the length of the whole sarcomere of *Drosophila* IFM. This is achieved by sequential elongation of different domains downstream of kettin (Fukuzawa *et al.*, 2001). The C-terminus of crayfish SLS protein, which has Fn repeats, is attached to the end of the A-band and the I-band can be stretched until the elastic part of the molecule reaches its contour length. It seems likely that other, shorter, isoforms of SLS protein will be found in crayfish muscles that shorten less than the claw muscle.

These two arthropod examples, insect and crustacean, demonstrate the adaptability of SLS protein. By expression of different modules, the elasticity is tuned to a stiff muscle like IFM in which the sarcomere is stretched 100 nm or to the extensible claw muscle where the sarcomere is stretched 5 μm. IFM has the additional sophistication of having more than one isoform within a sarcomere.

The genomic sequence of *C. elegans* might be expected to have sequence downstream of kettin but this is not present. In view of the long sarcomeres in the obliquely striated muscle and the potential for considerable extension, it seemed likely that the extensible region of SLS protein would be needed to protect the sarcomeres from excessive stress, as in crayfish claw muscle. However, the linkage of dense bodies to the cuticle may prevent the sarcomeres stretching beyond the extent allowed by the compliance of the cuticle. Thus SLS protein expression is adapted to the needs of muscles attached to the hard exoskeleton of arthropods or to the flexible cuticle of nematodes.

Stretchin-MLCK

A transcription unit in the *Drosophila* genome has been identified which codes for a 926 kDa protein, stretchin-MLCK, predicted to have tandem Ig domains, PEVK and unique sequence, with a kinase domain near the C-terminus (Champagne *et al.*, 2000 and Figure 2c). As well as the full length transcript, an isoform called stretchin, lacking the C-terminal part of the molecule (including the kinase domain) and other isoforms

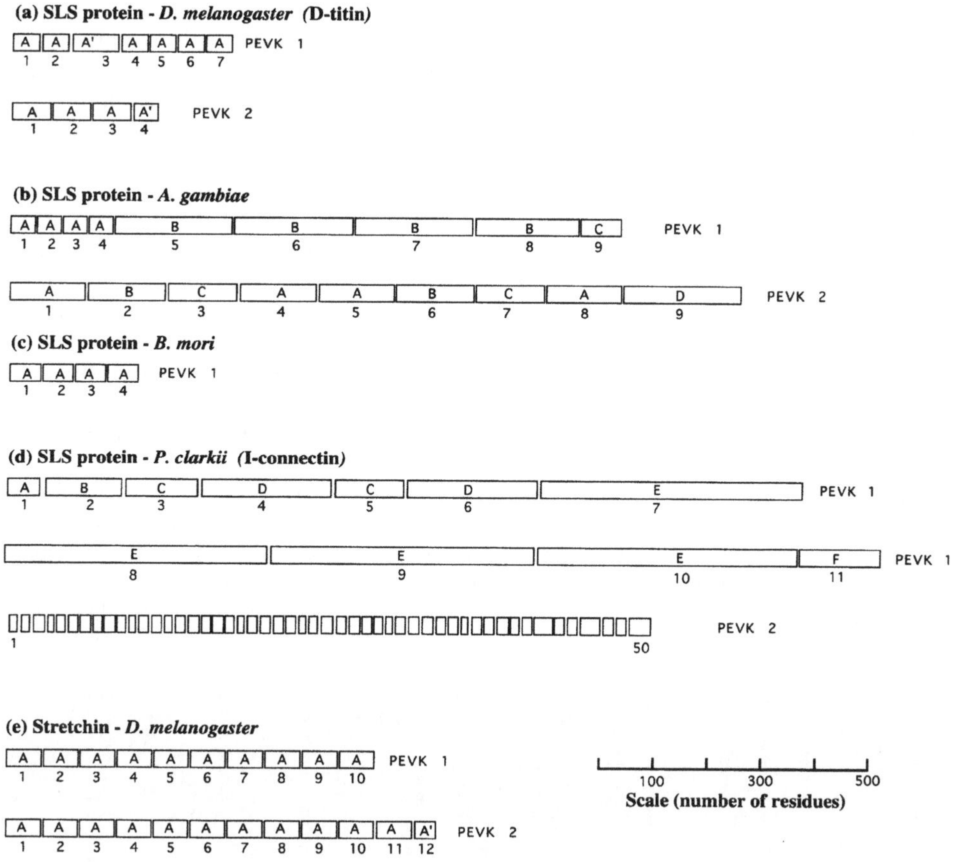

Fig. 4. Repeats in the PEVK sequence of invertebrate elastic proteins. In each case, the sequences are divided into PEVK 1 and 2 regions as shown in Figure 2. Domains are numbered and labelled A, B, C etc. to denote different types of repeat within each group. CLUSTAL X alignments for the PEVK repeats are given on the web site www.krembl.org/SLS (a) SLS protein – *D. melanogaster* (D-titin). PEVK 1 has 6 homologous tandem repeats of approximately 48 residues and a longer one (A′) of 91 residues. PEVK 2 has three highly homologous (>90% identity) repeats of 71 residues and a shorter one (A′) of 45 residues. (b) SLS protein – *A. gambiae*. PEVK 1 has 4 short homologous repeats of 45 residues that are similar to the repeats in PEVK 1 of *D. melanogaster*. These are followed by four longer repeats and a final short repeat. The B-type 216 residue repeats, numbered 5, 6 and 7, have 96, 98 and 98% identity to each other. PEVK 2 has eight repeats of 130–140 residues followed by a ninth 200 residue repeat. There is a pattern of similarity so that repeat 1 is more homologous to 4, 5 and 8, repeat 2 to repeat 6 and repeat 3 to repeat 7, giving the sequence ABCAABCA. (c) SLS protein – *B. mori*. This sequence is incomplete so there are only data for PEVK 1. This has four short repeats (56 residues) that are highly homologous (>90% identity). (d) SLS protein – *P. clarkii* (I-connectin) PEVK 1 has a complex pattern of repeats extending over nearly 2600 residues; six different types of repeat are present. C and D occur twice, E occurs four times. The E repeats are highly homologous despite being about 470 residues in length. E8, E9 and E10 are more than 99% identical at both the protein and DNA level. PEVK 2 has a large number (approximately 50) of short homologous repeats which range in length from 11 to 40 residues. (e) Stretchin – *D. melanogaster*. PEVK 1 and 2 are made up of 21 homologous 64 residue repeats, 10 in PEVK 1 and 11 in PEVK 2; A′ is a shorter repeat. The repeats in PEVK 1 have 90% identity and those in PEVK 2 have 84%.

thought to be myosin light chain kinases (MLCKs) are predicted products of the gene. The only evidence that any of the transcripts is in muscle is that cDNA from the second tandem Ig region codes for partial sequence in a 225 kDa protein, A(225), the antibody to which labels the A-band of *Drosophila* IFM (Patel and Saide, 2001).

Repeating structure of PEVK domains

The SLS proteins are notable for the high proportion of PEVK sequence in two separate domains downstream of kettin; stretchin also has two PEVK domains (Figure 2). Projectin, like vertebrate titin, has one PEVK domain. Since this region is likely to be important in determining the elasticity of the proteins, we have analysed the repeating patterns that are seen in all the longer PEVK domains (but not in projectin). Details of the repeats are given in Figure 4. The pattern in both PEVK domains of *Drosophila* SLS protein is relatively simple; PEVK 2 contains repeats identified previously (Machado *et al.*, 1998). In contrast, the pattern in the *Anopheles* protein is more complicated and most of the repeats are longer. Repeats in PEVK domains of crayfish SLS protein (I-connectin) were noted by Fuzukawa *et al.* (2001). Further analysis reveals a varied pattern of repeats that extends over nearly 2600 residues in PEVK 1 and shorter repeats of different lengths in PEVK 2. The entire PEVK 2 is short compared with that of the other species (Figure 2b), but there is an extra domain immediately upstream which has another type of repeating structure not present in other SLS proteins (Fukuzawa *et al.*, 2001). Stretchin has a simple pattern of repeats in both PEVK domains (Champagne *et al.*, 2000).

There is a large variation in the length of sequence making up a repeat, which ranges from around 11–480 residues. The high degree of homology between repeats is remarkable: three of the 480 residue repeats in the crayfish protein are more than 99% identical at both the protein and the DNA level. This suggests recent duplication of sequence within the gene, because genetic drift would be expected to produce some random sequence variation. Such sequence multiplication may be a mechanism for rapidly extending the length of extensible region in these proteins and changing the elasticity. The designation of domains as 'PEVK' is somewhat subjective and there is variation in the proportion of these residues in a domain: for example, the PEVK domain of soleus titin has 36% proline while stretchin PEVK 1 and 2 have 4.6 and 1.5% proline respectively. CLUSTAL X alignments for the PEVK repeats are given on the web site www.krembl.org/SLS

The diversity in the size and repeating pattern of PEVK repeats in invertebrate elastic proteins is in contrast to the makeup of vertebrate titin, where the single PEVK domain has one type of repeat of 28 residues (Greaser, 2001). This difference is probably due to the large range of sarcomere lengths in invertebrate muscles compared to vertebrate muscles, where sarcomere lengths vary little in different muscle types and in general are less than 2.5 μm. Isoforms of widely differing length and elasticity could be expressed from the *sls* gene.

The origin of passive stiffness in IFM

The structure of the IFM sarcomere

The stiffness of IFM is usually attributed to connections between the ends of the thick filaments and the Z-disc, spanning the short I-bands, the so-called C filaments (Reedy, 1971; Trombitas and Tigyi-Sebes, 1974; White, 1983; Trombitas, 2000). These filaments can be seen

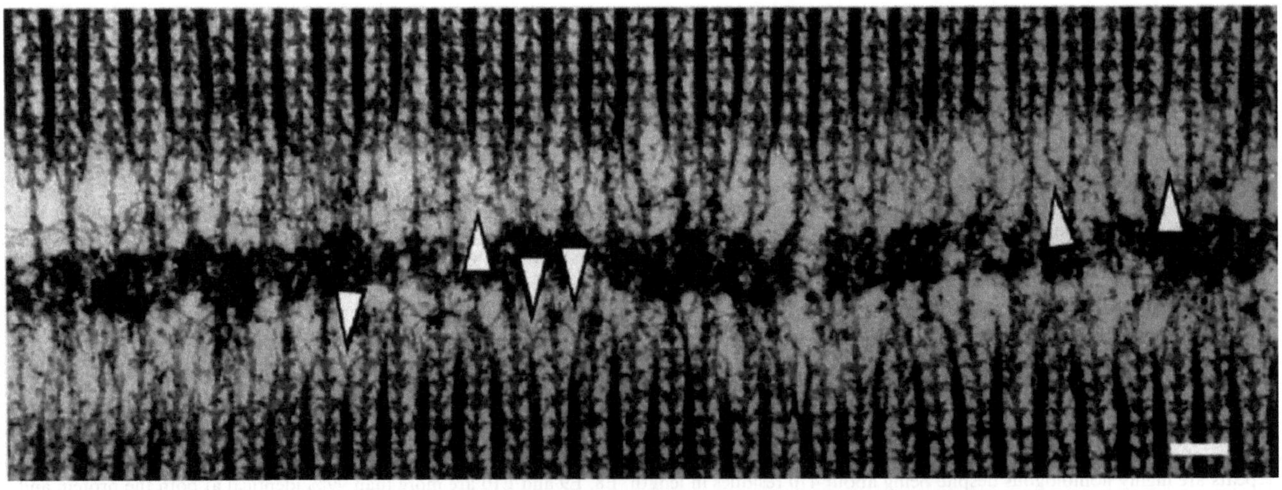

Fig. 5. Elastic filaments in *Lethocerus* IFM. Electron micrograph of a longitudinal section of a *Lethocerus* IFM fibre swollen at low ionic strength in rigor. Arrowheads point to filaments splitting from the ends of the thick filaments and joining actin outside the Z-disc. In less stretched regions, there are dense blobs where the fine filaments join actin. The electron micrograph is by Mary Reedy. Scale bar 100 nm.

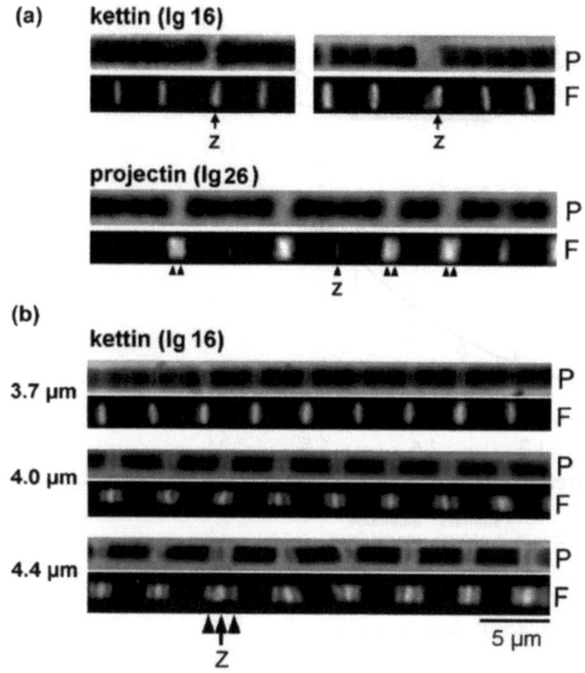

Fig. 6. The position of kettin and projectin in *Drosophila* IFM. (a) Single myofibrils stretched in relaxing conditions, labelled with monoclonal antibody to kettin Ig16 (top) or projectin Ig26 (bottom), followed by fluorescent second antibody. Where sarcomeres are disrupted and thin filaments broken, the kettin epitope is near the Z-disc and the projectin epitope is pulled out from the end of the A-band. (b) Myofibrils from which actin has been extracted with gelsolin at different sarcomere lengths, labelled with anti-kettin Ig16. Both the Z-disc (arrow) and the edge of the A-band (arrowheads) are labelled in stretched myofibrils.

clearly in electron micrographs of fibres stretched in rigor so that thin filaments are broken away from the Z-disc and do not confuse the picture. Projectin is in this region of the IFM sarcomere and was identified as a component of C filaments (Bullard *et al.*, 1977; Saide, 1981; Lakey *et al.*, 1990). Better preservation of *Lethocerus* IFM during preparation for electron microscopy has meant that the connections between thick filaments and Z-disc can be seen without removing thin filaments. Electron micrographs show strands emerging from thick filaments, which diverge and join the thin filaments before entering the Z-disc (Trombitas, 2000 and Figure 5). There is a dense blob where the strands join the thin filament and this is less pronounced when thick filaments are further away from the Z-disc, as though material were unravelling from it (Figure 5). This arrangement is rather different from the classical picture of C filaments joined directly to the Z-disc and will affect the interpretation of passive stiffness measurements.

Elastic proteins and IFM stiffness

An estimate can be made of the contribution of the two elastic proteins, projectin and SLS protein, to the passive stiffness of single IFM myofibrils. The main isoforms of SLS protein in *Drosophila* IFM are kettin and kettin spliced to sequence from the C-terminal region of the gene. There are also minor amounts of an isoform containing PEVK sequence (Kulke *et al.*, 2001). The main isoform of projectin in *Drosophila* IFM is about 900 kDa (Kulke *et al.*, 2001).

The position of kettin and projectin in stretched myofibrils can be followed with fluorescently labelled antibody, which shows that kettin remains close to the Z-disc in disrupted sarcomeres where thin filaments are broken (Figure 6a). Projectin is exposed and labels more strongly at the edge of the A-band in stretched sarcomeres, suggesting the labelled epitope, Ig26, is pulled from the A-band by part of the molecule which is still attached to the Z-disc (Kulke *et al.*, 2001 and Figure 6a). When kettin is freed from thin filaments by extracting actin with gelsolin, the sarcomeres extend uniformly and while some of the kettin Ig16 epitope remains near the Z-disc, some is on the end of the A-band and moves with it on increasing stretch (Figure 6b). Therefore, the C-terminal part of kettin is attached directly or indirectly to the A-band. The position of projectin is not affected by removing actin, and projectin appears to remain intact when the extracted myofibrils are stretched. Similar results are obtained when kettin is selectively cleaved by μ-calpain. Myofibrils extend easily because thin filaments can be pulled from the overlap region when not restrained by kettin, and the Ig16 epitope is detected at the edge of the A-band after it is cleaved from the rest of the molecule (Kulke *et al.*, 2001). Thus, kettin restricts sarcomere extension, and when kettin is rendered inoperable, projectin bears the remaining stress.

The force developed by relaxed IFM myofibrils at increasing sarcomere length is a measure of passive stiffness. Figure 7a shows that single *Drosophila* IFM myofibrils are much stiffer than vertebrate skeletal or cardiac myofibrils (Kulke *et al.*, 2001). The myofibrils give after an extension of only about 5% (Figure 7a), which was suggested to be due to breakage of actin and/or loosening up the connections between elastic filaments and their anchorage points at the A-band edge or Z-disc. Alternatively, it is possible that an elastic element unfolds. Ig domains could unfold *in situ* at high physiological sarcomere lengths (Li *et al.*, 2002).

A different way to look at stiffness is to oscillate the single myofibril using a micromotor and measure the force response at the force-transducer end (Figure 7b). The average peak-to-valley amplitude of force oscillations indicates myofibrillar stiffness. The passive stiffness of single *Drosophila* IFM myofibrils under control relaxing conditions (SL = 3.7 μm) decreased slowly during a 30-min experiment (Figure 7b, open circles and fit), indicating normal force relaxation. In contrast, stiffness dropped immediately and greatly upon exposure of myofibrils to a Ca^{2+}-independent gelsolin fragment to extract actin (Figure 7b, squares and fit curve). A similar magnitude of stiffness decrease was seen when myofibrils were treated with μ-calpain to specifically degrade kettin (Kulke *et al.*, 2001). At time

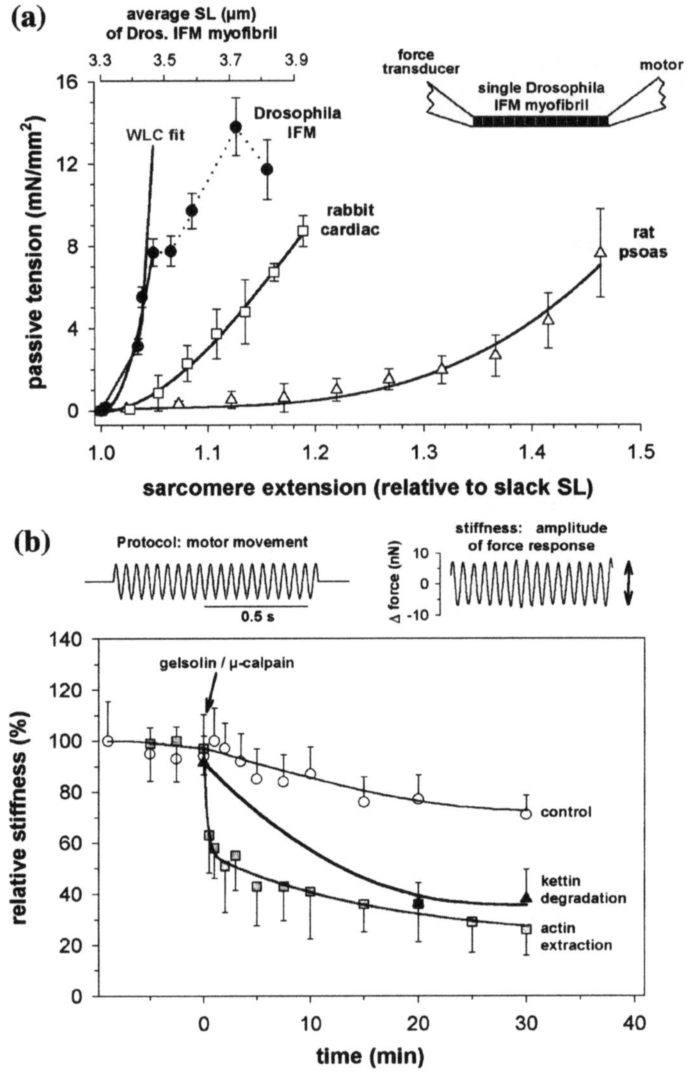

Fig. 7. Force measurements on single, non-activated, myofibrils. (a) Passive-tension *versus* sarcomere-extension relationship of *Drosophila* IFM myofibrils ($n = 24$), in comparison with that of rabbit heart ($n = 17$) and rat psoas ($n = 14$) myofibrils. Pooled data points (mean ± SEM) represent quasi steady-state tension recorded after decay of most of the viscous/viscoelastic force. Black solid curves are third-order polynomial fits. The tension-SL curve of *Drosophila* IFM myofibrils shows several steps at extensions >5% (dotted line). The grey solid curve ('WLC fit') is a fit to the steep initial portion of the *Drosophila* IFM data, calculated using the WLC model of entropic elasticity (Eq. (1)) (parameters: A, 10 nm; L, 126 ± 0.5 nm; z of elastic polymer at slack sarcomere length, 30 nm; scaling factor to obtain force of single molecule, $2.5 \times 10^9/\text{mm}^2$ cross-sectional area; for further details, see text). (b) Force response of single *Drosophila* IFM myofibrils to 20 Hz sinusoidal motor oscillations (taken as an indicator of stiffness) under different experimental conditions. Stiffness was measured at SL = 3.7 µm (pooled data points are mean ± SD, $n = 3$ myofibrils, for each condition). Fit curves are simple exponential decay functions.

point 30 min, average stiffness of control myofibrils was 71.0 ± 7.7%, statistically significantly higher ($P < 0.001$ in Student's t-test) than that of kettin-extracted (38.0 ± 11.5%) or actin-depleted (26.0 ± 10.1%) myofibrils; differences between the latter two types of specimen were statistically insignificant. Thus, both actin extraction and kettin proteolysis result in a large drop in myofibrillar stiffness. The 'residual' stiffness may be due to the remaining projectin molecules. In *Lethocerus* IFM, which has a 700 kDa isoform of SLS protein (kettin), the myofibrils are less stiff than in *Drosophila* IFM (Kulke *et al.*, 2001); removal of actin with gelsolin also results in a drop in passive stiffness in *Lethocerus* IFM fibres (Granzier and Wang, 1993).

Wormlike-chain behaviour of IFM elastic proteins

We wanted to know whether the force-extension curve of single *Drosophila* IFM myofibrils can be approximated by the wormlike-chain (WLC) model of entropic polymer elasticity, which has been successfully applied to explain the elasticity of titin molecules in vertebrate muscle sarcomeres (e.g. Linke *et al.*, 1998). This model states (Bustamante *et al.*, 1994; Marko and Siggia, 1995) that external force (f) is related to fractional extension (z/L) of a polymer chain by

$$f = \left(\frac{k_B T}{A}\right)\left[\frac{1}{4(1-z/L)^2} - \frac{1}{4} + \frac{z}{L}\right] \qquad (1)$$

where A is the persistence length, k_B is the Boltzmann constant, T is absolute temperature (~300 K in our experiments), z is end-to-end length (extension), and L is the chain's contour length. An entropic chain undergoes thermally-induced bending movements that tend to shorten its end-to-end length. Stretching the chain reduces its conformational entropy and thus, requires an external force. Entropic compliance results when $L \gg A$, due to the numerous configurations the polymer may adopt. The larger the persistence length (at a given contour length), the stiffer the polymer and the smaller the external forces required for stretching out (straightening) the molecule.

The idea here was to calculate the contour length of the polymer chain that makes up the freely extensible molecular segment in the short I-band of *Drosophila* IFM myofibrils, by taking into consideration the myofibrillar force–extension curve and the persistence-length value for a chain consisting solely of serially linked Ig-domains; the latter is known to be 10 nm (Li *et al.*, 2002). Then, the calculated contour length should give a hint at how many Ig-domains could, in theory, make up the extensible segment, since one Ig-domain has a length of approximately 4–4.5 nm (Improta *et al.*, 1996; Liversage *et al.*, 2001).

Accordingly, we tried to fit the initial data points of the *Drosophila* IFM force–extension curve by using the WLC model (Equation (1)) with the following assumptions:

(1) The polymer chain consists of a certain number of serially linked Ig-domains.
(2) The persistence length of this poly-Ig-domain polymer is 10 nm (Li *et al.*, 2002).
(3) The end-to-end length of this polymer (=extension) at slack sarcomere length is between 15 and 30 nm. (These values were assumed; the centre-to-centre distance between thick and thin filaments in *Drosophila* IFM is 28 nm (Irving and Maughan, 2000) and the outside-to-outside distance is 14–15 nm; the extensible segment must span at least this distance.)
(4) The myofibrillar force measured at a given extension is scaled to the single molecule by using scaling factors between 1.25×10^9 and $2.5 \times 10^9/\text{mm}^2$ myofibrillar cross-sectional area (the number of thin filaments per mm^2 cross-sectional area is $1.07 \times 10^9/\text{mm}^2$; the stoichiometry between thin and elastic filaments is not known but likely no more than 1:1 to 1:3).

Using these assumptions, and after scaling up the calculated single-molecule data to the myofibril level (using the same scaling factors as above), we obtain a curve of the kind shown in Figure 7a (grey solid line), which reveals a contour length of 126 ± 0.5 nm. Varying the (scaled-down) force and/or the end-to-end length in the WLC model by a factor of two, did not dramatically change the calculated contour-length value. Similarly, variations in persistence length between 8 and 15 nm did not lead to dramatically different contour-length values. The range of contour lengths obtained, was approximately 100–130 nm. This would suggest that the flexible polymer chain of *Drosophila* IFM myofibrils consists of approximately 22–32 Ig-domains.

Examination of the sequences of SLS protein and projectin (Figure 2) in *Drosophila* shows there are four tandem Ig domains at the end of kettin and a further 18 downstream. If these were spliced together (the C-terminal Igs are known to be spliced to the end of kettin in one IFM isoform), there would be 22 tandem Igs; it is unlikely there is appreciable PEVK because there are no large isoforms of SLS protein in IFM. Thus, there are potentially enough tandem Igs to approximate the contour length of the flexible chain predicted from the force extension curve using the WLC model. There are a maximum of 14 tandem Igs at the N-terminus of projectin, and these may be responsible for the passive stiffness in myofibrils when kettin is broken (Figure 7b). An alternative scenario is that kettin and projectin are somehow linked in series, possibly at the position of the dot-like structures seen in electron micrographs. A possible model based on the micrograph of *Lethocerus* IFM is shown in Figure 8.

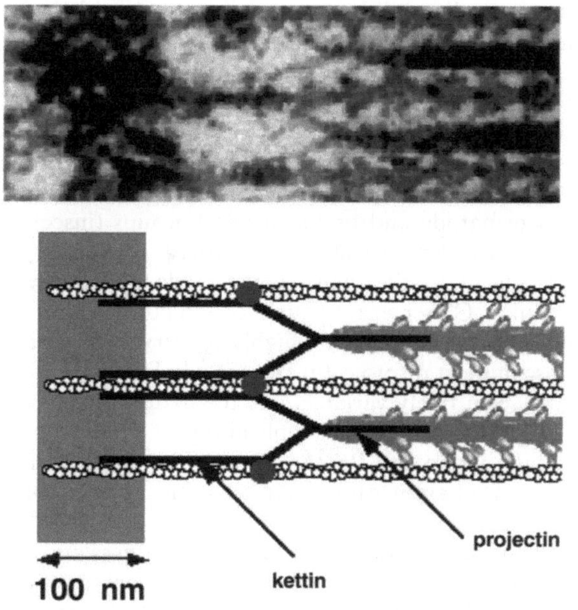

Fig. 8. Model of elastic filaments in *Lethocerus* IFM. Top, part of the electron micrograph in Figure 5. Bottom, suggested position of kettin and projectin. Kettin is bound to actin outside the Z-disc and projectin branches from thick filaments to join the thin filaments. The fine filaments may be made up of both proteins.

Future studies will be necessary to distinguish between these possibilities.

Conclusions

The analysis of sequence domains in the elastic proteins of a nematode and two arthropod groups (insects and crustacea) demonstrates the variety of forms these proteins may take, while being made up of common modules. Closer examination of structure within these modules reveals some are highly conserved, for example those in most of projectin and the whole of kettin; other parts seem freely variable, like the pattern of repeats in PEVK domains. The duplication of long sequence repeats within the PEVK domains is recent and may be a mechanism for increasing the length of the molecule.

Flight muscle needs well regulated stiffness and this is provided by projectin and kettin. The model for entropic polymer elasticity used to predict the elastic properties of titin also applies to IFM, which suggests the basic mechanism is the same in both cases, although IFM has two elastic components and vertebrate striated muscle only has one. IFM is particularly suitable for this study because the muscle has predominately the shorter isoforms of the elastic proteins, without the more complicated arrangement of sequence modules in larger isoforms.

This study of invertebrate proteins puts titin into perspective as an elastic protein adapted to the special requirements of vertebrate muscle. Some of the complexity has been lost and the functions of twitchin or projectin, and SLS protein have been combined in one molecule. Titin appears to have evolved from projectin with the loss of SLS protein in vertebrates. The sequence of twitchin or projectin and SLS protein in the annelids and molluscs, when available, will show if the conserved and highly variable regions that are a feature of the nematode and arthropod proteins, are present in these invertebrates. Comparison of domain sequences in elastic proteins may help to resolve the controversy over the relationship of arthropods to nematodes and annelids.

Acknowledgements

We are very grateful to Mary Reedy for the electron micrograph in Figure 5, and to Toby Gibson for helpful discussion.

Note added in Proof

Recently a *C. elegans* gene coding for an elastic protein of 2.2 MDa has been identified (Flaherty *et al.*, J Mol Biol (2002) **323**: 533–549). The protein (called Ce titin) has Ig and Fn domains, a PEVK-like domain and a kinase domain. The domain pattern differs from that of twitchin and SLS protein.

References

Adams MD, *et al.* (2000) The genome sequence of *Drosophila melanogaster*. *Science* **287:** 2185–2195.

Aguinaldo AM, Turbeville JM, Linford LS, Rivera MC, Garey JR, Raff RA and Lake JA (1997) Evidence for a clade of nematodes, arthropods and other moulting animals. *Nature* **387:** 489–493.

Akam M (2000) Arthropods: developmental diversity within a (super) phylum. *Proc Natl Acad Sci USA* **97:** 4438–4441.

Averof M (1997) Arthropod evolution: same Hox genes, different body plans. *Curr Biol* **7:** R634–R636.

Averof M and Cohen SM (1997) Evolutionary origin of insect wings from ancestral gills. *Nature* **385:** 627–630.

Ayme-Southgate A, Vigoreaux J, Benian G and Pardue ML (1991) *Drosophila* has a twitchin/titin-related gene that appears to encode projectin. *Proc Natl Acad Sci USA* **88:** 7973–7977.

Bang ML, Centner T, Fornoff F, Geach AJ, Gotthardt M, McNabb M, Witt CC, Labeit D, Gregorio CC, Granzier H and Labeit S (2001) The complete gene sequence of titin, expression of an unusual approximately 700-kDa titin isoform, and its interaction with obscurin identify a novel Z-line to I-band linking system. *Circ Res* **89:** 1065–1072.

Benian GM, Kiff JE, Neckelmann N, Moerman DG and Waterston RH (1989) Sequence of an unusually large protein implicated in regulation of myosin activity in *C. elegans*. *Nature* **342:** 45–50.

Benian GM, L'Hernault SW and Morris ME (1993) Additional sequence complexity in the muscle gene, unc-22, and its encoded protein, twitchin, of *Caenorhabditis elegans*. *Genetics* **134:** 1097–1104.

Blair JE, Ikeo K, Gojobori T and Hedges SB (2002) The evolutionary position of nematodes. *BMC Evol Biol* **2:** 7–13.

Bullard B, Hååg P, Brendel S, Benes V, Qiu F and Leonard K (2002) Mapping kettin and D-titin in the invertebrate sarcomere. *J Mus Res Cell Motil* **22:** 602.

Bullard B, Hammond KS and Luke BM (1977) The site of paramyosin in insect flight muscle and the presence of an unidentified protein between myosin filaments and Z-line. *J Mol Biol* **115:** 417–440.

Bullard B and Leonard K (1996) Modular proteins of insect muscle. *Adv Biophys* **33:** 211–221.

Bustamante C, Marko JF, Siggia ED and Smith S (1994) Entropic elasticity of lambda-phage DNA. *Science* **265:** 1599–1600.

Champagne MB, Edwards KA, Erickson HP and Kiehart DP (2000) *Drosophila* stretchin-MLCK is a novel member of the Titin/Myosin light chain kinase family. *J Mol Biol* **300:** 759–777.

Chan WP and Dickinson MH (1996) *In vivo* length oscillations of indirect flight muscles in the fruit fly *Drosophila virilis*. *J Exp Biol* **199 (12):** 2767–2774.

Conway Morris S (2000) The Cambrian 'explosion': slow-fuse or megatonnage? *Proc Natl Acad Sci USA* **97:** 4426–4429.

Daley J, Southgate R and Ayme-Southgate A (1998) Structure of the *Drosophila* projectin protein: isoforms and implication for projectin filament assembly. *J Mol Biol* **279:** 201–210.

de Rosa R, Grenier JK, Andreeva T, Cook CE, Adoutte A, Akam M, Carroll SB and Balavoine G (1999) Hox genes in brachiopods and priapulids and protostome evolution. *Nature* **399:** 772–776.

Fukuzawa A, Shimamura J, Takemori S, Kanzawa N, Yamaguchi M, Sun P, Maruyama K and Kimura S (2001) Invertebrate connectin spans as much as 3.5 micron in the giant sarcomeres of crayfish claw muscle. *Embo J* **20:** 4826–4835.

Funabara D, Kinoshita S, Watabe S, Siegman MJ, Butler TM and Hartshorne DJ (2001) Phosphorylation of molluscan twitchin by the cAMP-dependent protein kinase. *Biochemistry* **40:** 2087–2095.

Fyrberg CC, Labeit S, Bullard B, Leonard K and Fyrberg E (1992) *Drosophila* projectin: relatedness to titin and twitchin and correlation with *lethal(4) 102 CDa* and *bent-dominant* mutants. *Proc Roy Soc Lond B Biol Sci* **249:** 33–40.

Granzier HL and Wang K (1993) Interplay between passive tension and strong and weak binding cross-bridges in insect indirect flight muscle. A functional dissection by gelsolin-mediated thin filament removal. *J Gen Physiol* **101:** 235–270.

Greaser M (2001) Identification of new repeating motifs in titin. *Proteins* **43:** 145–149.

Hakeda S, Endo S and Saigo K (2000) Requirements of kettin, a giant muscle protein highly conserved in overall structure in evolution, for normal muscle function, viability, and flight activity of *Drosophila. J Cell Biol* **148:** 101–114.

Hu DH, Matsuno A, Terakado K, Matsuura T, Kimura S and Maruyama K (1990) Projectin is an invertebrate connectin (titin): isolation from crayfish claw muscle and localization in crayfish claw muscle and insect flight muscle. *J Muscle Res Cell Motil* **11:** 497–511.

Improta S, Politou AS and Pastore A (1996) Immunoglobulin-like modules from titin I-band: extensible components of muscle elasticity. *Structure* **4:** 323–337.

Irving TC and Maughan DW (2000) In vivo X-ray diffraction of indirect flight muscle from *Drosophila melanogaster. Biophys J* **78:** 2511–2515.

Kennison JA and Tamkun JW (1988) Dosage-dependent modifiers of *polycomb* and *antennapedia* mutations in *Drosophila. Proc Natl Acad Sci USA* **85:** 8136–8140.

Kiff JE, Moerman DG, Schriefer LA and Waterston RH (1988) Transposon-induced deletions in *unc-22* of *C. elegans* associated with almost normal gene activity. *Nature* **331:** 631–633.

Kolmerer B, Clayton J, Benes V, Allen T, Ferguson C, Leonard K, Weber U, Knekt M, Ansorge W, Labeit S and Bullard B (2000) Sequence and expression of the kettin gene in *Drosophila melanogaster* and *Caenorhabditis elegans. J Mol Biol* **296:** 435–448.

Kreuz AJ, Simcox A and Maughan D (1996) Alterations in flight muscle ultrastructure and function in *Drosophila* tropomyosin mutants. *J Cell Biol* **135:** 673–687.

Kulke M, Neagoe C, Kolmerer B, Minajeva A, Hinssen H, Bullard B and Linke WA (2001) Kettin, a major source of myofibrillar stiffness in *Drosophila* indirect flight muscle. *J Cell Biol* **154:** 1045–1057.

Labeit S and Kolmerer B (1995) Titins: giant proteins in charge of muscle ultrastructure and elasticity. *Science* **270:** 293–296.

Lakey A, Labeit S, Gautel M, Ferguson C, Barlow DP, Leonard K and Bullard B (1993) Kettin, a large modular protein in the Z-disc of insect muscles. *Embo J* **12:** 2863–2871.

Lakey A, Ferguson C, Labeit S, Reedy M, Larkins A, Butcher G, Leonard K and Bullard B (1990) Identification and localization of high molecular weight proteins in insect flight and leg muscle. *Embo J* **9:** 3459–3467.

Li H, Linke WA, Oberhauser AF, Carrion-Vazquez M, Kerkvliet JG, Lu H, Marszalek PE and Fernandez JM (2002) Reverse engineering of the giant muscle protein titin. *Nature Lond* **418:** 998–1002.

Linke WA, Ivemeyer M, Mundel P, Stockmeier MR and Kolmerer B (1998) Nature of PEVK-titin elasticity in skeletal muscle. *Proc Natl Acad Sci USA* **95:** 8052–8057.

Liversage AD, Holmes D, Knight PJ, Tskhovrebova L and Trinick J (2001) Titin and the sarcomere symmetry paradox. *J Mol Biol* **305:** 401–409.

Machado C and Andrew DJ (2000) D-TITIN: a giant protein with dual roles in chromosomes and muscles. *J Cell Biol* **151:** 639–652.

Machado C, Sunkel CE and Andrew DJ (1998) Human autoantibodies reveal titin as a chromosomal protein. *J Cell Biol* **141:** 321–333.

Maki S, Ohtani Y, Kimura S and Maruyama K (1995) Isolation and characterization of a kettin-like protein from crayfish claw muscle. *J Muscle Res Cell Motil* **16:** 579–585.

Marko JF and Siggia ED (1995) Statistical mechanics of supercoiled DNA. *Physical Review. E. Statistical Physics, Plasmas, Fluids, and Related Interdisciplinary Topics* **52:** 2912–2938.

Moerman DG, Benian GM, Barstead RJ, Schriefer LA and Waterston RH (1988) Identification and intracellular localization of the *unc-22* gene product of *Caenorhabditis elegans. Genes Dev* **2:** 93–105.

Nave R and Weber K (1990) A myofibrillar protein of insect muscle related to vertebrate titin connects Z band and A band: purification and molecular characterization of invertebrate mini-titin. *J Cell Sci* **95:** 535–544.

Patel SR and Saide J (2001) A(225), a novel A-band protein of *Drosophila* indirect flight muscle. *Biophys J* **80:** 71a.

Peckham M, Molloy JE, Sparrow JC and White DC (1990) Physiological properties of the dorsal longitudinal flight muscle and the tergal depressor of the trochanter muscle of *Drosophila melanogaster. J Muscle Res Cell Motil* **11:** 203–215.

Pringle J (1978) Stretch activation of muscle: function and mechanism. *Proc Roy Soc Lond* B **201:** 107–130.

Reedy MK (1971) Electron microscope observations concerning the behavior of the cross-bridge in striated muscle. In: Podolsky RJ (ed.) *Contractility of Muscle Cells and Related Processes.* (pp. 229–246) Prentice Hall Inc., Englewood Cliffs, NJ.

Saide JD (1981) Identification of a connecting filament protein in insect fibrillar flight muscle. *J Mol Biol* **153:** 661–679.

Saide JD, Chin-Bow S, Hogan-Sheldon J and Busquets-Turner L (1990) Z-band proteins in the flight muscle and leg muscle of the honeybee. *J Muscle Res Cell Motil* **11:** 125–136.

Southgate R and Ayme-Southgate A (2001) Alternative splicing of an amino-terminal PEVK-like region generates multiple isoforms of *Drosophila* projectin. *J Mol Biol* **313:** 1035–1043.

Trinick J (1994) Titin and nebulin: protein rulers in muscle? *Trends Biochem Sci* **19:** 405–409.

Trombitas K (2000) Connecting filaments: a historical prospective. *Adv Exp Med Biol* **481:** 1–23.

Trombitas K and Tigyi-Sebes A (1974) Direct evidence for connecting C filaments in flight muscle of honey bee. *Acta Biochim Biophys Acad Sci Hung* **9:** 243–253.

van Straaten M, Goulding D, Kolmerer B, Labeit S, Clayton J, Leonard K and Bullard B (1999) Association of kettin with actin in the Z-disc of insect flight muscle. *J Mol Biol* **285:** 1549–1562.

Vibert P, Edelstein SM, Castellani L and Elliott BW, Jr. (1993) Mini-titins in striated and smooth molluscan muscles: structure, location and immunological crossreactivity. *J Muscle Res Cell Motil* **14:** 598–607.

Vigoreaux JO, Saide JD and Pardue ML (1991) Structurally different *Drosophila* striated muscles utilize distinct variants of Z-band-associated proteins. *J Muscle Res Cell Motil* **12:** 340–354.

Waterston RH, Thomson JN and Brenner S (1980) Mutants with altered muscle structure of *Caenorhabditis elegans. Dev Biol* **77:** 271–302.

Waterston RH (1988) Muscle. In: Wood WB (ed.) *The Nematode Caenorhabditis elegans.* (pp. 281–335) Cold Spring Harbor Laboratory Press, New York, NY.

White DC (1983) The elasticity of relaxed insect fibrillar flight muscle. *J Physiol* **343:** 31–57.

Zhang Y, Featherstone D, Davis W, Rushton E and Broadie K (2000) *Drosophila* D-titin is required for myoblast fusion and skeletal muscle striation. *J Cell Sci* **113:** 3103–3115.

Journal of Muscle Research and Cell Motility **23**: 449–453, 2002.

Single-molecule measurement of elasticity of Serine-, Glutamate- and Lysine-Rich repeats of invertebrate connectin reveals that its elasticity is caused entropically by random coil structure

ATSUSHI FUKUZAWA[1,†], MICHIO HIROSHIMA[2,3,†], KOSCAK MARUYAMA[1], NAOTO YONEZAWA[1], MAKIO TOKUNAGA[2,3,4] and SUMIKO KIMURA[1,*]
[1]*Department of Biology, Faculty of Science, Chiba University, Chiba, 263-8522, Japan;* [2]*Structural Biology Center, National Institute of Genetics, Mishima, Shizuoka, 411-8540, Japan;* [3]*Research Center for Allergy and Immunology, RIKEN, Yokohama, Kanagawa, 230-0045, Japan;* [4]*Department of Genetics, Graduate University for Advanced Studies, Mishima, Shizuoka, 411-8540, Japan*

Abstract

Invertebrate connectin (I-connectin) is a 1960 kDa elastic protein linking the Z line to the tip of the myosin filament in the giant sarcomere of crayfish claw closer muscle (Fukuzawa *et al.*, 2001 EMBO J **20:** 4826–4835). I-Connectin can be extended up to 3.5 μm upon stretch of giant sarcomeres. There are several extensible regions in I-connectin: two long PEVK regions, one unique sequence region and Ser-, Glu- and Lys-rich 68 residue-repeats called SEK repeats. In the present study, the force measurement of the single recombinant SEK polypeptide containing biotinylated BDTC and GST tags at the N and C termini, respectively, were performed by intermolecular force microscopy (IFM), a refined AFM system. The force vs. extension curves were well fit to the wormlike chain (WLC) model and the obtained persistence length of 0.37 ± 0.01 nm ($n = 11$) indicates that the SEK region is a random coil along its full length. This is the first observation of an entropic elasticity of a fully random coil region that contributes to the physiological function of I-connectin.

Introduction

The giant filamentous protein, titin/connectin (3000–3700 kDa) contributes to organizing the contracting sarcomere by extending from the Z line over the thick filament up to the M line at the center of the sarcomere in vertebrate striated muscle (Maruyama, 1997; Linke, 2000; Granzier and Labeit, 2002; Tskhovrebova and Trinick, 2002). A molecule of titin/connectin is very flexible and approximately 1 μm long in resting sarcomeres, and can be passively stretched resulting in tension generation.

The complete sequence of vertebrate titin/connectin was elucidated by Labeit and Kolmerer (1995). It consists of the N terminal Z line-binding region (ca. 80 kDa), I band region (ca. 800–1500 kDa) and myosin-binding region (ca. 2100 kDa). The extensible I band region mainly consists of a number of tandem immunoglobulin (Ig) domains and one PEVK domain. Within the physiological range of sarcomere lengths, entropic elasticity of both the Ig domains (straightening of interdomains) and the PEVK region is responsible for reversible extension. The latter mainly contributes to the passive tension generation upon stretch (cf. Linke, 2000). The size of the PEVK region is muscle-type dependent: cardiac, 163 residues; psoas, 1400 residues; and soleus, 2174 residues (Labeit and Kolmerer, 1995; Linke *et al.*, 1996). The N2B unique sequence of cardiac titin/connectin isoform has been also shown to be elastic (Helmes *et al.*, 1999; Linke *et al.*, 1999; Trombitas *et al.*, 2001).

Giant sarcomeres of crayfish claw closer muscle the length of which is 8.3 μm long at rest can be elongated to 14 μm upon stretch. Manabe *et al.* (1993) suggested that a 3000 kDa connectin-like protein is responsible for tension development when sarcomeres are stretched to up to 14 μm in length. The distance to be covered by this protein is as long as 4 μm even if it attaches to the tip of the thick filament (titin binds to the myosin filament up to the M line region). In order to clarify what kind of primary structure makes such a remarkable extension possible, we sequenced the cDNA encoding the 3000 kDa protein in crayfish claw closer muscle, and reported that the protein invertebrate connectin (I-connectin, actually 1960 kDa) in question has a novel unique sequence entirely different from that of vertebrate titin/connectin (Figure 1): (1) There is the complete sequence of kettin, 540 kDa Z line-binding protein (van Straaten *et al.*, 1999) at the very N terminal region. (2) The number of Ig C2 or fibronectin Type III (Fn3) domain, abundant in vertebrate connectin, is very small. Ig and Fn3 domains that constitute over 90% of vertebrate connectin (Labeit and Kolmerer, 1995) comprise only 20% of I-connectin and in particular,

*To whom correspondence should be addressed: Tel.: +81-43-290-2811; Fax: +81-43-290-2811; E-mail: sumiko@bio.s.chiba-u.ac.jp
†Authors contributed equally to this work.

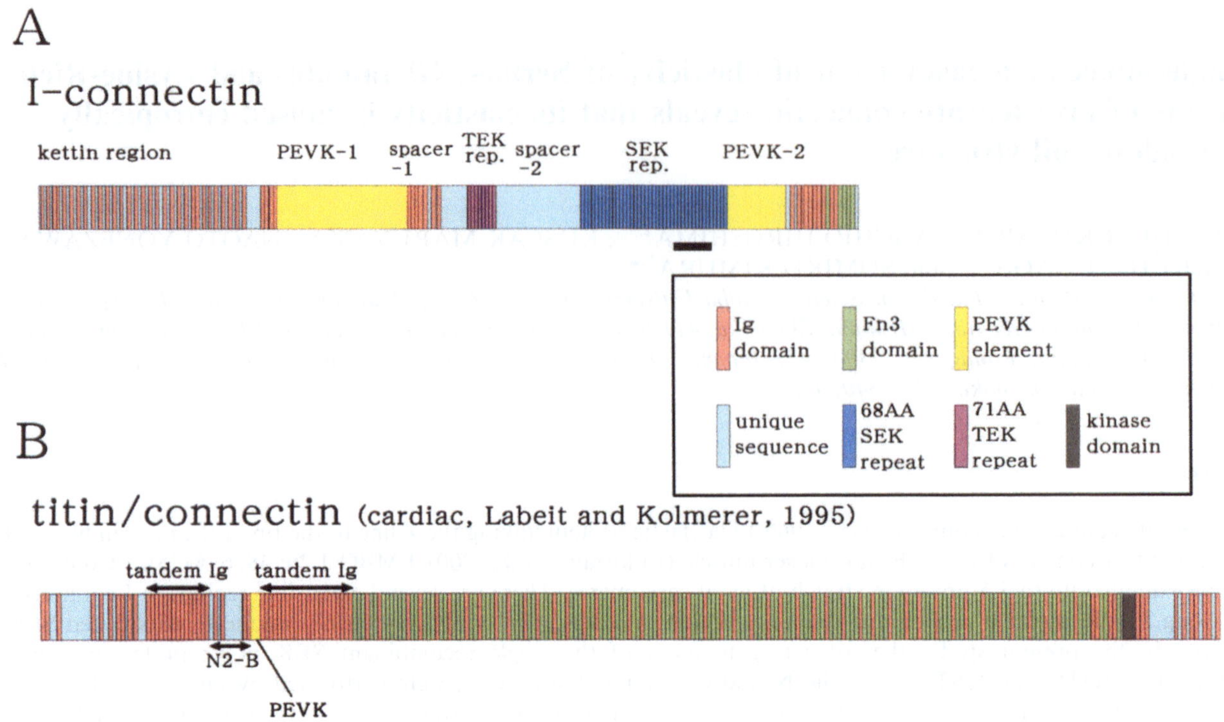

Fig. 1. Domain structure of I-connectin and human cardiac titin/connectin. A, In I-connectin. the N terminal 4822-residue kettin is followed by 3364-residue PEVK-1, seven Ig domains, 467-residue spacer-1, eight 71-residue TEK repeats, 1666-residue spacer-2, forty-one 68-residue SEK repeats, 1310-residue PEVK-2, ten Ig and five Fn3 domains. Total amino acids, 17,352 (1960 kDa). B, Human cardiac titin/connectin, see Labeit and Kolmerer (1995).

there are only five Fn3 domains at the very C terminal region (vertebrate titin/connectin has 132 copies of Fn3 domains). (3) There are two extensible PEVK domains consisting of 3364 and 1310 residues in I-connectin as compared to one PEVK region of titin/connectin in vertebrate striated muscle the size of which depends on muscle tissue-type (163–2174 residues). (4) There are novel repeating regions of 68 residues (called SEK repeat) and 71 residues (called TEK repeat). The extensibility of the novel SEK and TEK repeats and the two PEVK regions was demonstrated by immunofluorescence microscopy, using several site-specific antibodies. Furthermore, the SEK repeats and the two PEVK segments were shown to be involved in passive tension generation (Fukuzawa *et al.*, 2001).

Recently, single-molecule atomic force microscopy (AFM) has been applied to investigate dynamic mechanical properties of recombinant PEVK and N2B polypeptides of vertebrate striated muscle titin/connectin (Li *et al.*, 2001; Watanabe *et al.*, 2002). The results reported are consistent with those predicted from a wormlike chain (WLC) model, suggesting that the PEVK and N2B polypeptide behaves as an entropic spring. It is of much interest to elucidate whether the elastic portions of I-connectin mechanically behave just as titin/connectin's PEVK and N2B domains or not. Here we describe preliminary results with the serine-, glutamate- and lysine-rich (SEK) repeats of I-connectin studied by intermolecular force microscopy (IFM).

Materials and methods

Recombinant protein expression and purification

The SEK polypeptide with biotinylated BDTC tag at the N terminus and GST tag at the C terminus was expressed in *E. coli* and purified by use of two affinity columns and an ion exchange column.

The cDNA coding SEK region (EMBL/GenBank/DDBJ accession number: AB055861, 40510–42687; Fukuzawa *et al.*, 2001) was constructed in-frame gene into pinpoint-Xa protein expression vector (Promega) carrying a segment encoding a part of the 1.3 S subunit of *Propionibacterium shermanii* transcarboxylase peptide (Murtif *et al.*, 1985). Next, the cDNA coding GST tag was inserted into the above construct following the SEK region. The cDNA fragment was amplified by PCR using a pair of appropriate primers derived from pGEX6P-1 GST protein expression vector (EMBL/GenBank/DDBJ accession number: U78872 (Amersham Pharmacia)) and pGEX6P-1 vector as templates. The identity of the derived construct was verified by DNA sequencing.

The recombinant protein (BDTC-SEK-GST) was expressed in *E. coli* BL21 [DE3] pLysS. The BDTC tag was post-translationally biotinylated *in vivo*. The bacteria were harvested by centrifugation. Pellet was suspended with R buffer (140 mM NaCl, 2.7 mM KCl, 10 mM Na_2HPO_4, 1.8 mM KH_2PO_4, pH 7.4, 1 mM EDTA, 0.2 mM PMSF) and sonicated (TOMY UR-200P, output 1.5, 30 s, twice). After centrifugation, supernatant containing the soluble recombinant protein was applied to Glutathione Sepharose 4B (Amersham Pharmacia) column. After the resin was extensively washed with R buffer, the bound protein was eluted by a solution consisting of 50 mM Tris-HCl (pH8.0), 75 mM NaCl, 20 mM reduced glutathione, 1 mM EDTA and 0.2 mM PMSF. The fractions containing BDTC-SEK-GST were collected and applied to DEAE Sepharose FF (Amersham Pharmacia) column. In this step, several contaminants were adsorbed and BDTC-SEK-GST was eluted in the flow through fraction. Then, the fractions were applied to SoftLink™ Soft Release Avidin Resin (Promega) column. BDTC-SEK-GST was eluted by a biotin containing buffer (5 mM biotin, 50 mM Tris-HCl, pH8.0, 75 mM NaCl, 1 mM EDTA and 0.2 mM PMSF). Finally, BDTC-SEK-GST was dialyzed against 10 mM PIPES (pH 7.0) and 0.4 mM KCl. The purity and the concentration of BDTC-SEK-GST were examined by SDS gel electrophoresis and by BCA Protein Assay Kit (PIRECE).

Intermolecular force microscopy

For the extension measurement of the SEK polypeptide, we used a novel microscopy, IFM (Aoki *et al.*, 1997; Tokunaga *et al.*, 1997). IFM is a refined system of conventional AFM. Use of flexible handmade cantilevers with the spring constant of 0.1 [pN/nm] has achieved subpiconewton force resolution. The cantilever is more than 50-fold more sensitive and has higher signal-to-noise ratio than commercial ones. Such flexible cantilevers undergo large thermal bending motion of ~40 nm peak-to-peak amplitude. In order to reduce the large fluctuation, we used the feedback system which modulates laser radiation pressure on the cantilever opposite to the direction of bending motion. This system enabled us to control the cantilever position with nanometer accuracy.

IFM is appropriate for investigating a correlation between extension force of a single molecule and its microstructure because of the high sensitivity and the high signal-to-noise ratio. We therefore applied the IFM to extension measurements of the single SEK peptide.

Modification of substrate and IFM probe

The substrate used for the extension measurement was prepared as following: coverslips (18 mm × 18 mm) were cleaned in 0.1 M KOH for 1 h by ultrasonication and then rinsed for several times in milli-Q water. They were soaked in 10 mM HCl to remove KOH thoroughly and washed with water and then with ethanol. These cleaned coverslips were epoxy-silanized in ethanol containing 5% (v/v) silane coupling agent, 3-glycidoxypropyltriethoxysilane (LS-4668, Shin-etsu Chemicals), 0.3% (v/v) triethylamine at 80 for 1 h. After the coupling process, they were rinsed in water and in ethanol for several times and then dried in air. The epoxy-silanized substrate was fixed on a small dish (0.35 mm) with adhesive and stored in a desiccator until the extension measurement.

At the tip of the cantilever, ZnO whisker (needle crystal, Matsushita AMTEC) was adhered as probe. The whisker was biotinylated for binding to recombinant SEK-peptide through streptavidin. The first step of biotinylation was amino-silanizing with silane coupling agent, 3-aminopropyltriethoxysilane (LS-3150, Shin-etsu Chemicals). This step was similar to described above. The amino-silanized whisker was then immersed in dried DMF containing 1 mM biotin–$(AC_5)_2$–OSu at 37°C for 2 h, washed with ethanol and water, and then stored in ethanol until adhered to the cantilever with adhesive.

Sample preparation and force measurement

Before the extension measurement, 15 μl of solution containing anti-GST antibody (Sigma G7781) in 40.7 mM KCl and 10 mM HEPES-NaOH (pH 8.2) was dropped on the epoxy-silanized substrate and then a small coverslip (9 mm × 9 mm) was placed over the drop. Ten microliters of the antibody solution was infused to the narrow space between two coverslips and the dish was placed at 37°C for 20 h. After washout of non-adsorbed antibody, 10 μl of 1 μM recombinant SEK polypeptide was infused and incubated at room temperature for 30 min. Free polypeptide was then washed out and 1 mg/ml streptavidin was infused. After incubation for 15 min., the dish was then filled with extension solution containing 40.7 mM KCl, 10 mM PIPES-KOH (pH 7.0) and the small coverslip was removed. The solution was replaced several times by refluxing.

The cantilever with biotinylated probe was mounted to the IFM and immersed in extension solution in the dish. The probe position was controled by modulating the laser radiation pressure on the cantilever. The cantilever was moved at the velocity of 60–70 [nm/s]. In the repeated approach – detachment cycles, the biotinylated probe occasionally binds to streptavidin with the recombinant SEK polypeptide on the substrate and force signal of the peptide is detected while extending the molecule. The zero length point of the force-extension curves was determined as the point where the probe attached to the substrate. The magnitude of forces acting on the cantilever was read out from the modulation voltage of the laser for feedback. Obtained data of force-extension curves were analyzed by least square fitting based on WLC model.

Results and discussion

The SEK polypeptide with biotinylated BDTC tag at the N terminus and GST tag at the C terminus (BDTC-SEK-GST) (Figure 2A) was expressed. An SDS–PAGE of the purified sample is shown in Figure 2B. The purified sample was slightly contaminated with ~75 and ~20 kDa polypeptides. The molecular mass (MM) of BDTC-SEK-GST was estimated to be around 160–180 kDa. This value was much larger than the expected MM of 121 kDa (BDTC, 12.9 kDa; SEK, 78.5 kDa; GST, 25.7 kDa; connecting peptides, 3.9 kDa).

As reported by Gutierrez-Cruz *et al.* (2001), the calculated MW (51,467) of recombinant PEVK polypeptide is much smaller than the mobility-based MM (86 kDa). This is also true of recombinant SEK polypeptide (35,088 Da vs. 55 kDa) (Fukuzawa *et al.*, 2001). It is very likely that glutamic acid- and lysine-rich polypeptides show lower mobility on SDS polyacrylamide gels than that expected from the molecular sizes.

A typical example of force-extension curve of the SEK polypeptide is shown in Figure 3. The observed curves showed gradually rising force and then an abrupt decrease around 20–40 pN. It was probably induced by the detachment of the peptide from the substrate or the probe. The release curve of the peptide could not be observed because of the detachment, but the stretching curves obtained from different trials showed almost the same profiles. The persistence length and the contour length were obtained by fitting the curves with the WLC equation, $FA/k_BT = z/L + 1/(4(1 - z/L)^2) - 1/4$. ($F$, A, L, and z are observed force, persistence length, contour length, and end-to-end length of the stretching peptide, respectively) (Bustamante *et al.*, 1994). The thin continuous line superimposed on the stretch curve of Figure 3 is an example of a WLC fit.

The values of the persistence length and the contour length were 0.37 ± 0.01 nm and 188 ± 66 nm ($n = 11$), respectively. The calculated contour length (number of residues × 0.38 nm) of the recombinant polypeptide is as follows: 48.6 nm (128 residues biotinylated BDTC tag), 293.4 nm (722 residues SEK repeats) and 83.6 nm (220 residues GST tag). The total length, 425.6 nm, is much longer than the obtained value, 188 nm. Even the length of the SEK repeats, 293.4 nm, is considerably longer. This short length makes it unlikely that multimolecular complexes were stretched. Instead a few recombinant peptides may have adsorbed on the substrate or probe tip non-specifically (i.e. not via antibody – GST or biotin – streptavidin binding), and therefore not attached at their ends, explaining why the measured contour length was shorter than the calculated contour length.

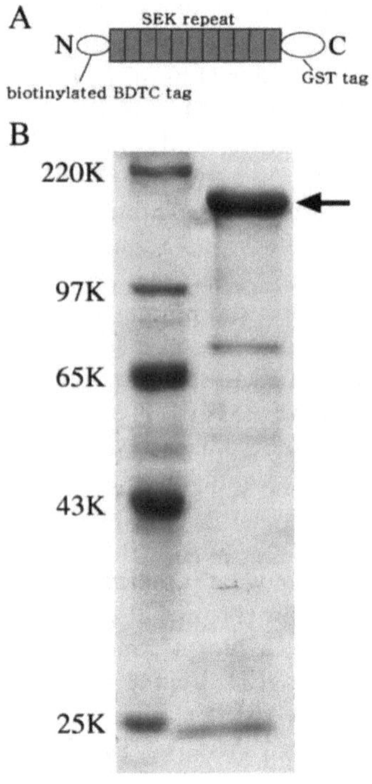

Fig. 2. A, Schematic of domain structure of the recombinant SEK repeats. BDTC, N terminal biotinylated tag; GST, C terminal GST tag. B, SDS polyacrylamide gel electrophoresis pattern of the recombinant SEK polypeptide. Left lane, molecular weight markers. Right lane, the recombinant SEK repeats (arrow). 10% polyacrylamide gel was used.

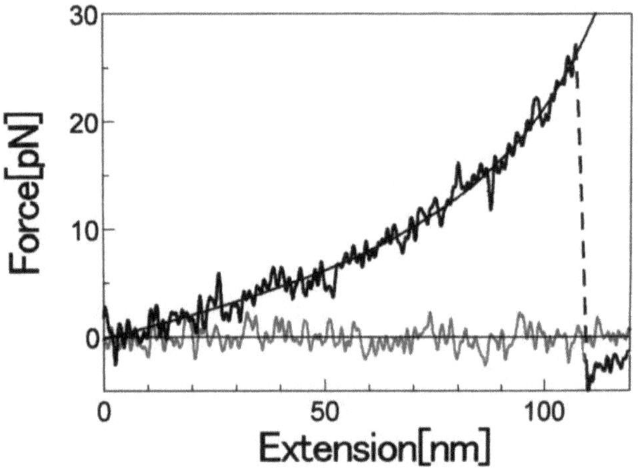

Fig. 3. A typical force-extension curve of recombinant SEK polypeptide as revealed by IFM. Curves of approach (gray) and stretch (black) with the WLC fit (thin line) are shown. A recombinant SEK polypeptide (1 μM in 40.7 mM KCl and 10 mM PIPES-NaOH buffer, pH 7.0) on the substrate was allowed to adsorb onto a probe surface through biotin – streptavidin bond. The peptide was picked up at random by cantilever probe (spring constant: 0.16 [pN/nm] (=0.00016 [N/m])) and then stretched. At ~25 pN, the peptide detached from the substrate or the probe. The rate of extension was 67 nm/s. WLC fitting yielded a persistence length of 0.37 ± 0.01 nm, which is almost the same as the repeat of amino acid residues in a fully extended random coil (0.38 nm).

The obtained persistence length of the SEK value, 0.37 ± 0.01 nm is considerably short in comparison to reported values of other peptides in titin/connectin. The values of the PEVK peptide are 0.45 nm for triplets, 0.8 nm for doublets and 1.45 nm for singlets (Watanabe *et al.*, 2002). Li *et al.* (2001) also reported the values of 0.4–2.5 nm. The cardiac muscle isoform, N2B titin/connectin, contains an elastic segment called unique N2B sequence (N2B-Us) in addition to the short PEVK segment (cf. Helmes *et al.*, 1999). The persistence length of N2B-Us is reported to be 0.3 nm (doublet) and 0.65 nm (singlet) (Watanabe *et al.*, 2002). When a doublet, two parallel peptides, is extended, the persistence length becomes half of that of the singlet. In the present experiments, a few force-extension curves of the doublet were found. In the force-extension curve of the doublet, force fell down to the zero level in a double step, whereas force in the curve of the singlet fell down in a single step as shown in Figure 3. The doublet curve yielded the persistence length of 0.19 nm (data not shown). Thus, we concluded that the persistence length of 0.37 nm is that of the singlet. The obtained persistence length of the SEK polypeptide (0.37 nm) is almost the same as the repeat of an amino acid residue in an unfolded and fully extended polypeptide (0.38 nm). It suggests that the SEK peptide forms random coil along nearly its full length and that the elasticity of the SEK peptide is caused entropically by the random coil structure. The peptides of titin/connectin, including the SEK repeat region, are thought to be entropic springs. The shortest persistence length of the SEK polypeptide indicates that the polypeptide has the largest entropic elasticity. It means that larger force is needed to extend the SEK polypeptide to a given fractional extension than other regions. This characteristics of the SEK peptide is consistent with our previous observations that the extent of extensibility is in the order of spacer 2 > PEVK > SEK repeats (Fukuzawa *et al.*, 2001).

The force measurements of the single SEK peptide have shown that the elasticity of the SEK peptide can be explained by the entropic elasticity of random coil. This study for the first time indicates that a complete random coil plays a unique role in the physiology of I-connectin.

Acknowledgements

This work was supported by Grants-in-Aid for Scientific Research of the Ministry of Education, Culture, Sports, Science and Technology of Japan, and a grant from Toray Science Foundation.

References

Aoki T, Hiroshima M, Kitamura K, Tokunaga M and Yanagida T (1997) Non-contact scanning probe microscopy with sub-piconewton force sensitivity. *Ultramicroscopy* **70:** 45–55.

Bustamante C, Marko JF, Siggia ED and Smith S (1994) Entropic Elasticity of λ-Pharge DNA. *Science* **265:** 1599–1600.

Fukuzawa A, Shimamura J, Takemori S, Kanzawa N, Yanaguchi M, Sun P, Maruyama K and Kimura S (2001) Invertebrate connectin spans as much as 3.5 μm in the giant sarcomeres of crayfish claw muscle. *EMBO J* **20:** 4826–4835.

Granzier H and Labeit S (2002) Cardiac titin: an adjustable multi-functional spring. *J Physiol* **541:** 335–342.

Gutierrez-Cruz G, Van Heerden AH and Wang K (2001) Modular motif, structural folds and affinity profiles of the PEVK segment of human fetal skeletal muscle titin. *J Biol Chem* **276:** 7442–7449.

Helmes M, Trombitas K, Centner T, Kellermayer M, Labeit S, Linke WA and Granzier H (1999) Mechanically driven contour-length adjustment in rat cardiac titin's unique N2B sequence: titin is an adjustable spring. *Circ Res* **84:** 1339–1352.

Labeit S and Kolmerer B (1995) Titins: giant proteins in charge of muscle ultrastructure and elasticity. *Science* **270:** 293–296.

Li H, Oberhauser AF, Redick SD, Carrion-Vazquez M and Erickson HP (2001) Multiple conformations of PEVK proteins detected by single-molecule techniques. *Proc Natl Acad Sci USA* **98:** 10,682–10,686.

Linke WA, Ivemeyer M, Olivieri N, Kolmerer B, Ruegg JC and Labeit S (1996) Towards a molecular understanding of the elasticity of titin. *J Mol Biol* **261:** 62–71.

Linke WA, Rudy DE, Centner T, Gautel M, Witt C, Labeit S and Gregorio CC (1999) I-band titin in cardiac muscle is a three-element molecular spring and is critical for maintaining thin filament structure. *J Cell Biol* **146:** 631–644.

Linke WA (2000) Stretching molecular springs: elasticity of titin filaments in vertebrate striated muscle. *Histol Histopathol* **15:** 799–811.

Manabe T, Kawamura Y, Higuchi H, Kimura S and Maruyama K (1993) Connectin, giant elastic protein, in giant sarcomeres of crayfish claw muscle. *J Muscle Res Cell Motil* **14:** 654–665.

Maruyama K (1997) Connectin/titin, giant elastic protein of muscle. *FASEB J* **11:** 341–345.

Murtif VL, Bahler CR and Samols D (1985) Cloning and expression of the 1.3 S biotin-containing subunit of transcarboxylase *Proc Natl Acad Sci USA* **82:** 5617–5621.

Tokunaga M, Aoki T, Hiroshima M, Kitamura K and Yanagida T (1997) Subpiconewton intermolecular force microscopy. *Biochem Biophys Res Commun* **231:** 566–569.

Trombitas K, Wu Y, Labeit D, Labeit S and Granzier H (2001) Cardiac titin isoforms are coexpressed in the half-sarcomere and extend independently. *Am J Physiol Heart Circ Physiol* **281:** H1793–H1799.

Tskhovrebova L and Trinick J (2002) Role of titin in vertebrate striated muscle. *Philos Trans R Soc Lond B Biol Sci* **357:** 199–206.

van Straaten M, Goulding D, Kolmerer B, Labeit S, Clayton J, Leonard K and Bullard B (1999) Association of kettin with actin in the Z-disc of insect flight muscle. *J Mol Biol* **285:** 1549–1562.

Watanabe K, Nair P, Labeit D, Kellermayer MS, Greaser M, Labeit S and Granzier H (2002) Molecular mechanics of cardiac titin's PEVK and N2B spring elements. *J Biol Chem* **277:** 11,549–11,558.

The Elastic Vertebrate Muscle Protein Titin

Journal of Muscle Research and Cell Motility **23**: 457–471, 2002.

Titin as a modular spring: emerging mechanisms for elasticity control by titin in cardiac physiology and pathophysiology

HENK GRANZIER[1,*], DIETMAR LABEIT[2], YIMING WU[1] and SIEGFRIED LABEIT[2]
[1]*Department VCAPP, Washington State University, Pullman, WA 99164-6520, USA;* [2]*Anesthesiology and Intensive Operative Medicine, University Hospital Mannheim, 68135, Germany*

Abstract

Titin is a giant elastic protein that functions as a molecular spring that develops passive muscle stiffness. Here we discuss the molecular basis of titin's extensibility, how titin's contribution to passive muscle stiffness may be adjusted and how adjustment of titin's stiffness may influence muscle contraction. We also focus on ligands that link titin to membrane channel activity, protein turnover and gene expression.

Introduction

Titin is a giant elastic protein located in the striated-muscle sarcomere where it is associated with the thin and thick filaments that underlie muscle contraction. A large part of the titin molecule is extensible and functions as a molecular spring that develops passive muscle stiffness when sarcomeres are stretched. Here we review our current knowledge on the molecular basis of titin's extensibility (Section 1). Importantly, recent work suggests that titin's contribution to passive muscle stiffness may be adjusted by a variety of mechanisms (Sections 2–5). Calcium/S100 may adjust passive stiffness within single contractile cycles (Section 2). Phosphorylation of cardiac titin's N2B spring elements reduces titin-based passive stiffness in cardiac muscle and involves signaling pathways that require minutes or longer to become mechanically relevant (Section 3). Differential splicing of titin's spring elements is a long-term mechanism of adjustment which plays a significant role in diastolic stiffness modulation in heart disease (Sections 4 and 5). Finally, we focus on how adjustment of titin's stiffness may influence muscle contraction and ligands that link titin to membrane channel activity, protein turnover and gene expression (Section 6).

Titin's extensible region: serially linked and mechanically distinct springs

Titin forms a striated muscle-specific myofilament that develops passive force in response to sarcomere stretch. (For recent reviews and original citations see Wang, 1984; Wang, 1996; Maruyama, 1997; Gregorio *et al.*, 1999; Linke, 2000; Granzier and Labeit, 2002; Tskhovrebova and Trinick, 2002). Titin's force is generated by an extensible segment located in the I-band region of the sarcomere (Figure 1). The extensible region comprises tandemly arranged immunoglobulin(Ig)-like domains (~90-amino-acid domains with a seven-strand β-barrel structure (Improta *et al.*, 1996)) making up a proximal tandem Ig segment near the Z-line and a distal tandem Ig segment near the A-band (Figure 2A). Both tandem Ig segments are separated by another extensible region of titin, the PEVK domain, so named because it is rich in proline (P), glutamate (E), valine (V) and lysine (K) residues (Labeit and Kolmerer, 1995). The PEVK region of titin and the N- and C-terminally located tandem Ig segments are spring elements found in both cardiac and skeletal muscle titins, whereas the N2B-unique sequence (N2B-Us) is cardiac-specific and forms a third spring element (Figure 2B).

Immunoelectron microscopy on human soleus fibers has shown that in slack sarcomeres (no external force) the tandem Ig segments and the PEVK are in a 'contracted' state. Upon sarcomere stretch the extension of tandem Ig segments dominates initially, followed by a dominating PEVK extension (see Figure 2A). It is likely that tandem Ig extension is primarily due to unbending of linkers between folded Ig domains (Linke *et al.*, 1998; Trombitas *et al.*, 1998b). The tandem Ig segments exhibit wormlike-chain (WLC) behavior with a persistence length (measure of chain's bending rigidity) that is relatively long (~15 nm based on a recent single-molecule study (Tskhovrebova and Trinick, 2001), explaining why tandem Ig extension dominates in moderately stretched sarcomeres where passive force is low.

When a continuous stretch is imposed on slack sarcomeres, tandem Ig segments of soleus muscle titin extend rapidly to a near constant length (Figure 2A, bottom). Further extension is halted despite the generation of high levels of passive force (in this regime PEVK extension dominates). Although unfolding of a few domains cannot be excluded (Minajeva *et al.*, 2001), the constant length of the tandem Ig segments in highly stretched sarcomeres of soleus muscle (Figure 2A) that

* To whom correspondence should be addressed: Tel.: +1-509-335-3390; Fax: +1-509-335-4650; E-mail: granzier@wsunix.wsu.edu

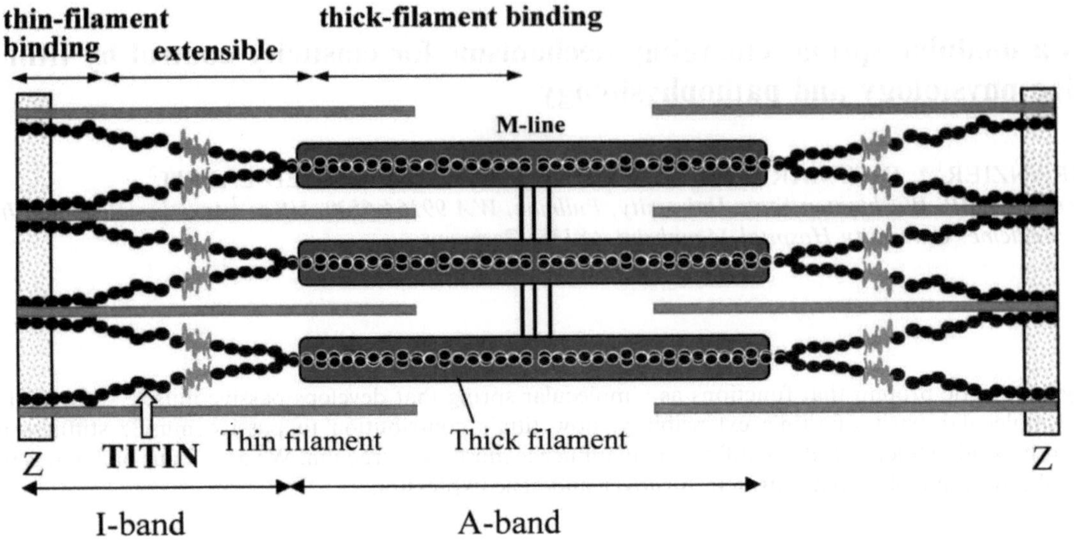

Fig. 1. (A) Layout of titin in sarcomere. A single titin polypetide runs from the Z-disc (Z) to the middle of A-band (M-line). In and near the Z-disc titin is inextensible, whereas titin's remaining I-band portion extends as sarcomeres are stretched.

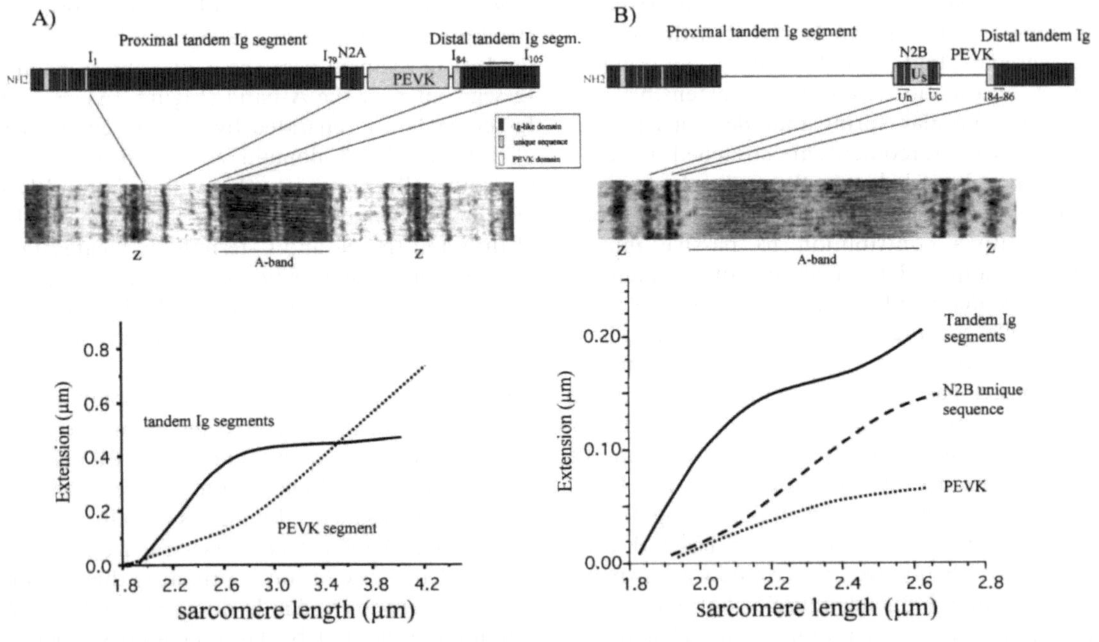

Fig. 2. Top: Domain structure of titin's I-band region of skeletal muscle titin (left) and cardiac titin (N2B isoform) (right). Middle left: Electron micrograph of human soleus sarcomere labeled with anti-titin antibodies that demarcate tandem Ig and PEVK segments. Middle right: Electron micrograph of rat cardiac sarcomere labeled with anti-titin antibodies that demarcate the unique N2B sequence (Us) and the PEVK. Note that in the sarcomere both Us and the PEVK are extended and that Us extension is largest. Bottom left: extension vs. sarcomere length of tandem Ig segments and PEVK in human soleus sarcomere. Bottom right: extension vs. sarcomere length of tandem Ig segments, PEVK, and N2B-Us in rat cardiac sarcomeres. (A: Adapted from the Journal of Cell Biology (Trombitas *et al.*, 1998b) by copyright permission of the *Rockefeller University Press* and B: Based on data in Trombitas *et al.* (2000) reproduced with permission of the *Biophysical Society*.)

can be accommodated by only folded domains indicates that large-scale Ig-domain unfolding is unlikely to take place under physiological conditions (Trombitas *et al.*, 1998b). The extensibility provided by the N2B-Us of cardiac titins (Figure 2B, bottom) reduces the necessity for Ig domain unfolding within the physiological sarcomere length range of cardiac muscle (see also below). The PEVK extends at relatively high forces (Trombitas *et al.*, 1998a, b) and this process is likely to result from straightening of random coil structures and polyproline II helices that comprise this element (Li *et al.*, 2001; Ma *et al.*, 2001).

A model has emerged for extensibility of skeletal muscle titin in which the PEVK segment (acting largely as an unfolded polypeptide) and the tandem Ig segments (containing folded Ig domains) behave as serially linked entropic springs with low and high persistence length, respectively (Linke and Granzier, 1998). In short sarcomeres these springs are in a high entropy state (PEVK and tandem Ig segments are 'contracted') and upon sarcomere extension the springs straighten, lowering their conformational entropy and resulting in a force, known as entropic force. This force increases in a non-linear fashion with the segment's fractional extension (end-to-end length divided by the contour length; for details see Bustamante *et al.*, 1994; Marko and Siggia, 1994) and is inversely proportional to the chain's persistence length. Using persistence lengths of 15 and 2 nm for the tandem Ig and PEVK segments, respectively, allowed us to model the extensible region of soleus titin and predict the measured extension of these segments in the sarcomere (Trombitas *et al.*, 1998b). Although minor deviations from measured extension are present (they may result from the arrangement and environment of titin within the sarcomere structure that limits the conformational space (Tskhovrebova and Trinick, 2002), the predictions and measurements in soleus fibers are close. Whether this model can be applied to cardiac titin, however, requires a better understanding of the mechanical properties of the N2B-Us, titin's third molecular spring (Helmes *et al.*, 1999; Linke *et al.*, 1999; Trombitas *et al.*, 1999).

Molecular mechanics of cardiac titin's spring elements

The single-molecule mechanical properties of titin and other elastic proteins have been studied in recent years by using laser tweezers (Kellermayer *et al.*, 1997, 1998, 2000, 2001; Tskhovrebova *et al.*, 1997) and atomic force microscopy (for review and original citations see Carrion-Vazquez *et al.*, 2000). To explore the mechanical properties of the N2B-Us we performed atomic force microscopy on this element and compared its properties with those of the fragment I91–I98 (comprising eight Ig domains from the distal tandem Ig segment), and the 188-residue PEVK domain (Watanabe *et al.*, 2002a, b). Force–extension curves were measured by using an atomic force microscope ('Molecular Force Probe', MFP) specialized for stretching individual molecules (Figure 3A).

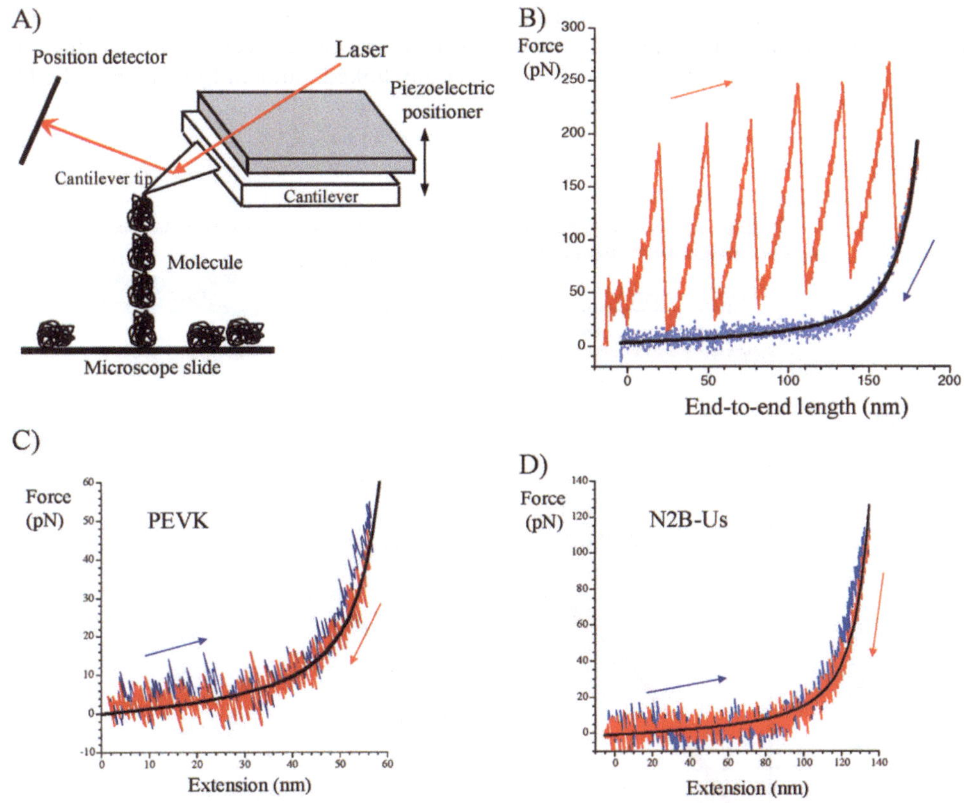

Fig. 3. (A) Schematic of force-measuring AFM. A molecule is attached between the tip of the cantilever and a glass microscope slide. Extension of the molecule by retraction of the piezoelectric positioner generates force in the molecule that deflects the cantilever. Cantilever deflection is measured from which the force generated in the molecule can be calculated by using the experimentally determined cantilever stiffness. (B–D) examples of stretch–release curves of I91–I98 (B), PEVK (C) and N2B-Us (D) of human cardiac titin. WLC curve fitted to release (black line). Note that I91–I98 displays during stretch regularly spaced (~28 nm) force peaks and that force during release is much less than during stretch (i.e., force–extension curve displays hysteresis). In contrast, the stretch and release curves largely overlap in case of the PEVK (C) and N2B-Us (D). (Based on Figures 2, 3 and 5 of Watanabe *et al.* (2002b). Reproduced with permission of *The American Society for Biochemistry and Molecular Biology*.)

An example of a stretch–release curve of I91–I98 is shown in Figure 3B. Consistent with earlier findings (Rief *et al.*, 1997; Carrion-Vazquez *et al.*, 1999a, b; Marszalek *et al.*, 1999) the stretch curve displays sawtooth-like force peaks, indicating that repetitive structural transitions occurred during stretch with each sudden force drop representing the unfolding of a β-barrel Ig domain. The unfolding force varies with stretch rate: at a rate of 50 nm/s the mean unfolding force is 180 pN, and at a rate of 1000 nm/s the value is 250 pN (Watanabe *et al.*, 2002a). The stretch-rate dependence of unfolding force allowed us to deduce the rate constant of unfolding at zero external force (K_u^0) and the location of unfolding barrier along the unfolding reaction coordinate (width of the unfolding potential, ΔX_u). Results were $K_u^0 = 6.1 \times 10^{-5}\ s^{-1}$ and $\Delta X_u = 0.25$ nm. These values were used to ascertain whether unfolding of Ig domains has to be taken into account when modeling the extensible region of cardiac titin. Results indicate that in repeated stretch–release cycles the probability of domain unfolding is low and that at the upper range of the physiological force regime (~15 pN) of all titin molecules in the sarcomere only ~10% might contain a single unfolded domain (Watanabe *et al.*, 2002b). Our results suggest that the extensible region of cardiac titin is more realistically simulated with Ig domains in the folded state.

AFM-based stretch–release force–extension curves of both the PEVK and N2B-Us displayed little hysteresis: the stretch and release data nearly overlapped (Figure 3C and D). The force–extension curves closely followed WLC behavior. Thus, on the time scale of the experiments, both the PEVK and N2B-Us behave as entropic springs with presumed random coil structure. Considering the limited force resolution of the AFM (~5 pN), these findings do not exclude deviations from WLC behavior at low force. For example, the secondary structure predicted for the N2B-Us (Watanabe *et al.*, 2002b) indicates that ~1/3 of the sequence may form α-helical structures (the remaining sequences are unstructured). Stretching these α-helical structures may cause sequential breakage of intra-helix hydrogen bonds at forces below the resolution of the AFM (Carrion-Vazquez *et al.*, 2000).

By fitting the release curves of stretch–release cycles with the WLC equation we determined the persistence length (measure of chain's bending rigidity) of the PEVK and N2B-Us. Experimental statistics indicate that the persistence lengths of the single molecules are ~1.4 and ~0.65 nm for the PEVK and the N2B-Us, respectively (Watanabe *et al.*, 2002b).

Using the above-described molecular characteristics, we modeled the extensible region of cardiac titin (N2B isoform) as three mechanically distinct springs that are serially linked: tandem Ig segments, the N2B-Us and the

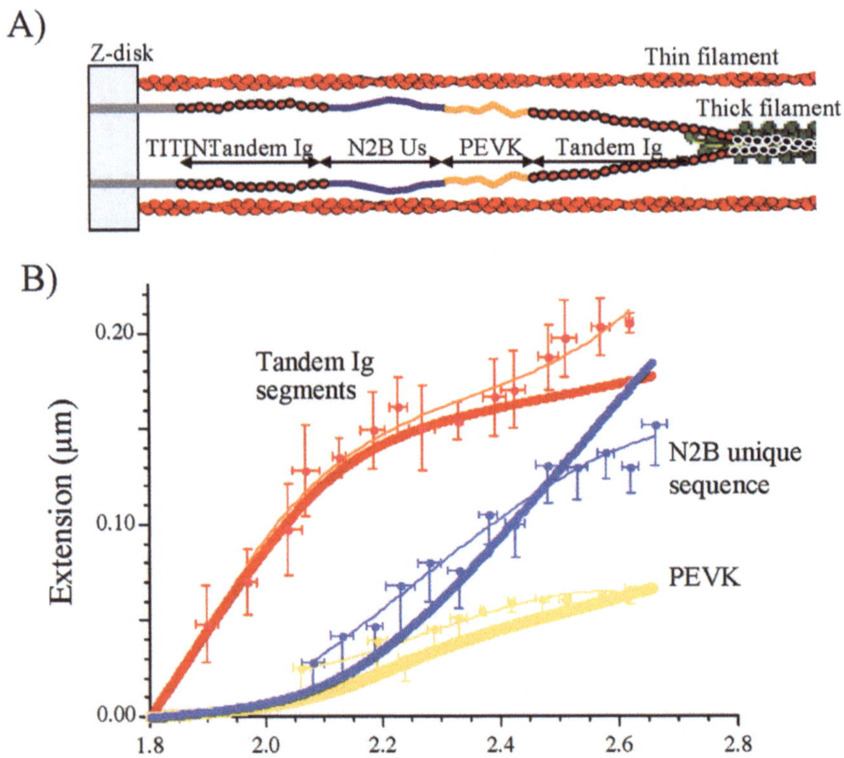

Fig. 4. (A) Model of extensible region of cardiac titin (N2B isoform) in sarcomere. (B) Predicted and measured extensions of tandem Ig segments (proximal and distal segments are summed), N2B-Us and PEVK. Predictions are shown as thick solid lines, mean values of measurements as symbols and standard errors of the mean as thin vertical and horizontal lines. (Based on Figure 6 of Watanabe *et al.* (2002b). Reproduced with permission of *The American Society for Biochemistry and Molecular Biology*.)

PEVK. We calculated the extension of the various spring elements as function of sarcomere length. In the physiological sarcomere length range (~1.8–2.4 μm), predicted values were within experimental error of those measured (Figure 4B) suggesting that the model parameters used in the simulations approximate those *in situ* well. Because the model calculations assume that domain unfolding is absent, these findings also support the view that unfolding is not prominent under physiological conditions. Only at sarcomere lengths greater than ~2.4 μm does the measured tandem Ig segment extension exceed predictions (Figure 4B). Thus, beyond the physiological sarcomere length range, Ig unfolding probably occurs (calculations reveal that unfolding of less than two domains can explain the results). By having Ig domains in their folded state, the tandem Ig segments attain a long persistence length allowing them to extend under low force (Figure 4B) and to set the sarcomere length at which extension of the PEVK and N2B-Us begins to dominate.

In summary, AFM studies on N2B cardiac titin reveal that the PEVK and N2B-Us both behave as entropic springs, but with different persistence lengths. Using the obtained persistence lengths a model of the extensible region of cardiac titin can be constructed containing three serially linked springs (tandem Igs, PEVK and N2B-Us) that correctly simulates the complex extension of titin in the sarcomere.

Titin–actin interaction modulates muscle stiffness

In the sarcomere titin's extensible region is located in close proximity to the actin-based thin filaments, and several earlier studies have suggested that interactions between titin and actin occur (reviewed in Yamasaki *et al.* (2001)). Recently we expressed recombinant fragments representing the sub-domains comprising the extensible region of cardiac N2B titin (tandem Ig segments, the N2B splice element, and the PEVK domain), and assayed them for binding to F-actin (Yamasaki *et al.*, 2001). The PEVK region bound F-actin (Figure 5A), while no binding was detected for the other constructs. The results of Kulke *et al.* (2001) are consistent with these findings. Comparison with a skeletal-muscle PEVK fragment revealed that only the cardiac PEVK binds actin at physiological ionic strengths (Yamasaki *et al.*, 2001).

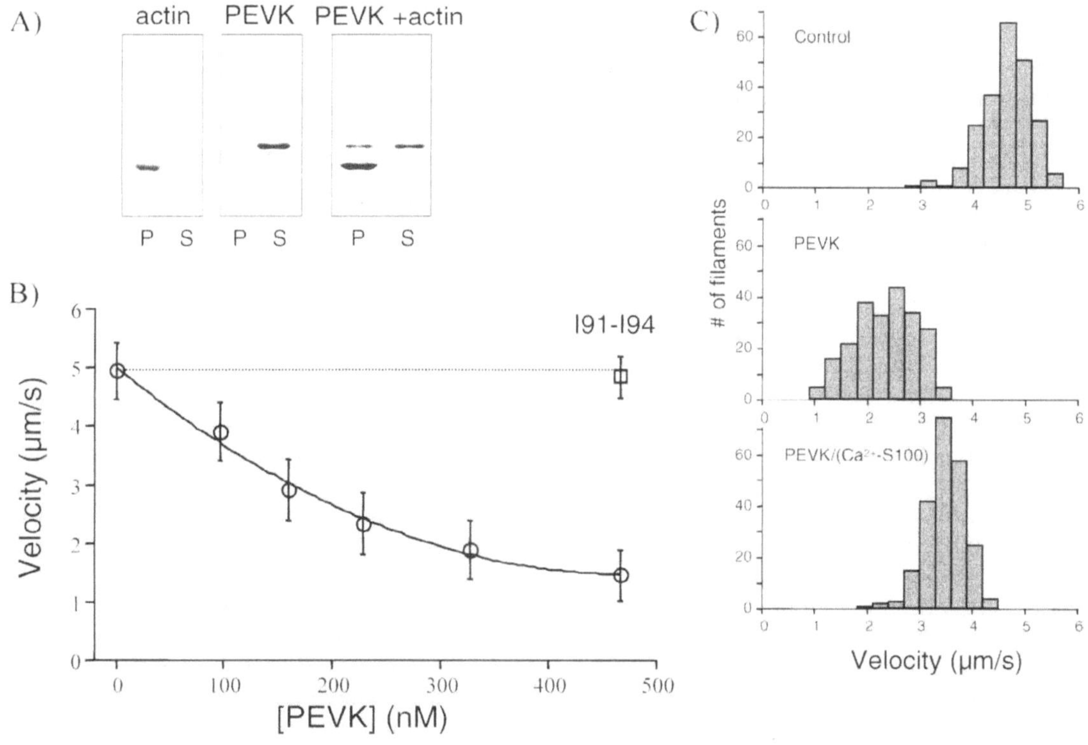

Fig. 5. (A) Co-sedimentation assay reveals PEVK (N2B isoform)–actin interaction. In control experiments, the PEVK is found in the supernatant and F-actin in the pellet, while when the two are mixed, the PEVK fragment is also detected in the pellet. (B) *In vitro* motility assay. Circles: F-actin sliding velocities were measured over surfaces coated with either HMM (0 nm PEVK) or a mixture of HMM and varying concentrations of PEVK (100–475 nM). PEVK inhibited velocities in a concentration-dependent manner, with a 475 nm concentration of the fragment resulting in a ~70% reduction in velocity. Square: A four Ig-like domain fragment (I91–I94) from titin's distal tandem Ig segment did not affect velocities when added at the highest PEVK concentration used (475 nM). (C) Calcium/S100A1 alleviates PEVK-based inhibition of *in vitro* motility. F-actin velocities were measured over surfaces coated with either HMM (top) or a mixture of HMM and PEVK (middle and bottom) in the presence of 0.1 mM $CaCl_2$ (bottom). PEVK inhibited sliding velocities by ~50% relative to the control. Calcium/S100A1 alleviated the PEVK-based motility inhibition by ~50%. (Based on Figures 2, 4 and 9 of Yamasaki *et al.* (2001). Reproduced with permission of the *Biophysical Society*.)

F-actin contains a large patch of negatively charged residues on its exposed surface (Kabsch *et al.*, 1990), and several actin-binding proteins are known to bind to this region via basic charge clusters (e.g., Friederich *et al.*, 1992; Pfuhl *et al.*, 1994; Fulgenzi *et al.*, 1998). The results of our binding studies suggest that PEVK–actin interaction also includes an electrostatic component. Both the skeletal and cardiac PEVK fragments exhibited decreased binding in response to increased ionic strength, with the skeletal PEVK exhibiting weaker binding at intermediate ionic strength as well as a greatly enhanced ionic strength sensitivity (Yamasaki *et al.*, 2001). Only the cardiac N2B PEVK bound F-actin at physiological ionic strengths. These results are consistent with those of Gutierrez-Cruz *et al.* (2000), who observed binding between a fetal skeletal muscle PEVK fragment and F-actin at intermediate ionic strengths, and an absence of binding at elevated (physiological) ionic strength. The differences in actin-binding affinity may be explained by the different charge characteristics of the cardiac and skeletal PEVK fragments, which have theoretical isoelectric points (pI) of 9.7 and 8.0, respectively. At physiological pH, the cardiac PEVK therefore carries a greater net positive charge, which likely facilitates its interaction with the negatively charged actin filament.

Assuming that the actin-binding propensity of the PEVK domain is indeed related to its charge, it is possible to make predictions about the binding properties of the PEVK domains expressed in different titin isoforms. The majority of the PEVK sequences in all titin isoforms are comprised of basic (pI $\sim$ 9–10) 27–28-residue repeats, termed PPAK or PEVK repeats, that are rich in proline (Freiburg *et al.*, 2000; Greaser, 2001). In addition, the larger versions of the PEVK domain expressed in skeletal titins and the cardiac N2BA isoform contain highly acidic (pI $\sim$ 4.0) regions rich in glutamic acid residues (E-rich regions) which are interspersed with the PPAK repeats (Greaser, 2001). The presence of the E-rich regions gives the larger PEVK isoforms an overall acidic character. For example, the soleus muscle PEVK has a predicted pI of 5.1. In contrast, the cardiac N2B PEVK lacks E-rich segments and has a pI (9.70) that is characteristic of the PPAK repeats which comprise $>75\%$ of its primary structure. The cardiac N2B PEVK may therefore be unique among the PEVK isoforms in its ability to bind actin.

In addition to their net positive charge, the PPAK repeats are also rich in proline. Proline-rich regions (PRRs) often assume extended conformations, such as the polyproline II (PPII)-helix that was recently predicted for skeletal muscle PEVK fragments (Gutierrez-Cruz *et al.*, 2000; Ma *et al.*, 2001). Due to the conformational restrictions that proline residues impose on the polypeptide backbone, PRRs experience a relatively small reduction of entropy upon ligand binding (Williamson, 1994). This property makes them energetically favorable sites for protein–protein interactions, and there are numerous examples of PRRs that perform binding functions *in vivo* (for review, see Williamson, 1994). The binding of actin by the PPAK-rich N2B PEVK may therefore be both electrostatically and thermodynamically favorable.

The significance of interaction between actin and the cardiac PEVK (N2B isoform) was investigated using an *in vitro* motility assay technique. The findings indicate that, as the thin filament slides relative to titin in the motility assay, a dynamic interaction between the PEVK domain and F-actin retards filament sliding (Figure 5B). We also investigated the effect of calcium on PEVK–actin interaction. Although calcium alone had no effect, S100A1, a soluble calcium-binding protein found at high concentrations in the myocardium, inhibited PEVK–actin interaction in a calcium-sensitive manner. *In vitro* motility results indicate that S100A1–PEVK interaction alleviates the force that arises as F-actin slides relative to the PEVK domain (Figure 5C). We speculate that S100A1 may provide a mechanism to free the thin filament from titin and reduce the sarcomere's passive tension during active contraction. Thus, a dynamic interaction between titin and actin contributes to passive stiffness of the sarcomere and the interaction varies with the physiological state of the myocardium.

Post-translational modulation of titin's spring properties

We recently showed (Yamasaki *et al.*, 2002) that cardiac titin can be phosphorylated by protein kinase A (PKA) and that the phosphorylation site is the N2B-Us (Figure 6A–C). Thus, in addition to the well-characterized myofibrillar PKA substrates MyBP-C and TnI, titin also contains a PKA-responsive domain expressed only in cardiac muscle. Interestingly, PKA-based phosphorylation of titin results in a reduction of passive tension of cardiac myocytes (Strang *et al.*, 1994; Yamasaki *et al.*, 2002) that varies inversely with sarcomere length (Figure 6D). This reduction may be explained by assuming that phosphorylation destabilizes native structures within the N2B element (for example, α-helical structures, see above) causing it to extend and lower its fractional extension (via increased contour length). This would give the tandem Ig segments and PEVK domain lower fractional extensions at a given SL (via reduced end-to-end length), leading to a decrease in passive tension. At longer SLs, α-helical structures may be denatured due to stretch, and phosphorylation will therefore have less impact on the extension of the N2B element and, consequently, on passive tension.

It is well established that activation of the β-adrenergic pathway increases phosphorylation of TnI and MyBP-C (Solaro *et al.*, 1976; Hartzell and Glass, 1984; Strang and Moss, 1995). Whether phosphorylation of titin can also be enhanced by β-adrenergic agonists was investigated by using back-phosphorylation experiments. Cell suspensions were subjected to either the β-receptor antagonist propranolol (prop) or the β-receptor agonist isoproterenol (iso), followed by rapid skinning and incubation with ^{32}P-ATP and PKA. Cell suspen-

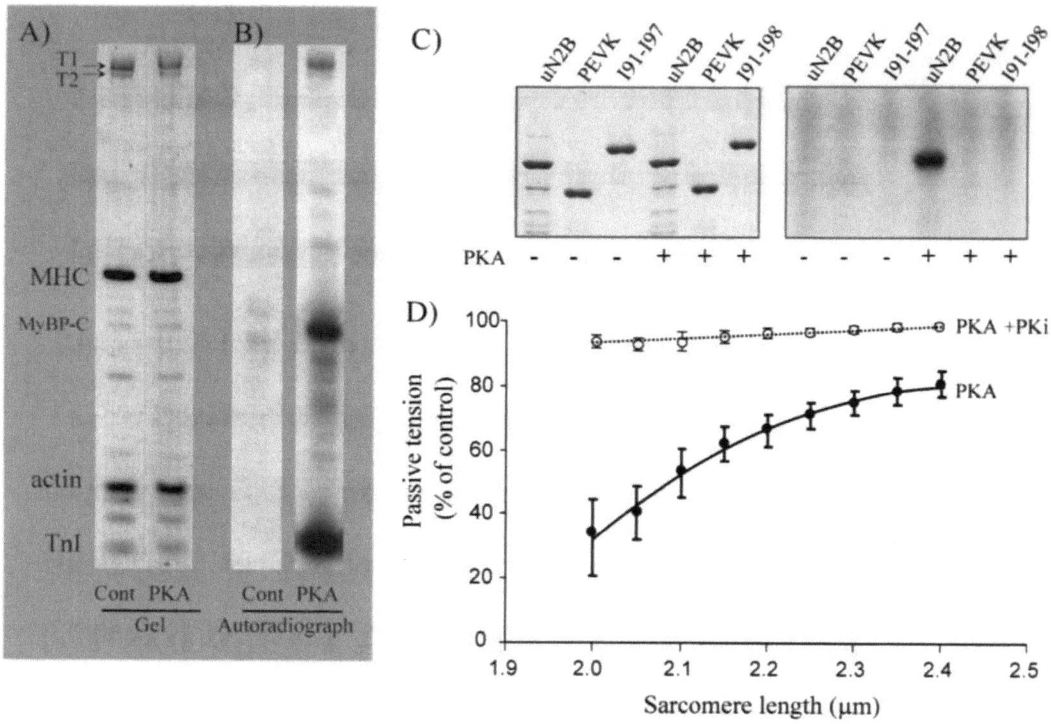

Fig. 6. (A) Coomassie–stained gel of solubilized cardiac myocyte suspensions incubated with ATP-γ ^{32}P in the absence of exogenous PKA (Cont), or in the presence of PKA (PKA). (B) 24-h autoradiographic exposure showing phosphorylation of MyBP-C and TnI and titin. (C) *In vitro* PKA phosphorylation survey of titin's extensible segment. We used recombinant fragments representing the 572-residue unique sequence within the N2B element (N2B-Us), the PEVK domain (PEVK), and the tandem Ig segments (I91–I98). (left) Coomassie–stained gel of recombinant fragments incubated with ATP-γ ^{32}P in the absence (−) and presence (+) of PKA (right). Autoradiogram showing phosphorylation of N2B-Us by PKA. (D) Mean passive tension values relative to the control as a function of SL. Cells were either treated with PKA alone or PKA + PKi (PKA inhibitor). Error bars are SEM ($n = 8$). (Based on Figures 1, 4 and 5 of Yamasaki *et al.* (2002). Reproduced with permission of *Lippincott Williams and Wilkins*.)

sions were then solubilized, and the samples were subjected to electrophoresis. The gels were stained with Coomassie blue and analyzed with autoradiographic techniques. As observed for MyBP-C and Tn-I, isoproterenol-treatment resulted in reduced ^{32}P incorporation (Yamasaki *et al.*, 2002). Relative to propranolol, isoproterenol treatment decreased ^{32}P incorporation by 52 ± 12% ($P < 0.01$). These findings indicate that titin is phosphorylated by PKA in response to β-adrenergic stimulation in intact myocytes. Considering that the activation of PKA via β-adrenergic stimulation constitutes a major regulatory pathway in the heart, the PKA responsive element of cardiac titins may allow modulation of diastolic function *in vivo*. We speculate that when β-adrenergic stimulation enhances the heartbeat frequency and the rate of contraction and relaxation, the reduction in titin's force resulting from N2B phosphorylation allows for more rapid and complete ventricular filling.

Modulation of titin's elasticity by differential splicing of spring elements

Titin is encoded by a single gene (Figure 7) located on chromosome 2 in humans and in the mouse. The recent genomic analysis of human titin revealed 363 exons that code for a total of 38,138 amino-acid residues (Bang *et al.*, 2001). This includes 105 exons which code for conserved ~28-residue PEVK/PPAK-repeats, possibly corresponding to structural spring units (Freiburg *et al.*, 2000; Greaser, 2001). Interestingly, two titin exons, M10 and novex-III, function as alternative C-termini (Bang *et al.*, 2001), giving rise to truncated (Z1-Z2 to novex-III; Mr ~620 kDa), and full-length titins (Z1 to M10; Mr >3000 kDa). The truncated novex-3 titin isoform can integrate into the Z-line lattice but is too short to reach the A-band, and is expressed in both skeletal and cardiac muscles (Bang *et al.*, 2001). We have speculated that co-expression of truncated and full-length titins may adjust the titin filament system to both three- and twofold symmetries of thick and thin filaments, respectively (Bang *et al.*, 2001).

The I-band region of full-length titins (Figure 8), comprising I50 to I84, undergoes extensive differential splicing (Freiburg *et al.*, 2000; Bang *et al.*, 2001). Thereby, highly variable isoforms of different spring composition are generated. The soleus muscle expresses the largest titin isoform observed so far (M_r 3.7 MDa), while psoas muscle expresses a smaller titin (M_r 3.35 MDa) with shorter proximal tandem Ig and PEVK

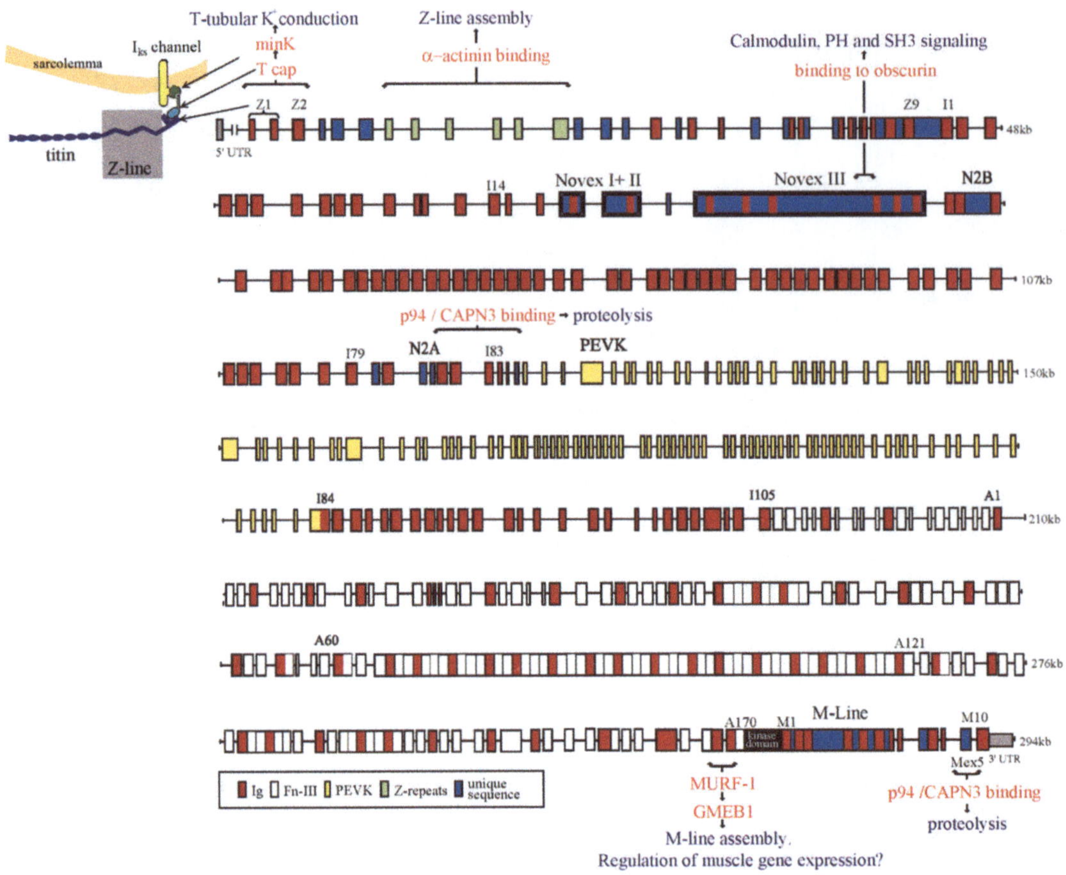

Fig. 7. Exon–intron structure of human titin gene. Note novel exons I–III. Binding sites of known titin ligands are indicated. Top left: example of a signaling complex in which titin may contribute to regulation of ion channel activity. (Based on Figure 1 of Bang *et al.* (2001). Reproduced with permission of *Lippincott Williams and Wilkins.*)

segments (Figure 9A and B). Passive tension increases steeper with sarcomere length in psoas than in soleus fibers (Figure 9C). The differences in tandem Ig and PEVK segment length of the different muscle types provide a molecular framework to understand these mechanical differences. The effect of tandem Ig- and PEVK-segment lengths on passive force can be evaluated by considering the model of passive force generation in which the tandem Ig segments (containing folded Ig domains) and the PEVK segment (acting largely as an unfolded polypeptide) behave as serially linked entropic springs (see above). In short sarcomeres these springs are in a high-entropy state, and upon sarcomere extension the springs become stretched. This results in a force that is inversely proportional to the chain's persistence length and that increases in a non-linear fashion with the segment's fractional extension. The 'serially linked entropic springs model' of passive-force development may be applied to titin isoforms by adapting the entropic forces to the fractional extensions multiplied by the contour lengths of the isoform's tandem Ig and PEVK segments. As a result of sequence differences (Figure 9B) contour lengths of tandem Ig and PEVK segments in psoas are ~100 and ~400 nm shorter, respectively, than in soleus titin. (This assumes folded Ig domains with a maximal 5 nm repeat per domain and an unfolded PEVK sequence with a maximal residue spacing of 3.8 Å.) Thus, for a given sarcomere length, the sequence differences predict that the fractional extension is higher in psoas than in soleus muscle. It follows that entropic force is predicted to be much higher in psoas than in soleus fibers, consistent with the measured passive tension differences (Figure 9C).

Exon 49 (containing the N2B sequence) is excluded in skeletal muscle titins, but is present in all cardiac titin isoforms. The major cardiac N2B isoform results from splicing of exons 49/50 to 225, thereby excluding exons 51–224. This cardiac-specific N2B isoform is co-expressed with larger cardiac-specific isoforms, the N2BA titins. N2BA cardiac titins contain in addition to exon 49 (N2B) also the exons 101–111 (coding for the so-called N2A segment). N2BA titins also contain a longer PEVK segment and more Ig domains than N2B titins (Figure 8). The expression level of N2B and N2BA cardiac titins varies considerably. Small rodents express predominantly N2B titins, whereas large mammals (including humans) co-express both isoforms at intermediate levels in their ventricles and predominantly N2BA titin in their atria (Cazorla *et al.*, 2000). Passive

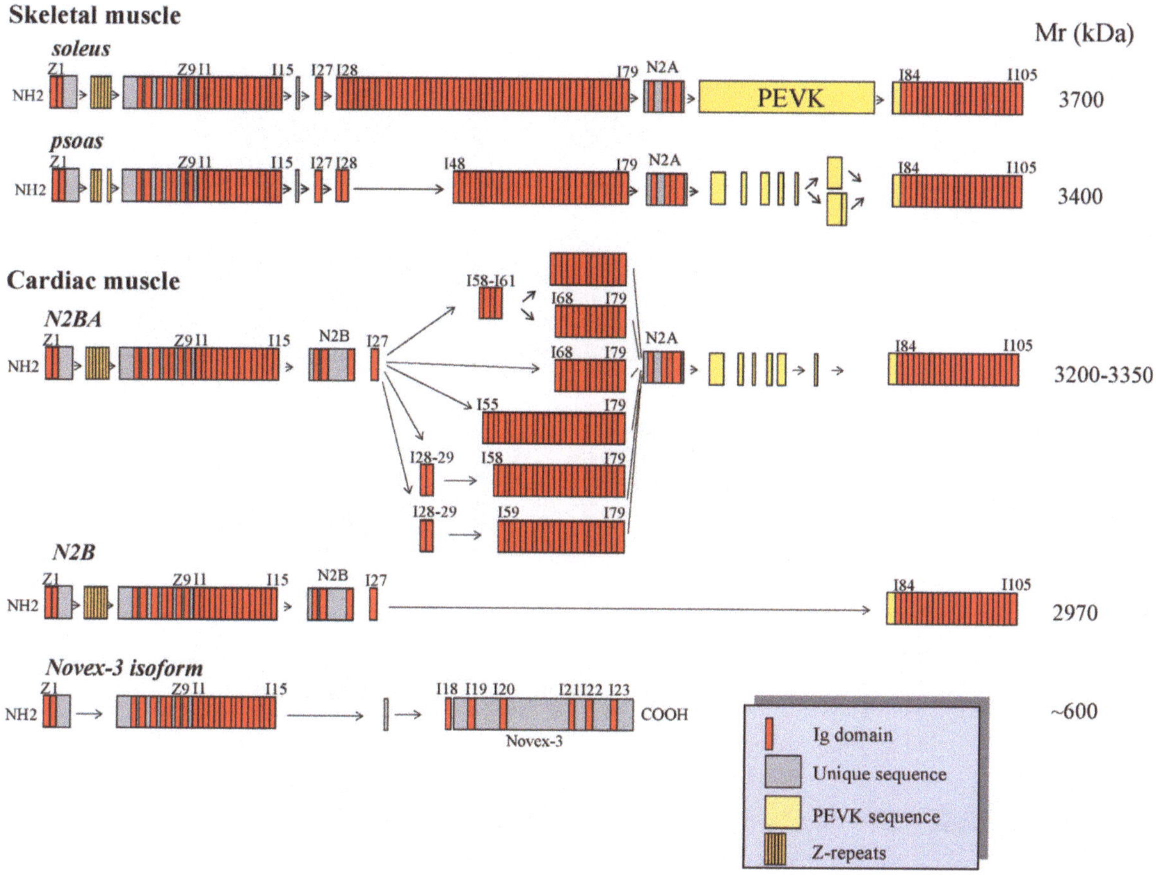

Fig. 8. Splice isoform diversity in titin's I-band regions. Skeletal muscles express N2A-based titins that vary in size in different muscle types (two of many possible splice pathways are shown). Heart muscles express large N2BA titins and small N2B titins. All striated muscles express novex-3 titin. (Based on Figure 3 of Freiburg *et al.* (2000) and Figure 3 of Bang *et al.* (2001). Reproduced with permission of *Lippincott Williams and Wilkins.*)

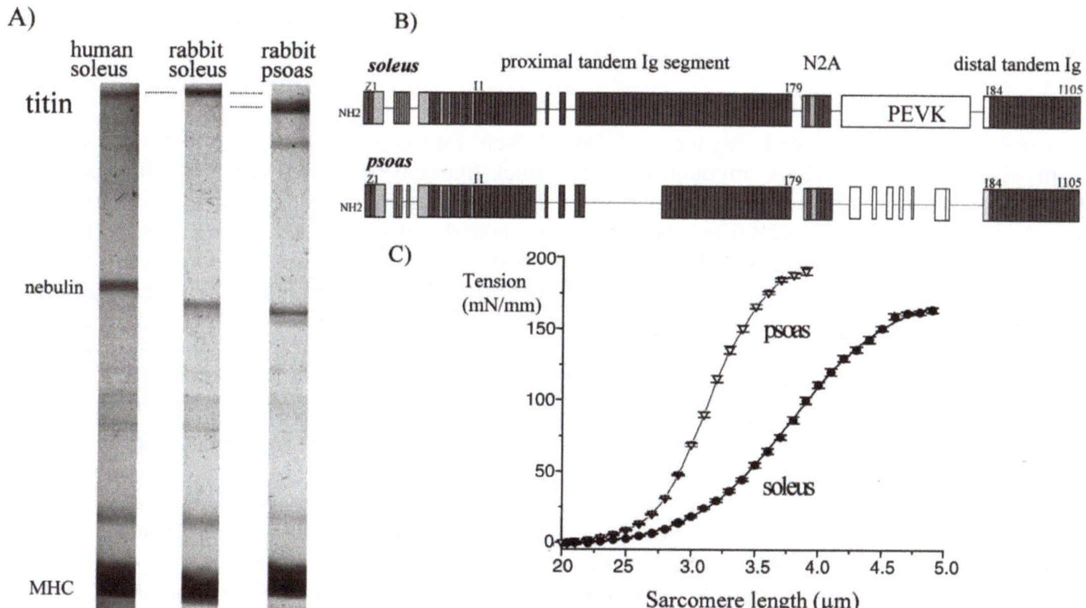

Fig. 9. (A) SDS–PAGE of soleus and psoas muscle. Titin of psoas muscle has a greater mobility than soleus titin, reflecting its shorter I-band segment. (B) Domain structure of the I-band segment of soleus and psoas titins. (C) Passive tension–sarcomere length relations of mechanically skinned fibers of psoas and soleus muscles. (Based on Figures 4 and 6 of Freiburg *et al.* (2000). Reproduced with permission of *Lippincott Williams and Wilkins.*)

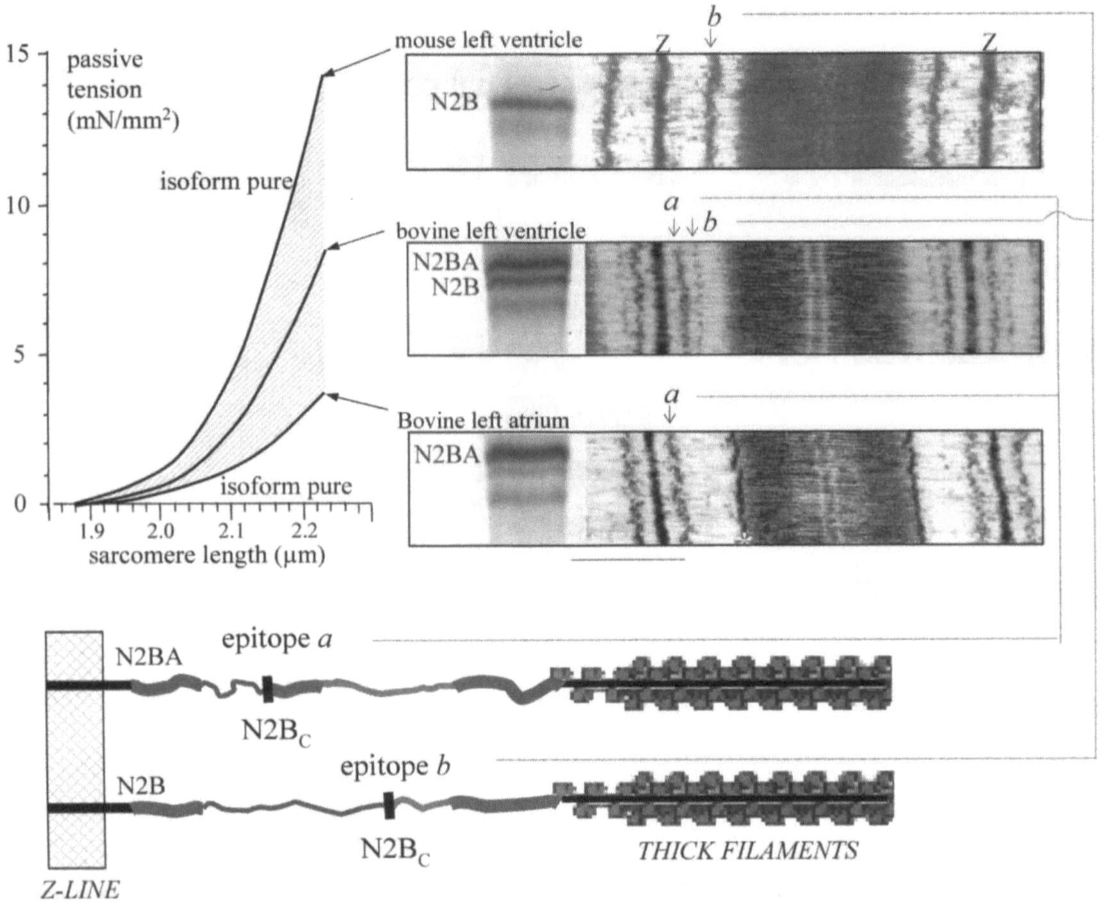

Fig. 10. Co-expression of cardiac isoforms and passive tension modulation. Micrographs: sarcomeres of cells labeled with N2B$_C$ antibody that binds C-terminal of the N2B-Us. To the left of each micrograph an electrophoresis result is shown. N2B-expressing cells (top) have a single N2B$_C$ epitope near the A-band (epitope *b*) and develop high passive tensions (left) whereas N2BA-expressing cells (bottom) have a single N2B$_C$ epitope near the Z-line (epitope *a*) and develop low passive tensions. The two I-band epitopes in the middle micrograph suggest co-expression of N2B and N2BA isoforms at the level of the half-sarcomere. These cells develop intermediate force levels. (Based on Figures 2 and 7 of Trombitas *et al.* (2001). Reproduced with permission of the *American Physiological Society.*)

stiffness of cardiac myocytes is much higher in N2B expressing myocytes than in N2BA myocytes (Figure 10). The higher passive stiffness is likely to result from the shorter I-band segment of N2B titin which, at a given sarcomere length, results in an extensible segment with a fractional extension (end-to-end length divided by the contour length) that is higher than that in N2BA titin (Trombitas *et al.*, 2000).

The high titin-based passive stiffness provided by N2B titin may allow rapid and stable determination of the end-diastolic volume at the high beat frequencies encountered in small mammals (where N2B titin dominates). Titin is expected to play a role in centering the A-band within the sarcomere (Horowits and Podolsky, 1987), and the high N2B-based passive stiffness may rapidly reset the A-band location during each diastole. It is also worthwhile to consider titin's contribution to restoring forces (Helmes *et al.*, 1996). The segment of titin near the Z-line binds the thin filament and can withstand compressive forces (Linke *et al.*, 1997; Trombitas and Granzier, 1997; Granzier *et al.*, 2000). When sarcomeres shorten to below slack length, the thick-filament tip moves into the stiff region of titin near the Z-line, and titin's extensible region is stretched in a reverse direction of that during stretch (Granzier *et al.*, 2000). This gives rise to so-called restoring forces (pushing Z-lines away from each other) and these forces are expected to be highest in N2B expressing myocytes. Considering that titin's restoring force may contribute to the early diastolic suction force that aids ventricular filling, expressing high levels of N2B titin may be relevant for achieving rapid diastolic filling.

Co-expression of cardiac titin isoforms

The ventricular myocardium of large mammals co-expresses isoforms at the level of the half-sarcomere, and co-expression results in passive tension levels intermediate between that of N2B and N2BA pure myocytes (Figure 10). Such intermediate tensions, in theory, could be achieved by varying the number of titin molecules per thick filament. However, this would also influence functions performed by titin's inextensible regions, such as thick-filament length control and construction and maintenance of Z-lines and M-lines (Gregorio *et al.*, 1999). When passive force levels are tuned via variation in the isoform expression ratio, the inextensible regions are not impacted because these regions are the same in different isoforms (Labeit and Kolmerer, 1995). Thus, co-expressing isoforms at various ratios may provide effective means for tuning passive properties, and any force level intermediate between that of isoform-pure cells may be obtained (see shaded area of passive tension curve in Figure 10).

Titin and heart disease

Several earlier studies reported that the amount of intact titin is reduced in patients with heart failure (Hein *et al.*, 1994; Morano *et al.*, 1994; Hein and Schaper, 1996), suggesting that titin is involved in the altered ventricular compliance associated with certain heart conditions. Furthermore, titin has been implicated in familial forms of dilated cardiomyopathy (DCM) by genetic linkage studies (Siu *et al.*, 1999). Recent sequencing of the titin gene in several families with DCM indeed identified titin mutations in and near its Z-line region as well as in the N2B exon (Gerull *et al.*, 2002; Itoh-Satoh *et al.*, 2002), whereas a 2 bp insertion in titin's 3′ region in another large family resulted in a truncated titin protein (Gerull *et al.*, 2002).

To better understand titin's role in diastolic dysfunction we have studied the canine tachycardia-induced model of DCM (Wu *et al.*, 2002b). We investigated animals paced for 4 weeks, and showed that relative to controls the total amount of titin was unchanged but that the ratio of N2BA to N2B isoforms was reduced (Figure 11). The functional consequence of this change was assessed by mechanical studies on muscle. This revealed that titin-based passive tension is significantly elevated in tachycardia-induced DCM (Figure 11C). That adjustment in cardiac isoform expression is not restricted to animal models but may also occur in human is revealed by Figure 12, which shows myocardium from a healthy individual and of patients with DCM or mitral regurgitation (MR). Control myocardium co-expresses N2B and N2BA isoforms at similar levels whereas in disease states the isoform expression ratios are modified. Interestingly, in the two examined samples the adjustment is in the opposite direction, with elevated N2B expression in DCM and elevated N2BA expression in MR (see Figure 12). Although more detailed studies are required that focus on, for example, possible location-specific changes in titin expression, the present data clearly reveal plasticity in the expression of cardiac titin isoforms.

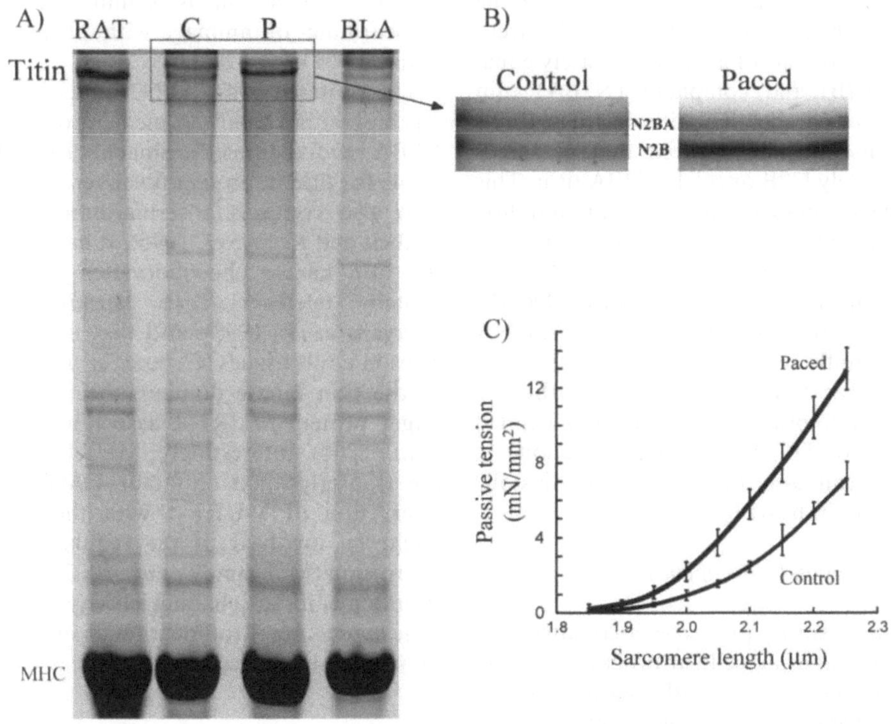

Fig. 11. (A) SDS–PAGE of left ventricular (LV) myocardium of control dogs (C) and dogs exposed to 4-week-long rapid pacing (P). Rat and bovine left atrium (BLA) were co-electrophoresed to provide standards for N2B titin (upper band in rat) and N2BA titin (upper band in BLA). (B) Enlargement of top of gel. Note that N2B titin is upregulated. (C) Titin-based passive tension–sarcomere length relations of skinned myocardium from control and paced animals. Titin-based passive tension is significantly ($P < 0.05$) elevated in the myocardium of paced animals. Results are from five control and six paced animals. Shown are mean and SEM values. (Based on Figures 1 and 3 of Wu *et al.* (2002b). Reproduced with permission of *Lippincott Williams and Wilkins*.)

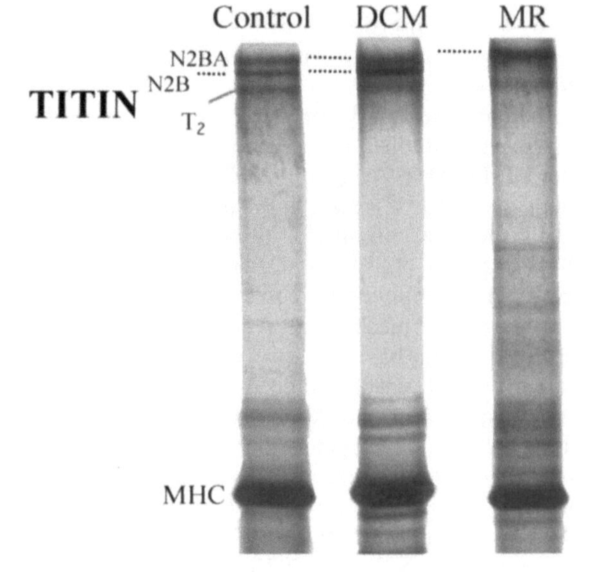

Fig. 12. SDS–PAGE of left ventricular myocardium from a healthy patient (Control), and two patients with cardiac diseases (DCM and MR. Control expresses similar level of N2B and N2BA titins. In DCM, N2B titin is up-regulated and in MR, N2BA titin dominates. Based on Figure 7 of Wu *et al.* (2002a). Reproduced with permission of *Churchill Livingstone*).

It is worth highlighting that both the canine and human myocardium express similar levels of compliant N2BA and stiff N2B titins under normal conditions. Considering that the isoforms are co-expressed at the level of the half-sarcomere (Figure 10) each half-sarcomere in control tissue consists of an approximately equal number of stiff (N2B) and compliant (N2BA) titin molecules arranged in parallel. The titin-based stiffness will therefore be intermediate between that of sarcomeres that express solely N2B or solely N2BA titin. This intermediate stiffness allows for maximal adaptability. Thus, sarcomeres can either greatly increase compliance by increasing the N2BA/N2B expression ratio (MR) or greatly increase stiffness by reducing this ratio (DCM). We conclude that the processing of the titin pre-mRNA must be subject to subtle regulatory mechanisms which control entry into either N2B or N2BA splice pathways. Therefore, in addition to primary titin defects associated with familial forms of DCM, acquired alterations in isoform expression can also play a role in diastolic dysfunction occurring in heart failure.

Active force modulation and titin ligands

Modification of titin's spring properties may affect not only passive stiffness but other cellular processes as well, and any of these processes may be affected by differential expression of isoforms. For example, recent evidence suggests that titin may play a role in regulating active force. Cazorla *et al.* (2001) reported that titin-based passive force enhances length-dependent activation of cardiac myocytes, and Fukuda *et al.* (2001) reported a length-dependent effect of titin on maximal active tension in rat cardiac trabeculae. We proposed, as underlying mechanism, an effect of titin on the distance between the myosin heads and their acting-bind sites, via either influencing the spacing between thin and thick filaments and/or the mobility of the myosin head (Cazorla *et al.*, 2001). Thus, the adjustment of titin's elastic properties that results from differential splicing may also influence titin-based modulation of active force.

Finally, during the last few years a diverse family of titin binding proteins has been identified that suggest the participation of the titin filament system in the regulation of ion channels, protein turnover, and gene expression (see Figure 7). At the edge of the Z-line, titin's N-terminal domains Z1-Z2 interact with the 19-kDa protein Tcap (Gregorio *et al.*, 1999), which in turn associates with the minK β-subunit of the stretch-regulated IsK potassium channel (Furukawa *et al.*, 2001). Thus, titin's passive force might be transmitted to ion channels and possibly influence their activity. In the I-band, the novex-3 titin isoform (Figures 2 and 3) interacts with obscurin (Bang *et al.*, 2001), a large protein containing several types of signaling domains (Young *et al.*, 2001). Because the novex-3 titin/obscurin complex extends in stretched sarcomeres (Bang *et al.*, 2001), the complex may have signaling properties that respond to sarcomere stretch and that are involved in stretch-initiated sarcomeric restructuring that occurs during muscle adaptation and disease. Consistent with this view, obscurin is significantly up-regulated in myocardium of animals with pacing-induced cardiac failure (Wu *et al.*, 2002b).

The protease p94/CAPN3 binds near titin's C-terminus and within the N2A element of skeletal muscle and N2BA cardiac titins (Sorimachi *et al.*, 1995), suggesting a role for titin in protein turnover. Near the C-terminus, titin also contains a serine/threonine kinase domain (Labeit and Kolmerer, 1995). It has been speculated that the titin kinase phosphorylates the Z-line associated protein telethonin/Tcap during myofibrillogenesis (Mayans *et al.*, 1998), and that it may have additional roles in adult tissues (Centner *et al.*, 2001). In proximity to the titin kinase domain, the muscle specific RING finger protein MURF-1 associates with titin (Centner *et al.*, 2001). Interestingly, MURF-1 also interacts with the transcriptional co-factor GMEB1. Therefore, the interaction of MURF-1 with the titin kinase region could be involved in the regulation of muscle gene expression (McElhinny *et al.*, 2002). In sum, ligands link titin to membrane channel activity, protein turnover and gene expression. Modification of titin's elastic properties may affect these processes.

Summary

The giant elastic protein titin contains an extensible segment that underlies the majority of physiological passive muscle stiffness. The extensible segment comprises mechanically distinct, serially linked spring elements: the tandem Ig segments, the PEVK and the cardiac-specific N2B-Us. Under physiological conditions the tandem Ig segments are likely to largely consist of folded Ig domains, whereas the N2B-Us and PEVK are largely unfolded and behave as WLCs with different persistence lengths. The mechanical characteristics of titin's extensible region may be tuned to match changing mechanical demands placed on muscle via mechanisms that operate on different time scales. The tuning mechanisms include post-transcriptional and post-translational processes. PKA reduces titin-based passive tension by phosphorylating the N2B sequence, a mechanism that is unique to cardiac muscle. We also provided evidence that differential splicing is a mechanism that is utilized to alter passive stiffness during heart disease. Finally, recent work suggests that titin performs roles that extend beyond passive stiffness generation. In contracting muscle, titin may modulate actomyosin interaction by a titin-based alteration of the distance between myosin heads and actin. Furthermore, novel ligands have been identified that link titin to membrane channels, protein turnover and gene expression. Thus, titin is a versatile and adjustable spring with a range of important functions in passive and contracting muscle and titin significantly contributes to the altered stiffness of the cardiomyopathic heart.

Acknowledgements

Supported by National Institutes of Health (HL61497 and HL62881 to HG) and Deutsche Forschungsgemeinschaft (La La668/6-2 and 7-1 to SL). We thank Dr Miklos Kellermayer for review of this manuscript. Use of the Advanced Photon Source was supported by the U.S. Department of Energy, Basic Energy Sciences, Office of Energy Research, under Contract No. W-31-109-ENG-38.BioCAT is a U.S. National Institutes of Health-supported Research Center RR08630.

References

Bang ML, Centner T, Fornoff F, Geach AJ, Gotthardt M, McNabb M, Witt CC, Labeit D, Gregorio CC, Granzier H, *et al.* (2001) The complete gene sequence of titin, expression of an unusual ~700 kDa titin isoform and its interaction with obscurin identify a novel Z-line to I-band linking system. *Circ Res* **89:** 1065–1072.

Bustamante C, Marko JF, Siggia ED and Smith S (1994) Entropic elasticity of lambda-phage DNA. *Science* **265:** 1599–1600.

Carrion-Vazquez M, Marszalek PE, Oberhauser AF and Fernandez JM (1999a) Atomic force microscopy captures length phenotypes in single proteins. *Proc Natl Acad Sci USA* **96:** 11288–11292.

Carrion-Vazquez M, Oberhauser AF, Fisher TE, Marszalek PE, Li H and Fernandez JM (2000) Mechanical design of proteins studied by single-molecule force spectroscopy and protein engineering. *Prog Biophys Mol Biol* **74:** 63–91.

Carrion-Vazquez M, Oberhauser AF, Fowler SB, Marszalek PE, Broedel SE, Clarke J and Fernandez JM (1999b) Mechanical and chemical unfolding of a single protein: a comparison. *Proc Natl Acad Sci USA* **96:** 3694–3699.

Cazorla O, Freiburg A, Helmes M, Centner T, McNabb M, Wu Y, Trombitas K, Labeit S and Granzier H (2000) Differential expression of cardiac titin isoforms and modulation of cellular stiffness. *Circ Res* **86:** 59–67.

Cazorla O, Wu Y, Irving TC and Granzier H (2001) Titin-based modulation of calcium sensitivity of active tension in mouse skinned cardiac myocytes. *Circ Res* **88:** 1028–1035.

Centner T, Yano J, Kimura E, McElhinny AS, Pelin K, Witt CC, Bang ML, Trombitas K, Granzier H, Gregorio CC, *et al.* (2001) Identification of muscle specific ring finger proteins as potential regulators of the titin kinase domain. *J Mol Biol* **306:** 717–726.

Freiburg A, Trombitas K, Hell W, Cazorla O, Fougerousse F, Centner T, Kolmerer B, Witt C, Beckmann JS, Gregorio CC, *et al.* (2000) Series of exon-skipping events in the elastic spring region of titin as the structural basis for myofibrillar elastic diversity. *Circ Res* **86:** 1114–1121.

Friederich E, Vancompernolle K, Huet C, Goethals M, Finidori J, Vandekerckhove J and Louvard D (1992) An actin-binding site containing a conserved motif of charged amino acid residues is essential for the morphogenic effect of villin. *Cell* **70:** 81–92.

Fukuda N, Sasaki D, Ishiwata S and Kurihara S (2001) Length dependence of tension generation in rat skinned cardiac muscle: role of titin in the Frank–Starling mechanism of the heart. *Circulation* **104:** 1639–1645.

Fulgenzi G, Graciotti L, Granata AL, Corsi A, Fucini P, Noegel AA, Kent HM and Stewart M (1998) Location of the binding site of the mannose-specific lectin comitin on F-actin. *J Mol Biol* **284:** 1255–1263.

Furukawa T, Ono Y, Tsuchiya H, Katayama Y, Bang ML, Labeit D, Labeit S, Inagaki N and Gregorio CC (2001) Specific Interaction of a minK Channel Subunit with the Sarcomeric Protein T-cap Suggests a T-Tubule – Myofibril Z-line Linking System. *J Mol Biol* **213:** 775–784.

Gerull B, Gramlich M, Atherton J, McNabb M, Trombitas K, Sasse-Klaassen S, Seidman JG, Seidman C, Granzier H, Labeit S, *et al.* (2002) Mutations of TTN, encoding the giant muscle filament titin, cause familial dilated cardiomyopathy. *Nat Genet* **30:** 201–204.

Granzier H and Labeit S (2002) Cardiac titin: an adjustable multi-function spring. *J Physiol (Lond)* **541:** 335–342.

Granzier H, Helmes M, Cazorla O, McNabb M, Labeit D, Wu Y, Yamasaki R, Redkar A, Kellermayer M, Labeit S, *et al.* (2000) Mechanical properties of titin isoforms. *Adv Exp Med Biol* **481:** 283–300.

Greaser M (2001) Identification of new repeating motifs in titin. *Proteins* **43:** 145–149.

Gregorio CC, Granzier H, Sorimachi H and Labeit S (1999) Muscle assembly: a titanic achievement? *Curr Opin Cell Biol* **11:** 18–25.

Gutierrez-Cruz G, Van Heerden AH and Wang K (2000) Modular motif, structural folds and affinity profiles of PEVK segment of human fetal skeletal muscle titin. *J Biol Chem* **276:** 7442–7449.

Hartzell HC and Glass DB (1984) Phosphorylation of purified cardiac muscle C-protein by purified cAMP-dependent and endogenous Ca2+-calmodulin-dependent protein kinases. *J Biol Chem* **259:** 15587–15596.

Hein S and Schaper J (1996) Pathogenesis of dilated cardiomyopathy and heart failure: insights from cell morphology and biology. *Curr Opin Cardiol* **11:** 293–301.

Hein S, Scholz D, Fujitani N, Rennollet H, Brand T, Friedl A and Schaper J (1994) Altered expression of titin and contractile proteins in failing human myocardium. *J Mol Cell Cardiol* **26:** 1291–1306.

Helmes M, Trombitas K, Centner T, Kellermayer M, Labeit S, Linke A and Granzier H (1999) Mechanically driven contour-length adjustment in rat cardiac titin's unique N2B sequence: titin is an adjustable spring. *Circ Res* **84:** 1139–1352.

Helmes M, Trombitas K and Granzier H (1996) Titin develops restoring force in rat cardiac myocytes. *Circ Res* **79:** 619–626.

Horowits R and Podolsky RJ (1987) The positional stability of thick filaments in activated skeletal muscle depends on sarcomere length: evidence for the role of titin filaments. *J Cell Biol* **105:** 2217–2223.

Improta S, Politou AS and Pastore A (1996) Immunoglobulin-like modules from titin I-band: extensible components of muscle elasticity. *Structure* **4:** 323–337.

Itoh-Satoh M, Hayashi T, Nishi H, Koga Y, Arimura T, Koyanagi T, Takahashi M, Hohda S, Ueda K, Nouchi T, *et al.* (2002) Titin mutations as the molecular basis for dilated cardiomyopathy. *Biochem Biophys Res Commun* **291:** 385–393.

Kabsch W, Mannherz HG, Suck D, Pai EF and Holmes KC (1990) Atomic structure of the actin: DNase I complex. *Nature* **347:** 37–44.

Kellermayer MS, Smith SB, Granzier HL and Bustamante C (1997) Folding–unfolding transitions in single titin molecules characterized with laser tweezers. *Science* **276:** 1112–1116.

Kulke M, Fujita-Becker S, Rostkova E, Neagoe C, Labeit D, Manstein DJ, Gautel M and Linke WA (2001) Interaction between PEVK-titin and actin filaments: origin of a viscous force component in cardiac myofibrils. *Circ Res* **89:** 874–881.

Kellermayer MS, Smith SB, Bustamante C and Granzier HL (1998) Complete unfolding of the titin molecule under external force. *J Struct Biol* **122:** 197–205.

Kellermayer MS, Smith S, Bustamante C and Granzier HL (2000) Mechanical manipulation of single titin molecules with laser tweezers. *Adv Exp Med Biol* **481:** 111–126; discussion 127–128.

Kellermayer MS, Smith SB, Bustamante C and Granzier HL (2001) Mechanical fatigue in repetitively stretched single molecules of titin. *Biophys J* **80:** 852–863.

Labeit S and Kolmerer B (1995) Titins: giant proteins in charge of muscle ultrastructure and elasticity. *Science* **270:** 293–296.

Li H, Oberhauser AF, Redick SD, Carrion-Vazquez M, Erickson HP and Fernandez JM (2001) Multiple conformations of PEVK proteins detected by single-molecule techniques. *Proc Natl Acad Sci USA* **98:** 10682–10686.

Linke W, Rudy D, Centner T, Gautel M, Witt C, Labeit S and Gregorio C (1999) I-band titin in cardiac muscle is a three-element molecular spring and is critical for maintaining thin filament strucuture. *J Cell Biol* **146:** 631–644.

Linke WA (2000) Stretching molecular springs: elasticity of titin filaments in vertebrate striated muscle. *Histol Histopathol* **15:** 799–811.

Linke WA and Granzier H (1998) A spring tale: new facts on titin elasticity. *Biophys J* **75:** 2613–2614.

Linke WA, Ivemeyer M, Labeit S, Hinssen H, Ruegg JC and Gautel M (1997) Actin-titin interaction in cardiac myofibrils: probing a physiological role. *Biophys J* **73:** 905–919.

Linke WA, Stockmeier MR, Ivemeyer M, Hosser H and Mundel P (1998) Characterizing Titin's I-Band Ig Domain Region as an Entropic Spring. *J Cell Sci* **111:** 1567–1574.

Ma K, Kan L and Wang K (2001) Polyproline II helix is a key structural motif of the elastic PEVK segment of titin. *Biochemistry* **40:** 3427–3438.

Marko JF and Siggia ED (1994) Fluctuations and supercoiling of DNA. *Science* **265:** 506–508.

Marszalek PE, Lu H, Li H, Carrion-Vazquez M, Oberhauser AF, Schulten K and Fernandez JM (1999) Mechanical unfolding intermediates in titin modules. *Nature* **402:** 100–103.

Maruyama K (1997) Connectin/titin, giant elastic protein of muscle. *Faseb J* **11:** 341–345.

Mayans O, van der Ven PF, Wilm M, Mues A, Young P, Furst DO, Wilmanns M and Gautel M (1998) Structural basis for activation of the titin kinase domain during myofibrillogenesis. *Nature* **395:** 863–869.

McElhinny AS, Sorimachi H, Labeit S and Gregorio CC (2002) Dual roles of MURF-1 in the regulation of sarcomeric M-line structure and in the potential modulation of gene expression via its interaction with titin and GMEB-1. *J Cell Biol* **57:** 125–136.

Minajeva A, Kulke M, Fernandez JM and Linke WA (2001) Unfolding of titin domains explains the viscoelastic behavior of skeletal myofibrils. *Biophys J* **80:** 1442–1451.

Morano I, Hadicke K, Grom S, Koch A, Schwinger RH, Bohm M, Bartel S, Erdmann E and Krause EG (1994) Titin, myosin light chains and C-protein in the developing and failing human heart. *J Mol Cell Cardiol* **26:** 361–368.

Pfuhl M, Winder SJ and Pastore A (1994) Nebulin, a helical actin binding protein. *Embo J* **13:** 1782–1789.

Rief M, Gautel M, Oesterhelt F, Fernandez JM and Gaub HE (1997) Reversible unfolding of individual titin immunoglobulin domains by AFM. *Science* **276:** 1109–1112.

Siu BL, Niimura H, Osborne JA, Fatkin D, MacRae C, Solomon S, Benson DW, Seidman JG and Seidman CE (1999) Familial dilated cardiomyopathy locus maps to chromosome 2q31. *Circulation* **99:** 1022–1026.

Solaro RJ, Moir AJ and Perry SV (1976) Phosphorylation of troponin I and the inotropic effect of adrenaline in the perfused rabbit heart. *Nature* **262:** 615–617.

Sorimachi H, Kinbara K, Kimura S, Takahashi M, Ishiura S, Sasagawa N, Sorimachi N, Shimada H, Tagawa K, Maruyama K, *et al.* (1995) Muscle-specific calpain, p94, responsible for limb girdle muscular dystrophy type 2A, associates with connectin through IS2, a p94- specific sequence. *J Biol Chem* **270:** 31158–31162.

Strang KT and Moss RL (1995) Alpha 1-adrenergic receptor stimulation decreases maximum shortening velocity of skinned single ventricular myocytes from rats. *Circ Res* **77:** 114–120.

Strang KT, Sweitzer NK, Greaser ML and Moss RL (1994) Beta-adrenergic receptor stimulation increases unloaded shortening velocity of skinned single ventricular myocytes from rats. *Circ Res* **74:** 542–549.

Trombitas K and Granzier H (1997). Actin removal from cardiac myocytes shows that near Z-line titin attaches to actin while under tension. *Am J Physiol* **273:** C662–C670.

Trombitas K, Freiburg A, Centner T, Labeit S and Granzier H (1999) Molecular dissection of N2B cardiac titin's extensibility. *Biophys J* **77:** 3189–3196.

Trombitas K, Greaser M, French G and Granzier H (1998a) PEVK extension of human soleus muscle titin revealed by immunolabeling with the anti-titin antibody 9D10. *J Struct Biol* **122:** 188–196.

Trombitas K, Greaser M, Labeit S, Jin JP, Kellermayer M, Helmes M and Granzier H (1998b) Titin extensibility in situ: entropic elasticity of permanently folded and permanently unfolded molecular segments. *J Cell Biol* **140:** 853–859.

Trombitas K, Redkar A, Centner T, Wu Y, Labeit S and Granzier H (2000) Extensibility of isoforms of cardiac titin: variation in contour length of molecular subsegments provides a basis for cellular passive stiffness diversity. *Biophys J* **79:** 3226–3234.

Trombitas K, Wu Y, Labeit D, Labeit S and Granzier H (2001) Cardiac titin isoforms are coexpressed in the half-sarcomere and extend independently. *Am J Physiol Heart Circ Physiol* **281:** H1793–H1799.

Tskhovrebova L and Trinick J (2001) Flexibility and extensibility in the titin molecule: analysis of electron microscope data. *J Mol Biol* **310:** 755–771.

Tskhovrebova L and Trinick J (2002) Role of titin in vertebrate striated muscle. *Philos Trans R Soc Lond B Biol Sci* **357:** 199–206.

Tskhovrebova L, Trinick J, Sleep JA and Simmons RM (1997) Elasticity and unfolding of single molecules of the giant muscle protein titin. *Nature* **387:** 308–312.

Wang K (1984) Cytoskeletal matrix in striated muscle: the role of titin, nebulin and intermediate filaments. *Adv Exp Med Biol* **170:** 285–305.

Wang K (1996) Titin/connectin and nebulin: giant protein rulers of muscle structure and function. *Adv Biophys J* **33:** 123–134.

Watanabe K, Muhle-Goll C, Kellermayer MS, Labeit S and Granzier H (2002a) Different molecular mechanics displayed by titin's constitutively and differentially expressed tandem Ig segments. *J Struct Biol* **137:** 248–258.

Watanabe K, Nair P, Labeit D, Kellermayer MS, Greaser M, Labeit S and Granzier H (2002b) Molecular mechanics of cardiac titin's PEVK and N2B spring elements. *J Biol Chem* **277:** 11549–11558.

Williamson MP (1994) The structure and function of proline-rich regions in proteins. *Biochem J* **297:** 249–260.

Wu Y, Bell SP, LeWinter MM, Labeit S and Granzier H (2002a) Titin: an endosarcomeric protein that modulates myocardial stiffness. *J Cardiac Failure* **8:** S276–S286.

Wu Y, Bell SP, Trombitas K, Witt CW, Labeit S, LeWinter MM and Granzier H (2002b) Changes in titin isoform expression in pacing-induced cardiac failure give rise to increased passive muscle stiffness. *Circulation*, in press.

Yamasaki R, Berri M, Wu Y, Trombitas K, McNabb M, Kellermayer MS, Witt C, Labeit D, Labeit S, Greaser M, *et al.* (2001) Titin-actin interaction in mouse myocardium: passive tension modulation and its regulation by calcium/s100a1. *Biophys J* **81:** 2297–2313.

Yamasaki R, Wu Y, McNabb M, Greaser M, Labeit S and Granzier H (2002) Protein kinase A phosphorylates titin's cardiac-specific N2B domain and reduces passive tension in rat cardiac myocytes. *Circ Res* **90:** 1181–1188.

Young P, Ehler E and Gautel M (2001) Obscurin, a giant sarcomeric Rho guanine nucleotide exchange factor protein involved in sarcomere assembly. *J Cell Biol* **154:** 123–136.

Journal of Muscle Research and Cell Motility **23**: 473–482, 2002.

Species variations in cDNA sequence and exon splicing patterns in the extensible I-band region of cardiac titin: relation to passive tension

MARION L. GREASER*, MUSTAPHA BERRI†, CHAD M. WARREN and PAUL E. MOZDZIAK‡
University of Wisconsin-Madison, Madison, Wisconsin 53706, USA

Abstract

Titin is believed to play a major role in passive tension development in cardiac muscle. The cDNA sequence of cardiac titin in the I-band sarcomeric region was determined for several mammalian species. Contiguous sequences of 3749, 12,230, 6602, and 11,850 base pairs have been obtained for the rat N2B, rat N2BA, dog N2B, and dog N2BA isoforms respectively. The length of the PEVK region of the N2B isoform did not correlate with rest tension properties since the only species showing an altered length was the dog that expressed a shorter form. No differences were found between the N2B PEVK lengths in ventricular and atrial muscle. New N2BA splicing pathways in the first tandem Ig region were found in human and dog cardiac muscle. Most of the rat and dog sequences were 85–95% identical with the reported human sequence. However, the N2B unique amino acid sequences of rat and dog were only 51 and 67% identical to human. The rat N2B unique sequence was 526 amino acids in length compared to 572 in human. The difference in length was due to deletion of amino acid segments from six different regions of the N2B unique domain. Patterns of PEVK exon expression were also much different in the dog, human, and rat. Six separate dog N2BA PEVK clones were sequenced, and all had different exon splice combinations yielding PEVK lengths ranging from 703 to 900 amino acids. In contrast a rat N2BA clone had a PEVK length of 525 amino acids, while a human clone had an 908 amino acid PEVK segment. Thus, in addition to the higher proportion of the shorter N2B isoform found in rat compared with dog cardiac muscle observed previously, shorter N2B unique and N2BA PEVK segments may also contribute to the greater passive tension in cardiac muscle from rats.

Introduction

Most of the passive force, or resting tension, of heart and skeletal muscles is now thought to arise from stretching long, elastic filaments composed of titin (Horowits and Podolsky, 1987; Funatsu *et al.*, 1990; Wang *et al.*, 1991; Granzier and Irving, 1995; Wu *et al.*, 2000). This extremely large protein (Wang *et al.*, 1979), which is also known as connectin (Maruyama, 1976), spans each half of the sarcomere from the Z-line to the M-line (Furst *et al.*, 1988; Labeit and Kolmerer, 1995) (see also Wang, 1996; Maruyama, 1997; Gregorio *et al.*, 1999; Trinick and Tskhovrebova, 1999; Granzier and Labeit, 2002 for more extensive reviews). While the segment of the titin molecule found in the I-band behaves elastically when the sarcomeres are stretched, the A-band segment is bound along the thick filament and is therefore inelastic. An unusual amino acid composition region (with nearly 80% proline, glutamic acid, valine and lysine) and termed 'PEVK' is found in the central I-band region (Labeit and Kolmerer, 1995). In skeletal muscle increasing tension extends titin in three phases: an initial straightening of the collapsed Ig domain chain, a lengthening of the PEVK region, and finally an unraveling of individual Ig domains (Linke *et al.*, 1996; Linke *et al.*, 1998; Trombitas *et al.*, 1998). The latter stage requires forces unlikely to occur under physiological conditions (Linke and Granzier, 1998).

Titin is expressed in different length versions in cardiac, psoas and soleus muscles by means of alternative splicing, and the lengths correlate with the passive tension properties of the different muscle types (Labeit and Kolmerer, 1995). The different lengths result from expression of different numbers of I-band Ig domains and different lengths of the PEVK region. The N2B unique region (expressed only in cardiac muscle) has also been shown to lengthen during passive stretch (Helmes *et al.*, 1999; Linke *et al.*, 1999; Trombitas *et al.*, 1999, 2000; Watanabe *et al.*, 2002). Different isoforms of cardiac I-band titin have been identified, and these fall into two distinct classes termed 'N2B' and 'N2BA' (Freiburg *et al.*, 2000) (see Figure 1). The former has a short PEVK region and fewer Ig domains; the latter has more Igs and a somewhat longer PEVK region. Both N2B and N2BA isoforms contain the extensible N2B unique sequence. Both isoform types

†Current address: INRA de Tours-Nouzilly, Nouzilly, France
‡Current address: North Carolina State University, Raleigh, North Carolina
*To whom correspondence should be addressed: Tel.: +608-262-1456; Fax: +1 608-265-3110; E-mail: mgreaser@facstaff.wisc.edu

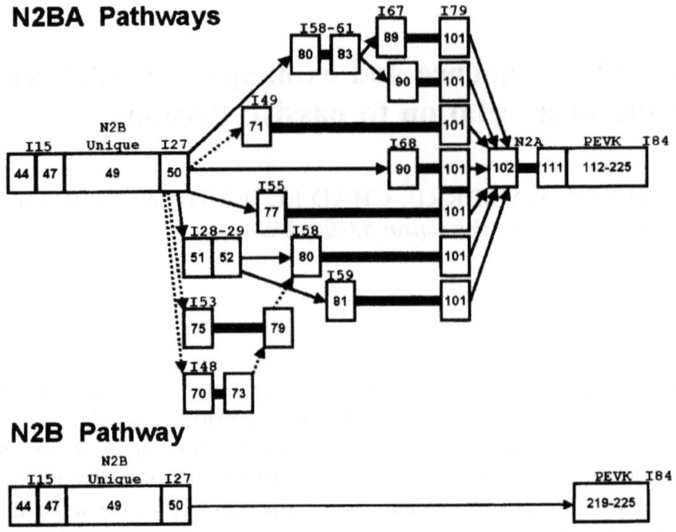

Fig. 1. Splicing pathways for cardiac titin isoforms. The figure is redrawn and modified from that of Freiburg *et al.* (2000). The boxes indicate numbered exons from human cardiac titin, GenBank accession number AJ277892. The Ig domains, the N2B unique, N2A region, and PEVK regions are labeled above the appropriate exons with the Ig nomenclature of Freiburg *et al.* (2000). The wide dense black lines indicate numerically continuous exons between the connected boxes. Arrows indicate discontinuous exon splice patterns. Numbers inside the N2BA PEVK box indicate the full range of exon numbers; only a small proportion (<30) are expressed in any one isoform. The arrows with dotted lines (upper-rat; two lower-human) indicate splicing patterns found in cardiac titin that have not previously been described.

may co-exist in the same sarcomere (Trombitas *et al.*, 2001).

Large passive tension differences have been shown between cardiac myocytes of different species (Fabiato and Fabiato, 1978; Cazorla *et al.*, 2000). Passive tension might be altered due to (1) different length PEVK segments, (2) different elasticity properties because of amino acid sequence changes, (3) different numbers of Ig domains expressed, or (4) different ratios of titin isoform types with different lengths. The present study applies RT-PCR and DNA sequencing to characterize central I-band regions of titin from several species. Larger regions of the rat and dog cardiac N2B and N2BA titin isoforms have been sequenced, and the results suggest multiple combinations of these potential alterations could affect passive tension.

Methods

RNA preparation

Immediately following euthanasia, hearts were removed, trimmed, and rinsed free of blood with phosphate buffered saline (PBS). The tissues were frozen in liquid nitrogen and stored at −70°C prior to RNA preparation. Total RNA was prepared using TRIzol reagent (Invitrogen Life Technologies, Carlsbad, CA) from the ventricular and atrial muscles of adult rabbit (New Zealand White), pig (crossbred), rat (Sprague Dawley), dog (mongrel and Beagle), and human (left ventricular tissue only) using the procedure of Chomczynski (1993).

cDNA synthesis, PCR procedures, and DNA sequencing

The cDNA synthesis and PCR were performed using the thermoscript RT-PCR system (Invitrogen). The cDNA was primed using random hexamers at 50°C for 50 min. After transcription, RNA was removed by digestion with RNase H at 37°C for 20 min. A PCR reaction (Invitrogen protocol) consisted of 5 μL 10× PCR buffer minus $MgCl_2$, dNTP mix to give a final concentration of 0.2 mM each in the reaction mixture, 0.2 μM of forward and reverse primers, 2 units of Platinium *Taq* DNA polymerase, 2 μL cDNA, and DEPC-treated water to 50 μL. The total $MgCl_2$ concentration was varied between 1.82 and 3.64 mM in separate tubes to optimize PCR amplification. Primers (listed in Table 1) were constructed on the basis of the human titin sequences (Labeit and Kolmerer, 1995; accession numbers X90568 and X90569) and new titin sequence from the rat and dog obtained during the course of the work. Temperature parameters for the reactions were: step 1 – 94°C 2 min; step 2 – 94°C 30 s; step 3 – temperature of 5° below the melting temperature of the primer pair 30 s; step 4 – 68°C 1 min/Kb; step 5 – repeat steps 2–5 a total of 34 times, and step 6 – 68°C 10 min.

PCR products were electrophoretically separated on 1% agarose gels containing ethidium bromide. Bands of interest were excised, and the DNA purified using a gel extraction kit (Qiagen, Valencia, CA). Blunt ended PCR products were ligated into Promega's pGEM®-T Easy vector (Promega, Madison, WI). The ligated DNA was then transformed into JM109 competent cells using heat shock. Plasmids were purified using Qiagen kits, and sequenced at the University of Wisconsin Biotechnology

Table 1. Forward and reverse primers for PCR amplification of titin clones from rat, dog, and human cardiac muscle

Rat clones	5′ Forward primer 3′	5′ Reverse primer 3′
2 Rat	TCAAGCCATCTCGGTTCTT	TCTTGATCACTGGGGCAG
Rat I24	GTGCATGGCCAGTAATGA	CTACTTCCAAGGGCTCAA
Rat K25	AAAGACAGCCCAAACGTA	CCTCTGTCAAAACGATGTC
Rat A31	GAGGTGAACTGGTATTTC	CAGAGCGTTGTTCTTGAA
Rat C35	TACAATGCCAAGTTGCTGGA	TTTCAGCTTGCGGATGAAGG
Rat C43	GCACTCCTGACTGTCCTA	GCTTTCTTGAATTTCCACCTC
Rat J26B	AAAGACAGCCCAAACGTA	GGTTTCTTCTCTTCTGGA
Rat N33	TACAAAGGACCGACAGAG	CCTCTGTCAAAACGATGTC
Rat A47A	TGAGCCCTTGGAAGTAGC	GGCAATAACATCTATGGACACTG
Dog clones		
Dog B7	CCAGAAGATGAAGGAACTTAC	AGTTAGCGTGGCCGTACAG
Dog A16	ACTCCTTCTGCTGACTACA	TTTTGGAGGTGGCTCTGGTA
Dog F23	AGCACCATTTTGAAAGCTGC	TTTTGGAGGTGGCTCTGGTA
Dog 615 & 760	TGCGGCGAGTATACATGCAAAGC	CCGTCTTTCATCCAGGTTGT
Dog L43	CCAATGAGTATGGCAGCG	GCTGGCTTCATTGATGGTAA
Dog K43	CCAATGAGTATGGCAGCG	GCTGGCTTCATTGATGGTAA
Dog D36	CTTCAAAGTATGTCTCCACC	GTAAGCTCAGGCTCAAAG
Dog 27B	AAGCCTCCAATGAGTATGGCA	TCACTTCCAGCCGCATTTGA
Dog 23	AAAGCAGGCATGAACGACGCTGG	TTCAAAGATCGCAGAAGACCCAACGGA
Dog 499	GATGGTTGTTCCTCCTATTCCTCTG	GAGAGGTATTGCTGGCTTAATTTCAG
Dog 599	CCATTCGAAGAACCTTACTATGAAGAG	ATATTCTTCTTCCCGTTGTACTGAAAC
Dog 699	GTACAAAAGAAAACTGTTACCGAAGAG	GATCTCTTTCACAAACTTCAGTGGTAC
Dog F34	CCTTCATCCGCAAGCTGAAA	CGACACTGAGGTTGTGCATT
Dog A46A-F	GTACAACTTCAGGAATGATGGT	GCCTTTAGCTGGTAGGTGA
Human clones		
B45C	CCAATGAGTATGGCATGT	ACTAGCTTCATTGATCGTAA
B45A	CCAATGAGTATGGCATGT	ACTAGCTTCATTGATCGTAA
C38	ATACAACTTCAGGAATGATGGC	ACTTCCTCCTTTCGGCTTCTA

Center. In most cases sequences from both strands were obtained.

Sequence processing and analysis

Sequences were identified and compared with Genbank sequences using the NCBI Blast (Altschul *et al.*, 1990), translated with DNA Strider (Marck, 1988), and analyzed using the GCG Wisconsin Package SeqWeb Version 2 from Accelrys. Identification of expressed exons was conducted using the 'Blast 2 sequences' method (Tatusova and Madden, 1999) and the human titin genomic sequence (accession number AJ277892, Freiburg *et al.*, 2000).

Results

Characterization of the titin N2B PEVK region in different species

We initially hypothesized that the length of the PEVK in the N2B isoform might be a primary factor affecting passive tension. We therefore used PCR with primers based on the human sequence (accession number X90568, Labeit and Kolmerer, 1995) corresponding to positions in Ig27 and Ig84 (flanking the N2B type of PEVK, Figure 1) (Ig nomenclature of Freiburg *et al.*, 2000 here and throughout the paper) to amplify cDNAs from different species. The major PCR product using rabbit, pig, or rat ventricle was a single major cDNA band with an expected size (based on the human sequence) of 760-bp (Figure 2 left). However, two fragments (a major 615-bp and a minor 760-bp) were amplified from dog RNA. Similar sized PCR fragments were obtained using atrial muscles from each species

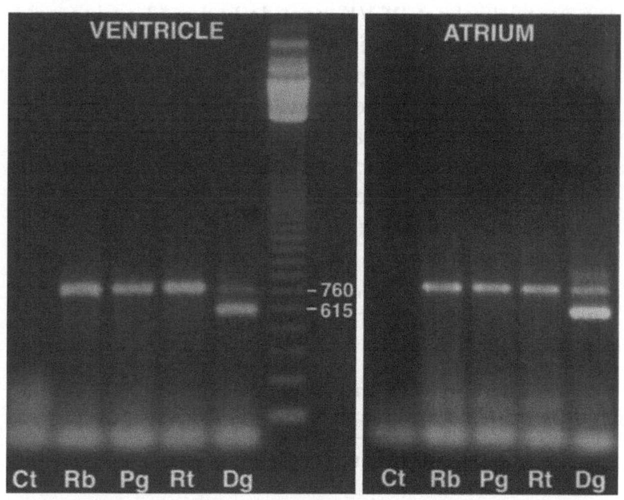

Fig. 2. One percent agarose gel electrophoresis of products obtained after RT-PCR of ventricular and atrial (right) total RNA using primers from Ig27 and Ig84 (N2B PEVK region). Ct – control, no cDNA template; Rb – rabbit; Pg – pig; Rt – rat; Dg – dog; right lane – 123-bp DNA ladder. Numbers represent the approximate base pair sizes of the major PCR products.

```
           1                                                 50
Human      GSVSCTATLT VTVPGGEKKV RKLLPERKPE PKEEVVLKSV LRKRPEEEEP
Dog760     GSVSCTATLT VTVPGAEKKV RKLLPEPKPQ PKEEVVLKSV LRKRPEEEEP
Dog615     GSVSCTATLT VTV~~~~~~~ ~~~~~~~~~~ ~~~~~~~~~~ ~~~~~~~~~~
Rat        GSVSCTATLT VTVPTAEKKV RKLLPEPKPQ PKEEVVLKSV LRKRPEEEEP
Pig        GSVSCTATLT VTVPGGEKKV RKLLPEPKPE PKEEVVLKSV LRKRPEEEEP
Rabbit     GSVSCTATLT VTVPGGEKKV RKLLPEPKPQ PKEEVVLKSV LRKRPEEEEP

           51                                               100
Human      KVEPKKLEKV KKPAVPEPPP PKPVEEVEVP ..TVTKRERK IPEPTKVPEI
Dog760     KVEPKKLEKI KKP.VPE.PP PKAVEEAEVP PAPVTKRERK IPEPVKVPEI
Dog615     ~~~~~~~~~~ ~~~~~PE.PP PKAVEEAEVP PAPVTKRERK IPEPVKVPEI
Rat        KVEPKKVEKV KKPEEP.PPP PKAV.EVEAP PEPKPK.ERK VPEPTKVPEI
Pig        KVEPKKLEKI KKPAVPE.PP PKAVEEVEAP PAAVPKRERK VPEPTKVPEI
Rabbit     KVEPKKLEKI KKPAAPE.PP PKAVEEVEVP PATVEKKERK IPEPTKVPEI

           101                                              150
Human      KPAIPLPAPE PKPKPEAEVK TIKPPPVEPE PTPIAAPVTV PVVGKKAEAK
Dog760     KPAIPLPGPE PKPKPEPEVK VIKPPPVEPA PTPIAAPVTV PVVGKKAEAK
Dog615     KPAIPLPGPE PKPKPEPEVK VIKPPPVEPA PTPIAAPVTV PVVGKKAEAK
Rat        KPAIPLPGPE PKPKPEPEVK TMKAPPIEPA PTPIAAPVTA PVVGKKAEAK
Pig        KPAIPLPGPE PKPKPEPEVK VIKPPPVEPP PAPIAAPVTV PVVGKKAEAK
Rabbit     KPAIPLPGPE PKVKPEAEVK TIKPPPVEPA PTPIAAPVTA PVVGKKTEAQ

           151                                              200
Human      APKEEAAKPK GPIKGVPKKT PSPIEAERRK LRPGSGGE KPPDEAPFTYQL
Dog760     APKEEAAKPK GPIKGIPKKT PSPIEVERKK LRPGSGGE KPPDEAPFTYQL
Dog615     APKEEAAKPK GPIKGIPKKT PSPIEVERKK LRPGSGGE KPPDEAPFTYQL
Rat        .PKDEAAKPK GPIKGVAKKT PSPIEAERKK LRPGSGGE KPPDEAPFTYQL
Pig        APKEEAAKPK GPIKGVAKKT PSPIEAERKK LRPGSGGE KPPDEAPFTYQL
Rabbit     .PKEEAAKPK GPIKGVAKKT PSPIEAERKK LRPGSGGE KPPDEAPFTYQL

           201                        229
Human      KAVPLKFVKE IKDIILTESE FVGSSAIFE
Dog760     KAVPLKFVKE IKDIVLTEAE SVGSSAIFE
Dog615     KAVPLKFVKE IKDIVLTEAE SVGSSAIFE
Rat        KAVPLKFVKE IKDIVLTEAE SVGSSAIFE
Pig        KAVPLKFVKE IQDIVLTEAE SVGSSAIFE
Rabbit     KAVPLKFVKE IKDIVLKEAE SVGSSAIFE
```

Fig. 3. Translated amino acid sequences of the N2B PEVK region for ventricular titins from various species. The amino acid sequences from human (GenBank accession number CAA62188, Labeit and Kolmerer, 1995), dog 615 (U82221), rabbit (U82222), pig (U82223), and rat (U82224) are compared with identities shown in bold. The underlined portions correspond to the N2B PEVK region (between segments from Ig27 and Ig84). Note that the 53 amino acid deletion near the amino terminal end of the dog 615 PEVK sequence coincides exactly with the beginning of the human cardiac PEVK segment.

(Figure 2 right). Blast searches revealed that the nucleotide sequences were >85% identical to the human cardiac N2B titin PEVK sequence. The cDNA sequences obtained from ventricular and atrial muscle total RNA were 100% identical within each species (data not shown). The amino-acid sequences were deduced from the nucleotide sequences and are shown in Figure 3. The dog ventricle short PEVK peptide had a 53 amino acid residue deletion coinciding with the amino terminal end of the N2B PEVK region. The rest of the sequence, however, is highly homologous with segments from the other species. There was a high degree of homology between human cardiac titin N2B PEVK amino acid sequences and the corresponding region of rat, rabbit, pig, dog 615, and dog 760 (Table 2). The GenBank accession numbers for the dog 615 (U82221), rabbit (U82222), rat (U82223), and pig (U82224) N2B PEVK regions have been reported previously.

Table 2. Comparisons of different N2B amino acid sequences from the PEVK region of different species

Species	Percent identity to human	Deletions, insertions vs. human
Rat	84	2 Deletions
Rabbit	90	2 Deletions, 2 insertions
Dog 615	89	53 + 1 Deletions, 2 insertions
Dog 760	90	2 Deletions, 2 insertions
Pig	91	1 Deletion, 2 insertions

Sequence of the dog and rat N2BA PEVK region

Since there was a difference in resting tension between dog and rat myocytes, we compared the N2BA-PEVK region in cardiac muscle of these two species using PCR amplification with primers corresponding to the N2BA isoforms. The N2BA PEVK of human ventricular muscle (GenBank accession number AF525413) was also determined since it had not previously been reported in the GenBank database. The dog sequence was first determined, but it initially proved to be very difficult to PCR amplify. Several cloning steps were necessary to bridge the entire PEVK region. A contig was finally constructed using the dog 23, 499, 599, and 699 clones (see Figure 4). PCR products were later obtained from the full rat (N33) and human (C38) PEVK regions amplified as single clones (Figures 5, 6). Normal Blast comparisons of these cDNAs and the translated amino acid sequences verified that they were from the PEVK region. However, the alignments were poor, and it was evident that there were different exon expression patterns among the different species products. The sequences were therefore compared individually against the human genomic titin contig (accession number AJ277892 – Freiburg *et al.*, 2000) to determine which exons were being expressed. The exon combinations for human, dog, and rat PEVK are shown in Figure 7. The identification of the human exons in clone C38 was straightforward since the sequences were

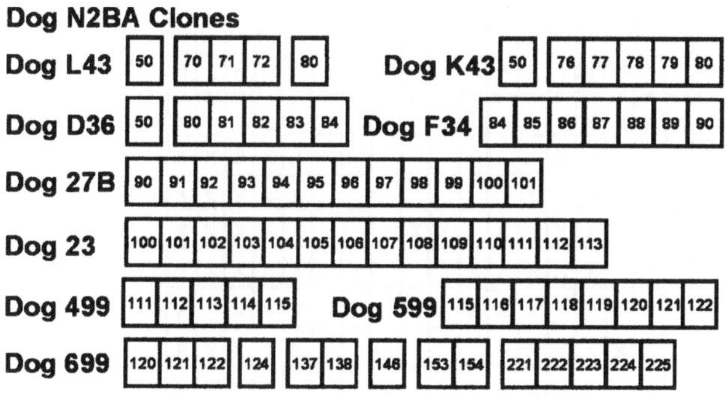

Fig. 4. Exon patterns of dog N2BA isoform PCR clones. Exon numbers correspond to those of human titin, GenBank accession number AJ277892 (Freiburg *et al.*, 2000). The 5′ and 3′ exons in each clone are usually only partial. The small gaps between some of the exons highlight splicing regions that are not numerically continuous.

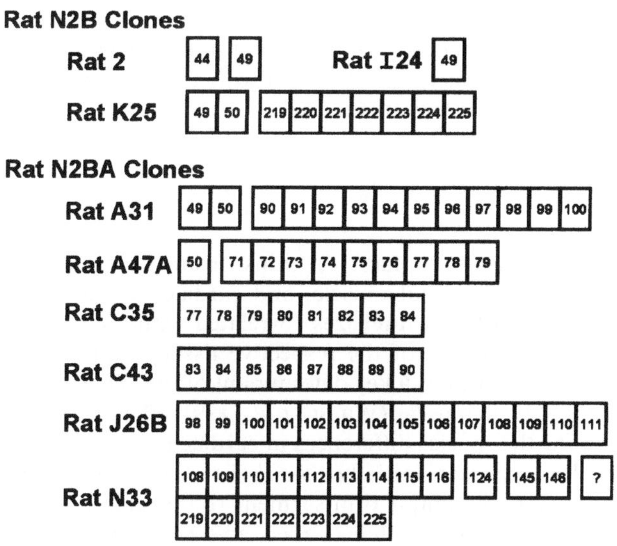

Fig. 5. Exon patterns of rat N2B and N2BA PCR clones.

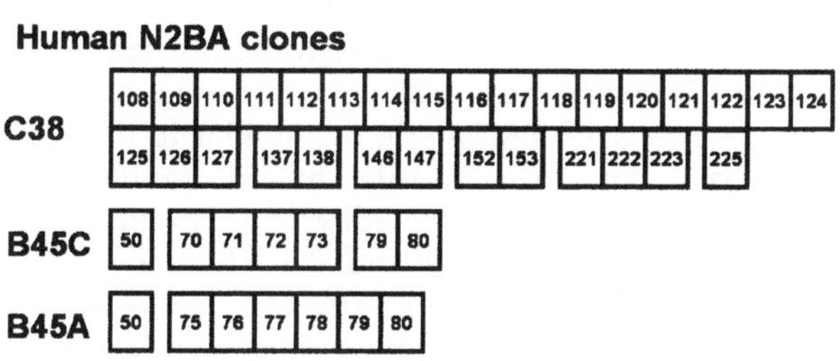

Fig. 6. Exon patterns in human N2BA PCR clones.

identical to those in the database. Fortunately the assignment of most dog and rat exons was also uncomplicated. Both species had essentially identical exon–intron borders to those in human titin. Each segment with the highest identity to the human sequence was aligned in the expected order in the clone sequences. However, the pattern of exon expression was highly inconsistent between species. Most of the exons expressed in common among the three species are at the 5′ and 3′ ends; more variable patterns occurred in the center.

After successfully obtaining full-length N2BA PEVK clones from human and rat, we again attempted to amplify this region in dog cardiac titin. PCR products of the expected size were produced, but they showed a blurred band upon agarose electrophoresis, suggesting multiple-sized species. The DNA was cloned, and plasmids from six individual colonies were sequenced. The exon expression patterns from these clones (A46A-F) are shown in Figure 7. All six clones had different exon patterns. The PEVK lengths were 703, 788, 894, 900, 703, and 819 amino acids for clones A–F respectively.

Sequence from the N2B unique region

Sequence was also obtained from the N2B unique region since this titin segment has been shown to extend under passive stretch. It is coded by the very large exon 49 that also includes coding for Ig domains 24, 25, and 26. The sequence for the dog titin in this region was obtained as a contig from clones B7, A16, and F23 (Figure 8). With rat the full exon 49 was obtained by alignment of clones Rat 2, I24, and K25. A comparison of human, dog, and rat N2B unique region amino acid sequences revealed striking differences (Figure 9). Blast comparisons showed that the rat and dog amino acid sequences were 51 and 67% identical with human respectively. In addition there were six sequence segments (5, 5, 6, 9, 10, 12, 7 amino acids each) that were apparently deleted in the rat compared to the human. Several other deletions and insertions of one or two amino acids also were found. The resulting segment lengths were 572, 573, and 526 for human, dog, and rat respectively. The aligned rat N2B sequences yielded a contig from exon 44 (Ig15) through the short PEVK region and beginning of Ig84, a total of 3749 bp (GenBank accession number AF525412). A contig of the dog N2B clones covering the same region plus additional exons in the second tandem Ig region was 6602 bp in length (GenBank accession number AY136512). The overall pattern of exon expression was similar to that shown previously with human titin (Labeit and Kolmerer, 1995).

Species comparisons of the N2A unique sequences

In light of the large differences in the N2B unique sequences, the N2A unique regions were also compared. Amino acid sequences of this segment from human, dog, and rat titins were highly homologous. Both the dog and rat N2A unique regions (106 amino acids each with no deletions or insertions) were 94% identical to the corresponding region in human titin (not shown).

Dog and rat N2BA clones and splicing pathways

A number of different alternative splicing patterns have been reported with human and rabbit N2BA cardiac

Exon No.	112	113	114	115	116	117	118	119	120	121	122	124	125	126	127	136	137	138	145	146	147	152	153	154	?	?	?	?	?	219	220	221	222	223	224	225
Repeat	P	P	P	E	P	P	P	P	P	P	P	P	P	P	P	P	E	P	P	P	P	E	P	P	P	P	P	P	P	P	P	P	P	P	P	P
Human C38	X	X	X	X	X	X	X	X	X	X	X	X	X	X	X		X	X	X	X	X	X	X							X	X	X	X	X		X
Dog Contig	X	X	X	X	X	X	X	X	X	X	X	X				X	X	X	X	X		X	X	X								X	X	X	X	X
Dog A46A	X	X	X	X	X	X	X					X				X	X	X	X	X		X	X	X								X	X	X	X	X
Dog A46B	X	X	X	X	X	X	X					X				X	X	X		X		X	X	X	X	X	X	X				X	X	X	X	X
Dog A46C	X	X	X	X	X	X	X	X	X	X	X	X				X	X	X		X		X	X	X	X	X				X	X	X	X	X	X	X
Dog A46D	X	X	X	X	X	X	X	X	X	X	X	X				X	X	X		X		X	X	X	X	X	X	X				X	X	X	X	X
Dog A46E	X	X	X	X	X	X	X					X				X	X	X		X	X	X	X	X								X	X	X	X	X
Dog A46F	X	X	X	X	X							X	X			X	X	X	X	X		X	X	X	X	X	X	X	X			X	X	X	X	X
Rat	X	X	X	X	X							X							X	X					X					X	X	X	X	X	X	X

Fig. 7. Exon expression patterns in the PEVK region of human, dog, and rat N2BA titin isoforms. The type of repeat coded by each exon is identified as either 'P' for PPAK or 'E' for polyglutamic acid (Greaser, 2001). Note that the splicing patterns are different between species and that all six of the individual dog clones had different exon compositions.

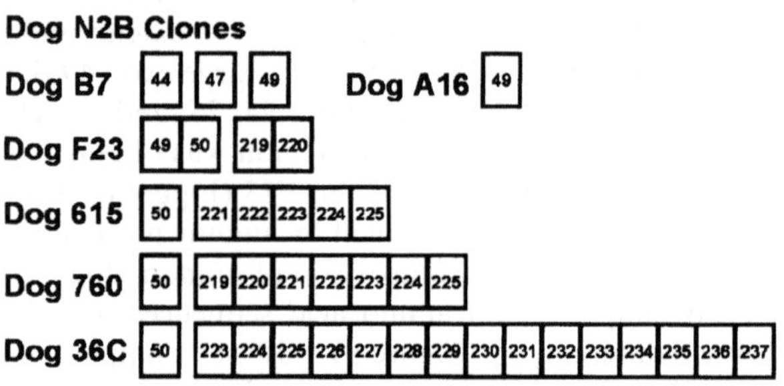

Fig. 8. Exon patterns in dog N2B PCR clones.

titin. We therefore obtained cDNA sequence for the titin segment spanning between exon 50 (I27) and the N2BA type PEVK. Several clones with different splicing pathways were found (Figures 4, 5). A contig for the rat N2BA isoform type has been assembled from rat A47A (exon 50; Ig27), C35, C43, A31, J26B, and N33 (exon 225, Ig84) with a total length of 12230 bp (GenBank accession number AF525411). A dog N2BA contig of 11850 bp has been assembled from K43 (exon 50; Ig27), D36, F34, 27B, dog 23, and A46C (exon 225, Ig84) (GenBank accession number AY136513).

New tandem Ig splicing pathways

Two new types of splicing pathways were initially found in the dog. The L43 clone goes from exon 50 (I27) to a group of three exons (70–72, Ig48–50) and then to exon 80 (Ig58). Another clone obtained using the same PCR primers shows connection between exon 50 and exon 75 (Ig53). Two similar but different new pathways have also been found in human cardiac titin (Figure 6, clones B45C and B45A). The B45C has exon 50 connected to exons 70–73 and then to exons 79 and 80; B45A has exon 50 spliced to the exon group from 75 to 80. These splice paths are shown as the lower two with dotted arrows in Figure 1. A further unique splice pathway has been found in the rat. The A47A clone has splicing between exon 50 and a continuous group between exons 71 and 79. These consecutive exons would presumably extend to exon 101 although there is no experimental evidence for this as yet. This pathway is shown with the upper dotted arrow in Figure 1.

Discussion

The observations that (1) titins from slow skeletal muscle, fast skeletal muscle, and cardiac muscle differ in size and (2) that the sizes correlated with the passive tension properties of the respective tissues support the concept that titin is a key determinant of the extensibility properties of muscle (Wang *et al.*, 1991). The explanation for the species differences in cardiac passive tension was less apparent. Initially it was believed that the PEVK region was the major titin segment responsible for titin extension during stretch (Labeit and Kolmerer, 1995). Thus it seemed logical that a shorter length PEVK region (of the N2B isoform) would be the reason for the greater passive tension in rat vs. dog cardiac muscle. However, the length of the short PEVK in rat was similar to human, rabbit, and pig even though the rest tension of these muscle types was different. The dog had two different types of PEVK length, one similar to the other species, but the other shorter by 53 amino acids (Figure 3). The shorter form (dog 615) resulted from the exclusion of exons 219 and 220 (Figure 7). This smaller sized PEVK was opposite of what would be expected from rest tension measurements; dog cardiac myocytes have low passive tension while a shorter PEVK should result in higher tension. The short PEVK found in dog has not been described before in cardiac tissue from other species (Higgins *et al.*, 1994; Witt *et al.*, 1998).

The recognition of titin isoforms of the N2BA and N2B type brought additional complexity to the interpretation of titin's role in passive tension (Freiburg *et al.*, 2000; Granzier and Labeit, 2002). N2BA type titins contain more Ig domains than the ones of the N2B isoform. In addition the PEVK region is longer in the N2BA class of isoform (Freiburg *et al.*, 2000). Thus

```
             1                                                    50
N2B_dog  DMTDALRKAK PTPEATKDFP EMSLKGPAVE ASDSEQEIAA FVNSTILKAA
N2B_hum  DMTDTPCKAK STPEAPEDFP QTPLKGPAVE ALDSEQEIAT FVKDTILKAA
N2B_rat  DVPDVSCKEE STSGVPGDFL ETSARGLSVQ AFDSEQEITA FA.....KGA

             51                                                  100
N2B_dog  SVAEESQQLS HEHIVQTSEL SSQVTLVAEQ LQSTVTLEQN MPDAESTRAR
N2B_hum  LITEENQQLS YEHIAKANEL SSQLPLGAQE LQS..ILEQD KLTPESTREF
N2B_rat  FVAEEKLQLP YEHKVDSSEL GTGVTLGAQK LQPVI..... MSTLQGTGEL

             101                                                 150
N2B_dog  LSINGTIHVQ PIKEPSPNLQ LQVIQSQKTL PKEDVLMLQE PESQVILSET
N2B_hum  LCINGSIHFQ PLKEPSPNLQ LQIVQSQKTF SKEGILMPEE PETQAVLSDT
N2B_rat  PSIDGAVHTQ PGKEPPPTLH LQTVQSQRTL PKEDTLQFEE P.........

             151                                                 200
N2B_dog  EKIFPSAMSV EEISSLTVES MKTFLAEPEG SYPQSSIEEA PAPSYSTSVA
N2B_hum  EKIFPSAMSI EQINSLTVEP LKTLLAEPEG NYPQSSI.EP PMHSYLTSVA
N2B_rat  EELCPGASSA AQVSSVTVKP LITLTAEPKG KSPQSS.TAA PDGSLLSSVT

             201                                                 250
N2B_dog  EEVLLSKEKT VSAMDKEQSL PPPKQEAQSA PILSQSLAEG NMESLKGADV
N2B_hum  EEVLSLKEKT VSDTNREQRV TLQKQEAQSA LILSQSLAEG HVESLQSPDV
N2B_rat  EETLHLGEKM ISEVDKEQR. .........A LLLSQSIAEG CVESLEALDV

             251                                                 300
N2B_dog  IISEVHCEPQ VPSEHTCTE. DEIVMESADE LESAGQDFAA RTEEGKSLRF
N2B_hum  MISQVNYEPL VPSEHSCTEG GKILIESANP LENAGQDSAV RIEEGKSLRF
N2B_rat  AVSEVRSEPQ VPSQHTCIEE GKTLMASADT LESTGQNVAL RTEEGKSLSF

             301                                                 350
N2B_dog  PLAYEEKQVL LKEEHSDNLA VPLNQTSEYK KEPMVINGVP EVQASDTLSK
N2B_hum  PLALEEKQVL LKEEHSDNVV MPPDQIIESK REPVAIKKVQ EVQGRDLLSK
N2B_rat  PLALEEKQVL LKEEQSEVMA VPTSQTSKPK KEPEAVKGVK EVQESDLLSK

             351                                                 400
N2B_dog  ESLLSGIPEE QRLNLKTQIR RALQAAVASE QPSLFSEWLR NVEKVEVKAI
N2B_hum  ESLLSGIPEE QRLNLKIQIC RALQAAVASE QPGLFSEWLR NIEKVEVEAV
N2B_rat  ETLFSSIPKE QRLHLKTQVR RALQAAVARE QASLFSEWPR NIDKVEVTAV
N2B_rab  ~~~~~~~~~~ ~~~~~~~~~~ ~~~~~~~~~~ ~PSLFSEWLR NIETVEIKVV

             401                                                 450
N2B_dog  HFTQEPKRVM CTYLITSVNT LTEELTIAIE GIDPQMADLT TELKDALCSV
N2B_hum  NITQEPRHIM CMYLVTSAKS VTEEVTIIIE DVDPQMANLK MELRDALCAI
N2B_rat  NFTQEPKRIL CTYLITSVKS LTEELTLTIE DIDPQMANLE TELKDALCSI
N2B_rab  SFTQEPKCIM CTYLITTVKS LMEEVTITLE GVDPQMANLN IELKDALCSI

             451                                                 500
N2B_dog  ICEEINILTA EDPRIQKGAT TGLPEEMGPS SESQKVEAIW EPEAESKYLV
N2B_hum  IYEEIDILTA EGPRIQQGAK TSLQEEMDSF SGSQKVEPIT EPEVESKYLI
N2B_rat  VCEERNILMA EDPRFQEEDK IGVQAGRGHL SDAQKVETVI EAEVDSKYLV
N2B_rab  ICEEINIFTA EDPRIQKGAQ TGLQEEMDSC SDAQMVEAIM EPEVEPNYLV

             501                                                 550
N2B_dog  SVEEVSCFHV ESQVKVGDTT GVPDGVTPAV SPDEKQAEIP KPSEEKEKLP
N2B_hum  STEEVSYFNV QSRVKYLDAT PVTKGVASAV VSDEKQDESL KPSEEKEESS
N2B_rat  PKEEVSWLNV ESHFKDGDTD EVP....... .....QTETL KVAEEGD...
N2B_rab  PAEEVSCFSV ESQVKDIDTP TGTDELAPAA VSAETQEESL KSSEEEEKSS

             551                   575
N2B_dog  G.GGAGEGGV GEAQEAEGSA VGEDG
N2B_hum  SESGTEEVAT VKIQEAEGGL IKEDG
N2B_rat  ....TQKTST VRSQEEAEGP LADLC
N2B_rab  K.CGTKEAGT RKIQDTEASL TKEAG
```

Fig. 9. Comparison of the N2B unique regions from dog, human, and rat. Note the six clusters of amino acid deletions in the rat vs. the dog or human (depicted by dots in the diagram). The human and rabbit sequences are from GenBank accession numbers CAA62188 (Labeit and Kolmerer, 1995) and CAB85900 (Freiburg *et al.*, 2000) respectively.

cardiomyocytes with a higher proportion of this longer titin would be expected to have lower passive tension. The elegant work of Cazorla *et al.* (2000) showed that the rest tension of cardiomyocytes from species that contained high proportions of the N2BA isoforms (such as bovine) was lower than with rats that show a preponderance of the N2B titin isoform. Thus, the isoform proportions appear to play a key role in determining rest tension.

The reasons for the large number of titin isoform splicing pathways remain obscure. The general patterns of splicing reported previously with human and rabbit were also found in the dog and rat with minor differences. Thus the dog splicing pathways from exon 50 (Ig27) to exons 76 (Ig54) (dog K43 clone) or 80 (Ig58) (dog D36 clone) were similar to the previously reported paths of Ig27–Ig55 or Ig58 in rabbit (Freiburg *et al.*, 2000). Rat N2BA titins, although found in only small proportions relative to the N2B isoform, are clearly present since cDNAs for all the Ig coding exons from 71 to 111 (Ig55–Ig83) could be PCR amplified (Figure 5). Two distinctly new pathways were found in the dog with splicing of exon 50 (Ig27) to a cluster of three exons (70–72, Ig48–50) and then to exon 80 (Ig58) (Figure 4, clone L43) and exon 50 spliced to exon 76 (clone K43). No evidence for cardiac expression of Ig domains in the range between Ig29 and Ig55 has been reported previously. A similar pathway has also been found in humans (Figure 6, clone B45C). A major difficulty in delineating splice pathways is the large size of titin. Thus, although procedures for long PCR continue to improve, it remains challenging to amplify cDNAs longer than 3–5 kb. The exon containing the N2B unique sequence (no. 49) alone contains nearly 2800 bp.

The highly variable expression of PEVK exons between species was an unexpected finding (Figure 7). Regular Blast comparisons were hopelessly confusing because of the different splicing patterns in this titin region. Also the only N2BA type pattern in the database is from soleus muscle that expresses a much larger PEVK segment. In the current work cardiac sequences were compared with the human genomic contig with GenBank accession number AJ277892 (Freiburg *et al.*, 2000). Identification of expressed exons in most cases proved to be less difficult than first anticipated. The exon–intron splice borders were virtually identical among the human, dog, and rat species. In spite of the high homology of the PPAK repeats (Greaser, 2001), there was enough difference between the cDNAs to distinguish each exon. The only exon that could not be definitely assigned in the rat was one between 146 and 219. This exon is located in one of three virtually identical segments (A,B,C) in the titin gene (Bang *et al.*, 2001). Thus this exon could be either 175, 184, or 193 since all have identical sequences in human. The exons near both the 5′ and 3′ ends of the PEVK region (exons 112–116 and 221–225) appeared more likely to be expressed in the three species. An exception to this pattern was exon 224 in human clone C38. This exon is apparently expressed in both the human cardiac N2B and the soleus skeletal muscle isoforms (Labeit and Kolmerer, 1995).

Several of the N2BA PEVK dog clones contained exons from the genomic ABC region. Many of the PPAK repeats in this area are very similar to each other even within a single genomic segment, so attempts to identify the specific exons were impossible (these are labeled with question marks in Figure 7). However, it is clear that at least some of these exons must be ones for which there has been no experimental evidence for expression in the human soleus. Thus there appear to be some PEVK exons expressed specifically in the cardiac N2BA isoform.

The large variation in exon expression between different dog N2BA PEVK clones was another unexpected observation. Six different clones were sequenced, and the sequences of each were different. Even the two clones with the same PEVK length (A46A, A46E) did not contain both exons 145 and 147. Since the clones were selected by chance, it seems likely that additional types of exon expression pattern still remain to be characterized. It has been previously noted that the PEVK contains two types of amino acid composition regions along its length, the PPAK repeats and the polyE segments rich in glutamic acid residues (Greaser, 2001). There are three of the polyE exons in the N2BA PEVK (labeled ‘E’ in the repeat line of Figure 7), and all three are expressed in human and the six dog clones that have been sequenced. However, the rat clone contains only the first one of these segments. The biological significance, if any, of these observations is unclear. It also remains to be seen how many kinds of human PEVK-related isoforms are expressed and whether all the human N2BA type isoforms exclude exon 224.

Previous studies have shown that passive tension is related to the ratios of the N2B to N2BA isoforms in different species (Cazorla *et al.*, 2000). The current work identifies two potential additional factors that relate to species differences in passive tension. First, there appears to be a significant difference in the character and/or size of the N2B unique region between species. This protein segment is now known to occur in both the N2B and N2BA classes of isoform (Freiburg *et al.*, 2000). The N2B unique segment is especially important in the N2B titin isoform because it provides an extensible region in addition to the very short PEVK. Previous studies have demonstrated extension of this region by antibody labeling of cells with different sarcomere lengths (Helmes *et al.*, 1999; Linke *et al.*, 1999; Trombitas *et al.*, 2000). Recent work has also shown that the N2B unique region elongates using single molecule atomic force measurements (Watanabe *et al.*, 2002). The large divergence in sequence identity in this region was surprising. The amino acid sequence of the rat was only 51% identical with the human (determined by Blast comparisons), while the identity between dog and human was 67%. Comparisons of the individual Ig and PEVK exons of rat to those from human cardiac titin ranged between 85 and 95% identity throughout the approximately 16 kb of sequence reported here. Dog and human titin comparisons yielded similar homologies. Even the N2A unique regions of dog and rat were 94% identical in amino acid sequence with human. The only other mammalian titin sequence in the database that contains the N2B unique region is from rabbit, and it just contains the latter half of this region (Freiburg *et al.*, 2000; accession number Y14853). It too has low homology with human or rat titin N2B unique sequences (57 and 63% identity respectively). The rabbit sequence also lacked deletions in the positions near the carboxyl end of the rat N2B unique region. We examined whether the apparent rat mutation rate was also high in the Ig domains 24, 25, and 26 found in exon 49. The amino acid identities with human titin were 74, 77, and 67% in these domains respectively; such values are somewhat lower than those found in other Ig domains in this region. Thus the apparent mutation rate is much higher in exon 49 than in other parts of the titin sequence in the exon range between 44 and 225.

If lower mutation rates occur in protein regions with essential structure or function, then it is logical that the N2B unique region has several segments of amino acid sequence that are not essential. Using the same sort of arguments regarding the PEVK, Ig domains, and the N2A unique regions, each of these latter regions appears to have resisted large mutation changes and have sequences important to their function. The PPAK repeats (Greaser, 2001) were found to have higher identity between species than the poly glutamic acid regions, suggesting that the former have a more important structural role. If a shorter N2B unique region were a benefit physiologically to the rat, the shortening of this domain from the 572 residues in human to 526 would be tolerated. The fact that the segments deleted are scattered through the rat sequence (Figure 9) also suggests that this protein domain may have little ordered structure.

A second possible factor that may affect passive tension is the length of the PEVK domain. A longer segment should lead to less passive tension at a given sarcomere length. The rat (N33) and human (C38) clones sequenced in the current study that contained a full-length PEVK segment of the N2BA type had different lengths. The human N2BA PEVK was 908 amino acid residues in length (measured from the beginning of exon 112 through the sequence...QLKAV in exon 225). In contrast the rat sequence was only 525 amino acids long. The dog sequences varied from 703 to 900 amino acids in length. There is apparently more than one type of human N2BA type of PEVK isoform, since the PEVK length has been quoted as being approximately 600 residues (Freiburg *et al.*, 2000). Clearly more information is necessary to determine the types and properties of the differing N2BA PEVK isoforms as well as their proportions if more than one type is present.

A limitation of the current studies is the lack of quantitative information about the amount of message of the various types. Products of regular PCR are often biased toward shorter amplified cDNA pieces. Thus one cannot conclude that greater PCR band intensity indicates more message. We therefore cannot conclude anything about the relative proportions of the short and long form of the dog N2B PEVK (Figure 2). Also the rat N2BA isoforms, although present as message in the ventricles, may still be at such low concentrations that they have little effect on passive tension. However, the shorter N2B unique region in rats, since it is found in all N2B and N2BA isoforms, would likely give rise to greater passive tension at a given sarcomere length than titin from humans or dogs even if all other isoform ratios were similar.

In summary, multiple types of changes in titin can modulate the passive tension differences between species. Different ratios of longer and shorter titin isoforms of the N2B and N2BA types can be expressed, but there may also be differences in the length of the N2B unique region and changes in the average length of the N2BA PEVK segment.

Acknowledgements

This work was supported by the College of Agricultural and Life Sciences, University of Wisconsin-Madison and by grants from the National Institutes of Health (HL47053; HL62466).

References

Altschul SF, Gish W, Miller W, Meyers EW and Lipman DJ (1990) Basic local alignment search tool. *J Mol Biol* **215:** 403–410.

Bang ML, Centner T, Fornoff F, Geach AJ, Gotthardt M, McNabb M, Witt CC, Labeit D, Gregorio CC, Granzier H and Labeit S (2001) The complete gene sequence of titin, expression of an unusual approximately 700-kDa titin isoform, and its interaction with obscurin identify a novel Z-line to I-band linking system. *Circ Res* **89:** 1065–1072.

Cazorla O, Freiburg A, Helmes M, Centner T, McNabb M, Wu Y, Trombitas K, Labeit S and Granzier H (2000) Differential expression of cardiac titin isoforms and modulation of cellular stiffness. *Circ Res* **86:** 59–67.

Chomczynski P (1993) A reagent for the single-step simultaneous isolation of RNA, DNA and protein from cell and tissue samples. *Bio Techniques* **15:** 532–537.

Fabiato A and Fabiato F (1978) Myofilament-generated tension oscillations during partial calcium activation and activation dependence of the sarcomere length-tension relation of skinned cardiac cells. *J Gen Physiol* **72:** 667–699.

Freiburg A, Trombitas K, Hell W, Cazorla O, Fougerousse F, Centner T, Kolmerer B, Witt C, Beckmann JS, Gregorio CC, Granzier H and Labeit S (2000) Series of exon-skipping events in the elastic spring region of titin as the structural basis for myofibrillar elastic diversity. *Circ Res* **86:** 1114–1121.

Funatsu T, Higuchi H and Ishiwata S (1990) Elastic filaments in skeletal muscle revealed by selective removal of thin filaments with plasma gelsolin. *J Cell Biol* **110:** 53–62.

Furst DO, Osborn M, Nave R and Weber K (1988) The organization of the titin filaments in the half-sarcomere revealed by monoclonal antibodies in immunoelectron microscopy: a map of ten nonrepetitive epitopes starting at the Z-line extends close to the M-line. *J Cell Biol* **106:** 1563–1572.

Granzier HL and Irving TC (1995) Passive tension in cardiac muscle: contribution of collagen, titin, microtubules, and intermediate filaments. *Biophys J* **68:** 1027–1044.

Granzier H and Labeit S (2002) Cardiac titin: an adjustable multi-functional spring. *J Physiol (Lond)* **541:** 335–342.

Greaser M (2001) Identification of new repeating motifs in titin. *Proteins* **43:** 145–149.

Gregorio CC, Granzier H, Sorimachi H and Labeit S (1999) Muscle assembly: a titanic achievement? *Curr Opin Cell Biol* **11:** 18–25.

Helmes M, Trombitas K, Centner T, Kellermayer M, Labeit S, Linke WA and Granzier H (1999) Mechanically driven contour-length adjustment in rat cardiac titin's unique N2B sequence: titin is an adjustable spring. *Circ Res* **84:** 1339–1352.

Higgins DG, Labeit S, Gautel M and Gibson TJ (1994) The evolution of titin and related giant muscle proteins. *J Mol Evol* **38:** 395–404.

Horowits R and Podolsky RJ (1987) The positional stability of thick filaments in activated skeletal muscle depends on sarcomere length: evidence for the role of titin filaments. *J Cell Biol* **105:** 2217–2223.

Labeit S and Kolmerer B (1995) Titins: giant proteins in charge of muscle ultrastructure and elasticity. *Science* **270:** 293–296.

Linke WA and Granzier H (1998) A spring tale: new facts on titin elasticity. *Biophys J* **75:** 2613–2614.

Linke WA, Ivemeyer M, Olivieri N, Kolmerer B, Ruegg JC and Labeit S (1996) Towards a molecular understanding of the elasticity of titin. *J Mol Biol* **261:** 62–71.

Linke WA, Rudy DE, Centner T, Gautel M, Witt C, Labeit S and Gregorio CC (1999) I-band titin in cardiac muscle is a three-element molecular spring and is critical for maintaining thin filament structure. *J Cell Biol* **146:** 631–644.

Marck C (1988) 'DNA Strider': a 'C' program for the fast analysis of DNA and protein sequences on the Apple Macintosh family of computers. *Nucleic Acids Res* **16:** 1829–1836.

Maruyama K (1976) Connectin, an elastic protein from myofibrils. *J Biochem* (*Tokyo*) **80:** 405–407.

Maruyama K (1997) Connectin/titin, a giant elastic protein of muscle. *FASEB J* **11:** 341–345.

Tatusova TA and Madden TL (1999) BLAST 2 Sequences, a new tool for comparing protein and nucleotide sequences. *FEMS Microbiol Lett* **174:** 247–250.

Trinick J and Tskhovrebova L (1999) Titin: a molecular control freak. *Trends Cell Biol* **9:** 377–380.

Trombitas K, Freiburg A, Centner T, Labeit S and Granzier H (1999) Molecular dissection of N2B cardiac titin's extensibility. *Biophys J* **77:** 3189–3196.

Trombitas K, Greaser M, Labeit S, Jin JP, Kellermayer M, Helmes M and Granzier H (1998) Titin extensibility in situ: entropic elasticity of permanently folded and permanently unfolded molecular segments. *J Cell Biol* **140:** 853–859.

Trombitas K, Redkar A, Centner T, Wu Y, Labeit S and Granzier H (2000) Extensibility of isoforms of cardiac titin: variation in contour length of molecular subsegments provides a basis for cellular passive stiffness diversity. *Biophys J* **79:** 3226–3234.

Trombitas K, Wu Y, Labeit D, Labeit S and Granzier H (2001) Cardiac titin isoforms are coexpressed in the half-sarcomere and extend independently. *Am J Physiol Heart Circ Physiol* **281:** H1793–H1799.

Wang K (1996) Titin/connectin and nebulin: giant protein rulers of muscle structure and function. *Adv Biophys* **33:** 123–134.

Wang K, McClure J and Tu A (1979) Titin: major myofibrillar components of striated muscle. *Proc Natl Acad Sci USA* **76:** 3698–3702.

Wang K, McCarter R, Wright J, Beverly J and Ramirez-Mitchell R (1991) Regulation of skeletal muscle stiffness and elasticity by titin isoforms: a test of the segmental extension model of resting tension. *Proc Natl Acad Sci USA* **88:** 7101–7105.

Watanabe K, Nair P, Labeit D, Kellermayer MS, Greaser M, Labeit S and Granzier H (2002) Molecular mechanics of cardiac titin's PEVK and N2B spring elements. *J Biol Chem* **277:** 11549–11558.

Witt CC, Olivieri N, Centner T, Kolmerer B, Millevoi S, Morell J, Labeit D, Labeit S, Jockusch H and Pastore A (1998) A survey of the primary structure and the interspecies conservation of I-band titin's elastic elements in vertebrates. *J Struct Biol* **122:** 206–215.

Wu Y, Cazorla O, Labeit D, Labeit S and Granzier H (2000) Changes in titin and collagen underlie diastolic stiffness diversity of cardiac muscle. *J Mol Cell Cardiol* **32:** 2151–2162.

Journal of Muscle Research and Cell Motility **23**: 483–497, 2002.

Cardiac titin: molecular basis of elasticity and cellular contribution to elastic and viscous stiffness components in myocardium

WOLFGANG A. LINKE[1,*] and JULIO M. FERNANDEZ[2]
[1]*Institute of Physiology and Pathophysiology, University of Heidelberg, Im Neuenheimer Feld 326, D-69120 Heidelberg, Germany;* [2]*Department of Biological Sciences, Columbia University, New York, NY, USA*

Abstract

Myocardium resists the inflow of blood during diastole through stretch-dependent generation of passive tension. Earlier we proposed that this tension is mainly due to collagen stiffness at degrees of stretch corresponding to sarcomere lengths (SLS) ≥2.2 μm, but at shorter lengths, is principally determined by the giant sarcomere protein titin. Myocardial passive force consists of stretch-velocity-sensitive (viscous/viscoelastic) and velocity-insensitive (elastic) components; these force components are seen also in isolated cardiac myofibrils or skinned cells devoid of collagen. Here we examine the cellular/myofibrillar origins of passive force and describe the contribution of titin, or interactions involving titin, to individual passive-force components. We construct force–extension relationships for the four distinct elastic regions of cardiac titin, using results of *in situ* titin segment-extension studies and force measurements on isolated cardiac myofibrils. Then, we compare these relationships with those calculated for each region with the wormlike-chain (WLC) model of entropic polymer elasticity. Parameters used in the WLC calculations were determined experimentally by single-molecule atomic force-microscopy measurements on engineered titin domains. The WLC modelling faithfully predicts the steady-state-force *vs.* extension behavior of all cardiac-titin segments over much of the physiological SL range. Thus, the elastic-force component of cardiac myofibrils can be described in terms of the entropic-spring properties of titin segments. In contrast, entropic elasticity cannot account for the passive-force decay of cardiac myofibrils following quick stretch (stress relaxation). Instead, slower (viscoelastic) components of stress relaxation could be simulated by using a Monte-Carlo approach, in which unfolding of a few immunoglobulin domains per titin molecule explains the force decay. Fast components of stress relaxation (viscous drag) result mainly from interaction between actin and titin filaments; actin extraction of cardiac sarcomeres by gelsolin immediately suppressed the quickly decaying force transients. The combined results reveal the sources of velocity sensitive and insensitive force components of cardiomyofibrils stretched in diastole.

Introduction

A wealth of information collected over the past decade has provided us with exciting new insight in the elasticity of titin (initially described as connectin; Maruyama *et al.*, 1977), the giant muscle protein (Wang, 1996; Trinick and Tskhovrebova, 1999). Among the well-known 'stretchy' biomolecules (Alper, 2002) – many of which are covered in this issue – titin arguably still is a 'newcomer', considering that its amino-acid sequence was elucidated only some 7 years ago (Labeit and Kolmerer, 1995). However, within a relatively short time, the concerted efforts of several research groups have led to a mechanical characterization of titin in such a detail that by now it can be said that the molecular mechanisms of titin elasticity are no secret anymore. The importance of a detailed understanding of titin mechanics is illuminated by the fact that changes in the elastic function of cardiac-titin regions accompany severe heart failure in humans (Neagoe *et al.*, 2002). In myocardium, titin acts together with collagen to determine the passive tension (PT) of the ventricular wall during diastolic filling. A brief update of an earlier report discussing the role of titin and collagen in cardiac muscle (Linke *et al.*, 1994) is provided here (Figure 1).

Titin is a ≥1 μm long and slender protein of 3–3.7 MDa size (Labeit and Kolmerer, 1995) that is present in all sarcomeres. A titin molecule extends over half of a sarcomere from the Z-disk to the M-line (cf., Figure 2A; Fürst *et al.*, 1988; Itoh *et al.*, 1988). Only the I-band titin is functionally elastic (except a 100-nm-wide region on either side of the Z-disk, which is tightly associated with the thin filaments; Linke *et al.*, 1997; Trombitas *et al.*, 1997). Like A-band titin (Labeit *et al.*, 1992), the elastic I-band titin region is assembled in a modular fashion (Figure 2A). Many individually folded domains of the immunoglobulin (Ig) type (Politou *et al.*, 1995; Improta *et al.*, 1996) are interspersed with unique sequences (Labeit and Kolmerer, 1995). The composition of I-band titin is regulated by alternative splicing,

*To whom correspondence should be addressed: Tel.: +49-6221-544130; Fax: +49-6221-544049; E-mail: wolfgang.linke@urz.uni-heidelberg.de

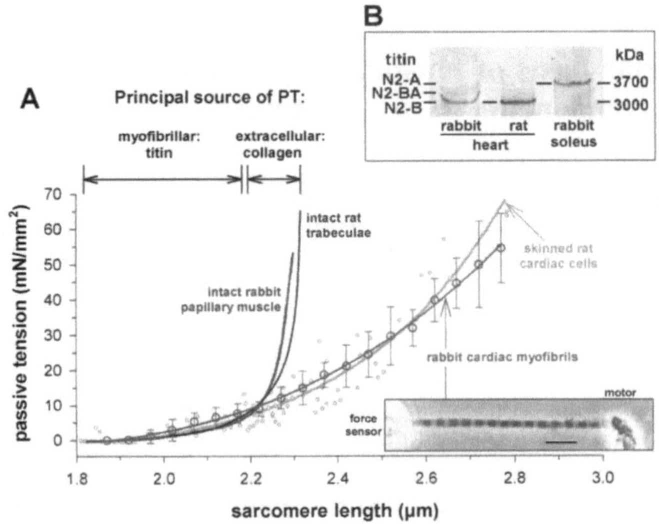

Fig. 1. Sources of PT in various cardiac-muscle preparations stretched under relaxing conditions. (A) Comparison of passive SL–tension curves of intact rabbit (Julian *et al.*, 1976) and intact rat (Kentish *et al.*, 1986) cardiac fibers with those of skinned rat cardiomyocytes (small grey symbols; Weiwad *et al.*, 2000) and isolated rabbit cardiac myofibrils (large symbols and error bars; Linke, 2000). Arrows in upper left corner indicate which protein dominates PT development in the left ventricle at which SL (Linke *et al.*, 1994). (B) 2% SDS–polyacrylamide gel to detect titin isoforms (for methodological description, cf., Neagoe *et al.*, 2002). Both rabbit and rat heart express almost exclusively N2B-titin. Rabbit soleus-titin band (N2A-titin isoform) is shown for comparison.

giving rise to muscle type-specific isoforms (Freiburg *et al.*, 2000). The titin isoform with the shortest I-band region is found in cardiac-muscle sarcomeres (N2B-isoform) and has a molecular mass of 2970 kD (Labeit and Kolmerer, 1995). This N2B-isoform makes up ~70% of all titin isoforms in normal human left ventricle, the remainder being various isoforms of a so-called N2BA type (Neagoe *et al.*, 2002). The elastic I-band region of N2B-titin can be subdivided into four structurally distinct regions (Figure 2): (1) a proximal Ig region containing 15 tandem-Ig domains; (2) a middle N2-B segment that contains a 572-residue amino-acid sequence of unknown structure; (3) a 186 amino-acid-long segment rich in proline (P), glutamate (E), valine (V) and lysine (K) residues, named the PEVK region; and (4) a distal Ig region containing 22 Ig-domain repeats. Whereas the middle N2-B segment is specific to titin in cardiac muscle, alternatively spliced isoforms of titin in other muscle tissues add varying numbers of Ig modules to the proximal Ig region and additional residues to the PEVK domain (Freiburg *et al.*, 2000). For example, human soleus titin (a so-called N2A-isoform, like in all skeletal muscles) contains an additional 53 proximal Ig domains and a PEVK region of 2174 residues (Labeit and Kolmerer, 1995). Thus, by adjusting the length of titin's extensible region, a muscle can vary its elastic properties (Linke *et al.*, 1996b).

Thanks to a combination of two novel approaches, protein engineering and single-molecule force spectroscopy using the atomic force microscope (AFM) (Carrion-Vazquez *et al.*, 2000), we can now study the contribution of the individual building blocks of titin, one by one, to the protein's overall elasticity (Li *et al.*, 2002). A complex mechanical behavior of cardiac titin is apparent, which originates in the structural heterogeneity of this finely tuned molecular spring. In this study, we investigate how the force–extension relationships of the four distinct N2B cardiac-titin regions measured *in situ* (Linke *et al.*, 1999) compare to those calculated with the wormlike-chain (WLC) model of entropic polymer elasticity, which has been successfully applied to explain the elastic behavior of single titin molecules (Kellermayer *et al.*, 1997; Rief *et al.*, 1997; Tskhovrebova *et al.*, 1997). Importantly, the parameters established by AFM mechanics on engineered titin domains (Li *et al.*, 2002) could be used to feed the WLC model. This way we probed the idea that entropic (WLC-like) elasticity of the individual titin segments underlies the elastic behavior of whole cardiac I-band titin.

Another aspect is that cardiac muscle works under nonequilibrium conditions – mammalian heart is stretched (and released) one or many times per second. Diastolic force development is not purely elastic, but has long been known to include stretch velocity-sensitive components of viscous and/or viscoelastic nature (Noble, 1977; Chiu *et al.*, 1982; de Tombe and ter Keurs, 1992; Bartoo *et al.*, 1997). The question of how titin might contribute to these force components, was another focus of the present analysis. Specifically, we asked whether titin itself, or interactions involving the elastic titin segment (Kulke *et al.*, 2001a), could play a role for the viscous force decay (stress relaxation) following stretch of cardiac myofibrils. Altogether, we provide a detailed characterization of the molecular events within cardiac myofibrils relevant to the elastic and viscoelastic behavior of myocardium in diastole.

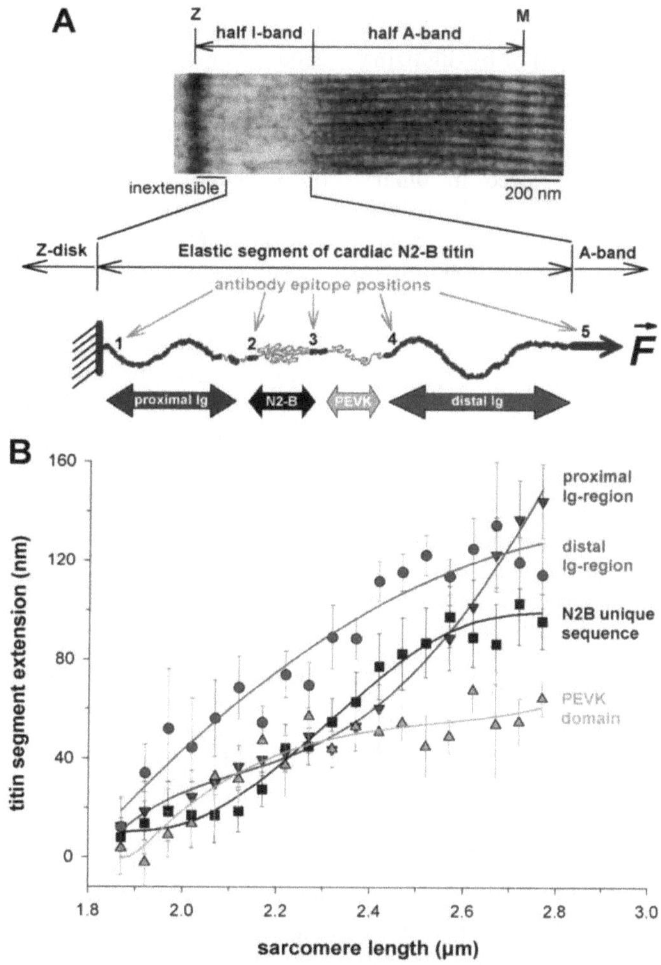

Fig. 2. Extension of cardiac I-band titin (N2B-isoform) *in situ*. (A) Four distinct elastic regions make up the extensible titin segment in each half I-band. Top picture shows electron micrograph of a half-sarcomere. The scheme below also indicates the epitope positions of titin-specific antibodies (1–5) to sites flanking the four elastic regions; these antibodies were used to measure the stretch-dependent extension of titin regions (Linke *et al.*, 1999). (B) Relationship between titin-segment extension and SL, compiled from the original results of immunoelectron/immunofluorescence microscopical analyses of rabbit cardiac sarcomeres (Linke *et al.*, 1999). Symbols show mean ± error estimate; fit curves are third-order regressions.

Materials and methods

Myofibril mechanics

Myofibrils were isolated from freshly excised rabbit cardiac muscle as described (Linke *et al.*, 1996b). Briefly, thin muscle strips were dissected, tied to thin glass rods and skinned in buffer solution ('rigor', composed of (in mM): KCl 75, Tris 10, $MgCl_2$ 2, EGTA 2, pH 7.1) containing 0.5% Triton X-100 for ≥4 h. The skinned strips were minced and homogenized in rigor buffer. All steps were performed at 4°C. A drop of the suspension was placed on a cover slip on the stage of an inverted microscope (Zeiss Axiovert 135). Glass microneedles attached to a piezoelectric micromotor (Physik Instrumente, Waldbrunn, Germany) and a home-built force transducer (sensitivity, ~5 nN, resonant frequency, ~700 Hz), respectively, were used to pick up single myofibrils or small bundles containing 2–4 myofibrils. To firmly anchor the specimen ends, the needle tips were coated with a silicone adhesive in a 1:1 (v/v) mixture of Dow Corning 3145 RTV and 3140 RTV. Water-hydraulic micromanipulators (Narishige, Japan) were used to control the position of both needles. Experiments were performed at room temperature in relaxing solution of 200 mM ionic strength, pH 7.1. Relaxing solution (for composition, see Linke *et al.*, 1997) contained 20 mM 2,3-butanedione monoxime to suppress any, possibly remaining, contractile activity. All solutions were supplemented with the protease inhibitor leupeptin to minimize titin degradation (Linke *et al.*, 1998b). In some experiments, a Ca^{2+}-independent gelsolin fragment (kindly provided by Dr H. Hinssen) was added to the relaxing buffer (final concentration, 0.2 mg/ml) to extract actin (Linke *et al.*, 1997).

Force data collection and motor control were done with a PC, data acquisition (DAQ) board (PCI-MIO-16-E1, National Instruments, Austin, TX) and custom-written LABVIEW software (Linke *et al.*, 1998a, b). Sarcomere length (SL) was measured either with a color

CCD camera (Sony) and frame grabber board including image processing software (Scion Image) or by digitizing and analyzing the myofibril image, using a 2048-element linear photodiode array (Sony), the PC, DAQ board, and LABVIEW algorithms. Force sampling rate typically was 5 kHz. Force data were stored in binary format and were median-filtered off-line. If it was desirable to compare force data from different experiments, the force was related to cross-sectional area inferred from the diameter of the specimens as described (Linke *et al.*, 1997).

Analysis of the stretch force-dependent in situ extension of cardiac titin

Immunoelectron and immunofluorescence microscopy with monoclonal or polyclonal antibodies flanking the four distinct regions of the elastic I-band titin (see Figure 2A) was used previously to establish the extensibility of titin in rabbit cardiac-muscle sarcomeres (Linke *et al.*, 1999). Here, the technical names of the antibodies (T12, I17, I18, I20/22, MIR) were replaced for simplicity by numbers 1 to 5, with 1 being closest to the Z-disk and 5 being located at the A-band/I-band junction (see Figure 2A). For each antibody type, the epitope-mobility data obtained over a range of SLs from 1.8 to 2.8 μm was pooled in SL bins of 50 nm. For each SL bin, the extension of a given titin segment was measured as the distance flanked by two nearest antibody epitopes: proximal Ig region, epitope 1 to epitope 2; N2B unique sequence, 2 to 3; PEVK, 3 to 4; distal Ig region, 4 to 5. Titin segment extension was then plotted against SL. The corresponding stretch force, *F*, was determined from mechanical recordings of the (steady-state) PT of isolated rabbit cardiac myofibrils (Linke, 2000).

AFM of engineered titin polyproteins

Single-molecule AFM has been described elsewhere (e.g., Carrion-Vazquez *et al.*, 2000; Fisher *et al.*, 2000). The cantilevers of the force-measuring unit are standard Si_3N_4 cantilevers from Digital Instruments, Santa Barbara, CA (spring constant, 100 mN/m) or TM Microscopes, Sunnyvale, CA (spring constant, 12 mN/m). Cantilevers were calibrated in solution using the equipartition theorem. AFM experiments described in this study used cloned polyproteins containing either eight identical titin-Ig domains, $I27_8$ (Carrion-Vazquez *et al.*, 1999) or I27 domains interspersed with cardiac PEVK-domains, $(I27\text{-}PEVK)_3$ (Li *et al.*, 2001) or a construct of the type ($I27_3$-N2B-$I27_3$) (Li *et al.*, 2002), where N2B is the 572-residue human cardiac N2B unique sequence. Engineering of these polyproteins has been described in detail (Carrion-Vazquez *et al.*, 2000; Li *et al.*, 2000, 2001). We note that I27 refers to the original nomenclature of titin domains proposed by Labeit and Kolmerer (1995). Following the discovery of additional Ig-domains in the human gene sequence of titin, the numbering of I27 was changed to I91 (Freiburg *et al.*, 2000; Bang *et al.*, 2001). In a typical AFM stretch experiment, 3–10 μl of protein sample concentrated at 10–100 μg/ml were deposited onto freshly evaporated gold coverslips, and the protein was allowed to adsorb onto the gold surface. Force–extension measurements were carried out in PBS buffer at an ionic strength of 200–300 mM.

Monte Carlo simulations

The time course of stress relaxation after quick stretch of cardiac myofibrils was reproduced by a Monte Carlo (MC) technique based on entropic elasticity theory (WLC model) and the kinetic parameters of titin-Ig domain unfolding/refolding obtained in AFM stretch experiments with single polyproteins (Carrion-Vazquez *et al.*, 1999). Details of the MC approach used by us were described previously (Rief *et al.*, 1998). The WLC model of entropic elasticity (Bustamante *et al.*, 1994; Marko and Siggia, 1995) predicts the relationship between the relative extension of a polymer (z/L) and the entropic restoring force (f) through

$$f = \left(\frac{k_B T}{A}\right)\left[\frac{1}{4(1-z/L)^2} - \frac{1}{4} + \frac{z}{L}\right] \tag{1}$$

where k_B is the Boltzmann constant, T is absolute temperature (300 K), A is the persistence length (a measure of the bending rigidity of the chain), z is the end-to-end length, and L is the chain's contour length.

In these MC simulations, the force decay following a stretch was considered to be due to the unfolding of titin-Ig domains (Minajeva *et al.*, 2001). As refolding is not observed in the presence of a force, the kinetics of domain refolding have no impact on the results of the simulation (Oberhauser *et al.*, 1998; Carrion-Vazquez *et al.*, 1999). However, external force greatly affects domain unfolding. The probability of observing the unfolding of any module P_u was calculated as

$$P_u = (k_u^0 \Delta t)(\exp(f \Delta x_u / k_B T)) \tag{2}$$

where k_u^0 is the Ig-domain unfolding rate at zero force, Δt is the polling interval, f is applied force, and Δx_u is the unfolding distance (Carrion-Vazquez *et al.*, 1999). The simulations were performed in a LABVIEW (National Instruments, Austin, TX) software environment. A feature of the custom-written software allowed selection of a desired number of iterations.

Results

Titin and collagen are the main determinants of myocardial stiffness

Within the normal 'working' SL range of the heart (1.7–2.3 μm SL; Allen and Kentish, 1985), two main structural elements have been identified, which are principally important for the elastic properties of the ventricular wall: titin and collagen (Linke *et al.*, 1994). Experimental evidence showed that, when the passive length–tension curves of single isolated cardiac myofibrils are compared to those of intact multicellular cardiac preparations, such as trabeculae or papillary muscles (Figure 1A), the curves are similar up to ~2.2 μm SL (Linke *et al.*, 1994). This finding holds true also when it is taken into account that myofibrils occupy only about 50–70% of a cardiomyocyte's cross-sectional area. Furthermore, the passive SL–tension curves of single cardiac myofibrils are readily comparable to those of chemically skinned (demembranated) cardiac cells (Figure 1A; see Weiwad *et al.*, 2000), indicating that most or all of the structures responsible for the cell's PT development lie inside the sarcomeres. Thus, it was proposed that, up to ~2.2 μm SL, the dominant factor for myocardial PT is titin, whereas at SL >2.2 μm, the high stiffness of the ventricular wall is caused mainly by collagen (Linke *et al.*, 1994). Intermediate filaments, such as desmin, contribute nothing or very little to PT development (Wu *et al.*, 2000; Anderson *et al.*, 2002).

A comparison of PT between different cardiac preparations must consider that species differ in their expression of cardiac-titin isoforms (Freiburg *et al.*, 2000). For example, rat and rabbit heart can be compared, as they express almost exclusively the shortest titin isoform, the N2B variant (Figure 1B). Low-percentage SDS–PAGE performed on left ventricular wall tissue from these species detects only trace amounts of the larger N2BA-isoform, with a tendency for rabbit heart to show a slightly higher incidence/degree of N2BA-titin expression than rat heart. However, the passive SL–tension curves of rat and rabbit cardiac myofibrils are virtually indistinguishable (Linke *et al.*, 1996b; and Figure 1A). Thus, in both species, titin determines most of the PT over much of the heart's physiological range of extensions, whereas extracellular elements (collagen) are responsible for the steep increase in PT towards the high end of this range. It is likely that this conclusion can also be extended to many other mammalian species. Below we describe our current view of the molecular mechanisms underlying titin elasticity/viscoelasticity in mammalian heart.

In situ *extensibility of cardiac titin segments*

The functionally elastic I-band segment of cardiac titin contains four distinct extensible regions: proximal Ig-domain region, N2B-unique sequence, PEVK domain, and distal Ig-domain region (Figure 2A). In a previous work (Linke *et al.*, 1999), we used titin-specific antibodies against epitopes flanking all four regions (numbered 1–5 in Figure 2A) and measured the stretch-dependent extension of each segment by immunoelectron microscopy and immunofluorescence microscopy on rabbit cardiac sarcomeres. A compilation of these data is presented in Figure 2B, showing titin-segment extension plotted against SL. It appears that each titin segment extends in a nonlinear fashion. Also, Ig-domain regions begin to extend before titin's unique sequences. On the other hand, the PEVK domain already stretches before the N2B-unique sequence starts to extend. A differential segment-extension behavior was recognized previously also for skeletal muscle titin (Gautel and Goulding, 1996; Linke *et al.*, 1996b). Why do titin segments exhibit these differences in their extensibility?

Single-molecule AFM of titin polyproteins

This question can be answered by performing mechanical measurements directly on titin's molecular subsegments (Li *et al.*, 2002). By combining protein engineering and single-molecule AFM approaches (Carrion-Vazquez *et al.*, 2000; Fisher *et al.*, 2000), interpretational limitations associated with previous single-molecule mechanical studies on whole proteins or large protein fragments (Kellermayer *et al.*, 1997; Tskhovrebova *et al.*, 1997) can be overcome. For instance, it is possible to study the mechanical properties of individual Ig-domains contained within the elastic I-band titin segment (Figure 3A). An example is shown in Figure 3B, C: a single titin-Ig domain (here, I27 according to nomenclature by Labeit and Kolmerer, 1995; I91 according to nomenclature by Freiburg *et al.*, 2000) from the elastic titin region is expressed and multiplied to obtain a polyprotein containing eight serially linked I27 (=I91) modules, $I27_8$ (Figure 3B). Stretching this polyprotein in the AFM gives a characteristic 'fingerprint' (Figure 3C): (1) the force–extension curve shows a sawtooth pattern, in which each peak corresponds to the unfolding of one Ig module; (2) the unfolding force of each peak is ~200 pN (Li *et al.*, 2000); and (3) the peaks are spaced at a constant interval. The increase in contour length of the polyprotein when one Ig domain unfolds can be measured with the WLC model of entropic elasticity (Equation (1)). Fitting this model to each sawtooth peak in the measured force trace reveals that each unfolding event increases the contour length of the polyprotein by 28.1 nm (Figure 3C, red curves). Similar experiments have provided us with important new information about the mechanical stability of several other distal and proximal Ig-domains (Li *et al.*, 2002).

N2B unique sequence

A novel approach was used to mechanically study the unique sequences in titin, N2B and PEVK (Li *et al.*, 2002). Here, the protein fragment of interest is contained within an engineered polyprotein also containing a certain number of I27 (=I91) Ig domains. In the case of the N2B unique sequence, a heteropolyprotein of the type ($I27_3$-N2B-$I27_3$) was generated (Figure 3D, inset). The characteristic 'fingerprinting' of Ig domains when the polyprotein is stretched in the AFM setup, is then used to distinguish recordings obtained from multiple polyproteins from those obtained from a single

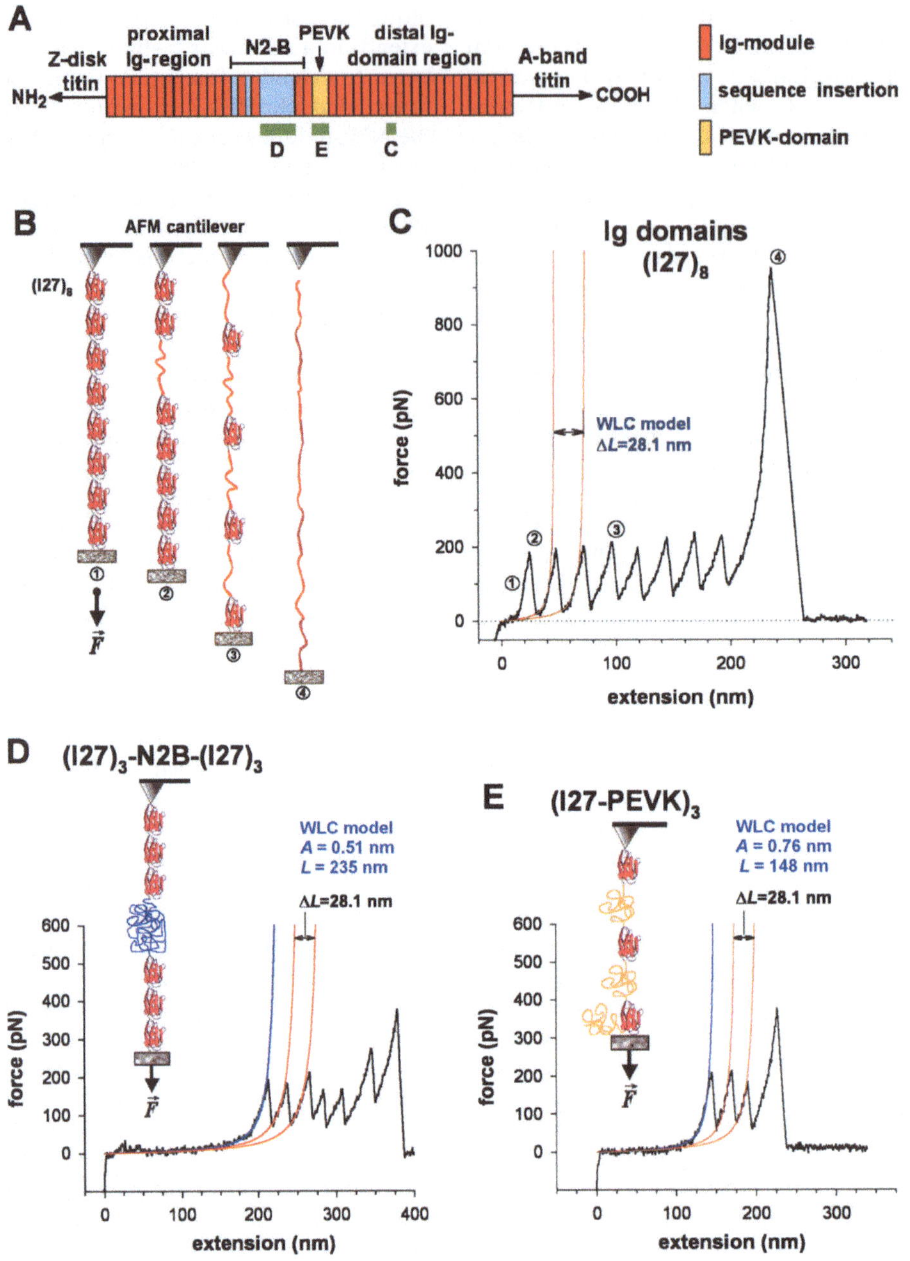

Fig. 3. Single-molecule AFM of engineered titin domains. (A) Domain architecture of the extensible I-band segment of N2B cardiac titin. Thick green bars below scheme indicate location of titin domains studied by AFM; capital letters refer to the panels in this figure showing AFM results for the particular domain. (B) Ig-module I27 (=I91 in nomenclature by Freiburg *et al.*, 2000) can be used as a 'gold standard' in the AFM mechanical measurements of titin domains. Shown is a schematic of how an engineered $I27_8$ ($=I91_8$) polyprotein is attached to the AFM cantilever and a gold-coated coverslip surface and is stretched by a force, *F*. Stages 1–4 correspond to the numbers above the force trace in (C). (C) Typical sawtooth pattern of the force–extension curve of $I27_8$ ($=I91_8$). Force peaks near 200 pN indicate Ig-domain unfolding events. The last peak (4) corresponds to rupture of the polymer from the sites of attachment. The red lines are WLC fits (Levenberg–Marquardt algorithm) based on Equation (1), to the second and third force peak. One unfolding event is calculated to increase the contour length, *L*, of the polyprotein by exactly 28.1 nm. (D) AFM force spectroscopy of a cloned heteropolyprotein consisting of titin's N2B unique sequence flanked by three I27 (=91) Ig domains on either side (Li *et al.*, 2002). The force trace shows a long, featureless, initial region corresponding to N2B extension. The WLC model (Equation (1)) measures the contour length and persistence length of N2B (blue curve). The first six force peaks (near 200 pN) indicate unfolding of all Ig domains within this tether. The spacing between force peaks is a constant 28.1 nm, indicating that a single molecule was stretched. (E) Representative result of AFM studies with engineered polyproteins of the type $(I27\text{-}PEVK)_3$, where 'PEVK' is the 186-residue cardiac PEVK domain of titin's N2B-isoform (Li *et al.*, 2001). Again, the WLC model was used to parameterize PEVK extension (blue curve) and to measure the spacing between Ig-domain unfolding peaks (red curves). Here, the single-molecule tether contained all three I27 Ig modules and two of the three PEVK domains.

polyprotein (for details, see Li *et al.*, 2001, 2002). Only recordings obtained from single polyproteins are included in the statistical analysis. A typical force–extension curve of ($I27_3$-N2B-$I27_3$) is shown in Figure 3D. The featureless force trace before the first unfolding peak corresponds to the extension of the N2B-unique sequence. In this example, the tether contained all six I27 Ig modules (six unfolding peaks spaced at 28.1 nm) plus the intervening N2B-sequence, which according to WLC calculation (blue curve in Figure 3D) had a contour length, L, of 235 nm. This value is close to the contour length of a fully extended N2B-unique sequence (as expected from the amino-acid sequence; Labeit and Kolmerer, 1995), i.e., the number of residues (572) multiplied by the average, maximally possible, spacing of ~0.4 nm. The average contour length of the N2B-unique sequence, measured from 48 recordings like the one shown in Figure 3D, was 232 nm. The persistence length, A, of the N2B-unique sequence, calculated by the WLC model, was 0.51 nm in the example of Figure 3D and 0.66 nm on average. We found a relatively narrow distribution of persistence lengths, from ~0.4 to 1.3 nm (Li *et al.*, 2002). Thus, it appears that the N2B-unique sequence is a random coil behaving like a purely entropic WLC.

PEVK-domain

Similar AFM experiments were performed with a heteropolyprotein of the type $(I27\text{-}PEVK)_3$ (Li *et al.*, 2001, 2002). The PEVK-domains contained within this construct (Figure 3E, inset) have a length of 186 nm (=PEVK domain of the cardiac N2B isoform; Freiburg *et al.*, 2000) and are interspersed with Ig I27 (=I91) domains to be used for 'fingerprinting'. Again, only recordings obtained from single polyproteins are included in the statistical analysis. Figure 3E shows a typical force–extension curve of $(I27\text{-}PEVK)_3$. No obvious features could be detected in the force trace before the first unfolding peak, which corresponds to the extension of PEVK-titin. The tether contained three I27 Ig modules (three unfolding peaks spaced at 28.1 nm) and two PEVK-domains, as judged from the contour length, $L = 148$ nm, calculated with the WLC model (blue curve in Figure 3E); half this value equals the expected contour length of a fully extended PEVK domain (186 residues times ~0.4 nm). The average contour length of one cardiac PEVK-domain was ~72 nm; the persistence length, A, was 0.76 nm in the example of Figure 3E and 0.91 nm on average (Li *et al.*, 2001, 2002). However, a relatively wide distribution of persistence lengths, from ~0.5 to 2.5 nm, was found. We suggested earlier that this wide range of persistence lengths could be due to multiple conformations of the PEVK domain (Li *et al.*, 2001). In any case, the cardiac PEVK domain can be considered a polymer chain exhibiting WLC properties.

Force–extension relationship of titin's unique sequences: A comparison of in situ *measurements and WLC calculations based on AFM data*

How does the extensibility of the N2B and PEVK sequences, measured by AFM on engineered titin constructs, compare to that of the native unique sequences in the intact cardiac-muscle sarcomere? To make this comparison, we first need to obtain the force–extension relationships of these cardiac-titin segments *in situ*. The (SL-dependent) extension of PEVK and N2B is taken from the plot in Figure 2B (symbols and fits). The stretch force corresponding to a given *in situ* extension is inferred from passive-force measurements on isolated rabbit cardiac myofibrils. Here, we take the purely elastic force component recorded under steady-state conditions, i.e., 2–3 min following a stretch to a new SL. The PT–SL curve used for this purpose is shown in Figure 1A (large symbols with error bars and fit labelled 'rabbit cardiac myofibrils'). These data explicitely do not include viscous and viscoelastic force components. The values of the measured PT are then scaled down to the level of the single titin molecule by assuming a value of 6×10^9 titins per mm^2 cross-sectional area (Li *et al.*, 2002). We also assume that, at a given SL, the (scaled-down) force measured at the ends of the myofibril equals the force acting on each elastic titin segment. This is a reasonable assumption, given that the individual segments in a titin molecule are connected in series and should experience the same stretch force at any time.

Combining the results of titin-segment extensibility studies and myofibrillar force measurements, we can now plot the *in situ* force–extension relationships of both the PEVK-domain and the N2B-unique sequence (Figure 4, symbols and error estimates). Then, we try to reproduce the force corresponding to a given extension (Figure 4, curves) by applying the WLC model of entropic elasticity (Equation (1)). Note that this is not a fitting procedure, but a comparison of measured and calculated data. However, for the calculations we use parameters established experimentally in atomic force spectroscopy measurements (Figure 3; and Li *et al.*, 2002). The end-to-end length (=extension), z, of each region is taken from the fit curves in Figure 2B.

N2B extensibility

To reproduce the force–extension relationship of the N2B-unique sequence, we feed the WLC model with the parameters $L = 230$ nm and $A = 0.66$ nm. Using these parameters, a striking correlation is seen between measured (Figure 4A, symbols and error estimates) and calculated data (Figure 4A, thick curve), at least for forces up to ~6 pN. Values of $A = 0.33$ nm or $A = 1.32$ nm clearly were insufficient to reproduce the *in situ* data (Figure 4A, thin curves). At forces >6 pN, the scatter in the measured data was high, and useful conclusions could not be drawn. Nevertheless, over much of the extension range studied, *in situ* data and AFM-derived reconstitutions gave identical results. For comparison, at physiologically relevant degrees of extension, the equilibrium (elastic) force per titin molecule may reach no more than 3–4 pN (Li *et al.*, 2002). This fact should be kept in mind when interpreting the graphs in Figures 4 and 5.

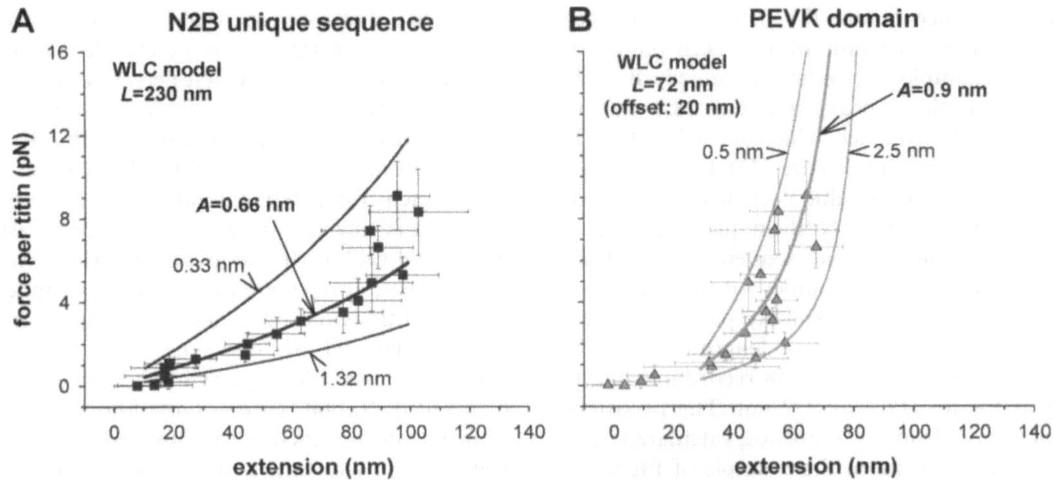

Fig. 4. Reconstitution of the *in situ* force–extension relationships of cardiac titin's unique sequences, based on results obtained in single-molecule AFM measurements. (A) Force-extension data for the N2B-unique sequence of rabbit cardiac sarcomeres (symbols: mean values and error bars). The lines are force–extension curves calculated with the WLC model (Equation (1)), which was fed with a fixed contour length of 230 nm and different persistence lengths (0.66; 0.33; and 1.32 nm). $A = 0.66$ nm produced best results. (B) Force-extension data for the PEVK-domain *in situ* (symbols: mean values and error bars). Lines are force–extension curves calculated with the WLC model (Equation (1)), which was fed with the parameters for persistence length (average: 0.9 nm; upper and lower extremes: 0.5 and 2.5 nm) and contour length (72 nm) established by AFM work. An offset of 20 nm was added to all extension values of these WLC curves, to account for the fact that the antibody epitopes 3 and 4 (Figure 2A) do not directly flank the PEVK-domain. For further details, see text.

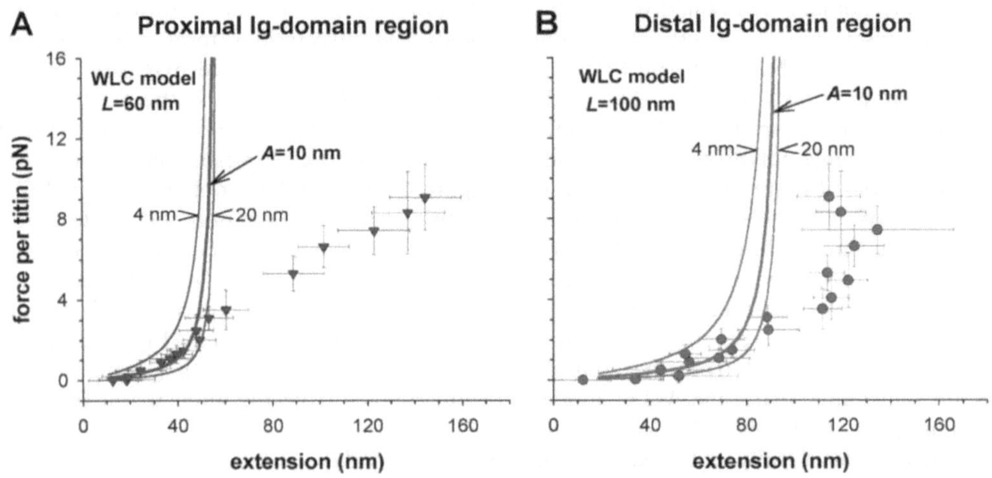

Fig. 5. Comparison of force–extension relationships of Ig-domain regions measured in cardiac sarcomeres with those obtained by WLC modelling. (A) Proximal tandem-Ig region; (B) Distal tandem-Ig-region. *In situ* data (symbols and error bars) were compared to WLC curves (lines) calculated using the indicated parameters for persistence length (A) and contour length (L). Note that physiological stretch forces may not exceed 3–4 pN per titin molecule (Li *et al.*, 2002). For further details, see text.

PEVK extensibility

A description of the PEVK force–extension relationship again uses the parameters previously established for the PEVK-domain in the AFM work: $A = 0.9$ nm and $L = 72$ nm. The thick line in Figure 4B shows that the WLC fit based on these parameters connects the majority of the data points. However, some data points deviate significantly from the modeled curve, especially at forces >4 pN. Using a persistence-length value of 0.5 nm in the WLC model (at constant L) leads to a better description of the high-force data, whereas with $A = 2.5$ nm, reproducibility of measured data is low (Figure 4B, thin lines). However, all data points measured *in situ* fall in between the WLC fit curves obtained with the lowest ($A = 0.5$ nm) and the highest ($A = 2.5$ nm) persistence-length value. Altogether, reconstituting the *in situ* force–extension relationship from AFM results appeared to be feasible.

WLC modelling of in situ *force–extension relationships of Ig-domain regions*

We also try to reproduce the force–extension behavior of the proximal and distal tandem-Ig regions (Figure 5, symbols and error estimates) by using the WLC model. Unfortunately, it is not possible to reliably determine contour-length and persistence-length values for (relatively short) Ig-domain regions with folded Ig modules by single-molecule AFM force spectroscopy (Li *et al.*, 2002). However, the persistence length, A, of an engineered titin construct containing 12 serially linked tandem-Ig modules ($I27_{12}$) was recently measured on electron microscopic images, and was found to be 10 nm (Li *et al.*, 2002). This is similar to values suggested earlier for whole titin (15 nm, Higuchi *et al.*, 1993; or 13.5 nm, Tskhovrebova and Trinick, 2001). We therefore use $A = 10$ nm as a standard in the WLC calculations. Furthermore, the contour length, L, of the proximal tandem-Ig region was taken to be 60 nm, because the region contains 15 tandem-Ig modules (Labeit and Kolmerer, 1995) and one Ig-module spans approximately 4–4.5 nm (Improta *et al.*, 1996). Corresponding values for the distal Ig-domain region were $L = 100$ nm (22 modules). As before, the fit curves in Figure 2B provide the end-to-end length (or extension), z, of each Ig-domain region.

Proximal Ig-region

Using the above values, we find a good correlation between measured and calculated data describing the proximal Ig-segment extension (Figure 5A), up to forces of ~3 pN per titin molecule. At higher forces, the WLC curve continues to rise steeply, whereas the measured force shows a breakpoint and then increases much more shallow. A clear deviation between measured and calculated data is obvious at forces ≥3.5 pN. This deviation most likely is related to an increased probability of unfolding of individual Ig domains, which is not taken into account in the WLC modelling. The correlation between measured and calculated data was not satisfactory when we used persistence-length values of 4 or 20 nm in the WLC model (Figure 5A, thin curves). In contrast, the good reproducibility of the low-force–extension data, using $A = 10$ nm, suggests that the proximal Ig-domain region (with folded modules) may behave as a purely entropic spring up to an extension of ~45 nm.

Distal Ig-region

Here it was difficult to decide which persistence-length value used in the WLC modelling resulted in best reproducibility of the measured data (Figure 5B). With $A = 10$ nm, the majority of data points up to forces of 2–3 pN could be connected (thick curve), but given the substantial scatter in the measured force–extension data, also the WLC curves using $A = 4$ nm or $A = 20$ nm (thin curves) crossed some points. The measured data clearly deviated from those calculated with the WLC model, at forces ≥3.5 pN. However, no obvious 'breakpoint' could be observed, unlike for the proximal Ig-domain region, and there was no levelling off of the measured force–extension relationship even at high forces. Thus, the graph suggests that unfolding of modules is rather unlikely to occur in the distal Ig-domain region.

Is titin involved in determining viscoelastic properties of cardiac myofibrils?

The analyses described above probed the steady-state (equilibrium) elastic properties of nonactivated cardiac myofibrils. However, if the myofibrils are stretched within the 'working' SL range at physiological stretch rates, they exhibit viscoelastic behavior (Figure 6A). This behavior manifests itself in hysteresis during a stretch-release cycle (Linke *et al.*, 1996a) and in stress relaxation (force decay at constant SL). The magnitude of the stress-relaxation response varies with the stretch rate (Bartoo *et al.*, 1997). Also SL is important: at longer SL, the force-decay amplitude is higher, and stress relaxation lasts longer, than at shorter SL (Figure 6A). In the present study, we asked whether the stretch velocity-sensitive force components could be related to some properties of the titin filaments.

Because PEVK and N2B behave like purely entropic springs (Figure 3), any contribution of titin to myofibrillar viscoelasticity must come from the other extensible titin segments, the Ig-domain regions. In particular, viscoelastic behavior could be due to Ig-domain unfolding. Therefore, our approach was to try to reproduce the force response of cardiac myofibrils to a stretch-hold protocol (Figure 6A) with a MC simulation that takes into account the entropic elasticity of titin regions (Equation (1)), but also the unfolding characteristics of Ig domains established by single-molecule AFM work (Li *et al.*, 2002). This kind of simulation has been attempted previously for skeletal myofibrils (Minajeva *et al.*, 2001). A fresh look into this issue is warranted considering the newly acquired information on the mechanical properties of proximal and distal Ig-domains of cardiac titin (Li *et al.*, 2002). For instance, the unfolding rate at zero force, k_u^0, of the mechanically weakest Ig domain characterized so far by AFM work, is $\sim 3.3 \times 10^{-3}\ s^{-1}$ (the I4 domain of the proximal tandem-Ig segment). This value is an order of magnitude higher than the previously reported value for I27 (=I91) from the distal Ig-segment (Carrion-Vazquez *et al.*, 1999). In the MC simulations presented here, we generally use $k_u^0 = 3.3 \times 10^{-3}\ s^{-1}$ to calculate the unfolding probability of an Ig domain, P_u (Equation (2)).

The modeling also considers the presence of the unique sequences in I-band titin, which are modelled as WLC springs with a combined contour length of 300 nm (PEVK = 70 nm; N2B = 230 nm) and a persistence length of 0.9 nm (Li *et al.*, 2002). A fixed value in the simulation is the number of Ig domains within the extensible section of one titin molecule, which is 40 in the N2B-isoform of cardiac titin (Freiburg *et al.*, 2000). For unfolding distance, we use a value of 0.25 nm (Carrion-Vazquez *et al.*, 1999) and for the contour-length gain upon unfolding of one Ig domain, 28.5 nm.

Stress relaxation and Ig-domain unfolding

Figure 6B shows simulated force traces. The stretch amplitudes were 100 nm in both step 1 and step 2, thus

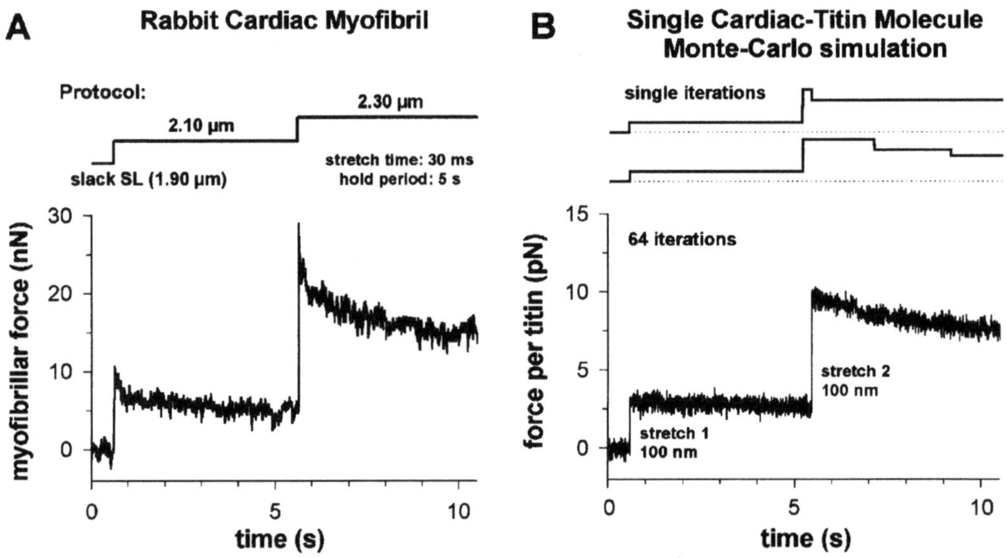

Fig. 6. Stress relaxation of cardiac titin: data measured *in situ* and simulated results using a MC approach. (A) Force response of a small bundle of nonactivated cardiac myofibrils to a stretch-hold protocol of the type shown above the panel. SL is indicated. The force trace is an average of the forces recorded in four consecutive recordings on the same specimen, using the same stretch protocol. (B) MC simulation of 'stress relaxation' of cardiac titin, using Equations (1) and (2), and parameters indicated in the main text. The top traces show two examples of single iterations where both stretch 1 and stretch 2 are 100 nm and each hold period is 5 s. Steps in the second hold period indicate unfolding of titin-Ig domains. The bottom graph shows an average of 64 iterations, to which white noise was added. Note that force relaxation exhibits a similar time course as the slow component of force relaxation in (A).

reproducing the amount of stretch imposed onto I-band titin in sarcomeres extended by $\Delta SL = 0.2$ µm from 1.9 and 2.1 µm SL, respectively (Figure 6A, 'Protocol'). The average force of 64 iterations was calculated, and noise was added to the force trace to make it look more 'real' (Figure 6B). Evidently, the simulated trace does not show strong similarity with the myofibril data. No fast components of force relaxation can be seen in the simulations. However, it appears that the slow components of force relaxation are quantitatively similar in Figure 6A and B. Thus, by feeding the MC simulation with parameters established experimentally in AFM work, part of the stress-relaxation response of cardiac myofibrils is explainable. The results suggest that titin Ig-domain unfolding could underlie the slow component of myofibrillar stress relaxation.

In situ *Ig domain unfolding is a relatively rare event*

To predict how many Ig domains can potentially unfold during stress relaxation, we generated 100 single iterations like the two examples shown in Figure 6B, top. In each iteration, we counted the number of unfolding events, detected as distinct force steps during hold periods. The MC simulation predicted that, over a 5-s-long hold period following a 100 nm/titin step from slack length (simulating a 1.90–2.10 µm SL increase), 0.29 Ig domains per titin molecule unfold on average. Following the second 100 nm/titin step (simulating a 2.10–2.30 µm SL increase), 1.36 Ig domains/titin were predicted to unfold. If the hold time was shortened to 0.5 s (of course, cardiomyocytes *in situ* are not held for long in the stretched state), the values changed to 0.04 (first step-hold) and 0.17 (second step-hold) unfolded Ig domains/molecule. Further, if the hold time was 0.5 s and the step size was modelled to be 200 nm (one step only) – mimicking a more physiological range of motion, from 1.9 to 2.3 µm SL – the MC simulations predicted unfolding of 0.21 Ig domains per titin. Thus, Ig-module unfolding (in the proximal tandem-Ig segment) may become relevant for the viscoelastic behavior of heart muscle when cardiac sarcomeres are stretched to greater physiological SLs.

Fast components of myofibrillar stress relaxation originate in actin–titin interactions

If the fast components of stress relaxation are not due to titin itself (Figure 6), what other factors could be responsible? Recently, we showed that a main determinant of the rapidly decaying viscous force following quick stretch of nonactivated cardiac myofibrils is actin–titin interactions (Kulke *et al.*, 2001a). Isolated rabbit cardiomyofibrils were quickly stretched (within 4 ms) in relaxing buffer from 2.1 to 2.3 µm SL, held in the stretched state, and stress relaxation was measured over a 10-s-long hold period (Figure 7A, left panel). We recorded peak force, as well as steady-state force at the end of the hold period (Figure 7A, panel 'control'). The force decline could be fitted with a three-order exponential-decay function, with decay time constants of 4–5 s, 50–60 ms, and ~30 s (Kulke *et al.*, 2001a). Then, the force decay after stretch was studied in relaxed myofibrils exposed to a Ca^{2+}-independent gelsolin fragment to extract actin. Thin filaments are rapidly removed by gelsolin, except the portion attached to titin in a ~100-nm-wide region adjoining the Z-disk (Linke

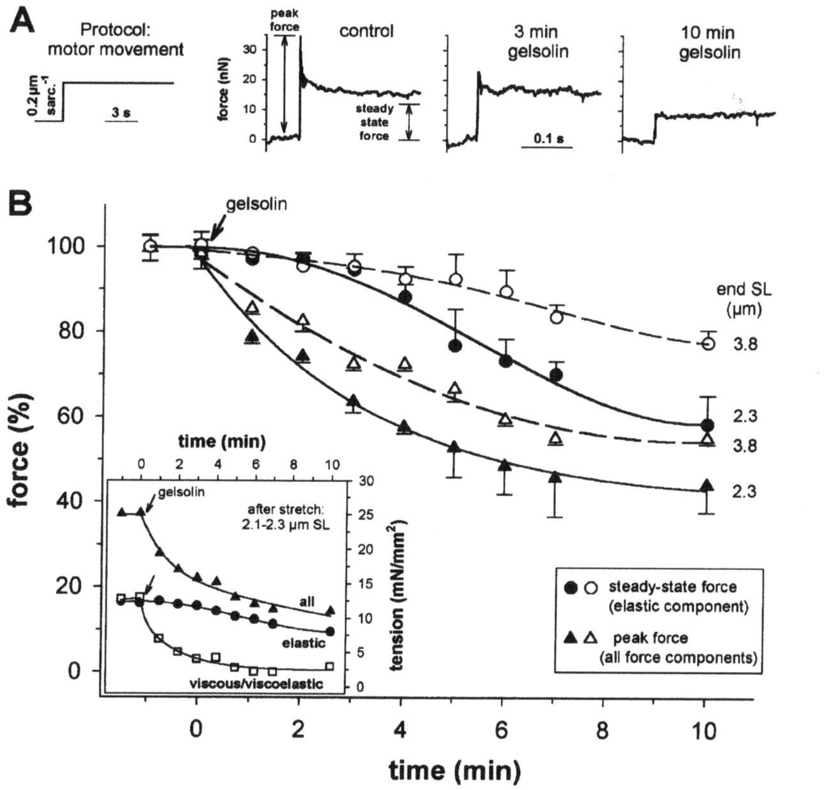

Fig. 7. Stress-relaxation measurements of nonactivated rabbit cardiac myofibrils before and during actin extraction using a Ca^{2+}-independent gelsolin fragment. (A) Motor stretch-hold ramp (stretch completed within 4 ms) and force responses: before actin extraction ('control'), 3 min, and 10 min, respectively, following application of gelsolin. Calibration bar ('0.1 s') applies to all force traces. Note that peak force is greatly decreased after 3-min gelsolin treatment, due to a major drop in the rapidly decaying viscous force component. (B) Comparison of summarized results of stress-relaxation measurements following 4-ms stretches from 2.1 to 2.3 µm SL (filled symbols) or from 3.6 to 3.8 µm SL (open symbols). Data are mean ± SD (n = 6 myofibrils). Curves are third-order regressions (fits to circles) or simple exponential decay functions (fits to triangles). Inset: Example showing absolute force changes related to myofibrillar cross-sectional area. Viscous/viscoelastic force is peak force ('all') minus steady-state force ('elastic').

et al., 1997). Analysis of stress relaxation during gelsolin treatment showed a drop in peak-force amplitude by ~35% within ~3 min (Figure 7A; 7B, filled triangles). During that time period, the velocity-sensitive force components could decrease by as much as 70% (example in Figure 7B, inset, open squares) – attributable almost exclusively to a decrease of the quickly decaying force component (Kulke *et al.*, 2001a). The slower-decay components and steady-state force were affected only >3 min following gelsolin application (Figure 7A, B). Some viscous/viscoelastic force decay remained after 10 min of gelsolin exposure (Figure 7B, inset), and most likely is due to Ig-domain unfolding. However, the results clearly suggest that interactions involving actin and titin play a main role for myofibrillar viscosity.

This conclusion was verified in several control experiments (Kulke *et al.*, 2001a). Most importantly, we tested whether weakly binding crossbridges in the actin–myosin overlap zone might contribute to the stretch-velocity-sensitive forces. Nonactivated cardiac myofibrils were stretched to zero overlap of actin and myosin filaments (3.6 µm SL), and the stretch-hold protocol again was applied before and during actin extraction. Figure 7B shows that results (open symbols) were qualitatively similar to those obtained at shorter SLs (closed symbols). Three minutes of gelsolin treatment decreased peak force by 26–28%, slightly less than the ~35% seen at short SLs. Thus, actin–myosin interactions may contribute to some degree to viscous force decay following quick stretch of relaxed muscle. However, most of the viscous drag originates in actin–titin interactions. Earlier, we have provided evidence that the only elastic I-band-titin region involved in (weak) binding to actin, is the PEVK-domain (Kulke *et al.*, 2001a; Linke *et al.*, 2002).

Discussion

The issues addressed in this study concern the molecular basis of (visco)elasticity in normal myocardium. Based on the results shown, the following questions will be discussed:

(1) How significant is titin, particularly in comparison to collagen, in determining the overall stiffness of the heart in diastole?

(2) What are the molecular mechanisms of elasticity of the individual titin regions that together, provide the

sarcomere with a unique passive length–tension relationship? How and why do the four distinct extensible titin regions differ in their (steady-state) elastic properties?
(3) Do cardiac titin filaments, or interactions involving the elastic titin segment, play a role in determining velocity-sensitive (viscous, viscoelastic) passive force components in myofibrils?

(1) Titin and collagen stiffness in myocardium

Compared to other proteins with known elastic function, cardiac titin was estimated to have a stiffness similar to that of elastin (Urry, 1984), but much lower than collagen (Linke *et al.*, 1994). In rabbit cardiac sarcomeres, predominant expression of the short N2B-titin leads to PT levels >10 times higher than those measured (at the same SL) in psoas-muscle sarcomeres expressing the longer N2A-titin isoform (Linke *et al.*, 1994; Kulke *et al.*, 2001b). Conversely, vertebrate cardiac N2B-titin is about an order of magnitude less stiff than the titin-like proteins present in the I-bands of *Drosophila* indirect flight muscle (Kulke *et al.*, 2001b).

The observation that single rabbit cardiac myofibrils and multicellular cardiac preparations generate about the same passive force per unit cross-sectional area, over much of the physiological SL range of myocardium (approximately 1.7–2.3 μm; Allen and Kentish, 1985), led to a model describing the role of titin and collagen in mammalian heart (Linke *et al.*, 1994). Accordingly, the myofibrillar titin-filament system is responsible for most of the diastolic stiffness of the whole heart at extensions corresponding to ≤2.2 μm SL. The extracellular collagen network becomes more important than titin in determining the PT level (wall stiffness), only at long physiological SLs above ~2.2 μm (Linke *et al.*, 1994). Here, we provided an updated account of our results relevant to this topic (Figure 1). Furthermore, passive force per unit cross-sectional area is independent of the presence of desmin – as found in mechanical measurements on skinned muscle fibers of desmin knockout mice (Anderson *et al.*, 2002) – and intermediate filaments are thus unlikely to make a significant contribution to passive stiffness. Similar conclusions have been drawn by others (Wu *et al.*, 2000). Both titin elasticity (Neagoe *et al.*, 2002) and collagen stiffness (Weber, 1997) can be altered in pathological situations, but it is the increased fibrosis and hence, the dramatic rise in collagen content, which mainly causes the much-elevated end-diastolic pressure levels in failing hearts (Neagoe *et al.*, 2002; Hein *et al.*, 2002). Nevertheless, titin can be considered the main determinant of myocardial passive stiffness in normal healthy hearts.

(2) Molecular basis of cardiac-titin elasticity

Recently we showed that the steady-state (equilibrium) force–extension relationship of nonactivated cardiac myofibrils can be fully reconstituted with a mathematical model that is fed with parameters determined in single-molecule AFM measurements on engineered titin domains (Li *et al.*, 2002). In the present study, we functionally 'dissected' the elastic I-band titin (cardiac N2B isoform) into its four structurally distinct regions (Figure 2) and constructed a force–extension curve for each region (*in situ*) by using previously established experimental data (Linke *et al.*, 1999). We then tested whether the regions can be described as purely entropic springs behaving accoding to WLC theory (Figures 4, 5). We found that within the physiological SL range, all four regions may indeed exhibit entropic elasticity. However, there are important differences in the mechanical behavior of these regions, which are highlighted below.

Ig-domain regions: At low stretch forces, titin's tandem-Ig-regions straighten out (Erickson, 1994), thus being responsible for most of the extensibility of sarcomeres at shorter physiological lengths (Linke *et al.*, 1996b). The Ig-regions essentially are semiflexible polymer chains with a low contour-length to persistence–length ratio of 6:1 to 10:1 (Figure 5). Polymer-elasticity theory predicts that little force is needed for the initial extension (straightening) of such polymer chains; they behave more like a leash than a spring (Erickson, 1997).

At higher stretch forces (>4 pN/titin molecule), the proximal Ig-domain region is more 'stretchable' than the distal Ig-region (Figure 5; also see Gautel *et al.*, 1996, and Linke *et al.*, 1999). Li *et al.* (2002) showed that a main difference between the two tandem-Ig-regions is that the proximal Ig-domains have dramatically higher unfolding probabilities than the distal Ig-domains. The distal Ig-domain region seems to be designed to avoid unfolding of its modules – a property that could be important for the proposed interaction between parallel titin strands in that region (Liversage *et al.*, 2001). In contrast, the proximal Ig-domain region clearly must unfold some of its modules to reach the extensions measured in highly extended cardiac sarcomeres (Linke *et al.*, 1996a). In conclusion, both Ig-domain regions act mainly as entropic springs, but it cannot be ruled out that Ig-domain unfolding in the proximal region contributes to the mechanical behavior of cardiac titin *in situ*. However, Ig-domain unfolding is energy-costly and thus, massive unfolding may not be part of the normal mechanism of titin elasticity. On the other hand, already the unfolding of only one Ig-domain per titin molecule drops force by a great amount (Minajeva *et al.*, 2001). Thus, by (reversibly) unfolding just a few modules, titin's proximal Ig-region can act as a 'shock absorber' to prevent damaging high stretch forces acting on a muscle.

Unique sequences: Both the N2B-PEVK domain and the N2B-unique sequence of cardiac titin behave as entropic WLCs *in vitro* (Figure 3) and apparently *in situ* (Figure 4). If these titin regions are indeed random coils, as proposed (Li *et al.*, 2002), none of them should be involved in contour-length adjustments of titin in the cardiac sarcomere (Helmes *et al.*, 1999). Whether PEVK-elasticity generally is purely entropic or also has an enthalpic component, is not entirely clear. Enthalpic contributions to (mainly) entropic elasticity were proposed earlier for skeletal PEVK-titin (Linke *et al.*, 1998a; Gutierrez-Cruz *et al.*, 2001). Differences in elastic properties between skeletal and cardiac PEVK-titins may be expected, due to differences in the modular structure of these domains (Greaser, 2001). We note that the PEVK-domain of N2B-cardiac titin studied here by AFM, in fact is constitutively expressed in all muscle types (Freiburg *et al.*, 2000). The conclusions reached in the present work may therefore extend to the C-terminal portion of the PEVK sequences of all titin isoforms (Linke *et al.*, 2002). In sum, titin's unique sequences can be viewed in principle as entropic springs, but some modulatory role might be played by 'enthalpic' elasticity mechanisms.

To explain why significant extension of the N2B-unique sequence begins only at longer SLs (Linke *et al.*, 1999), WLC elasticity theory can be used: the force needed to extend a WLC is higher, the greater the chain's ratio between contour length and persistence length. The L:A ratio measured for titin's N2B-unique sequence, is 232:0.66, or ~350:1 (Li *et al.*, 2002). In comparison, the L:A ratio determined for the cardiac PEVK-domain, is 72:0.91, or ~80:1. Therefore, the PEVK-domain will extend at lower forces than the N2B-unique sequence. Even lower forces are needed to extend (straighten) titin's tandem-Ig regions (L:A ratio, 6:1 to 10:1). These considerations provide a theoretical explanation for the differential extension of the four distinct cardiac-titin regions under equilibrium forces.

Finally, it remains to be seen, whether the thermal fluctuations of the entropic titin spring could be constrained significantly in the environment of the sarcomeric filament lattice. Some evidence suggesting that there is little constraint imposed on the Brownian motion of titin, comes from the observation that the steady-state elasticity of relaxed muscle fibers is not influenced by the presence of 4% dextran in the bathing medium (Ranatunga, 2001). As dextran shrinks the lattice spacing (Konhilas *et al.*, 2002), an effect on (equilibrium) elasticity of myofibrils would have been expected if steric hindrance of titin fluctuations were increased.

(3) Cardiac titin and myofibrillar viscoelasticity

If the mechanical properties of cardiac-titin regions were based entirely on entropic mechanisms, the force development upon stretch of titin should not exhibit any dependency on stretch velocity. However, when nonactivated single myofibrils are stretched at different rates, PT shows stretch velocity-sensitive components (Bartoo *et al.*, 1997; Minajeva *et al.*, 2001). This kind of viscous/viscoelastic behavior is seen also in intact or demembranated muscle fibers (de Tombe and ter Keurs, 1992; Wang *et al.*, 1993; Mutungi and Ranatunga, 1996, 1998). Until some time ago, the viscoelasticity of muscle had been attributed largely to the presence of weak interactions between actin and myosin filaments (weakly bound crossbridges; Hill, 1968; Proske and Morgan, 1999). A re-evaluation of this earlier concept became necessary with the general acceptance of titin as a principal passive force-bearing element in the sarcomere. In a previous study, we showed that unfolding of a relatively small number of titin-Ig-domains is sufficient to explain a significant portion of the viscoelastic force decay following stretch of skeletal myofibrils (Minajeva *et al.*, 2001). Here, using new data accumulated in mechanical studies on both cardiac-titin domains and isolated cardiac myofibrils, we made a prediction as to whether or not titin filaments play a role for the stretch velocity-sensitive passive-force components in cardiac muscle.

Results of MC simulations indicated that some viscoelastic force decay of stretched cardiac myofibrils might be due to unfolding of a minor fraction of proximal titin-Ig domains (Figure 6). Importantly, Ig-domain unfolding could explain only the slow phases of stress relaxation, but not the rapidly decaying forces. We caution that the evidence for a contribution of titin domains to slow viscoelastic force decay is only indirect. However, our evidence is based on real, measured AFM data on the unfolding probability of the mechanically weakest proximal Ig-module in cardiac titin (Li *et al.*, 2002). Similarly, in the MC approach, we specifically excluded that unique sequences in titin make a contribution to stress relaxation (PEVK and N2B were modelled as pure WLCs) – an assumption made based on the results of AFM force spectroscopy. In any case, the fact that slow stress relaxation is still seen in thin filament-extracted cardiac myofibrils containing only Z-disks, titin, and thick filaments (Figure 7B, inset), led us to argue that structural rearrangements within titin itself contribute to viscoelasticity of sarcomeres.

Mechanical measurements on actin-extracted cardiac myofibrils revealed that a large portion of the viscous/viscoelastic force decay – mainly the quickly decaying component – originates in actin–titin interactions (Figure 7; and Kulke *et al.*, 2001a). We estimated that approximately 80% of the stress relaxation following rapid stretch could be due to weak actin–titin binding, the remainder resulting from interactions between actin and myosin filaments (Figure 7B). Thus, weakly bound crossbridges could make a small contribution to passive muscle properties (Campbell and Lakie, 1998; Proske and Morgan, 1999), but much of the viscoelasticity is unrelated to actin–myosin interactions (Kulke *et al.*, 2001a). The fact that actin–titin interactions are the principal source of viscous drag in the sarcomere, was not taken into consideration in a recent calculation aimed at quantifying sarcomeric viscoelasticity (Ranatunga, 2001). Consequently, the values calculated fell short by a factor of at least five, of the experimental values known from mechanical measurements on relaxed muscle fibers (Mutungi and Ranatunga, 1996, 1998). We suggest that actin–titin interactions account for most, if not all, of the heretofore unexplained viscous effects.

Novel twists in titin viscoelasticity: By far not all elastic titin regions interact with the thin filaments. Besides a 100-nm-wide (functionally stiff) titin segment adjoining the Z-disk (Linke *et al.*, 1997; Trombitas *et al.*, 1997), the only I-band-titin domain found to bind to actin, is the PEVK domain (Kulke *et al.*, 2001a). Furthermore, the actin-PEVK titin interaction was shown to be modulated by Ca^{2+} (Kulke *et al.*, 2001a), which might give rise to lowered viscous drag. Future studies will focus on the exciting possibility that titin's viscoelastic properties can be adjusted *in situ* by regulatory mechanisms involving the action of Ca^{2+} ions (Stuyvers *et al.*, 1998) and/or other, still unknown, factors (Linke *et al.*, 2002).

Acknowledgements

We thank all members of the Linke and Fernandez laboratories for their important contributions to the works cited in this article. We acknowledge financial support of the Deutsche Forschungsgemeinschaft (grants Li 690/2-3 and Li 690/6-2). W.A.L. is supported by a Heisenberg fellowship from the Deutsche Forschungsgemeinschaft.

References

Allen DG and Kentish JC (1985) The cellular basis of the length–tension relation in cardiac muscle. *J Mol Cell Cardiol* **17:** 821–840.

Alper J (2002) Protein structure. Stretching the limits. *Science* **297:** 329–331.

Anderson J, Joumaa V, Stevens L, Neagoe C, Li Z, Mounier Y, Linke WA and Goubel F (2002) Passive stiffness changes in soleus muscles from desmin knockout mice are not due to titin modifications. *Pflügers Arch* **444:** 771–776.

Bang ML, Centner T, Fornoff F, Geach AJ, Gotthardt M, McNabb M, Witt CC, Labeit D, Gregorio CC, Granzier H and Labeit S (2001) The complete gene sequence of titin, expression of an unusual approximately 700-kDa titin isoform, and its interaction with obscurin identify a novel Z-line to I-band linking system. *Circ Res* **89:** 1065–1072.

Bartoo ML, Linke WA and Pollack GH (1997) Basis of passive tension and stiffness in isolated rabbit myofibrils. *Am J Physiol* **273:** C266–C276.

Bustamante C, Marko JF, Siggia ED and Smith S (1994) Entropic elasticity of λ-phage DNA. *Science* **265:** 1599–1600.

Campbell KS and Lakie M (1998) A cross-bridge mechanism can explain the thixotropic short-range elastic component of relaxed frog skeletal muscle. *J Physiol* **510:** 941–962.

Carrion-Vazquez M, Oberhauser AF, Fowler SB, Marszalek PE, Broedel SE, Clarke J and Fernandez JM (1999) Mechanical and chemical unfolding of a single protein: a comparison. *Proc Natl Acad Sci USA* **96:** 3694–3699.

Carrion-Vazquez M, Oberhauser AF, Fisher TE, Marszalek PE, Li H and Fernandez JM (2000) Mechanical design of proteins studied by single-molecule force spectroscopy and protein engineering. *Prog Biophys Mol Biol* **74:** 63–91.

Chiu Y-L, Ballou EW and Ford LE (1982) Internal viscoelastic loading in cat papillary muscle. *Biophys J* **40:** 109–120.

de Tombe P and ter Keurs HEDJ (1992) An internal viscous element limits unloaded velocity of sarcomere shortening in rat myocardium. *J Physiol* **454:** 619–642.

Erickson HP (1994) Reversible unfolding of fibronectin type III and immunoglobulin domains provides the structural basis for stretch and elasticity of titin and fibronectin. *Proc Natl Acad Sci USA* **91:** 10,114–10,118.

Erickson HP (1997) Stretching single protein molecules: titin is a weird spring. *Science* **276:** 1090–1092.

Fisher TE, Marszalek PE and Fernandez JM (2000) Stretching single molecules into novel conformations using the atomic force microscope. *Nature Struct Biol* **7:** 719–724.

Freiburg A, Trombitas K, Hell W, Cazorla O, Fougerousse F, Centner T, Kolmerer B, Witt C, Beckmann JS, Gregorio CC, Granzier H and Labeit S (2000) Series of exon-skipping events in the elastic spring region of titin as the structural basis for myofibrillar elastic diversity. *Circ Res* **86:** 1114–1121.

Fürst DO, Osborn M, Nave R and Weber K (1988) The organization of titin filaments in the half-sarcomere revealed by monoclonal antibodies in immunoelectron microscopy: a map of ten nonrepetitive epitopes starting at the Z line extends close to the M line. *J Cell Biol* **106:** 1563–1572.

Gautel M and Goulding D (1996) A molecular map of titin/connectin elasticity reveals two different mechanisms acting in series. *FEBS Lett* **385:** 11–14.

Gautel M, Lehtonen E and Pietruschka F (1996) Assembly of the cardiac I-band region of titin/connectin: expression of the cardiac-specific regions and their structural relation to the elastic segments. *J Muscle Res Cell Motil* **17:** 449–461.

Greaser M (2001) Identification of new repeating motifs in titin. *Proteins* **43:** 145–149.

Gutierrez-Cruz G, van Heerden AH and Wang K (2001) Modular motif, structural folds and affinity profiles of the PEVK segment of human fetal skeletal muscle titin. *J Biol Chem* **276:** 7442–7449.

Hein S, Gaasch WH and Schaper J (2002) Giant molecule titin and myocardial stiffness. *Circulation* **106:** 1302–1304.

Helmes M, Trombitas K, Centner T, Kellermayer M, Labeit S, Linke WA and Granzier H (1999) Mechanically driven contour-length adjustment in rat cardiac titin's unique N2B sequence: titin is an adjustable spring. *Circ Res* **84:** 1339–1352.

Higuchi H, Nakauchi Y, Maruyama K and Fujime S (1993) Characterization of beta-connectin (titin 2) from striated muscle by dynamic light scattering. *Biophys J* **65:** 1906–1915.

Hill DK (1968) Tension due to interaction between the sliding filaments in resting striated muscle. The effect of stimulation. *J Physiol* **199:** 637–684.

Improta S, Politou A and Pastore A (1996) Immunoglobulin-like modules from I-band titin: extensible components of muscle elasticity. *Structure* **4:** 323–337.

Itoh Y, Suzuki T, Kimura S, Ohashi K, Higuchi H, Sawada H, Shimizu T, Shibata M and Maruyama K (1988) Extensible and less-extensible domains of connectin filaments in stretched vertebrate skeletal muscle sarcomeres as detected by immunofluorescence and immunoelectron microscopy using monoclonal antibodies. *J Biochem* **104:** 504–508.

Julian FJ, Sollins MR and Moss RL (1976) Absence of a plateau in length-tension relationship of rabbit papillary muscle when internal shortening is prevented. *Nature* **260:** 340–342.

Kellermayer MSZ, Smith SB, Granzier HL and Bustamante C (1997) Folding-unfolding transitions in single titin molecules characterized with laser tweezers. *Science* **276:** 1112–1116.

Kentish JC, ter Keurs HE, Ricciardi L, Bucx JJ and Noble MI (1986) Comparison between the sarcomere length-force relations of intact and skinned trabeculae from rat right ventricle. Influence of calcium concentrations on these relations. *Circ Res* **58:** 755–768.

Konhilas JP, Irving TC and de Tombe PP (2002) Myofilament calcium sensitivity in skinned rat cardiac trabeculae: role of interfilament spacing. *Circ Res* **90:** 59–65.

Kulke M, Fujita-Becker S, Rostkova E, Neagoe C, Labeit D, Manstein DJ, Gautel M and Linke WA (2001a) Interaction between PEVK-titin and actin filaments: origin of a viscous force component in cardiac myofibrils. *Circ Res* **89:** 874–881.

Kulke M, Neagoe C, Kolmerer B, Minajeva A, Hinssen H, Bullard B and Linke WA (2001b) Kettin, a major source of myofibrillar stiffness in *Drosophila* indirect flight muscle. *J Cell Biol* **154:** 1045–1057.

Labeit S and Kolmerer B (1995) Titins, giant proteins in charge of muscle ultrastructure and elasticity. *Science* **270:** 293–296.

Labeit S, Gautel M, Lakey A and Trinick J (1992) Towards a molecular understanding of titin. *EMBO J* **11:** 1711–1716.

Li H, Linke WA, Oberhauser AF, Carrion-Vazquez M, Kerkvliet JG, Lu H, Marszalek PE and Fernandez JM (2002) Reverse engineering of the giant muscle protein titin. *Nature* **418:** 998–1002.

Li H, Oberhauser AF, Fowler SB, Clarke J and Fernandez JM (2000) Atomic force microscopy reveals the mechanical design of a modular protein. *Proc Natl Acad Sci USA* **97:** 6527–6531.

Li H, Oberhauser AF, Redick SD, Carrion-Vazquez M, Erickson HP and Fernandez JM (2001) Multiple conformations of PEVK proteins detected by single-molecule techniques. *Proc Natl Acad Sci USA* **98:** 10,682–10,686.

Linke WA (2000) Stretching molecular springs: elasticity of titin filaments in vertebrate striated muscle. *Histol Histopathol* **15:** 799–811.

Linke WA, Popov VI and Pollack GH (1994) Passive and active tension in single cardiac myofibrils. *Biophys J* **67:** 782–792.

Linke WA, Bartoo ML, Ivemeyer M and Pollack GH (1996a) Limits of titin extension in single cardiac myofibrils. *J Muscle Res Cell Motil* **17:** 425–438.

Linke WA, Ivemeyer M, Olivieri N, Kolmerer B, Rüegg JC and Labeit S (1996b) Towards a molecular understanding of the elasticity of titin. *J Mol Biol* **261:** 62–71.

Linke WA, Ivemeyer M, Labeit S, Hinssen H, Rüegg JC and Gautel M (1997) Actin-titin interaction in cardiac myofibrils: probing a physiological role. *Biophys J* **73:** 905–919.

Linke WA, Ivemeyer M, Mundel P, Stockmeier MR and Kolmerer B (1998a) Nature of PEVK-titin elasticity in skeletal muscle. *Proc Natl Acad Sci USA* **95:** 8052–8057.

Linke WA, Stockmeier MR, Ivemeyer M, Hosser H and Mundel P (1998b) Characterizing titin's I-band Ig domain region as an entropic spring. *J Cell Sci* **111:** 1567–1574.

Linke WA, Rudy DE, Centner T, Gautel M, Witt C, Labeit S and Gregorio CC (1999) I-band titin in cardiac muscle is a three-element molecular spring and is critical for maintaining thin filament structure. *J Cell Biol* **146:** 631–644.

Linke WA, Kulke M, Li H, Fujita-Becker S, Neagoe C, Manstein DJ, Gautel M and Fernandez JM (2002) PEVK domain of titin: an entropic spring with actin-binding properties. *J Struct Biol* **137:** 194–205.

Liversage AD, Holmes D, Knight PJ, Tskhovrebova L and Trinick J (2001) Titin and the sarcomere symmetry paradox. *J Mol Biol* **305:** 401–409.

Marko JF and Siggia ED (1995) Stretching DNA. *Macromolecules* **28:** 8759–8770.

Maruyama K, Murakami F and Ohashi K (1977) Connectin, an elastic protein of muscle. Comparative Biochemistry. *J Biochem* (*Tokyo*) **82:** 339–345.

Minajeva A, Kulke M, Fernandez JM and Linke WA (2001) Unfolding of titin domains explains the viscoelastic behavior of skeletal myofibrils. *Biophys J* **80:** 1442–1451.

Mutungi G and Ranatunga KW (1996) Tension relaxation after stretch in resting mammalian muscle fibers: stretch activation at physiological temperatures. *Biophys J* **70:** 1432–1438.

Mutungi G and Ranatunga KW (1998) Temperature-dependent changes in the viscoelasticity of intact resting mammalian (rat) fast- and slow-twitch muscle fibres. *J Physiol* **508:** 253–265.

Neagoe C, Kulke M, del Monte F, Gwathmey JK, de Tombe PP, Hajjar R and Linke WA (2002) Titin isoform switch in ischemic human heart disease. *Circulation* **106:** 1333–1341.

Noble MIM (1977) The diastolic viscous properties of cat papillary muscle. *Circ Res* **40:** 288–292.

Oberhauser AF, Marszalek PE, Erickson HP and Fernandez JM (1998) The molecular elasticity of the extracellular matrix protein tenascin. *Nature* **393:** 181–185.

Politou AS, Thomas DJ and Pastore A (1995) The folding and stability of titin immunoglobulin-like modules, with implications for the mechanism of elasticity. *Biophys J* **69:** 2601–2610.

Proske U and Morgan DL (1999) Do cross-bridges contribute to the tension during stretch of passive muscle? *J Muscle Res Cell Motil* **20:** 433–442.

Ranatunga KW (2001) Sarcomeric visco-elasticity of chemically skinned skeletal muscle fibres of the rabbit at rest. *J Muscle Res Cell Motil* **22:** 399–414.

Rief M, Gautel M, Oesterhelt F, Fernandez JM and Gaub HE (1997) Reversible unfolding of individual titin immunoglobulin domains by AFM. *Science* **276:** 1109–1112.

Rief M, Fernandez JM and Gaub HE (1998) Elastically coupled two-level systems as a model for biopolymer extensibility. *Phys Rev Lett* **81:** 4764–4767.

Stuyvers BD, Miura M, Jin JP and ter Keurs HE (1998) Ca^{2+}-dependence of diastolic properties of cardiac sarcomeres: involvement of titin. *Progr Biophys Mol Biol* **69:** 425–443.

Trinick J and Tskhovrebova L (1999) Titin: a molecular control freak. *Trends Cell Biol* **9:** 377–380.

Trombitas K, Greaser ML and Pollack GH (1997) Interaction between titin and thin filaments in intact cardiac muscle. *J Muscle Res Cell Motil* **18:** 345–351.

Tskhovrebova L and Trinick J (2001) Flexibility and extensibility in the titin molecule: analysis of electron microscope data. *J Mol Biol* **310:** 755–771.

Tskhovrebova L, Trinick J, Sleep JA and Simmons RM (1997) Elasticity and unfolding of single molecules of the giant muscle protein titin. *Nature* **387:** 308–312.

Urry DW (1984) Protein elasticity based on conformations of sequential polypeptides: the biological elastic fiber. *J Protein Chem* **3:** 403–436.

Wang K (1996) Titin/connectin and nebulin: giant protein rulers of muscle structure and function. *Adv Biophys* **33:** 123–134.

Wang K, McCarter R, Wright J, Beverly J and Ramirez-Mitchell R (1993) Viscoelasticity of the sarcomere matrix of skeletal muscles. The titin-myosin composite filament is a dual-stage molecular spring. *Biophys J* **64:** 1161–1177.

Weber KT (1997) Extracellular matrix remodeling in heart failure: a role for de novo angiotensin II generation. *Circulation* **96:** 4065–4082.

Weiwad WK, Linke WA and Wussling MH (2000) Sarcomere length-tension relationship of rat cardiac myocytes at lengths greater than optimum. *J Mol Cell Cardiol* **32:** 247–259.

Wu Y, Cazorla O, Labeit D, Labeit S and Granzier H. (2000) Changes in titin and collagen underlie diastolic stiffness diversity of cardiac muscle. *J Mol Cell Cardiol* **32:** 2151–2162.

Journal of Muscle Research and Cell Motility **23**: 499–511, 2002.

Stretching and visualizing titin molecules: combining structure, dynamics and mechanics

MIKLÓS S. Z. KELLERMAYER* and LÁSZLÓ GRAMA
Department of Biophysics, University of Pécs, Medical School Pécs, Szigeti ut 12, Pécs H-7624 Hungary

Abstract

The details of the global and local structure and function of titin, a giant filamentous intrasarcomeric protein are largely undiscovered. Here we discuss a combination of bulk-solution and novel single-molecule techniques that may lend unique insights into titin's molecular dynamic, structural and mechanical characteristics.

Introduction

Titin is a giant filamentous protein of modular construction, whose elastic properties define the passive mechanical properties of striated muscle (for recent reviews, see (Wang, 1996; Maruyama, 1997; Gregorio *et al.*, 1999; Trinick and Tskhovrebova, 1999; Granzier and Labeit, 2002; Tskhovrebova and Trinick, 2002)). A titin molecule is a tandem array of immunoglobulin (Ig) type C2 and fibronectin (FN) type III domains (~300/molecule) interspersed with unique sequences, most notably the proline-, glutamate-, valine- and lysine-rich PEVK segment, and the N2A and N2B segments (Labeit and Kolmerer, 1995; Freiburg *et al.*, 2000; Bang *et al.*, 2001). The I-band segment of titin acts as a molecular spring (Horowits *et al.*, 1986; Trombitás *et al.*, 1995; Granzier *et al.*, 1996; Linke *et al.*, 1996). Titin's A-band segment, on the other hand, is thought to function as a scaffold that defines structural regularity within the A-band (Trinick, 1996).

Due to its large size, structural complexity and elastic nature it is difficult to explore titin's properties by conventional bulk-solution methods. Single-molecule techniques, on the other hand, may provide unique information about titin's characteristics (reviewed in (Wang *et al.*, 2001)). Titin, given its conveniently large size, has even been used as a protein-folding model system in various recent single-molecule studies (Zhuang *et al.*, 1999, 2000). Single-molecule-mechanics experiments described titin as an entropic chain in which domain unfolding was inferred to occur at high forces and refolding at low forces (Kellermayer *et al.*, 1997, 1998; Rief *et al.*, 1997; Tskhovrebova *et al.*, 1997). However, the structural basis of titin's elasticity, the various regimes of elasticity, the spatial distribution of intramolecular structural events during titin's elongation, and the exact structural transitions occurring at different spatial and temporal scales remain to be understood. In the following we discuss the structure, dynamics, and mechanics of titin with an emphasis on recent attempts to image and manipulate individual, fluorescently labeled molecules.

* To whom correspondence should be addressed: Tel.: +36-72-536-271; Fax: +36-72-536-261; E-mail: miklos.kellermayer.jr@aok.pte.hu

Structure

Primary structure

Titin is encoded by a single gene structured into 363 exons coding for a total number of 38,138 amino acids (Bang *et al.*, 2001). Via alternative splicing various size isoforms are expressed in different types of striated muscle (Freiburg *et al.*, 2000). The molecular weight of the various titin isoforms ranges between 3.0 and 3.7 Mda (Freiburg *et al.*, 2000). Along titin's primary structure various amino acids with reactive side chains occur with different frequencies. These reactive side chains may be labeled with different probes, such as fluorescent dyes or spin labels, for the purpose of exploring titin's structure and dynamics. The relative frequencies of cysteine and lysine residues along the skeletal-muscle titin sequence (Labeit and Kolmerer, 1995) are shown in Figure 1a. The relative frequency (running average per 1000 residues) of cysteine along titin is fairly constant (15 cysteines/1000 residues ±7 SD), except for the PEVK segment where cysteines are absent. The relative frequency of lysine is 86/1000 (±21 SD), except for the PEVK segment where lysine is abundant (Figure 1a). Tskhovrebova and Trinick have used Cy3-dye (Tskhovrebova and Trinick, 1999, 2000) or TRITC (Tskhovrebova and Trinick, 2002) to label titin fluorescently. In our work we have mainly used fluorescein (iodoacetamide-fluorescein, IAF and fluorescein-isothiocyanate, FITC) and rhodamine derivatives (tetramethylrhodamine-iodoacetamide, TMRIA and tetramethylrhodamine-isothiocyanate TRITC) to label skeletal-muscle titin prepared from rabbit back muscle for the purpose of studying titin's global conformation, internal dynamics, diffusive properties and extensibility at the single-molecule level (Grama *et al.*, 2000; Kellermayer, 2000a, b; Kellermayer *et al.*,

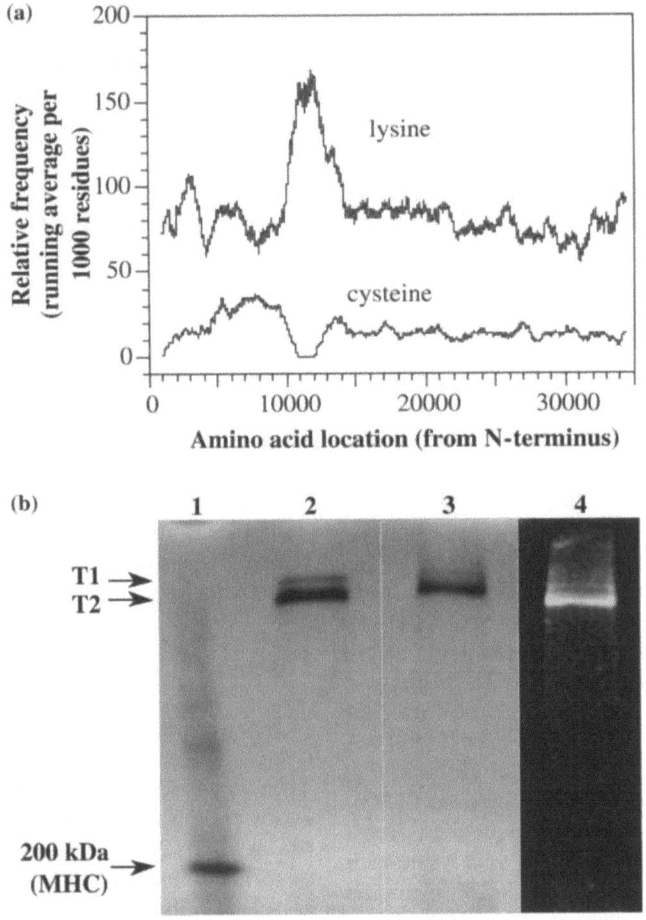

Fig. 1. (a) Distribution of cysteine and lysine residues along the skeletal-muscle titin sequence (Labeit and Kolmerer, 1995). Running average of the number of cysteines per 1000 residues is shown. (b) SDS–polyacrylamide electrophoretogram of purified titin prior to and following fluorescent modification with tetramethylrhodamine iodoacetamide (TMRIA). Titin was labeled overnight on ice with TMRIA in an 1650× molar excess. Typical labeling efficiencies were 100–150 TMRIA per titin. Molecular weight marker (lane 1), purified titin (lane 2), Coomassie-stained TMRIA-labeled titin (lane 3), and the fluorescence image of TMRIA-labeled titin (lane 4) are shown.

2000a, b; Grama *et al.*, 2001a, b, c). Typical labeling efficiencies obtained ranged between 100 and 150 dye molecules per titin. The SDS–PAGE pattern of TMRIA-labeled titin is shown in Figure 1b.

Secondary structure

Titin is a tandem array of various domains (Labeit and Kolmerer, 1995) which altogether form a ~1-μm-long chain in the native state (Nave *et al.*, 1989). The approximately 300 globular domains along titin may be grouped into immunoglobulin (Ig) C2 and fibronectin (FN) III types (Labeit *et al.*, 1992; Labeit and Kolmerer, 1995). The structure of titin's globular domains has been explored primarily with NMR spectroscopy (Politou *et al.*, 1994a, b, 1995, 1996; Improta *et al.*, 1996, 1998). These domains have a β-barrel structure formed from seven β-strands that run anti-parallel to each other. Neighboring domains are connected with an elastic linker region, which has been implicated to provide flexibility and extensibility to the chain (Improta *et al.*, 1998). Interspersed among the globular domains there are various unique sequences. Most of the unique sequences are found in titin's I-band section. Intensively studied unique sequences in titin are the proline(P)-, glutamate(E)-, valine(V)- and lysine(K)-rich PEVK segment and the N2A and N2B segments (Labeit and Kolmerer, 1995; Freiburg *et al.*, 2000). The exact structure of these elements is not known, although poly-proline helices have been implicated in the PEVK segment (Ma *et al.*, 2001).

Higher-order structure

Higher-order structure of titin concerns the axial arrangement of the various domains along the molecule, the global shape of titin, and supramolecular associations.

Axial arrangement of titin's domains: The globular domains and the unique sequences are arranged along the axis of the molecule according to titin's sarcomeric orientation and arrangement. Accordingly, the domain arrangement is thought to have physiological significance. Most notably, the I-band section of titin is partitioned into proximal and distal tandem Ig segments (i.e., proximal is closer, distal is farther away from the N-terminus), in between which unique sequences (PEVK, N2A, N2B) and extra Ig domains may be found depending on the particular isoform (muscle type) (Labeit and Kolmerer, 1995; Freiburg *et al.*, 2000). Since the presence and the size of the unique sequences varies – via alternative splicing – according to the muscle-isoform, these segments are thought to lend tissue specificity to titin (Freiburg *et al.*, 2000). The variation in the size of the unique sequences according to muscle type has been suggested to provide an adaptable elasticity to titin (Granzier *et al.*, 2000; Granzier and Labeit, 2002). In the A-band section, super-repeats of Ig and FN segments are found in a repeated arrangement which is thought to provide a template for the regular binding of sarcomeric proteins such as myosin and C-protein (Labeit and Kolmerer, 1995).

Global shape inferred from images of fluorescent, surface-adsorbed molecules: The global structure and shape of titin have been explored with various methods including morphological (electron microscopy (Trinick *et al.*, 1984; Wang *et al.*, 1984; Tskhovrebova *et al.*, 1997; Tskhovrebova and Trinick, 2001), atomic force microscopy, AFM (Hallett *et al.*, 1996; Tskhovrebova and Trinick, 1999; Kellermayer *et al.*, 2002b, submitted), fluorescence microscopy (Kellermayer, 2000a, b; Tskhovrebova and Trinick, 2000; Grama *et al.*, 2001a, b, c)), diffraction (Higuchi *et al.*, 1993) and hydrodynamic (Kurzban and Wang, 1988) techniques. While high-resolution imaging techniques (EM, AFM) allow the examination of titin's structure in detail, these methods may impose fixation and dehydration artefacts. Fluorescence imaging, on the other hand, allows individual titin molecules to be visualized under aqueous buffer conditions. We have used laser scanning confocal microscopy to image single, fluorescently labeled titin molecules (Grama *et al.*, 2000, 2001a, b, c; Kellermayer, 2000a, b; Kellermayer *et al.*, 2000a, b). When deposited by equilibration to a glass surface in a diffusion-driven process (Figure 2a), TMRIA-labeled titin molecules appeared as bright, diffraction-limited spots (Figure 2b). The sizes of the spots vary somewhat, probably due to a variation in labeling efficiency, variation in the local and global conformation of the molecule, and variation in the extent of rhodamine dimer formation (see below). The spots have a mean, apparently circular area of 0.11 μm^2 (± 0.11 SD, $n = 2334$). The mean diameter (d) of the bright spots is ~ 0.29 μm, which is comparable to the resolution limit of the microscope used ($d_{\min} \sim 0.25$ μm), suggesting that the dimensions of the surface area occupied by a molecule are less than or equal to the diffraction limit (Grama *et al.*, 2001a, b, c). The radius of gyration (R_G, root-mean-square distance of an array of atoms from their common center of gravity (Cantor and Schimmel, 1980)) of a surface-adsorbed titin molecule must therefore be comparable to or smaller than the radius of the smallest resolvable particle (~ 0.125 μm). Considering that R_G is related to the end-to-end distance (R) as (Cantor and Schimmel, 1980)

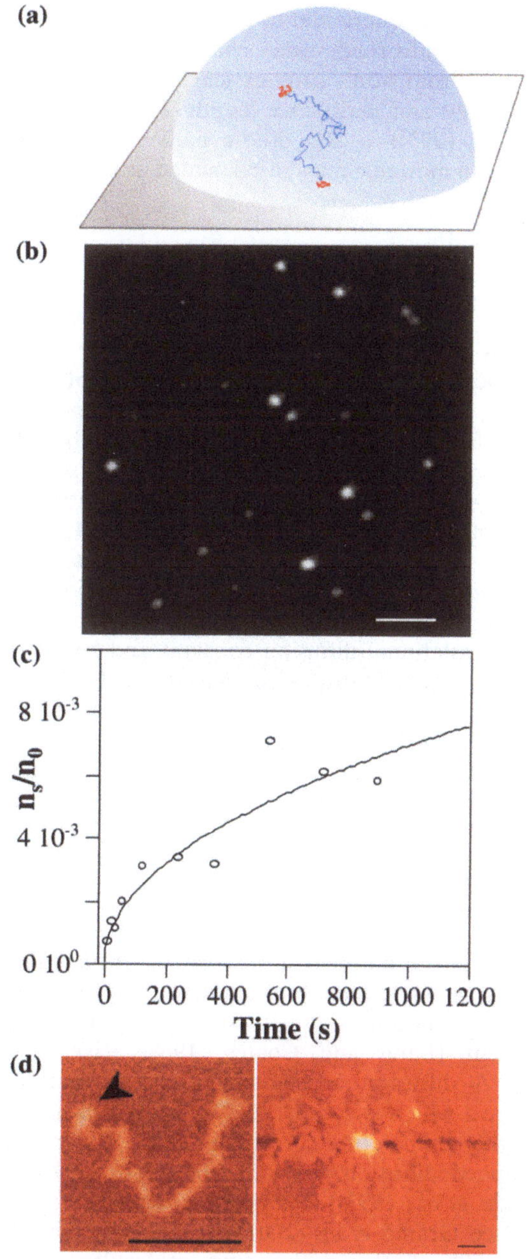

Fig. 2. (a) Deposition of fluorescently labeled titin molecules onto glass surface. The titin molecule is shown in red, and its diffusion trajectory in blue. (b) Confocal microscopic image of a glass surface equilibrated with TMRIA-labeled titin molecules. Scale bar, 2 μm. (c) Kinetics of surface adsorption of fluorescently labeled titin molecules. The ratio of surface-bound molecules (per cm^2) and initial concentration (molecules per cm^3 solution at $t = 0$) is shown as a function of incubation time. The data were fitted with Equation (3). A diffusion constant of 4.8×10^{-8} cm^2/s was calculated from the fit (modified from (Grama *et al.*, 2001a, b, c)). (d) AFM images of titin. Scale bar, 1 μm. Titin was allowed to bind to a mica surface for 10 min. Then the surface was washed with buffer, distilled water and dried under a gentle stream of nitrogen gas. Single titin molecule (left), and a titin oligomer (right) are shown (modified from (Kellermayer *et al.*, 2002, submitted)).

$$\langle R_G^2 \rangle = \frac{\langle R^2 \rangle}{6}, \tag{1}$$

the root-mean-square end-to-end distance $\left(\sqrt{\langle R^2\rangle}\right)$ of the surface-adsorbed titin chain may not exceed ~0.3 μm. From titin's contour length ($L \sim 1$ μm (Nave *et al.*, 1989)) and persistence length ($P \sim 15$ nm (Higuchi *et al.*, 1993)) $\sqrt{\langle R^2\rangle}$ of the native, surface-equilibrated titin molecule may be calculated as (Rivetti *et al.*, 1996)

$$\langle R^2\rangle_{2D} = 4PL\left(1 - \frac{2P}{L}\left(1 - e^{-\frac{L}{2P}}\right)\right). \tag{2}$$

$\sqrt{\langle R^2\rangle}$ is ~0.24 μm, which indeed falls below the upper limit calculated. Although the physical size of the area occupied by a titin molecule equilibrated or trapped to the glass surface cannot exactly be resolved by this method, the observations indicate that – in aqueous solution – titin, driven by its high flexibility, contracts into an area that is smaller than the optical resolution limit.

Global shape and size inferred from the kinetics of surface adsorption: From the kinetics of equilibration of fluorescently labeled titin molecules onto a glass surface titin's translational diffusion constant and hence shape parameters may be estimated (Rivetti *et al.*, 1996). We measured the fraction of molecules attached to the substrate surface (n_s/n_0) after varying periods of incubation time (t) (Figure 2c). The experimental data were fitted well ($r = 0.93$) with

$$\frac{n_s}{n_0} = \sqrt{\frac{4D}{\pi}}\sqrt{t}, \tag{3}$$

where n_s is the number of surface-adsorbed molecules per cm^2, n_0 is the number of molecules in unit volume (cm^3) of solution at $t = 0$, and D is the diffusion coefficient (Lang and Coates, 1968; Rivetti *et al.*, 1996). A diffusion constant of 4.8×10^{-8} cm^2/s was calculated which is comparable to but somewhat lower than the temperature-corrected value of the earlier-determined diffusion constant of titin (6×10^{-8} cm^2/s (Higuchi *et al.*, 1993)) (for possible reasons of discrepancy, see Grama *et al.* (2001a, b, c)). The persistence length (P) of the native titin molecule calculated according to (Higuchi *et al.*, 1993)

$$D = \frac{k_B T}{3\pi\eta L}\left(1 + 1.84\left(\frac{L}{2P}\right)^{\frac{1}{2}}\right), \tag{4}$$

using the diffusion constant (D) determined above and a contour length (L) of 1 μm (Nave *et al.*, 1989) is 26 nm. This persistence length exceeds that calculated earlier for native β-connectin (or T2) (15 nm (Higuchi *et al.*, 1993)), but can still be reconciled with the upper limit of the physical dimensions of the area occupied by the surface-adsorbed molecule. Possible reasons for the higher persistence lengths include contaminating aggregates (oligomers) of titin and the effect of fluorescent labeling. Conceivably, by using a more efficient method to prevent the aggregation of titin, or by precisely determining the titin-molecule concentration, measuring the kinetics of molecular deposition may, in the future, yield better estimates of the physical parameters of the titin molecule.

Global shape of titin molecules and oligomers explored with AFM: The shape of surface-adsorbed titin molecules can be explored with the high-resolution technique, AFM (Kellermayer *et al.*, 2002b, submitted). The AFM image of a single titin molecule equilibrated to a mica surface is shown in Figure 2d (left). The molecule displays a convoluted structure, which is determined by the molecule's bending rigidity. Frequently, a globular 'head' may be observed at one end of the chain (arrowhead), which corresponds to the M-line region of the molecule (Nave *et al.*, 1989). The persistence length, calculated based on Equation 2, is 18.7 nm, which corresponds well but slightly exceeds previously measured values (15 nm (Higuchi *et al.*, 1993), 13.5 nm (Tskhovrebova and Trinick, 2001)). In addition to single molecules typical samples also contain oligomers of titin. Oligomers range from bi-molecular to high-order complexes (Kellermayer *et al.*, 2002b, submitted). The AFM image of a high-order titin oligomer is shown in Figure 2d (right). In every case a globular head is observed through which the component titin molecules connect to each other. Notably, the globular head is the only point of attachment between the component titin molecules. The rest of the contour of the individual component titin molecules is apparently independent, and does not form connections between neighboring molecules within the oligomer. The surface-equilibrated titin oligomers display a point-symmetric global appearance, where the point of symmetry is the globular, presumably M-line-region head.

The titin molecules imaged with AFM apparently occupy an area greater than that seen with fluorescence microscopy (compare Figure 2b and d). Conceivably, in aqueous solution the surface-adsorbed titin molecules are more contracted and partially extend into the solution. Then, during buffer removal the titin chain becomes partially extended before becoming completely attached to the surface. One may speculate that this process is influenced by several factors, including surface tension, chain diameter and bending rigidity, excluded volume effects and the forces acting between the surface and the molecule. By better characterizing the details of the deposition process we may learn more about the structure and elasticity of titin.

Dynamics

With the help of fluorescent labeling it is possible to follow the dynamic, motional characteristics of titin across wide spatial and temporal scales. The spatial scale ranges from nanometers (explored with fluorescence resonance energy transfer (FRET)) to several micrometers (explored by fluorescence microscopy). The temporal scale ranges from nanoseconds (explored with time-resolved fluorescence spectroscopy) to tens of milliseconds (explored by video-rate imaging). So far the number of studies reported on the molecular dynamics of titin or its domains is limited and only a handful of papers on the fluorescent labeling of titin have been published. In the following we discuss observations on fluorescently labeled titin which bear relevance to the dynamics of this molecule.

Self-quenching in fluorescently labeled titin

Zhuang *et al.* observed that the fluorescence intensity of single, Oregon-Green 488-labeled titin molecules increases upon the addition of chemical denaturants, suggesting that by monitoring fluorescence the kinetics of protein folding and unfolding might be monitored at the single-molecule level (Zhuang *et al.*, 2000). The reduced fluorescence intensity in the native state was hypothesized to be due to self-quenching caused by the proximity of the dye molecules. The increase in fluorecence intensity upon the addition of denaturants has been explained by the reduced proximity, hence reduced self-quenching, of the fluorophores associated with the conformational changes occurring during unfolding. However, the exact mechanisms of fluorescence reduction in titin remained largely unknown. We have also observed reduced fluorescence in TMRIA-labeled titin, the intensity of which increased several-fold upon the addition of chemical denaturants (Grama *et al.*, 2001a, b, c). To explore the molecular mechanisms behind self-quenching, we carried out fluorescence and absorption spectroscopy of TMRIA-labeled titin in solution (Figure 3). In the absorption spectra a significant peak at 518 nm, absent from the excitation spectra, appeared in addition to the one at 555 nm (Figure 3b). While the 555 nm peak corresponds to rhodamine monomers, the peak at 518 nm is due to the presence of non-fluorescent rhodamine dimers (Levshin and Bocharov, 1961; Rohatgi and Singhal, 1966; Selwyn and Steinfeld, 1972). Since the probability of dimer formation increases with rhodamine concentration (Levshin and Bocharov, 1961; Baranova, 1965; Plant, 1986), the presence of a large dimer population suggests that titin has collected the associated fluorophores into a small volume. As a result, the dye molecules are in close proximity, allowing self-quenching. The mechanisms of such concentration quenching in rhodamine dimers are not precisely known; collisional quenching, dark complex formation and energy transfer have been previously implicated (Plant, 1986) suggesting that both dynamic and structural mechanisms may contribute to the phenomenon. Dimer formation is distance dependent (Packard *et al.*, 1996); therefore, dimers appear only if the spatial proximity between rhodamine monomers is sufficiently high, as in a concentrated solution. By measuring the ratio of dimers and monomers the average spatial proximity, which is analogous to the concentration, of the dye molecules may be estimated. From the average spatial proximity of chain-associated rhodamine dye molecules the global shape of the chain may be inferred (see below). In contrast to the explanation offered by Zhuang *et al.* (2000), our findings suggested that the increase in fluorescence intensity observed during chemical denaturation (with GdnHCl, urea, SDS) is the result of direct interaction between the denaturant and the rhodamine dimers rather than a marker of protein unfolding, and the unfolding of titin does not result in a significant dissociation of the chain-associated dimers.

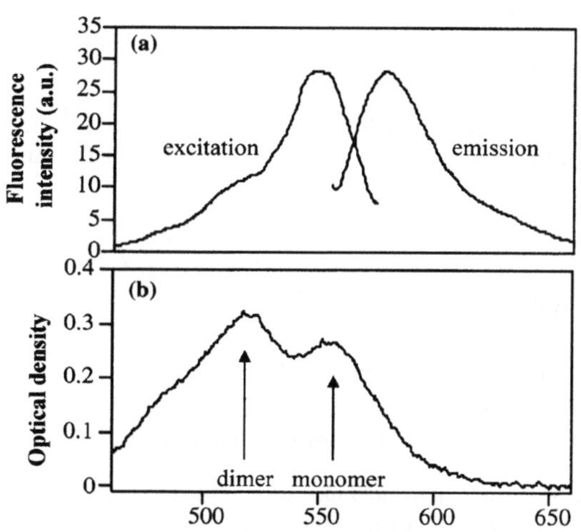

Fig. 3. Spectroscopic analysis of TMRIA-labeled titin. (a) Fluorescence excitation and emission spectra. Excitation spectra were recorded at an emission wavelength of 580 nm, while emission spectra at an excitation wavelength of 555 nm. (b) Absorption spectrum of TMRIA-labeled titin. Rhodamine monomer peak appears at 555 nm. In addition a peak is detected at 518 nm which is unseen in the excitation spectrum and corresponds to the rhodamine dimer (modified from (Grama *et al.*, 2001a, b, c)).

Global conformation of unfolded titin explored through rhodamine dimers

Considering that fluorescently labeled titin is studded with TMRIA molecules whose ensemble behavior is monitored, we may derive information about titin's global conformation if we make a few simplifying assumptions: (i) the fluorophores are more or less evenly distributed along the titin chain and (ii) they contribute equally to dimer formation. In native titin the spatial arrangement of the rhodamine molecules is determined by the domain structure. The dimers are formed by the interaction of rhodamine molecules that are either on the same domain or on closely spaced cysteines of neighboring domains, because the distance between point dipoles in the dimer ($R \approx 0.6$ nm (Packard *et al.*, 1996)) is smaller than the average interdomain spacing of Ig or FN domains (~4.6 nm (Trombitás *et al.*, 1998a, b)). In the completely unfolded titin the native three-dimensional structure is collapsed. In the first approximation titin is a linear chain with a nearly constant frequency of cysteines along its contour. Considering the labeling ratio obtained in this work and assuming an equal accessibility of the cysteines to TMRIA, approximately every fourth cysteine residue was labeled, If the segments of the unfolded titin chain are allowed to diffuse and interact with each other within the space allotted, or if they can freely rearrange during denaturation, then the ratio of dimers and monomers describes the average concentration of rhodamine in the volume occupied by a single titin molecule. The volume of the single titin molecule can then be calculated as

$$V = \frac{n}{MA}, \tag{5}$$

where n is the average number of TMRIA molecules on titin, M is the molar concentration of TMRIA (mol/l), and A is Avogadro's number (6.023×10^{23}). According to TMRIA concentration calibration (data not shown and (Burghardt *et al.*, 1996)), the dye concentration in the volume occupied by a single, acid-denatured titin molecule is in the order of millimolar. Thus, at 136 TMRIA molecules per titin and an assumed TMRIA concentration of 1 mM the mean volume of a single, acid-denatured titin molecule is 0.23×10^{-3} μm^3, which gives a mean radius of 0.054 μm for a spherical particle. For comparison, the equilibrium dimensions of a denatured titin molecule may be calculated from its statistical chain parameters. The three-dimensional mean-square end-to-end distance is obtained from the contour (L) and persistence (P) lengths as (Rivetti *et al.*, 1996)

$$\langle R^2 \rangle = 2PL\left(1 - \frac{P}{L}\left(1 - e^{-\frac{L}{P}}\right)\right). \tag{6}$$

For a contour length of 10 μm (Labeit and Kolmerer, 1995) and a persistence length of 1.6 nm (Kellermayer *et al.*, 2000a, b), the mean-square end-to-end distance of the unfolded titin molecule is $\sim 3 \times 10^{-2}$ μm^2, which gives a radius of gyration of ~0.07 μm according to Equation (1). It is likely that the small space occupied by the unfolded titin molecule, relative to the folded one, is due to an increased flexibility of the denatured protein chain. The high flexibility of unfolded domains or quasi-unfolded protein elements (such as the PEVK (Trombitás *et al.*, 1998a, b)), may have important implications for the passive mechanical behavior of striated muscle. The forces required to extend a flexible polymer chain are greater than those required to extend a rigid one. While physiologically occurring, repeated folding–unfolding transitions in titin are debated (Trombitás *et al.*, 1998a, b; Helmes *et al.*, 1999; Minajeva *et al.*, 2001), it is conceivable that unfolding of some of titin's globular domains does take place during pathological overextension of the sarcomere. The increased flexibility of the denatured domain, along with the contour-length gain, may provide a safety mechanism that prevents further denaturation along the molecule and facilitates domain refolding by contraction into a small volume once the molecule is allowed to relax.

Effect of fluorescent modification on titin mechanics

Rhodamine dimers are likely to appear for dynamic (chain diffusion and flexibility) and structural (rhodamine–rhodamine bond) reasons. Due to the dynamics of the titin chain the rhodamine molecules are brought together, resulting in an increased apparent local fluorophore concentration. In addition, weak interactions may hold the rhodamine dimer together. Such interactions may, in principle contribute to stabilizing the three-dimensional structure of the titin molecule. We investigated the effect of TMRIA-labeling on titin mechanics with single-molecule force spectroscopy using atomic force microscopy (AFM) (Figure 4). The unfolding force spectrum of TMRIA-labeled titin, recorded using AFM, was similar to that of the unlabeled

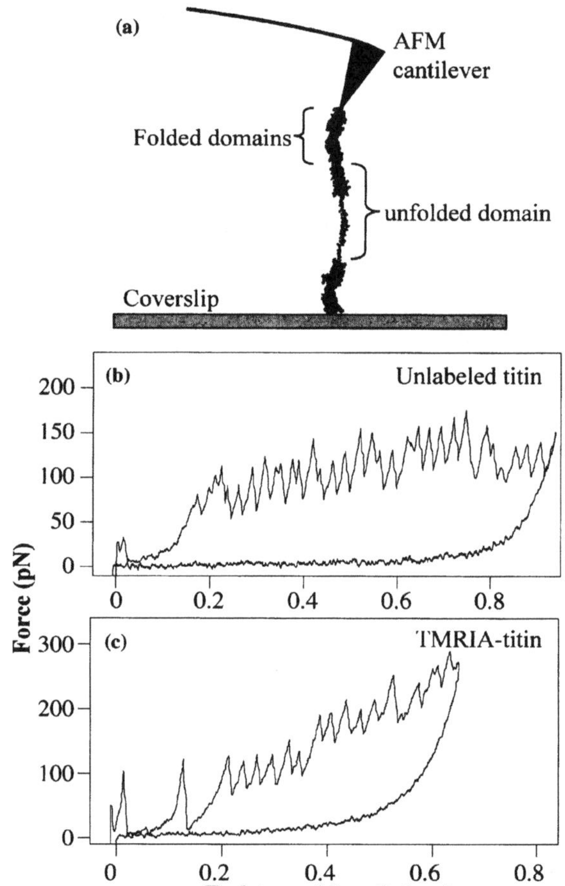

Fig. 4. Stretching titin molecules with the AFM. (a) Experimental arrangement. Titin molecules attached to the surface of a precleaned microscope slide were picked up with an AFM cantilever (#MSCT AUHW, ThermoMicroscopes). Titin was stretched by moving the cantilever away from the surface using an AFM dedicated to manipulating molecules (Asylum Research, Santa Barbara, CA). Cantilever stiffness was calibrated by the thermal method. Force was obtained by measuring cantilever bending. Extension was obtained by measuring cantilever displacement with linear voltage differential transformer (LVDT) and correcting for cantilever bending. (b) Force spectrum of unlabeled titin. (c) Force spectrum of TMRIA-labeled titin.

molecule (Figure 4). The domain unfolding forces were not significantly different in the case of the unlabeled and TMRIA-labeled titin, and we could not resolve significant additional force transitions in the case of TMRIA-titin. Since the AFM experiments did not reveal distinct rhodamine-dimer rupture events the dimer bond energies are much smaller than the free energy of titin domain unfolding (17–28k_BT (Soteriou *et al.*, 1993; Rief *et al.*, 1998)). Thus, we assume that the dimer bonds do not significantly alter the conformation of titin. Although dimer rupture events were undetected, it might be possible that fluorescent modification alters the elasticity of titin, considering that the persistence length of the polypeptide chain depends on the nature of the amino acid side chains (Cantor and Schimmel, 1980) and the modification thereof.

Mechanics

The idea that titin is an elastic molecule has been suggested shortly following the discovery of the protein (Wang *et al.*, 1984). The phenomenology of titin's extensibility has been described in a long series of excellent experiments using sequence-specific antibodies to demarkate specific segments of the molecule (Trombitás *et al.*, 1995, 1998a, b, 1999; Granzier *et al.*, 1996, 1997). These experiments have shown that titin is a multi-stage spring in which various segments extend at different levels of sarcomere stretch. These immunoelectron microscopic observations, however, provide static images and no force data. The extensibility of titin has also been measured using immunofluorescence methods (Linke *et al.*, 1996, 1998a, b, 1999) in combination with force measurements. However, the effect of antibody labeling on titin elasticity has been debated.

In order to measure the elastic behavior of titin, single-molecule manipulation techniques have been employed (for review see Wang *et al.* (2001)). In these experiments single titin molecules were stretched with either laser tweezers (Kellermayer *et al.*, 1997, 1998, 2000a, b, 2002b, submitted; Tskhovrebova *et al.*, 1997) or the atomic force microscope (Rief *et al.*, 1997, 1998). Titin was described as an entropic, wormlike chain in which force-driven unfolding events occur during stretch. Unfolding corresponds to the unfolding of globular domains along titin, and AFM experiments even resolved individual domain unfolding events (Rief *et al.*, 1997). The unfolding forces were measured to range from ~20 pN up to several hundred pN. Refolding has been recorded to occur at low forces (~5 pN) during the mechanical relaxation of the molecule, following a substantial chain shortening (Kellermayer *et al.*, 1997). Notably, a force hysteresis has been observed, which has been attributed to the kinetics of domain unfolding/refolding relative to the kinetics of the experiment. Accordingly, unfolding forces have been shown to vary with the rate of stretch (Rief *et al.*, 1997; Kellermayer *et al.*, 2000a, b). Furthermore, the mechanically driven domain unfolding has been successfully simulated with simple, two-state kinetic models based on the Boltzmann distribution with an added mechanical energy component (Kellermayer *et al.*, 1997; Rief *et al.*, 1997). Although the single-molecule-mechanics experiments contributed significantly to understanding titin elasticity, several intriguing questions remain to be answered: (1) The simple kinetic model of force-driven domain unfolding assumes that the domains are thermally (i.e., randomly) selected for unfolding. In a serially linked chain the mechanical stabilities of the individual domains may impose a mechanical hierarchy that determines the order of domain unfolding. It is unknown whether there is a specific spatial arrangement according to which the domains with different stabilities are arranged along the titin molecule. (2) Mechanical fatigue has been recently demonstrated in repetitively stretched single titin molecules (Kellermayer *et al.*, 1999, 2001). The responsible intramolecular component and the exact mechanisms, however, need to be further explored. (3) It is unclear what factors influence the elasticity and stability of titin *in situ*. The vicinity and binding of sarcomeric components (e.g., actin (Kellermayer and Granzier, 1996; Linke *et al.*, 2002)) may influence titin elasticity. Furthermore, the lack of significant domain unfolding *in situ* (Trombitás *et al.*, 1998a, b; Helmes *et al.*, 1999), compared with the kinetics of mechanically driven unfolding in single-molecule experiments, raises questions about possible domain-stabilizing factors. To sort out some of these questions, one may combine single-molecule manipulation techniques with imaging methods as discussed below.

Stretched, surface-attached titin strands

We have experimented with various techniques to stretch fluorescently labeled titin molecules and image them. First we pushed titin-coated, surface-adsorbed latex microspheres across a glass coverslip, which resulted in the extension of the anchoring titin strands (Figure 5). The image of a group of latex microspheres coated with fluorescently labeled titin is shown in Figure 5a(left). Prior to dissociating from the microsphere the titin strands bound lengthwise to the glass surface and their large-scale structure became stabilized. Stretched, surface-attached titin strands (Figure 5b) appeared as strings of bright beads interrupted with faint, low-fluorescence segments. The axial fluorescence intensity profile contained peaks and valleys corresponding to high fluorophore density in the bright regions and to low density in the faint ones, respectively (Figure 5c). The mean end-to-end length of the stretched titin strands was 2.1 μm (±0.98 SD, $n = 53$).

We have also deposited fluorescently labeled titin molecules directly to the surface from the solution and used meniscus force to stretch them (Tskhovrebova *et al.*, 1997). The force on a molecule caused by receding meniscus is

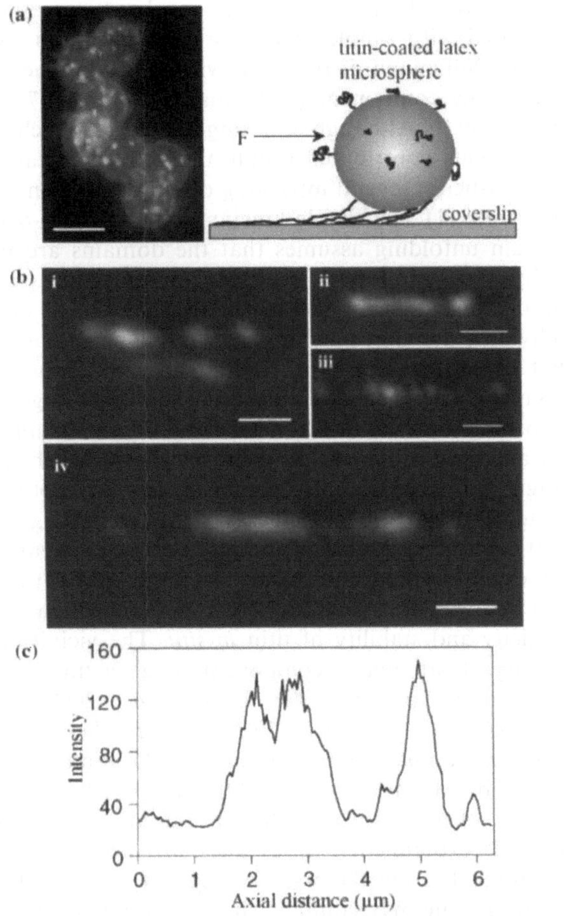

Fig. 5. Stretching titin by manipulating titin-coated latex microspheres. (a) Left: confocal microscopic image of a group of titin-coated microspheres. Scale bar, 3 μm. Images were made using a laser scanning confocal microscope (Bio-Rad, Hertfordshire, UK) with an Ar ion laser. The figure was generated by integrating the optical slices taken across the volume of the beads. Right: Scheme of the titin-stretching experiment. A microsphere carrying fluorescently labeled titin is adsorbed to a glass microscope coverslip, then it is moved laterally (with force F) by scraping the surface with a plastic pipette tip. (b) Images of stretched, surface-adsorbed TMRIA-titin strands. Scale bar, 1 μm. (c) Fluorescence intensity profile along the axis of the titin strand shown in b(iv).

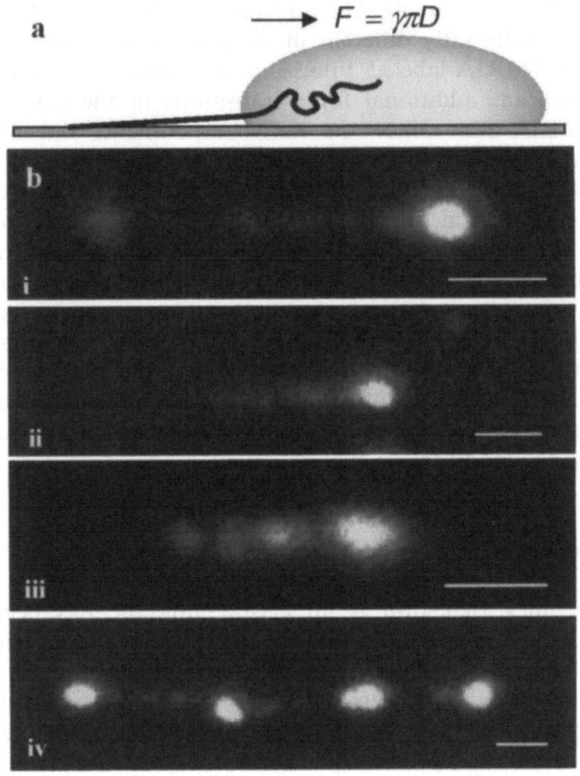

Fig. 6. Stretching fluorescently labeled titin molecules with meniscus force. (a) Scheme of the experimental arrangement. A 100-μl sample of TMRIA-labeled titin molecules in buffer containing 50% glycerol was pipetted onto a precleaned glass coverslip spinning at 5000 rpm. (b) Images of stretched titin molecules. Scale bar, 1 μm.

$$F = \gamma \pi D, \tag{7}$$

where γ is surface tension and D is the diameter of the molecule. The forces acting on titin may be up to several hundred piconewtons (Tskhovrebova *et al.*, 1997), although the force acts only for a brief period, until the meniscus travels along the length of the molecule (fraction of a second (Tskhovrebova *et al.*, 1997)). Images of titin filaments stretched with meniscus force are shown in Figure 6b. The filaments appear as strings of bright beads interrupted with low-fluorescence segments, similarly to those seen in Figure 5.

Single-titin-strand fluorescence resonance energy transfer (FRET)

To further explore the internal structure of the extended titin strand, titin was labeled with the FRET fluorophore pair of IAF (donor) and TMRIA (acceptor) (Figure 7). The overall structure of the extended titin strand revealed by the different fluorophores was similar (Figure 7b). By contrast, the axial fluorescence intensity profiles for IAF and TMRIA differed from each other in the local intensity values (Figure 7c, upper panel). Greater acceptor intensities were seen in the bright regions, while there was a greater donor fluorescence intensity in the interconnecting, faint region. The axially resolved FRET efficiency (E) was calculated as

$$E = [1 + \gamma I_d/I_a]^{-1}, \tag{8}$$

where I_d and I_a are the donor and acceptor intensities during donor excitation, respectively, at identical locations along the molecule axis (Ha *et al.*, 1999). The correction factor γ (Ha *et al.*, 1999) was experimentally determined to be 0.3. The axial distribution of E (Figure 7c, lower panel) contained maxima in the center of the bright regions and minima in the interconnecting faint regions. The structural heterogeneity inferred from the fluorescence images (Figures 5 and 6) was supported by the axial variation in FRET efficiency (E). Accordingly, the local reduction in E is caused by the separation, in short spatial scales (<10 nm), of the donor fluorophores from the acceptors due to the

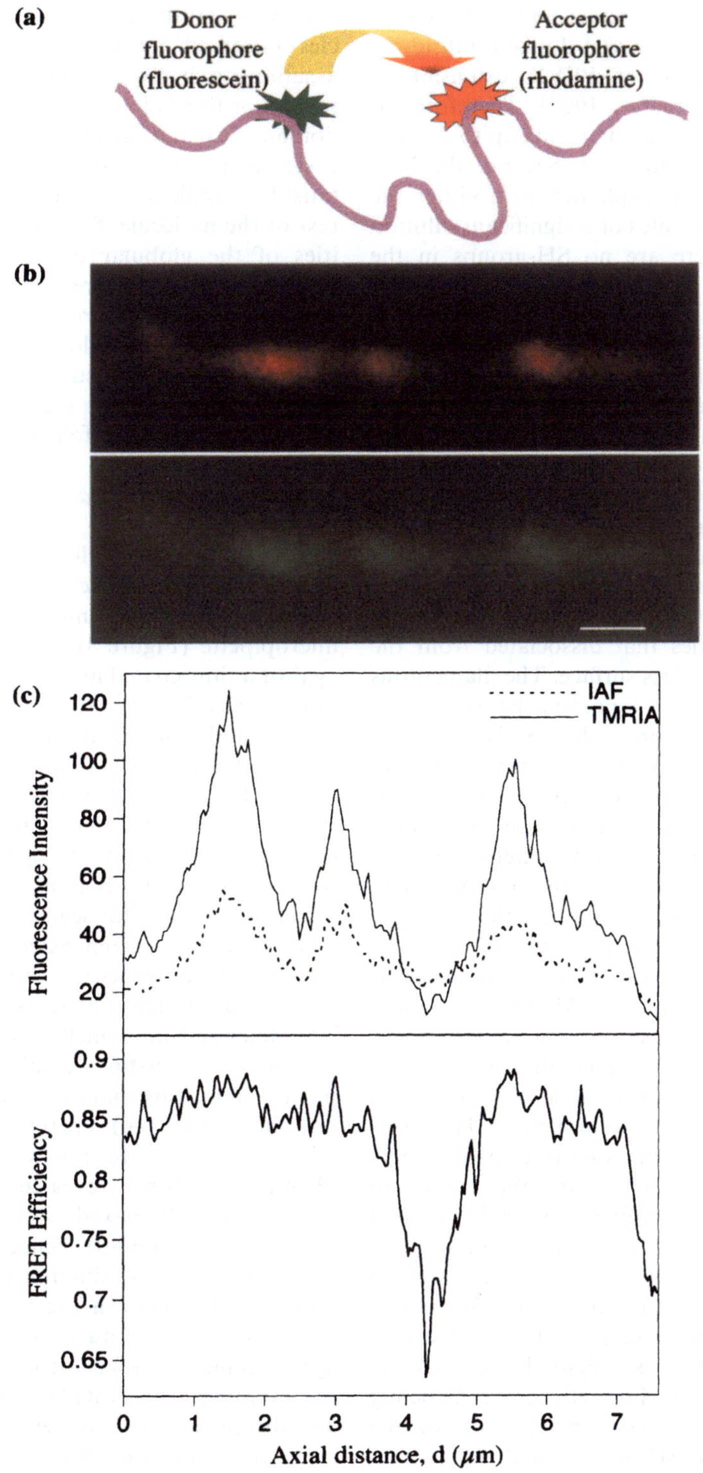

Fig. 7. Fluorescence resonance energy transfer (FRET) in single titin strand. (a) Principle of measurement. Titin was labeled overnight on ice in the presence of equal concentrations of iodoacetamido-fluoresceine (IAF, donor) and TMRIA (acceptor). (b) Donor (IAF) (upper panel) and acceptor (TMRIA) (lower panel) fluorescence images of the same, donor-excited (at 488 nm), stretched, surface-adsorbed titin strand. (c) Axially resolved donor and acceptor emission intensities (upper panel) and FRET efficiency (lower panel).

reduction in structural compactness of titin as a result of mechanical extension.

What structural elements might the high- and low-fluorescence regions correspond to? The faint, low-fluorescence segments are probably either the PEVK segment or unfolded globular domains or both, for at least two reasons: first, the filaments are longer than the contour length of the native titin molecule (~1 μm

(Nave *et al.*, 1989)). Such an overextension may occur at the expense of PEVK extension and domain unfolding. The maximum contribution of PEVK extension is ~0.8 μm (Labeit and Kolmerer, 1995). Therefore, to explain the extension of the titin filaments up to 2.1 μm, domain unfolding must be invoked. Second, the low-fluorescence regions may be explained with either the lack of fluorophore binding sites or a significant dilution of the fluorophores. There are no SH-groups in the PEVK segment (Labeit and Kolmerer, 1995), therefore the TMRIA-labeled and stretched PEVK segment is expected to appear as a dark region interrupting the image of the fluorescent titin molecules. Typically, however, there are several dark regions along the fluorescent filaments. These extra dark regions might be attributed to unfolded and extended globular domains. In such extended regions the fluorophores are axially diluted, displaying low fluorescence.

The alternating bright and faint regions in the filament images might also be explained by alternative mechanisms: (1) The bright spots might represent independent titin molecules that dissociated from the bead as it rolled along the glass surface. The filamentous appearance would then merely represent the path of the bead rather than a single titin molecule. This mechanism, however, is unlikely, because similar results are obtained when titin molecules are directly deposited on the surface, followed by centrifugation and stretching with meniscus force. (2) The filaments might represent more than one titin molecule. Since titin molecules may indeed form oligomers, it is possible that the filaments are composed of more than one molecule. Dimers and trimers have been frequently observed in preparations stretched with meniscus forces (Tskhovrebova *et al.*, 1997). The filament in Figure 6b (iv) may indeed be an oligomer of titin, based on its long length. Currently it is difficult to differentiate between single molecules and oligomers of titin with the fluorescent method. However, considering the point symmetry of the titin oligomers (Kellermayer *et al.*, 2002, submitted), the maximum number of molecules linked in series is probably limited to two. Therefore, even if more than one titin molecule is present in the filament, the number of molecules along the filament axis is unlikely to exceed two. (3) Finally, the heterogeneity in the fluorescence intensity distribution along the filament axis might be caused by variations in the local labeling efficiency. Labeling efficiency varies with the local concentration of the fluorophore binding sites (SH-groups) and the accessibility of the binding sites. Although the average axial cysteine distribution is fairly constant (except for the PEVK segment), it might be possible that some cysteine side chains are inaccessible and therefore unavailable for TMRIA binding. In order for a faint titin segment to be resolvable, its length must be greater than the resolution limit of the microscope at the wavelength used ($d_{min} \sim 0.25$ μm). Considering that the length of a globular domain is ~5 nm including the linker region (Trombitás *et al.*, 1998a, b), a 0.3 μm long titin segment contains ~60 domains. At the typical labeling efficiencies observed by us (~100–140 TMRIA per titin), assuming a homogeneous cysteine accessibility, every second or third globular domain is labeled based on the domain distribution of the cysteine residues. Thus, the cysteines in a stretch of ~60 consecutive domains in titin must be significantly more inaccessible than those in the rest of the molecule. Considering the structural similarities of the globular domains along titin, this is an unlikely scenario. Therefore, it is unlikely that the faint regions along the fluorescent titin filaments are folded, native but stretched globular domains with no fluorophores attached. In sum, there are good reasons to suppose that the faint regions are the PEVK segment and/or stretches of unfolded domains.

Real-time segmental elasticity

The real-time extensibility of titin was studied by visualizing a latex microsphere-adsorbed molecule while mechanically manipulating its free end with a glass micropipette (Figure 8). The micropipette was pressed against a titin-coated microsphere, and a titin strand was pulled away from the surface (Figure 8a). The stretched titin strand contained bright spots along its length interconnected with faint, low-fluorescence segments (Figure 8b). The time-resolved recording of titin stretch (Figure 8c) revealed that as the end-to-end length of the tether increased, the faint segments extended significantly, but the bright regions maintained their resolved dimensions. The faint segments separated by the bright, compact regions often extended in discord. The axial structural heterogeneity in titin persisted during repeated mechanical stretch and relaxation cycles. Notably, the latex microsphere, which was elastically coupled to the microscope coverslip, was displaced towards the micropipette, indicating that force was generated in titin (point no. 5 in Figure 8b–c). Therefore, while the faint segments connecting the bright spots were often difficult to discern directly, the force revealed the uninterrupted serial linkage. Serially linked titin segments often extended differently (Figure 8d). Since serially linked segments bear equal forces, differences in their extensibilities indicate differences in their elastic properties. The appearance of a plateau in segment extensibility implies approaching the segment contour length and thereby reaching an elastic limit (Trombitás *et al.*, 1998a, b), while an abrupt increase in segment extension points to structural transition. Accordingly, segment 4–5 in Figure 8d reaches a (transient) elastic limit at lower filament end-to-end lengths, therefore at lower forces, than segment 3–4, suggesting that, below a filament length of ~1.8 μm, segment 3–4 is more flexible than segment 4–5. Since unfolded titin is more flexible than the native (Kellermayer *et al.*, 1997; Rief *et al.*, 1997; Tskhovrebova *et al.*, 1997), segment 3–4 is possibly a preunfolded section of titin, while segment 4–5 is in the native state in this filament length range. Above a filament length of ~1.8 μm, however, a structural transition begins in

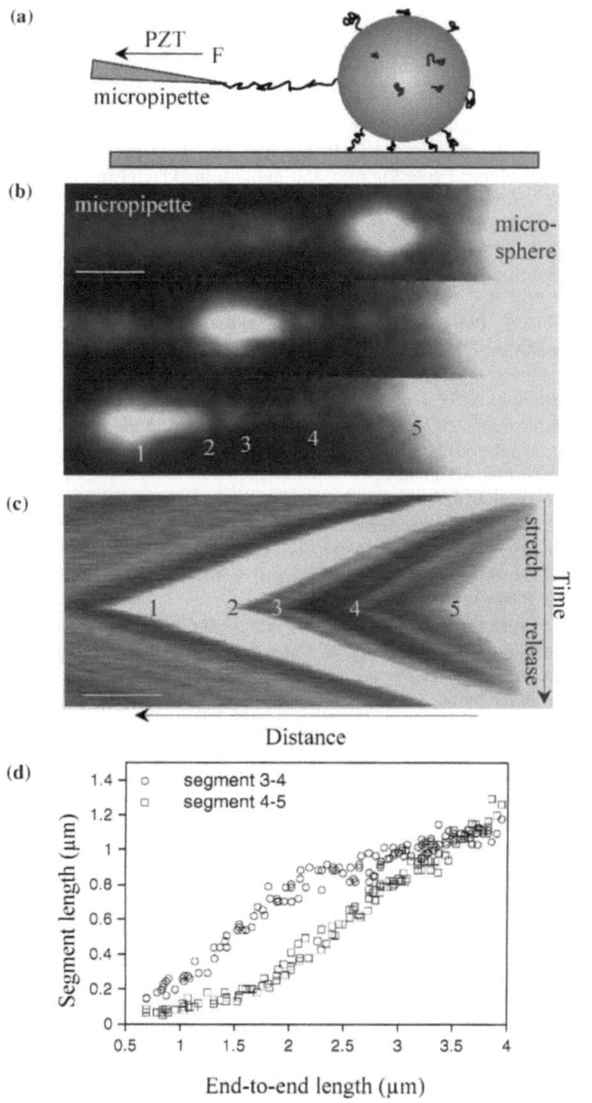

Fig. 8. Micromanipulation and real-time elasticity of fluorescently labeled titin. (a) The free end of a latex microsphere-adsorbed titin molecule was captured with a moveable glass micropipette, stretched, then relaxed. (b) Snapshots of a titin molecule during extension. Numbers identify measurement points along the molecule, which demarcate the ends of extensible segments. Scale bar, 1 μm. (c) Grayscale image reconstructed from the time-resolved fluorescence along the titin axis during stretch and release. Separation between consecutive scan lines, 0.5 s. Total time span, 58 s. Scale bar, 1 μm. (d) Segmental extensibility in titin. Length of segments 3–4 and 4–5 as a function of filament end-to-end length.

segment 4–5 which allows its extension beyond the transient contour length. The structural transition corresponds to the force-driven unfolding of titin's globular domains, leading to an axial dilution of TMRIA, which can be discerned as a local reduction in fluorescence intensity within segment 4–5 (Figure 8c).

Summary and perspectives

Single-molecule manipulation and imaging techniques have given a thrust to our understanding of the properties of the intriguingly interesting, giant titin molecule. There are many exciting questions and problems that remain to be addressed, particularly the exact spatial distribution of the mechanical hierarchy of titin's various sub-segments, the intramolecular dynamics related to repeated mechanical stretch (fatigue), and the effect of known and yet unknown titin-binding proteins on the elasticity and stability of the molecule. The combination of novel single-molecule techniques may give us further unique insights into the structure, dynamics and mechanics of titin (Figure 9). In a combination of high-sensitivity and high-resolution fluorescence imaging with force-measuring laser tweezers the dynamic changes in fluorescence can be correlated with molecular forces (Figure 9a). In the case of TMRIA-labeled titin, in such an arrangement stretching is expected to result in the axial dilution of the fluorophore, upon which rhodamine dimers dissociate, leading to a transiently increased fluorescence. The mutual recording of fluorescence and high-resolution surface topographic (AFM) images may help in precisely identifying the origins of the axial fluorescence variation seen in stretched titin molecules (Figure 9b). Furthermore, by detecting fluorescently modified, sequence-specific anti-titin antibodies, the mechanically driven intramolecular structural changes may be mapped onto titin's sequence at the single-molecule level.

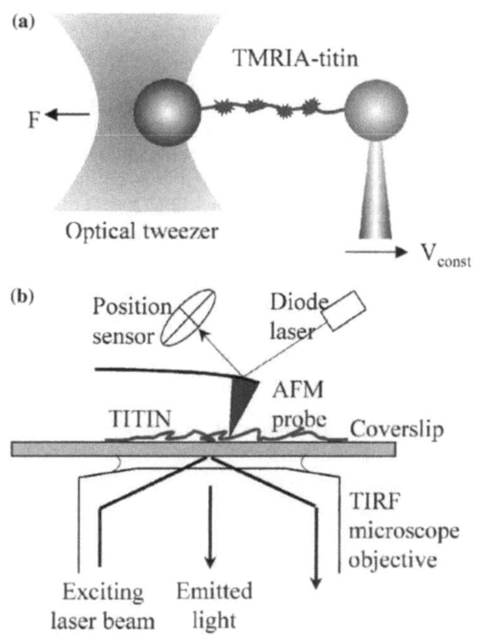

Fig. 9. Perspectives of experimentation on single, fluorescently labeled titin molecules. (a) Stretching single, TMRIA-labeled titin molecules with optical tweezers. Stretching is expected to result in an axial dilution of the fluorophore, upon which rhodamine dimers dissociate, leading to a transiently increased fluorescence. In such an arrangement the dynamic changes in fluorescence can be correlated with molecular forces. (b) Combined imaging of single titin molecules with AFM and TIRF microscopy. Such experiments may help in correlating molecular topography with underlying fluorescence intensity.

Acknowledgements

This work was supported by grants from the Hungarian Science Foundation (OTKA F025353) and the European Union (HPRN-CT-2000-00091) to MSZK. MSZK is a Howard Hughes Medical Institute International Research Scholar. We thank Dr A. Málnási-Csizmadia and A. Nagy for insightful reading of the manuscript.

References

Bang ML, Centner T, Fornoff F, Geach AJ, Gotthardt M, McNabb M, Witt CC, Labeit D, Gregorio CC, Granzier H and Labeit S (2001) The complete gene sequence of titin, expression of an unusual approximately 700-kDa titin isoform, and its interaction with obscurin identify a novel Z-line to I-band linking system. *Circ Res* **89(11):** 1065–1072.

Baranova EG (1965) On the concentration quenching of the luminescence of rhodamine 6G solutions. *Op Spectrosc* **18:** 230–234.

Burghardt TP, Lyke JE and Ajtai K (1996) Fluorescence emission and anisotropy from rhodamine dimers. *Biophys Chem* **59:** 119–131.

Cantor C and Schimmel PR (1980) *Biophysical Chemistry III. The Behavior of Biological Macromolecules.* W.H. Freeman and Company, San Francisco.

Freiburg A, Trombitas K, Hell W, Cazorla O, Fougerousse F, Centner T, Kolmerer B, Witt C, Beckmann JS, Gregorio CC, Granzier H and Labeit S (2000) Series of exon-skipping events in the elastic spring region of titin as the structural basis for myofibrillar elastic diversity. *Circ Res* **86(11):** 1114–1121.

Grama L, Somogyi B and Kellermayer MSZ (2000) *Global Configuration of Titin Revealed by Chain-Associated Rhodamine Dimer Formation.* III. International Conference on Molecular Recognition, University of Pécs, Pécs, Hungary.

Grama L, Somogyi B and Kellermayer MS (2001a) Global configuration of single titin molecules observed through chain-associated rhodamine dimers. *Proc Natl Acad Sci USA* **98(25):** 14,362–14,367.

Grama L, Somogyi B and Kellermayer MSZ (2001b) Direct visualization of surface-adsorbed single fluorescently labeled titin molecules. *Single Mol* **2:** 239–248.

Grama L, Somogyi B and Kellermayer MSZ (2001c) Global configuration and flexibility of fluorescently labeled titin molecules. *Biophys J* **80(1):** 277a.

Granzier H and Labeit S (2002) Cardiac titin: an adjustable multi-functional spring. *J Physiol* **541(Pt 2):** 335–342.

Granzier H, Helmes M, Cazorla O, McNabb M, Labeit D, Wu Y, Yamasaki R, Redkar A, Kellermayer M, Labeit S and Trombitas K (2000) Mechanical properties of titin isoforms. *Adv Exp Med Biol* **481:** 283–300.

Granzier H, Helmes M and Trombitás K (1996) Nonuniform elasticity of titin in cardiac myocytes: a study using immunoelectron microscopy and cellular mechanics. *Biophys J* **70(1):** 430–442.

Granzier H, Kellermayer M, Helmes M and Trombitás K (1997) Titin elasticity and mechanism of passive force development in rat cardiac myocytes probed by thin-filament extraction. *Biophys J* **73(4):** 2043–2053.

Gregorio CC, Granzier H, Sorimachi H and Labeit S (1999) Muscle assembly: a titanic achievement? *Curr Opin Cell Biol* **11(1):** 18–25.

Ha T, Ting AY, Liang J, Caldwell WB, Deniz AA, Chemla DS, Schultz PG and Weiss S (1999) Single-molecule fluorescence spectroscopy of enzyme conformational dynamics and cleavage mechanism. *Proc Natl Acad Sci USA* **96:** 893–898.

Hallett P, Tskhovrebova L, Trinick J, Offer G and Miles MJ (1996) Improvements in atomic force microscopy protocols for imaging fibrous proteins. *J Vac Sci Technol B* **14(2):** 1444–1448.

Helmes M, Trombitás K, Centner T, Kellermayer M, Labeit S, Linke WA and Granzier H (1999) Mechanically driven contour-length adjustment in rat cardiac titin's unique N2B sequence: titin is an adjustable spring. *Circ Res* **84(11):** 1339–1352.

Higuchi H, Nakauchi Y, Maruyama K and Fujime S (1993) Characterization of beta-connectin (titin 2) from striated muscle by dynamic light scattering. *Biophys J* **65(5):** 1906–1915.

Horowits R, Kempner ES, Bisher ME and Podolsky RJ (1986) A physiological role for titin and nebulin in skeletal muscle. *Nature* **323(6084):** 160–164.

Improta S, Krueger JK, Gautel M, Atkinson RA, Lefevre JF, Moulton S, Trewhella J and Pastore A (1998) The assembly of immunoglobulin-like modules in titin: implications for muscle elasticity. *J Mol Biol* **284(3):** 761–777.

Improta S, Politou AS and Pastore A (1996) Immunoglobulin-like modules from titin I-band: extensible components of muscle elasticity. *Structure* **4(3):** 323–337.

Kellermayer MSZ (2000a) *Mechanical Behavior of the Giant Protein Titin Explored by Imaging and Manipulating Single Molecules.* III. International Conference on Molecular Recognition, University of Pécs, Pécs, Hungary.

Kellermayer MSZ (2000b) The muscular break: molecular mechanisms of regulating sarcomeric extensibility by titin. *Biophys J* **78:** 3A.

Kellermayer MSZ and Granzier HL (1996) Calcium-dependent inhibition of in vitro thin-filament motility by native titin. *FEBS Lett* **380(3):** 281–286.

Kellermayer MSZ, Grama L and Somogyi B (2000a) Direct visualization of the extensibility of fluorescently labeled titin molecules. *Biophys J* **78:** 392A.

Kellermayer MSZ, Smith S, Bustamante C and Granzier HL (2000b) Mechanical manipulation of single titin molecules with laser tweezers. *Adv Exp Med Biol* **481:** 111–126.

Kellermayer MSZ, Smith SB, Bustamante C and Granzier HL (1999) A molecular segment in titin is an adjustable spring. *Biophys J* **76:** A9.

Kellermayer MSZ, Smith SB, Bustamante C and Granzier HL (1998) Complete unfolding of the titin molecule under external force. *J Struct Biol* **122(1–2):** 197–205.

Kellermayer MSZ, Smith SB, Granzier HL and Bustamante C (1997) Folding-unfolding transitions in single titin molecules characterized with laser tweezers. *Science* **276(5315):** 1112–1126.

Kellermayer MS, Smith SB, Bustamante C and Granzier HL (2001) Mechanical fatigue in repetitively stretched single molecules of titin. *Biophys J* **80(2):** 852–863.

Kellermayer MSZ, Smith BL, Bustamante C and Granzier HL (2002a) Force-driven folding and unfolding transitions in single titin molecules: single polymer strand manipulation. *NATO Science Series 3. High Technology* **87:** 311–326.

Kellermayer MSZ, Trombitás K and Granzier HL (2002b) Formation and elasticity of multimolecular bundles of titin. *Biophys J* **82:** 370.

Kurzban GP and Wang K (1988) Giant polypeptides of skeletal muscle titin: sedimentation equilibrium in guanidine hydrochloride. *Biochem Biophys Res Commun* **150(3):** 1155–1161.

Labeit S and Kolmerer B (1995) Titins: giant proteins in charge of muscle ultrastructure and elasticity. *Science* **270(5234):** 293–296.

Labeit S, Gautel M, Lakey A and Trinick J (1992) Towards a molecular understanding of titin. *Embo J* **11(5):** 1711–1716.

Lang D and Coates P (1968) Diffusion coefficient of DNA in solution at 'zero' concentration as measured by electron microscopy. *J Mol Biol* **36:** 137–151.

Levshin LV and Bocharov VG (1961) Study of concentration effects in solutions of certain organic compounds. *Op Spectrosc* **10:** 330–333.

Linke WA, Ivemeyer M, Mundel P, Stockmeier MR and Kolmerer B (1998a) Nature of PEVK-titin elasticity in skeletal muscle. *Proc Natl Acad Sci USA* **95(14):** 8052–8057.

Linke WA, Ivemeyer M, Olivieri N, Kolmerer B, Ruegg JC and Labeit S (1996) Towards a molecular understanding of the elasticity of titin. *J Mol Biol* **261(1):** 62–71.

Linke WA, Kulke M, Li H, Fujita-Becker S, Neagoe C, Manstein DJ, Gautel M and Fernandez JM (2002) PEVK Domain of Titin: an Entropic Spring with Actin-Binding Properties. *J Struct Biol* **137(1–2):** 194–205.

Linke WA, Rudy DE, Centner T, Gautel M, Witt C, Labeit S and Gregorio CC (1999) I-band titin in cardiac muscle is a three-element molecular spring and is critical for maintaining thin filament structure. *J Cell Biol* **146(3):** 631–644.

Linke WA, Stockmeier MR, Ivemeyer M, Hosser H and Mundel P (1998b) Characterizing titin's I-band Ig domain region as an entropic spring. *J Cell Sci* **111(Pt 11):** 1567–1574.

Ma K, Kan L and Wang K (2001) Polyproline II helix is a key structural motif of the elastic PEVK segment of titin. *Biochemistry* **40(12):** 3427–3438.

Maruyama K (1997) Connectin/titin, giant elastic protein of muscle. *Faseb J* **11(5):** 341–345.

Minajeva A, Kulke M, Fernandez JM and Linke WA (2001) Unfolding of titin domains explains the viscoelastic behavior of skeletal myofibrils. *Biophys J* **80(3):** 1442–1451.

Nave R, Fürst DO and Weber K (1989) Visualization of the polarity of isolated titin molecules: a single globular head on a long thin rod as the M band anchoring domain? *J Cell Biol* **109(5):** 2177–2187.

Packard BZ, Toptygin DD, Komoriya A and Brand L (1996) Protofluorescent protease substrates: intramolecular dimers described by the exciton model. *Proc Natl Acad Sci USA* **93:** 11,640–11,645.

Plant AL (1986) Mechanism of concentration quenching of a xanthene dye encapsulated in phospholipid vesicles. *Photochem Photobiol* **44(4):** 453–459.

Politou AS, Gautel M, Improta S, Vangelista L and Pastore A (1996) The elastic I-band region of titin is assembled in a 'modular' fashion by weakly interacting Ig-like domains. *J Mol Biol* **255(4):** 604–616.

Politou AS, Gautel M, Joseph C and Pastore A (1994a) Immunoglobulin-type domains of titin are stabilized by amino-terminal extension. *FEBS Lett* **352(1):** 27–31.

Politou AS, Gautel M, Pfuhl M, Labeit S and Pastore A (1994b) Immunoglobulin-type domains of titin: same fold, different stability? *Biochemistry* **33(15):** 4730–4737.

Politou AS, Thomas DJ and Pastore A (1995) The folding and stability of titin immunoglobulin-like modules, with implications for the mechanism of elasticity. *Biophys J* **69(6):** 2601–2610.

Rief M, Fernandez JM and Gaub HE (1998) Elastically coupled two-level systems as a model for biopolymer extensibility. *Phys Rev Lett* **81(21):** 4764–4767.

Rief M, Gautel M, Oesterhelt F, Fernandez JM and Gaub HE (1997) Reversible unfolding of individual titin immunoglobulin domains by AFM. *Science* **276(5315):** 1109–1112.

Rivetti C, Guthold M and Bustamante C (1996) Scanning force microscopy of DNA deposited onto mica: equilibration versus kinetic trapping studied by statistical polymer chain analysis. *J Mol Biol* **264:** 919–932.

Rohatgi KK and Singhal GS (1966) Nature of bonding in dye aggregates. *J Phys Chem* **70(6):** 1695–1701.

Selwyn JE and Steinfeld JI (1972) Aggregation equilibria of xanthene dyes. *J Phys Chem* **76(5):** 762–774.

Soteriou A, Clarke A, Martin S and Trinick J (1993) Titin folding energy and elasticity. *Proc R Soc Lond B Biol Sci* **254(1340):** 83–86.

Trinick J (1996) Cytoskeleton: Titin as a scaffold and spring. *Curr Biol* **6(3):** 258–260.

Trinick J and Tskhovrebova L (1999) Titin: a molecular control freak. *Trends Cell Biol* **9(10):** 377–380.

Trinick J, Knight P and Whiting A (1984) Purification and properties of native titin. *J Mol Biol* **180(2):** 331–356.

Trombitás K, Freiburg A, Centner T, Labeit S and Granzier H (1999) Molecular dissection of N2B cardiac titin's extensibility. *Biophys J* **77(6):** 3189–3196.

Trombitás K, Greaser M, French G and Granzier H (1998a) PEVK extension of human soleus muscle titin revealed by immunolabeling with the anti-titin antibody 9D10. *J Struct Biol* **122(1–2):** 188–196.

Trombitás K, Greaser M, Labeit S, Jin JP, Kellermayer M, Helmes M and Granzier H (1998b) Titin extensibility in situ: entropic elasticity of permanently folded and permanently unfolded molecular segments. *J Cell Biol* **140(4):** 853–859.

Trombitás K, Jin JP and Granzier H (1995) The mechanically active domain of titin in cardiac muscle. *Circ Res* **77(4):** 856–861.

Tskhovrebova L and Trinick J (1997) Direct visualization of extensibility in isolated titin molecules. *J Mol Biol* **265(2):** 100–106.

Tskhovrebova L and Trinick J (1999) Coordinated light, electron and atomic force microscopy of titin molecules. *Biophys J* **76:** A9.

Tskhovrebova L and Trinick J (2000) Extensibility in the titin molecule and its relation to muscle elasticity. *Adv Exp Med Biol* **481:** 163–173.

Tskhovrebova L and Trinick J (2001) Flexibility and extensibility in the titin molecule: analysis of electron microscope data. *J Mol Biol* **310(4):** 755–771.

Tskhovrebova L and Trinick J (2002) Role of titin in vertebrate striated muscle. *Philos Trans Roy Soc Lond B Biol Sci* **357(1418):** 199–206.

Tskhovrebova L, Trinick J, Sleep JA and Simmons RM (1997) Elasticity and unfolding of single molecules of the giant muscle protein titin. *Nature* **387(6630):** 308–312.

Wang K (1996) Titin/connectin and nebulin: giant protein rulers of muscle structure and function. *Adv Biophys* **33:** 123–134.

Wang K, Forbes JG and Jin AJ (2001) Single molecule measurements of titin elasticity. *Prog Biophys Mol Biol* **77(1):** 1–44.

Wang K, Ramirez-Mitchell R and Palter D (1984). Titin is an extraordinarily long, flexible, and slender myofibrillar protein. *Proc Natl Acad Sci USA* **81(12):** 3685–3689.

Zhuang X, Ha T, Kim HD, Centner T, Labeit S and Chu S (2000) Fluorescence quenching: a tool for single-molecule protein-folding study. *Proc Natl Acad Sci USA* **97(26):** 14,241–14,244.

Zhuang X, Ha T, Kim HD, Chu S and Labeit S (1999) Fluorescence and force study of folding/unfolding transitions in single titin molecules. *Biophys J* **76(1):** A9.

Journal of Muscle Research and Cell Motility **23**: 513–521, 2002.

Unfolding of titin domains studied by molecular dynamics simulations

MU GAO[1], HUI LU[2] AND KLAUS SCHULTEN[1,*]
[1]*Department of Physics and Beckman Institute, University of Illinois at Urbana-Champaign, IL 61801, USA;*
[2]*Department of Bioengineering, University of Illinois at Chicago, IL 60607, USA*

Abstract

Titin, a ~1 μm long protein found in striated muscle myofibrils, possesses unique elastic properties. The extensible behavior of titin has been demonstrated in atomic force microscopy and optical tweezer experiments to involve the reversible unfolding of individual immunoglobulin-like (Ig) domains. We have used steered molecular dynamics (SMD), a novel computer simulation method, to investigate the mechanical response of single titin Ig domains upon stress. Simulations of stretching Ig domains I1 and I27 have been performed in a solvent of explicit water molecules. The SMD approach provides a detailed structural and dynamic description of how Ig domains react to external forces. Validation of SMD results includes both qualitative and quantitative agreement with AFM recordings. Furthermore, combining SMD with single molecule experimental data leads to a comprehensive understanding of Ig domains' mechanical properties. A set of backbone hydrogen bonds that link the domains' terminal β-strands play a key role in the mechanical resistance to external forces. Slight differences in architecture permit a mechanical unfolding intermediate for I27, but not for I1. Refolding simulations of I27 demonstrate a locking mechanism.

Introduction

The giant muscle protein titin, also known as connectin, is a roughly 1 μm long macromolecule which spans over half of the muscle sarcomere and plays important roles in muscle contraction and passive elasticity as reviewed in Wang (1996), Erickson (1997), Maruyama (1997), Linke (2000), Granzier and Labeit (2002), Tskhovrebova and Trinick (2002). This string-like molecule is primarily composed of ~300 tandem repeats of two types of domains, immunoglobulin-like (Ig) domains and fibronectin type III (FnIII) domains, as well as of a unique region called PEVK that is rich in proline, glutamate, valine and lysine residues (Labeit and Kolmerer, 1995). Anchored at the Z-disk and at the M-line, titin develops a passive force which keeps sarcomere components uniformly organized during muscle contraction and restores sarcomere length when the muscle is relaxed. The I-band part of titin is believed to be responsible for titin extensibility and passive elasticity. It consists mainly of two tandem regions of Ig domains (proximal Ig region and distal Ig region), separated by the PEVK region. Each of the Ig domains possesses about 100 amino acids built in a β-sandwich architecture. The PEVK region, however, mostly contains coiled conformations which are easily elongated when the muscle is stretched (Linke *et al.*, 1998; Trombitas *et al.*, 1998; Ma *et al.*, 2001; Li *et al.*, 2001b). It has been proposed that individual Ig domains also unfold to provide the necessary extension for muscle (Soteriou *et al.*, 1993; Erickson, 1994). Recent studies have found that massive Ig domain unfolding cannot happen, rather, upon rapid stretching only a few of them unfold (Minajeva *et al.*, 2001).

Titin domains have been manipulated, using single molecule techniques, in atomic force microscopy (AFM) (Rief *et al.*, 1997; Carrion-Vazquez *et al.*, 1999, Marszalek *et al.*, 1999; Li *et al.*, 2002) and optical tweezer (Kellermayer *et al.*, 1997; Tskhovrebova *et al.*, 1997) experiments, and found to possess protection against strain-induced domain unfolding. We refer to proteins which are designed to respond to force application under physiological conditions as *mechanical proteins*. Other proteins, which may not encounter mechanical strain under physiological conditions, have been found to exhibit little resistance against strain-induced unfolding, as has been demonstrated through AFM experiments on spectrin (Rief *et al.*, 1999) and barnase (Best *et al.*, 2001).

The AFM force-extension profiles of multimers of Ig domains (from titin) and FnIII (from titin, and also from tenascin and fibronectin) display a regularly repeated sawtooth pattern (Rief *et al.*, 1997, 1998; Oberhauser *et al.*, 1998, 2002). The spacing between any two force peaks matches the extended length of one Ig or FnIII domain from its folded state, proving that, when these multi-domain proteins are stretched, their domains unfold one by one. The high magnitude of the force peaks, dependent on the pulling speed and type of protein, implies that the Ig and FnIII domains are designed to withstand significant stretching forces. At a pulling speed of 1 μm/s, for example, AFM unfolding of titin Ig domains requires about 200 pN while unfolding

* To whom correspondence should be addressed: E-mail: kschulte@ks.uiuc.edu

of FnIII domains requires a force ranging from 75 to 200 pN (Rief *et al.*, 1997; Oberhauser *et al.*, 2002). One remarkable property of these mechanical proteins is that they unfold reversibly. Refolding rates of around 1 s^{-1} were reported in AFM forced-unfolding experiments of titin I27 and FN-III_{10} (Rief *et al.*, 1997; Carrion-Vazquez *et al.*, 1999; Oberhauser *et al.*, 2002).

The experimental observations required an interpretation in terms of the architecture of the titin modules. Currently, only two experimental structures of titin I-band Ig domains are available, the 1st Ig domain (Mayans *et al.*, 2001) from the proximal Ig region and the 27th Ig domain (Improta *et al.*, 1996) (according to the new nomenclature (Freiburg *et al.*, 2000), I27 is now termed I91) from the distal Ig region. These domains adopt the typical I-frame immunoglobulin super-family fold (Politou *et al.*, 1995), consisting of two β-sheets packed against each other, as shown in Figure 1, with each sheet containing four or five β-strands. One sheet comprises strands A, B, E, and D, the other sheet comprises strands A′, G, F, C, and C′ (the latter in I1 only). All adjacent β-strands in both sheets are anti-parallel to each other, except for the parallel pair A′ and G. The β-strands A and A′ belong to different sheets but are part of the N-terminal strand. Both structures are stabilized by hydrophobic core interactions between the two β-sheets and by hydrogen bonds between β-strands. One major structural difference between I1 and I27, however, is that I1 contains a pair of cysteines which form a disulfide bridge under oxidizing conditions between two β-sheets, whereas I27 seems to lack the potential to form a disulfide bridge, even though I27 also contains two cysteine residues.

While single molecule experiments provide valuable dynamic force spectroscopy for mechanical proteins, the corresponding conformational changes of stretched proteins cannot be observed directly because conventional structure resolving methods such as X-ray and NMR are not applicable. SMD is the method that is well positioned to address this problem and unveil the atomic-level pictures behind the sawtooth like force profiles observed in AFM experiments. SMD, reviewed in (Isralewitz *et al.*, 2001), is a novel approach to the study of the dynamics of biomolecular systems, especially for mechanical proteins. Starting from an equilibrated X-ray or NMR structure, the dynamics of the protein are recorded during force-induced unfolding simulations, thus providing a better understanding of the structure-function relationship of the simulated protein. The advantage of SMD over conventional molecular dynamics is the ability to induce relatively large conformational changes in molecules on the nanosecond time scales accessible to computation. Two main pulling protocols have been used in simulating Ig domain unfolding: constant velocity pulling and constant force pulling. Constant velocity pulling mimics the AFM experiment and, hence, allows one to compare SMD data with AFM recordings and to provide interpretation for experiments. Constant force pulling allows extensive studies of the barrier crossing event

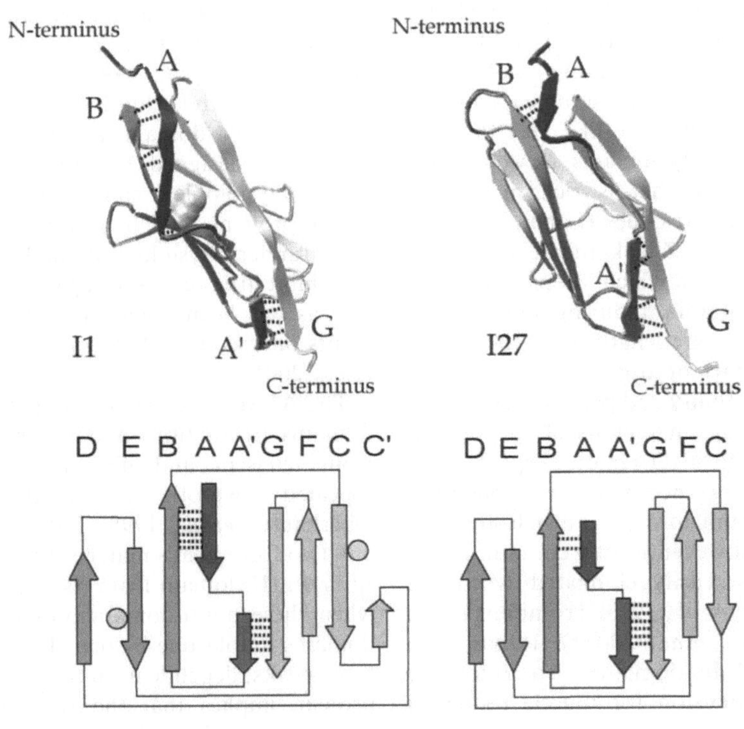

Fig. 1. Structures and topology diagrams of I1 and I27 domains. β-strand A and A′ are shown in black, other β-strands are shown in grey. Essential inter-strand hydrogen bonds for the mechanical stability of titin modules are represented as black dashed lines. The cysteine residues forming a disulfide bridge in I1 are represented as circles.

along the potential surface of the unfolding pathway. In combination with a statistical analysis, one can provide a quantitative description of the potential barrier separating the folded and unfolded domains and compare it to experimental observations.

In this paper we summarize the results of molecular dynamics simulations presented elsewhere (Lu *et al.*, 1998; Lu and Schulten, 1999a, b, 2002; Gao *et al.*, 2001, 2002b) to provide an overall view of the reversible unfolding of titin modules.

Methods

The Ig domain structures, I27 and I1, obtained from the protein data bank entry 1TIT and 1G1C, were employed in this modeling study. To simulate a solvent environment, the Ig domain structures were surrounded by a sphere of water molecules, which covered the molecular surface of the domain by at least four shells of water molecules, resulting in systems of 12,000–18,000 atoms.

The MD simulations were performed using the programs X-PLOR (Brünger, 1992) and NAMD (Kalé *et al.*, 1999) with the CHARMm22 force field (MacKerell *et al.*, 1998), assuming an integration time step of 1 fs and a uniform dielectric constant of one. The nonbonded van der Waals (vdW) and Coulomb interactions were calculated with a cut-off using a switching function starting at a distance of 10 Å and reaching zero at 13 Å. The TIP3P water model was employed for the solvent (Jorgensen *et al.*, 1983).

Two SMD protocols, constant-velocity and constant force stretching, have been applied to titin modules. SMD using constant-velocity moving restraints simulates the stretching of protein domains with a moving AFM cantilever. One terminus is fixed during the simulation, which prevents the protein from simply translating in space when external forces are applied, and this corresponds to attaching the protein to a fixed substrate in the AFM experiments. The other terminus is restrained to a point in space (restraint point) by an external, e.g., harmonic, potential. The restraint point is then shifted in a chosen direction (Grubmüller *et al.*, 1996; Israelewitz *et al.*, 1997; Lu *et al.*, 1998), forcing the restrained terminus to move from its initial position. Assuming a single reaction coordinate x and an external potential $U = k(x - x_0 - vt)^2/2$, where x_0 is the initial position of the restraint point moving with a constant-velocity v, and k is the stiffness of the restraint, the external force exerted on the system can be expressed as

$$F = k(x_0 + vt - x). \quad (1)$$

This force corresponds to a protein being pulled by a harmonic spring of stiffness k with one end attached to the restrained terminus and the other end moving with velocity v. The force applied in constant force SMD retains the same direction and magnitude regardless of the position of the restrained terminus.

SMD simulations with constant-velocity stretching were carried out on I1 and I27 by fixing the C_α atoms of the N-terminal residue, and by applying external forces to the C_α atom of the C-terminal residue. The value of k was set to 7 $k_B T/\text{Å}^2$, corresponding to spatial (thermal) fluctuations of the constrained C_α atom with variance $\delta x = \sqrt{k_B T/k} = 0.38$ Å at $T = 300$ K. SMD simulations with constant force stretching were implemented by fixing the N-terminus of the Ig domains and by applying a constant force to the C-terminus along the direction connecting the initial positions of the N-terminus to the C-terminus.

Results

The results of forced unfolding of I1 (oxidized and reduced) and I27 with constant-velocity stretching are compared in Figure 2a. I27 exhibits the major resistance to external forces in the extension from 12 to 15 Å, where a dominant force peak arises. At extensions larger than 20 Å the domain exhibits little resistance against stretching as is evident from the fact that only relatively weak forces are needed to increase extension. The force-extension profiles from previous work (Lu and Schulten, 1999a; Gao *et al.*, 2002b) have also shown that the lower the pulling speed, the lower the forces needed to extend the domain. A decrease of the speed from 0.5 to 0.01 Å/ps reduces the peak force from 2500 to 1500 pN for I27. Since the current accessible timescale to SMD simulations, nanosecond, is about six orders of magnitude shorter than a millisecond over which unfolding of Ig domains occurs in AFM experiments, the peak forces calculated in SMD simulations are larger than those measured in AFM experiments (Lu *et al.*, 1998; Lu and Schulten, 1999a). Despite this the simulated force-extension curves share the main qualitative feature with those observed in AFM stretching experiments: a single dominant force peak. I1 domains exhibit strong resistance against external forces in the extension range 5–16 Å, which is broader than the main burst region of I27. However, the unfolding force peak of I1 is lower than that of I27. For example, at 0.1 Å/ps pulling speed unfolding of I1 requires a peak force that is about 300 pN lower than that of unfolding I27. This implies that I1 is slightly less stable than I27. Similar to I27, unraveling I1 to extensions beyond the main force peak, which corresponds to the unfolding barrier, requires weaker and weaker forces until the protein backbone is completely extended when forces rise again. Since oxidized I1 contains a disulfide bond, the domain can only extend to ~220 Å as shown in Figure 2a, whereas reduced I1 can be stretched to ~300 Å, the length of a fully extended I1 domain.

Although constant velocity simulations reproduced a qualitatively similar force-extension profile as seen in AFM experiments, revealing the key factor that determines the unfolding barrier and obtaining a quantitative comparison to experimental data is difficult to

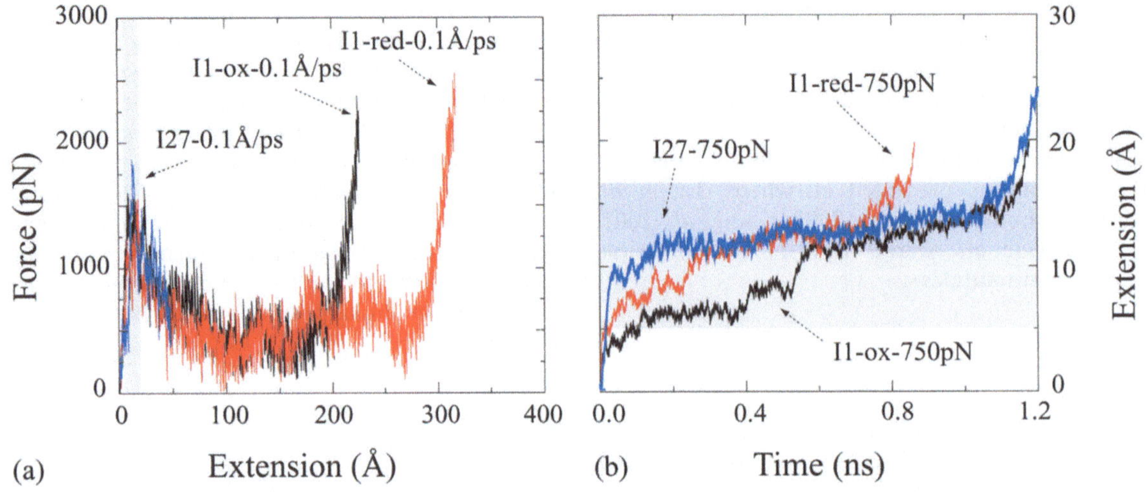

Fig. 2. (a) Force-extension profiles from SMD simulations with 0.1 Å/ps constant velocity stretching of oxidized I1 (black), reduced I1 (red) and I27 (blue). The gray area highlights the region where the force peak arises. (b) Extension-time profiles from SMD simulations with 750 pN constant force stretching of oxidized I1, reduced I1 and 27. The shaded areas correspond to key transition states for crossing barriers formed by hydrogen bonds between A- and B-strands (light gray) and by hydrogen bonds between A′ and G-strands (dark gray).

achieve from such approach because the proteins pass the transition state very quickly, typically in <100 ps. To address these issues, we have introduced constant force simulations in which forces much lower than the peak unfolding force calculated from constant velocity simulations were applied. During constant force simulations the proteins were hindered in front of the unfolding barrier for up to several ns, thus allowing us to capture the key events that characterize the barrier crossing (Lu and Schulten, 1999a). Moreover, with the mean first passage time theory we are capable to calculate the height of the unfolding barrier that can be compared with experimental results.

An analysis of the constant force simulation trajectories corresponding to the profiles shown in Figure 2b reveals that I27 unfolds in two key steps (Marszalek *et al.*, 1999; Lu and Schulten, 1999a). At the first step, an initial 12 Å extension takes place due to a straightening of the polypeptide chain near its terminal ends and disruption of the two hydrogen bonds between β-strands A and B. This extension, shown in Figure 2b for 750 pN stretching, continues until a plateau region (which is longer for weaker forces) is reached, at which point the six A′–G inter-strand hydrogen bonds come under mechanical strain. At the second step, the domain extensions fluctuate around a constant value during the plateau period, until the A′–G bonds concurrently break, after which the extension rapidly increases. Similarly, unfolding I1 also involves rupture of both A–B and A′–G hydrogen bonds (Gao *et al.*, 2002b). Compared to I27, however, disrupting the A–B bonds takes longer, for there are six A–B hydrogen bonds in I1. The main resistance of I1 is attributed to its six A–B bonds, especially for reduced I1 in which the disulfide bridge is absent. Under oxidizing conditions, the formation of the disulfide bond increases the stability of I1 by stabilizing A′–G bonds. After the A–B and A′–G bonds are broken, I1 and I27 unravel rapidly, with little resistance to the rupture of interstrand hydrogen bonds that occurs in a zipper-like, i.e., one-by-one, fashion.

The plateaus shown in Figure 2b correspond to barrier crossing processes, with τ_{barrier} denoting the mean first passage time spent at the plateau (Lu and Schulten, 1999a). The applied force effectively lowers the barrier such that stronger forces lead to faster barrier crossing than weaker forces. In all cases, the stretching motion gets temporarily 'stuck' in front of the barrier, which can then only be overcome by thermal fluctuations. This scenario can be described as Brownian motion governed by a potential which is the sum of the indigenous barrier and a linear potential accounting for the applied force. The mean time to cross the barrier can be evaluated using the expressions for the mean first passage time developed in Schulten *et al.* (1980, 1981), Szabo *et al.* (1980). By comparing the mean first passage times with the respective times τ_{barrier} for various forces one can estimate the height of the indigenous potential barrier.

To estimate the shape of the indigenous potential barrier we assume a linearly increasing ramp model for the energy $U(x)$ of the barrier: $U(x) = \Delta U(x-a)/(b-a)$ for $a \leq x \leq b$. The dynamics of barrier crossing is described as a Brownian motion with a reflective boundary condition at a, and an absorptive boundary condition at b. Here ΔU is the height of the barrier, and $b-a$ the barrier width. This linear potential function should yield approximately the same τ_{barrier} as results from the indigenous barrier. The simple form of the model potential permits an analytical expression for the mean first passage time (Izrailev *et al.*, 1997):

$$\tau(F) = 2\tau_d \delta(F)^{-2}[e^{\delta(F)} - \delta(F) - 1] \qquad (2)$$

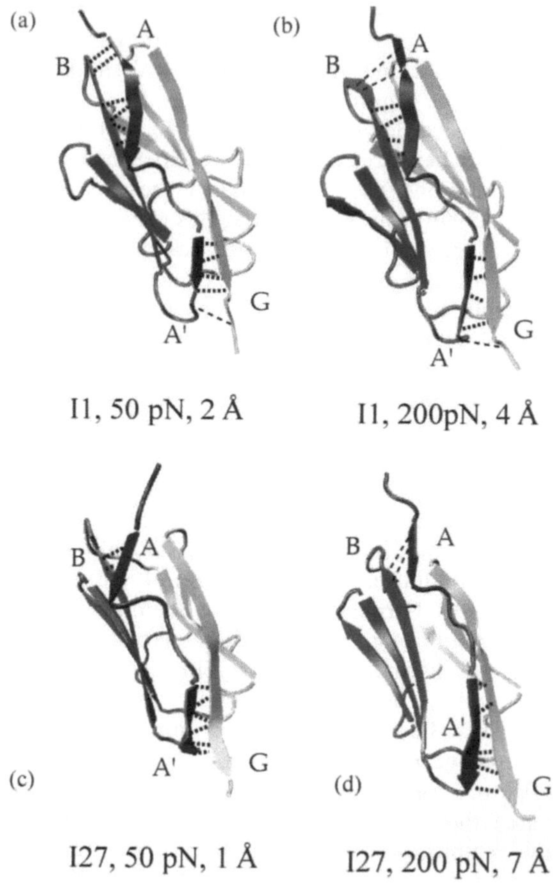

Fig. 3. Snapshots of titin modules from SMD simulations with constant forces of 50 and 200 pN. Two hydrogen bonds between A- and B-strands of I27 are disrupted with 200 pN force while I1's six hydrogen bonds between A- and B-strands are intact.

We have introduced here $\tau_d = (b - a)^2/2D$ and $\delta(F) = \beta[\Delta U - F(b - a)]$. Assuming a width of 3 Å, estimated from the fluctuation of the extension curves in Figure 2, a least square fit procedure produces a value of 20.3 kcal/mol for the height of the unfolding barrier of I27 (Lu and Schulten, 1999a), in close agreement with AFM and denaturing experiments that suggest 22.2 kcal/mol (Carrion-Vazquez *et al.*, 1999).

The constant force simulations also revealed an important mechanical difference between I1 and I27: I27 exhibits an intermediate of ~7 Å extension (Marszalek *et al.*, 1999), whereas I1 does not exhibit a similar intermediate (Gao *et al.*, 2002b). Figure 3 shows the extension of I1 and I27 in SMD simulations at constant forces of 50 and 200 pN applied for up to 10 ns. I1 appears less flexible than I27 as the force increased from 50 to 200 pN led to only ~2 Å additional extension. Although two hydrogen bonds between β-strands A and B broke upon a constant force of 200 pN, further extension of I1 was restricted by the remaining four intact A–B hydrogen bonds (Figure 3b). Obtaining additional extension for I1 requires breaking all six hydrogen bonds between A- and B-strands, which demands a larger force or longer stretching. In comparison, rupture of the two A–B hydrogen, bonds of I27 can result in ~6 Å additional extension (Figure 3c, d).

If Ig domains indeed unfold to fulfill their physiological roles, a rapid refolding rate is essential for their functions. Reversibly unfolding of Ig domains has been demonstrated in AFM experiments (Rief *et al.*, 1997; Carrion-Vazquez *et al.*, 1999). It takes about ~1 s for a fully stretched I27 domain to refold. The timescale, however, is too long to be reached in all-atom simulations. For practical purpose we simulate the refolding processes from structures that just crossed the main unfolding barrier (Gao *et al.*, 2001). Figure 4 presents reversible unfolding simulations of I27. The equilibrated protein was initially stretched with 750 pN constant force until all hydrogen bonds between A- and B-strands

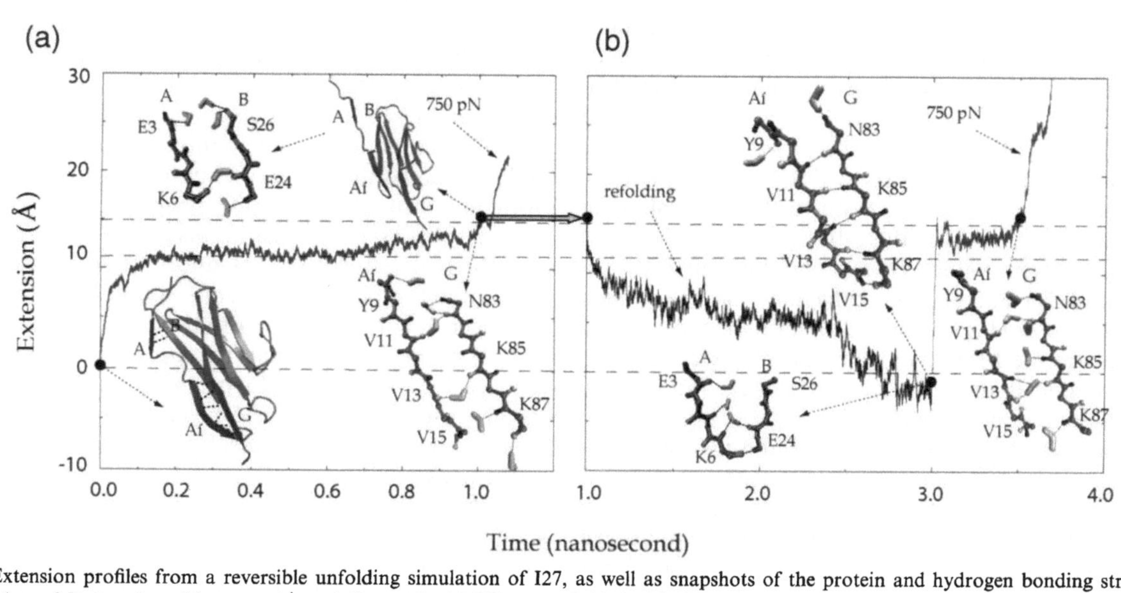

Fig. 4. Extension profiles from a reversible unfolding simulation of I27, as well as snapshots of the protein and hydrogen bonding structure between A- and B-strands and between A′- and G-strands. (a) I27 was unfolded with 750 pN constant force. (b) At 1 ns, the force was released and the module was allowed to spontaneously refold for additional 2 ns. The force was then resumed to stretch the partially refolded protein.

and between A′- and G-strands are broken (see snapshots on Figure 4a). The force was then released, and the partially unfolded I27 was subsequently allowed to spontaneously relax. During the 2 ns refolding simulation, the domain became more compact with its extension decreasing from over 15 to 0 Å, the extension of the native fold. At the end of the refolding simulation, five of the six hydrogen bonds connecting A′ and G reformed, except the bond between Y9(O) and N83(H) (see snapshots on Figure 4b). One of two A–B hydrogen bonds was also reformed. Finally, a 750 pN force was applied to stretch the refolded protein, resulting in a barrier passage time of about 500 ps, similar to the passage time of unfolding native I27, indicating that a refolding close to the native structure has been achieved.

Discussion

Multi-domain proteins like titin constitute a fascinating class of biopolymers with important cellular functions. Single molecule AFM experiments that probe the mechanical response of these proteins provide a unique source of information that becomes more valuable in combination with steered molecular dynamics simulations. The latter provide atomic-level pictures of the conformational changes governing the function of mechanical proteins which, however, need to be verified through comparison with experimental data.

Single molecule experiments (Kellermayer *et al.*, 1997; Rief *et al.*, 1997; Tskhovrebova *et al.*, 1997; Marszalek *et al.*, 1999; Li *et al.*, 2002) have elucidated the chief design requirements for titin Ig domains under extreme stretch conditions: they unravel one by one, and increase the length of titin at each unraveling event by about 280 Å without affecting the stability of those domains which still remain folded. At small extensions of titin, when the link regions are pulled taut to form a straight chain, the Ig domains are at their resting contour length of about 40 Å. Upon SMD force application, distal Ig domains such as I27 and I28 exhibit a pre-burst increase in contour length of about 7 Å, while proximal Ig domains such as I1 do not show this pre-burst intermediate state (Li *et al.*, 2002). With further extension, Ig domains continue to lengthen, but only after the force exceeds a given value. Our simulations provide an explanation for this bursting behavior. The links which must be ruptured to initiate the unfolding of the Ig domains are the backbone hydrogen bonds between β-strands A and B and between β-strands A′ and G. Due to the topology of the Ig domain (Figure 1), only when the A- and B-strands, as well as A′- and G-strands are separated, after breaking of all respective inter-strand hydrogen bonds, can the unfolding of an Ig domain continue, involving rupture of the inter-strand hydrogen bonds between the remaining β-strands. The AFM experiments have also found that unfolding distal Ig domains requires higher peak forces than unfolding proximal Ig domains (Li *et al.*, 2002). This is probably because distal Ig domains have more concurrently breaking inter-strand hydrogen bonds than proximal Ig domains (see Figure 1).

Ig domains, which exhibit high (>150 pN) dominant force peaks when stretched in AFM experiments and in SMD simulations, are called Class I proteins (Lu and Schulten, 1999b) (please note that this classification has nothing to do with the previous classification of the motifs of titin domains). Members belonging to this class exhibit strong resistance to forced unfolding; their fold topologies are such that interstrand hydrogen bonds must break in clusters in order to allow significant extension of the domain. Proteins like FN-III modules also belong to Class I. Simulations of FN-III modules revealed similar correlation between interstrand hydrogen bonds and mechanical stability of the modules (Krammer *et al.*, 1999; Craig *et al.*, 2001; Krammer *et al.*, 2002; Gao *et al.*, 2002a). Another example is ubiquitin, which exhibits multiple unfolding pathways (J. Fernandez, private communications). Other domains (e.g. all-helix domains or β proteins such as C2 domains) have topologies that can be extended while breaking backbone hydrogen bonds individually. They do not exhibit dominant force peaks when stretched in SMD simulations (Lu and Schulten, 1999b); we call these class II domains. The study suggested that AFM experiments on class II proteins should exhibit force-extension profiles different from those performed on proteins consisting of several class I domains. Either no dominant force peak (and thus no fingerprints like the sawtooth pattern) or much lower force peak values than those observed for class I domains are expected. The discernible-but-lower force peak values observed in class II domains might arise from hydrophobic effects, which do not play a dominant role in force-induced unfolding of class I domains. Both possibilities, i.e., non-discernible (Carrion-Vazquez *et al.*, 2000) and lower, discernible (Rief *et al.*, 1999) peaks have already been observed in recent AFM experiments.

The main obstacles to relating AFM measurements and MD simulations, namely the time scale discrepancy and the related discrepancy in unfolding forces, may be overcome through slower pulling speeds in the constant velocity SMD simulations. In this respect it is important to note that the slower (0.1 and 0.01 Å/ps) stretching simulations produce the same scenario of clustered hydrogen bond breaking as the faster 0.5 Å/ps simulation. On the other hand, the peak force values are reduced logarithmically. Figure 5 shows that by reducing the pulling velocity the peak force data obtained from SMD simulations merge with the extrapolated AFM peak force curve. Even though it is not expected that the main unfolding scenario revealed by SMD simulations depends on pulling speed, further simulations with lower pulling speed are of great value, as we can then compare the force peak value quantitatively between SMD and AFM recordings.

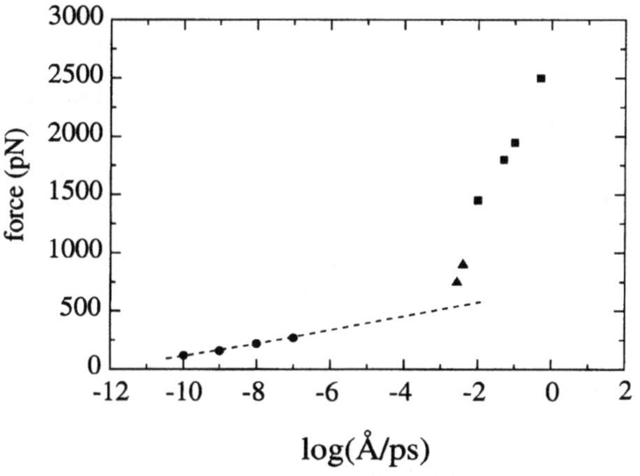

Fig. 5. Force spectroscopy of unfolding I27 from AFM experiments (●), constant velocity (■) and constant force SMD simulations (▲). The velocities corresponding to constant force pulling were estimated as the length of the reaction region, about 3 Å, divided by the mean first passage time. Since the SMD simulations and AFM experiments belong to different regime, they have different scaling properties. However, decreasing the pulling velocity in SMD simulations yielded SMD data that tend towards the extrapolated AFM peak force curve.

Several different theoretical studies on forced unfolding of protein domains have also been published recently, employing approaches and methods different from ours. A molecular mechanics approach on α-helix and β-hairpin stretching has been employed by Rohs *et al.* (1999). Stretching protein in reduced representation on a lattice model and an off-lattice model have been studied by Socci *et al.* (1999) and by Klimov and Thirumalai (1999, 2000). Evans and Ritchie (1999) modeled the Ig domain unfolding as a single bond breaking event using the Kramers-Smoluchowski theory that also underlies our calculation of mean first passage times. Paci and Karplus (1999, 2000) studied I27 and FnIII unfolding by means of biased molecular dynamics using an implicit solvent model to reduce computational effort; the authors suggested that the dominant barrier against unfolding is due to vdW interactions, and not due to hydrogen bonds. In contrast, in Lu and Schulten (2000) we have demonstrated that the key event in force-induced unfolding of I27 is hydrogen bond breaking assisted by bulk water, a feature neglected in implicit solvent models. Future SMD simulations will help clarify the differences among various theoretical approaches including those described above and will be aimed to provide a means to unite these approaches, while still relating them to experiments.

Our goal in modeling the stretching of titin domains is to obtain an atomic-level view of the process. Our resulting theoretical model should at least meet two criteria: first, the model should correspond closely to the experimental (AFM) situation, so as to provide a non-ambiguous check on the model's validity; second, the results should provide information on the process at atomic-level resolution that cannot be obtained from experiments. The SMD method has satisfied both criteria listed above. Force-extension curves produced by simulations can be directly compared to AFM data to examine the validity of the model. The SMD trajectories account for the unfolding process at atomic-level detail and, hence, reveal structure-dynamics-function relationships for protein elasticity properties not resolved by AFM.

SMD simulations can be used to help design proteins with unfolding barriers of desired strength. By mutating key residues involving forming A′–G hydrogen bonds of titin I27, for example, researchers have successfully created mutants that have fine tuned mechanical stabilities (Li *et al.*, 2001a). Additionally, modeling force-sensitive proteins with SMD can provide insights into novel processes. For example, simulations of $FnIII_{10}$ revealed a mechanosensitive switch of the protein (Krammer *et al.*, 1999, 2002). Mechanical forces can make conformational changes of a loop containing the ligand Asp-Gly-Arg peptide so that the binding to integrin receptors is weakened. With more structures of mechanical proteins becoming available and simulated, SMD can contribute fundamentally to the understanding of structure-function relationships of mechanical proteins. Even for the giant protein titin, for example, by systematically simulating individual domains or segments and piecing together the results one can eventually understand how nature designs the elastic spring.

Acknowledgements

This work was supported by the National Institutes of Health Research grants PHS5P41RR05969 and 1R01GM60946, as well as the National Science Foundation supercomputer time grant NRAC MCA93S028.

References

Best RB, Li B, Steward A, Daggett V and Clarke J (2001) Can non-mechanical proteins withstand force? stretching barnase by atomic force microscopy and molecular dynamics simulation. *Biophys J* **81:** 2344–2356.

Brünger AT (1992) X-PLOR, Version 3.1: A System for X-ray Crystallography and NMR. The Howard Hughes Medical Institute and Department of Molecular Biophysics and Biochemistry, Yale University.

Carrion-Vazquez M, Oberhauser A, Fowler S, Marszalek P, Broedel S, Clarke J and Fernandez J (1999) Mechanical and chemical unfolding of a single protein: a comparison. *Proc Natl Acad Sci USA* **96:** 3694–3699.

Carrion-Vazquez M, Oberhauser AF, Fisher TE, Marszalek PE, Li H and Fernandez JM (2000) Mechanical design of proteins studied by single-molecule force spectroscopy and protein engineering. *Prog Biophys Mol Biol* **74:** 63–91.

Craig D, Krammer A, Schulten K and Vogel V (2001) Comparison of the early stages of forced unfolding of fibronectin type III modules. *Proc Natl Acad Sci USA* **98:** 5590–5595.

Erickson H (1994) Reversible unfolding of fibronectin type III and immunoglobulin domains provides the structural basis for stretch and elasticity of titin and fibronectin. *Proc Natl Acad Sci USA* **91:** 10114–10118.

Erickson H (1997) Stretching single protein modules: titin is a weird spring. *Science* **276:** 1090–1093.

Evans E and Ritchie K (1999) Strength of a weak bond connecting flexible polymer chains. *Biophys J* **76:** 2439–2447.

Freiburg A, Trombitas K, Hell W, Cazorla O, Fougerousse F, Centner T, Kolmerer B, Witt C, Beckmann J, Gregorio C, Granzier H and Labeit S (2000) Series of exon-skipping events in the elastic spring region of titin as the structural basis for myofibrillar elastic diversity. *Circ Res* **86:** 1114–1121.

Gao M, Lu H and Schulten K (2001) Simulated refolding of stretched titin immunoglobulin domains. *Biophys J* **81:** 2268–2277.

Gao M, Craig D, Vogel V and Schulten K (2002a) Identifying unfolding intermediates of FN-III 10 by steered molecular dynamics. *J Mol Biol* **323:** 939–950.

Gao M, Wilmanns M and Schulten K (2002b) Steered molecular dynamics studies of titin I1 domain unfolding. *Biophys J*, **83:** 3435–3445.

Granzier H and Labeit S (2002) Cardiac titin: an adjustable multi-functional spring. *J Physiol* **541:** 335–342.

Grubmüller H, Heymann B and Tavan P (1996) Ligand binding and molecular mechanics calculation of the Streptavidin-Biotin Rupture Force. *Science* **271:** 997–999.

Improta S, Politou A and Pastore A (1996) Immunoglobulin-like modules from titin I-band: extensible components of muscle elasticity. *Structure* **4:** 323–337.

Isralewitz B, Gao M and Schulten K (2001) Steered molecular dynamics and mechanical functions of proteins. *Curr Op Struct Biol* **11:** 224–230.

Isralewitz B, Izrailev S and Schulten K (1997) Binding pathway of retinal to bacterio-opsin: a prediction by molecular dynamics simulations. *Biophys J* **73:** 2972–2979.

Izrailev S, Stepaniants S, Balsera M, Oono Y and Schulten K (1997) Molecular dynamics study of unbinding of the Avidin-Biotin complex. *Biophys J* **72:** 1568–1581.

Jorgensen WL, Chandrasekhar J, Madura JD, Impey RW and Klein ML (1983) Comparison of simple potential functions for simulating liquid water. *J Chem Phys* **79:** 926–935.

Kalé L, Skeel R, Bhandarkar M, Brunner R, Gursoy A, Krawetz N, Phillips J, Shinozaki A, Varadarajan K and Schulten K (1999) NAMD2: greater scalability for parallel molecular dynamics. *J Comp Phys* **151:** 283–312.

Kellermayer M, Smith S, Granzier H and Bustamante C (1997) Folding-unfolding transition in single titin modules characterized with laser tweezers. *Science* **276:** 1112–1116.

Klimov DK and Thirumalai D (1999) Stretching single-domain proteins: phase diagram and kinetics of force-induced unfolding. *Proc Natl Acad Sci USA* **96:** 1306–1315.

Klimov DK and Thirumalai D (2000) Native topology determines force-induced unfolding pathways in globular proteins. *Proc Natl Acad Sci USA* **97:** 7254–7259.

Krammer A, Craig D, Thomas WE, Schulten K and Vogel V (2002) A structural model for force regulated integrin binding to fibronectin's RGD-synergy site. *Matrix Biology* **21:** 139–147.

Krammer A, Lu H, Isralewitz B, Schulten K and Vogel V (1999) Forced unfolding of the fibronectin type III module reveals a tensile molecular recognition switch. *Proc Natl Acad Sci USA* **96:** 1351–1356.

Labeit S and Kolmerer B (1995) Titins, giant proteins in charge of muscle ultrastructure and elasticity. *Science* **270:** 293–296.

Li H, Linke W, Oberhauser AF, Carrion-Vazquez M, Kerkvliet JG, Lu H, Marszalek PE and Fernandez JM (2002) Reverse engineering of the giant muscle protein titin. *Nature* **418:** 998–1002.

Li H, Mariano CV, Oberhauser AF, Marszalek PE and Fernandez JM (2001a) Point mutations alter the mechanical stability of immunoglobulin modules. *Nature Struct Biol* **7:** 1117–1120.

Li H, Oberhauser AF, Redick SD, Carrion-Vazquez M, Erikson H and Fernandez JM (2001b) Multiple conformations of PEVK proteins detected by single-molecule techniques. *Proc Natl Acad Sci USA* **98:** 10682–10686.

Linke WA (2000) Stretching molecular springs: elasticity of titin filaments in vertebrate striated muscle. *Histol Histopathol* **15:** 799–811.

Linke WA, Ivemeyer M, Mundel P, Stockmeier MR and Kolmerer B (1998) Nature of PEVK-titin elasticity in skeletal muscle. *Proc Natl Acad Sci USA* **95:** 8052–8057.

Lu H, Isralewitz B, Krammer A, Vogel V and Schulten K (1998) Unfolding of titin immunoglobulin domains by steered molecular dynamics simulation. *Biophys J* **75:** 662–671.

Lu H and Schulten K (1999a) Steered molecular dynamics simulation of conformational changes of immunoglobulin domain I27 interpret atomic force microscopy observations. *Chem Phys* **247:** 141–153.

Lu H and Schulten K (1999b) Steered molecular dynamics simulations of force-induced protein domain unfolding. *Proteins Struct Func Gen* **35:** 453–463.

Lu H and Schulten K (2000) The key event in force-induced unfolding of titin's immunoglobulin domains. *Biophys J* **79:** 51–65.

Ma K, Kan L and Wang K (2001) Polyproline II helix is a key structural motif of the elastic PEVK segment of titin. *Biochemistry* **40:** 3427–3438.

MacKerell Jr. AD, Bashford D, Bellott M, Dunbrack Jr. RL, Evanseck J, Field MJ, Fischer S, Gao J, Guo H, Ha S, Joseph D, Kuchnir L, Kuczera K, Lau FTK, Mattos C, Michnick S, Ngo T, Nguyen DT, Prodhom B, Reiher IWE, Roux B, Schlenkrich M, Smith J, Stote R, Straub J, Watanabe M, Wiorkiewicz-Kuczera J, Yin D and Karplus M (1998) All-hydrogen empirical potential for molecular modeling and dynamics studies of proteins using the CHARMM22 force field. *J Phys Chem B* **102:** 3586–3616.

Marszalek PE, Lu H, Li H, Carrion-Vazquez M, Oberhauser AF, Schulten K and Fernandez JM (1999) Mechanical unfolding intermediates in titin modules. *Nature* **402:** 100–103.

Maruyama K (1997) Connectin/titin, a giant elastic protein of muscle. *FASEB J* **11:** 341–345.

Mayans O, Wuerges J, Canela S, Gautel M and Wilmanns M (2001) Structural evidence for a possible role of reversible disulphide bridge formation in the elasticity of the muscle protein titin. *Structure* **9:** 331–340.

Minajeva A, Kulke M, Fernandez JM and Linke WA (2001) Unfolding of titin domains explains the viscoelastic behavior of skeletal myofibrils. *Biophys J* **80:** 1442–1451.

Oberhauser A, Badilla-Fernandez C, Carrion-Vazquez M and Fernandez J (2002) The mechanical hierarchies of fibronectin observed with single molecule AFM. *J Mol Biol* **319:** 433–447.

Oberhauser AF, Marszalek PE, Erickson H and Fernandez J (1998) The molecular elasticity of tenascin, an extracellular matrix protein. *Nature* **393:** 181–185.

Paci E and Karplus M (1999) Forced unfolding of fibronectin type 3 modules: an analysis by biased molecular dynamics simulations. *J Mol Biol* **288:** 441–459.

Paci E and Karplus M (2000) Unfolding proteins by external forces and temperature: the importance of topology and energetics. *Proc Natl Acad Sci USA* **97:** 6521–6526.

Politou AS, Thomas D and Pastore A (1995) The folding and the stability of titin immunoglobulin-like modules, with implications for mechanism of elasticity. *Biophys J* **69:** 2601–2610.

Rief M, Gautel M, Oesterhelt F, Fernandez JM and Gaub HE (1997) Reversible unfolding of individual titin immunoglobulin domains by AFM. *Science* **276:** 1109–1112.

Rief M, Gautel M, Schemmel A and Gaub H (1998) The mechanical stability of immunoglobulin and fibronectin III domains in the muscle protein titin measured by AFM. *Biophys J* **75:** 3008–3014.

Rief M, Pascual J, Saraste M and Gaub H (1999) Single molecule force spectroscopy of spectrin repeats: low unfolding forces in helix bundles. *J Mol Biol* **286:** 553–561.

Rohs R, Etchebest C and Lavery R (1999) Unraveling proteins: a molecular mechanics study. *Biophys J* **76:** 2760–2768.

Schulten K, Schulten Z and Szabo A (1980) Reactions governed by a binomial redistribution process. The ehrenfest urn problem. *Physica* **100A:** 599–614.

Schulten K, Schulten Z and Szabo A (1981) Dynamics of reactions involving diffusive barrier crossing. *J Chem Phys* **74:** 4426–4432.

Socci N, Onuchic J and Wolynes P (1999) Stretching lattice models of protein folding. *Proc Natl Acad Sci USA* **96:** 2031–2035.

Soteriou A, Clarke A, Martin S and Trinick J (1993) Titin folding energy and elasticity. *Proc R Soc Lond B (Biol. Sci.)* **254:** 83–86.

Szabo A, Schulten K and Schulten Z (1980) First passage time approach to diffusion controlled reactions. *J Chem Phys* **72:** 4350–4357.

Trombitas K, Greaser M, Labeit S, Jin J, Kellermayer M, Helmes M and Granzier H (1998) Titin extensibility in situ: entropic elasticity of permanently folded and permanently unfolded molecular segments. *J Cell Biol* **140:** 853–859.

Tskhovrebova L and Trinick J (2002) Role of titin in vertebrate striated muscle. *Proc R Soc Lond B (Biol. Sci.)* **357:** 199–206.

Tskhovrebova L, Trinick J, Sleep J and Simmons R (1997) Elasticity and unfolding of single molecules of the giant protein titin. *Nature* **387:** 308–312.

Wang K (1996) Titin/connectin and nebulin: giant protein ruler of muscle structure and function. *Adv Biophys* **33:** 123–134.

Cytoskeletal Proteins

Journal of Muscle Research and Cell Motility **23**: 525–534, 2002.

Mechanical response of single filamin A (ABP-280) molecules and its role in the actin cytoskeleton

MASAHITO YAMAZAKI[1,2,*], SHOU FURUIKE[1] and TADANAO ITO[3]
[1]*Materials Science, Graduate School of Science and Engineering;* [2]*Department of Physics, Faculty of Science, Shizuoka University, Shizuoka 422-8529;* [3]*Department of Biophysics, Graduate School of Science, Kyoto University, Kyoto 606-8502, Japan*

Abstract

Filamin A produces isotropic cross-linked three-dimensional orthogonal networks with actin filaments in the cortex and at the leading edge of cells. Filamin A also links the actin cytoskeleton to the plasma membrane via its association with various kinds of membrane proteins. Recent new findings strongly support that filamin A plays important roles in the mechanical stability of plasma membrane and cortex, formation of cell shape, mechanical responses of cells, and cell locomotion. To elucidate the mechanical properties of the actin/filamin A network and the complex of membrane protein–filamin A–actin cytoskeleton, the mechanical properties of single human filamin A (hsFLNa) molecules in aqueous solution were investigated using atomic force microscopy. Ig-fold domains of filamin A can be unfolded by the critical external force (50–220 pN), and this unfolding is reversible, i.e., the refolding of the unfolded chain of the filamin A occurs when the external force is removed. Due to this reversible unfolding of Ig-fold domains, filamin A molecule can be stretched to several times the length of its native state. Based on this new feature of filamin A as the 'large-extensible linker', we describe our hypothesis for the mechanical role of filamin A in the actin cytoskeletons in cells and discuss its biological implications. In this review, function of filamin A in actin cytoskeleton, mechanical properties of single filamin A proteins, and the hypothesis for the mechanical role of filamin A in the actin cytoskeletons are discussed.

Function of filamin A in actin cytoskeleton

In cells, many kinds of cytoskeletons play important roles. One of the fundamental roles of cytoskeletons is to increase the mechanical stability of the plasma membranes and control cell shape. Bare cells without cytoskeletons are very weak. For example, recent studies on cell-size liposome (i.e., giant liposome) of phospholipid membranes show that a weak force induced by the growth of protein crystals breaks the liposome (Yamashita *et al.*, 2002). Thus, for the survival of cells, the cytoskeleton must control the mechanical stability and the shape of plasma membranes. For mechanical integrity, the actin cytoskeleton plays one of the most important roles. A large number of actin filament cross-linking proteins produce several kinds of assembly of actin filaments; tight parallel bundles (e.g., in filopodia), loose contractile bundles (e.g., in stress fibers), 70° branching network by Arp2/3 complex (Mullins *et al.*, 1998), and isotropic three-dimensional (3-D) networks (Stossel *et al.*, 2001). These various kinds of actin cytoskeletons can be assembled and disassembled dynamically in response to various kinds of signals, which is essential for cell locomotion, cell division, organelle transport, and various kinds of cell extensions such as filopodia.

Filamin A (ABP-280) is one of the actin filament cross-linking proteins (Flier and Sonnenberg, 2001; Stossel *et al.*, 2001). It can produce isotropic cross-linked 3-D orthogonal networks with actin filaments *in vivo* and *in vitro*, which is called actin/filamin A (actin/ABP-280) gel or actin/filamin A network (Figure 1). It is present at the leading edge and cortex of many kinds of nonmuscle cells such as fibroblasts, cerebral cortical neurons, platelets, and macrophages (Hartwig and Stossel, 1981; Hartwig and Shevlin 1986; Gorlin *et al.*, 1990; Fox *et al.*, 1998; Flier and Sonnenberg, 2001; Stossel *et al.*, 2001). Filamin A can also link the actin cytoskeleton to the plasma membrane via its association with integral membrane proteins (Figure 2). Recently, the number of these membrane proteins has increased rapidly; the list contains the integrin family of adhesion receptors such as integrin β_1, β_2, β_7 (Sharma *et al.*, 1995; Loo *et al.*, 1998; Calderwood *et al.*, 2001), glycoprotein Ibα in platelets (Fox, 1985; Meyer *et al.*, 1997), and several ion channel proteins such as dopamine D_2, D_3 receptor (Li *et al.*, 2000; Lin *et al.*, 2001), Kv4.2 potassium channel (Petrecca *et al.*, 2000), and metabotropic glutamate receptor (Enz, 2002) (Table 1).

Filamin A and the actin/filamin A network have been considered to play important roles in the mechanical

* To whom correspondence should be addressed: Tel./Fax: +81-54-238-4741; E-mail: spmyama@ipc.shizuoka.ac.jp

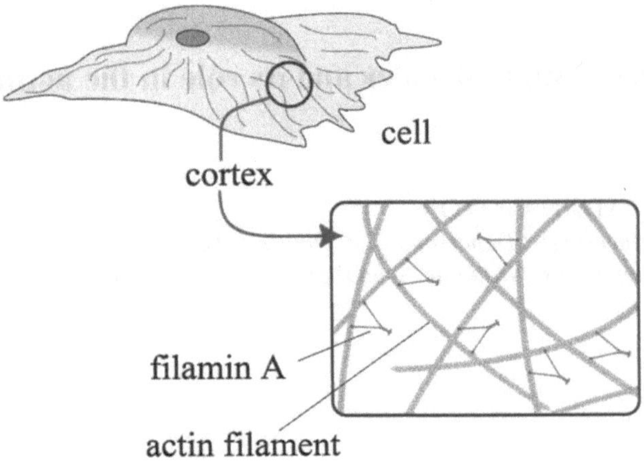

Fig. 1. Schematic model of 3-D actin/filamin A network that exists in the cortex of various nonmuscle cells.

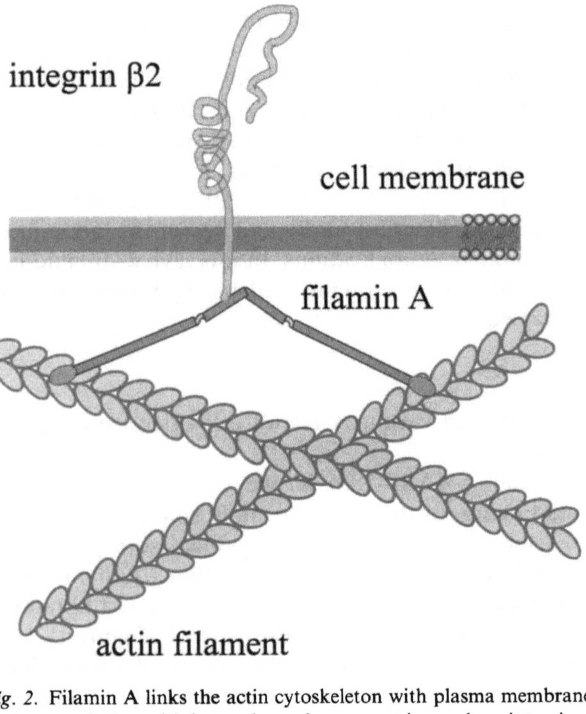

Fig. 2. Filamin A links the actin cytoskeleton with plasma membrane via its association with integral membrane proteins such as integrin.

Table 1. Proteins bound to filamin A

Protein	Reference
Membrane proteins	
GP-Ibα	Meyer *et al.* (1997)
β1A, β1D, β2, β3, β7 integrin	Calderwood *et al.* (2001), Loo *et al.* (1998), Sharma *et al.* (1995)
FcγRI	Ohta *et al.* (1991)
Tissue factor	Ott *et al.* (1998)
Dopamine D2, D3 receptor	Li *et al.* (2000), Lin *et al.* (2001)
Presenilin1, 2	Zhang *et al.* (1998)
Furin receptor	Liu *et al.* (1997)
Calvelolin 1	Stahlhut *et al.* (2000)
Potassium channel (Kv4.2)	Petrecca *et al.* (2000)
Glutamate receptor (type 7)	Enz (2002)
Calcium-sensing receptor	Hjälm *et al.* (2002)
Signalling proteins	
RalA, RhoA, Rac1, Cdc42	Ohta *et al.* (1999)
SEK1	Marti *et al.* (1997)
TRAF2	Leonardi *et al.* (2000)
Trio	Bellanger *et al.* (2000)
Others	
FAP52	Nikki *et al.* (2002)
cvHSP	Krif *et al.* (1999)
Granzyme B	Browne *et al.* (2000)
BRCA2	Yuan and Shen (2001)

response of cells, changing cell shape, and cell locomotion. Filamin A-deficient human melanoma cells (M2 cells) have impaired cell locomotion and exhibit circumferential blebbing of plasma membrane which results from the unstable cortex (Cunningham *et al.*, 1992; Cunningham, 1995). On the other hand, A7 cells, i.e., a clonal subline of M2 that has restored filamin A expression by transfection, showed normal locomotion and reduced membrane blebbing (Cunningham *et al.*, 1992). Both the cells (M2 and A7) contain similar amounts of Arp2/3, but M2 cells do not have 3-D network of actin filaments while A7 cells do, which indicates that Arp2/3 complex alone cannot form 3-D network of actin filaments (Flanagan *et al.*, 2001). In human periventricular heterotopica (PH), mutations in filamin A prevent long-range, directed migration of cerebral cortical neurons to the cortex (Fox *et al.*, 1998). Effects of mechanical force on function of cells and structures of cytoskeletons are investigated, and results show important roles for filamin A in mechanoprotection (Glogauer *et al.*, 1998; D'Addario *et al.*, 2001). Filamin A and actin filaments accumulate at membrane cortices that are under high membrane tension where they increase the strength of the cortex, which reduces the Ca^{2+}-influx through mechanosensitive ion channel proteins (Glogauer *et al.*, 1998). When mechanical force was applied on human fibroblasts cells using the fibronectin- or collagen-coated beads through the β_1 integrin, the production of filamin A increased (D'Addario *et al.*, 2001). These results clearly indicate that filamin A increase the mechanical stability of plasma membrane and cortex. Moreover, recently some researchers consider that in the focal adhesion which plays important roles in cell adhesion and locomotion, filamin A connects integrins with the actin cytoskeleton (Nikki *et al.*, 2002; Pokutta and Weis, 2002). FAP52, a focal adhesion-associated phosphoprotein in FAP52/PACSIN/syndapin family, binds the C-terminal region of filamin A, and these proteins are colocalized at the stress fiber-focal adhesion junction (Nikki *et al.*, 2002).

Filamin A can respond to signalling molecules and induce a change in actin assembly. Several signalling molecules such as Rho family GTPases (Rho, Rac, Cdc42, RalA) (Ohta *et al.*, 1999) and the stress-activated protein kinase (SAPK) activator SEK-1 (Marti *et al.*, 1997) were bound to the filamin A (Table 1). The binding of these signalling molecules with filamin A is an important step in signal transduction. For example, the binding of RalA with filamin A induces filopodia in

Swiss 3T3 cells (Ohta *et al.*, 1999). These results show that filamin A has multiple functions by interacting with many kinds of cellular proteins such as membrane proteins and various signalling molecules.

Human filamin A (hsFLNa) is a dimeric protein with equivalent 280-kDa subunits that associate with each other at their C-terminal domains. It has two N-terminal actin-binding domains (ABD) per dimer, and can thus cross-link actin filaments. Most of this protein (90%) is a semiflexible rod composed of 24 tandem repeats, each containing 96 amino acids on average, and its amino acid sequence predicts stretches of anti-parallel β-sheets in an immunoglobulin (Ig) fold (Figure 3a, b) (Gorlin *et al.*, 1990). There are many isoforms of hsFLNa. In human, there are three kinds of filamin; filamin A, filamin B, and filamin C (Flier and Sonnenberg, 2001). All of them have an N-terminal ABD, C-terminal self-association domain, and a semiflexible rod composed of 24 tandem repeats Ig-fold domains. They show high homology (about 70%) in their amino acid sequences, but in the region between C-terminal self-association domain and hinge-1(H1) there is large variation in the sequence (Stossel *et al.*, 2001; Flier and Sonnenberg, 2001). Binding sites of filamin A with most of membrane proteins and signalling molecules are located at one or several Ig-fold domains between the C-terminal domain and the hinge-1 (Table 1) (Flier and Sonnenberg, 2001). Chicken also has three kinds of filamins which have similar structures as human filamins (Tachikawa *et al.*, 1997; Ohashi *et al.*, private communication). On the other hand, filamins in some species have smaller number of Ig-fold domains; for example, *Dictyostelium* gelation factor (ABP-120) contains six tandem repeats of Ig-fold domains and an ABD (Noegel *et al.*, 1989). As for 3-D structure, only one structure of a part of filamin is known at present. The 3-D structure of the Ig-fold domain of segment 4 in the rod domain of the gelation factor in *Dictyostelium discoideum* was determined using NMR spectroscopy (Figure 3b) (Fucini *et al.*, 1997).

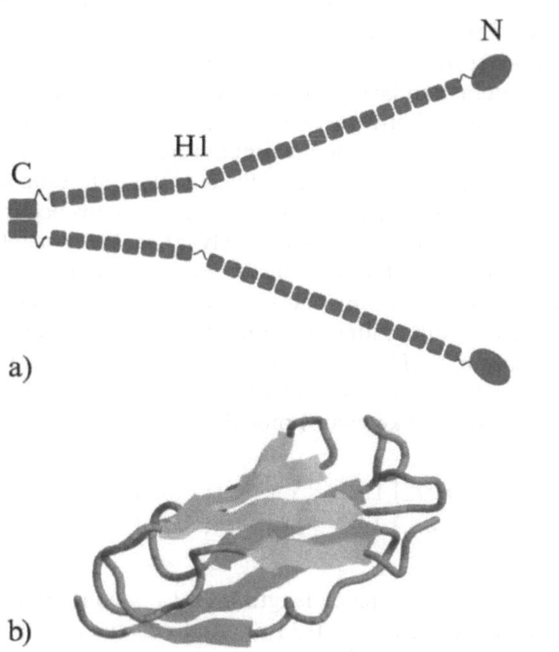

Fig. 3. (a) Schematic model of structure of dimer of filamin A. It has two N-terminal actin-binding domains (ABD) per dimer, C-terminal self-association domain, and semiflexible rod composed of 24 tandem repeats, which based on their amino acid sequence are likely to consist of anti-parallel β-sheets in an immunoglobulin (Ig) fold. H1 denotes hinge-1. (b) 3-D structure of the Ig-fold domain of segment 4 in the rod domain of the gelation factor in *Dictyostelium discoideum* (PDB ID; 1KSR). The structure was determined with NMR spectroscopy (Fucini *et al.*, 1997).

It is important to know the difference in functions of three kinds of filamins (A, B, C), their localization inside cells of various kinds of tissues, and the mechanism of its sorting, but at present we do not have enough information (Flier and Sonnenberg, 2001).

Mechanical property of single filamin A proteins

To elucidate mechanical properties (such as viscoelasticity) of actin/filamin A gel and complex of membrane protein–filamin A–actin cytoskeleton, it is important to investigate the mechanical property of their components; actin filament and filamin A. The mechanical behaviour of the actin filament is well known (Suzuki *et al.*, 1989, 1996; Käs *et al.*, 1996; Tsuda *et al.*, 1996). The actin filament is a semiflexible rod with a persistence length of ~1.8 μm (Käs *et al.*, 1996). Its longitudinal rigidity and its torsional rigidity are 1.8×10^9 N m^{-2} and 8.0×10^{-26} N m^2, respectively. The breaking force of the actin–actin bond in the actin filament was 600 pN under untwist and it decreased from 600 to 320 pN under twist (Tsuda *et al.*, 1996). This rupture force is much larger than that of avidin–biotin and that of antigen–antibody (Florin *et al.*, 1994; Hinterdorfer *et al.*, 1996). At semidilute concentration around 1.4 mg/ml, thermal motion of the actin filament was confined in the virtual tube formed around the filament; one motion is undulation motion of the filament within the tube and the other is the reptation along the tube (Käs *et al.*, 1996), which is in good agreement with the tube model developed by polymer physics (Doi and Edwards, 1986). In the presence of high concentrations of polymers such as poly(ethylene glycol) (PEG) or proteins which do not interact with actin filaments directly and are preferentially excluded from their neighbourhood, actin filaments are associated each other, aligning their axes parallel, to form bundle-type structures without cross-linking proteins (Suzuki *et al.*, 1989, 1996). This result suggests that actin filaments are prone to form parallel bundles in cells due to high concentrations of proteins. Therefore, to control assembly of actin filaments in cells for specific purpose, the cross-linking proteins must play indispensable roles.

On the other hand, information on the mechanical properties of filamin A has been lacking. Therefore, as a first step, we investigated the mechanical properties of hsFLNa molecules using atomic force microscopy (AFM) (Furuike *et al.*, 2001). Recent studies have shown that AFM can be useful to investigate mechanical property of biopolymers such as proteins during their extension by measuring external force in aqueous solutions (Ikai *et al.*, 1997; Rief *et al.*, 1997; Oberhauser *et al.*, 1998; Carl *et al.*, 2001; Watanabe *et al.*, 2002a, b).

We stretched single filamin A molecules in aqueous solution where they form dimers, and measured their force–extension relationship at room temperature. After adsorbing the protein molecules on the gold-coated mica substrate in the aqueous solution, the AFM tip was brought to the gold surface and kept there for 2 s with a force of 100 pN $\sim$ 1 nN to adsorb the segments of the filamin A protein, after which the tip was pulled away. In the retraction curves, we often observed a large attractive force between the tip and the gold surface to a large extension, sometimes more than 500 nm. This indicates that the filamin A molecule adsorbed on both the tip and the gold surface was stretched as the distance between the tip and the gold surface increased (i.e., extension). In most cases, we observed a periodic increase and decrease in force during the extension; i.e., so-called sawtooth pattern of the force–extension curve (Figure 4a, b). On occasion this pattern existed at extensions up to 700 nm (Figure 4a), but in most cases the sawtooth pattern suddenly disappeared at shorter distances (Figure 4b), which is due to the desorption of the filamin A molecule from the gold surface of the mica substrate or from the surface of the cantilever. At a pulling speed of 0.4 μm/s, the maximum force and the periodicity of the sawtooth patterns were 100–200 pN and 30 ± 5 nm, respectively.

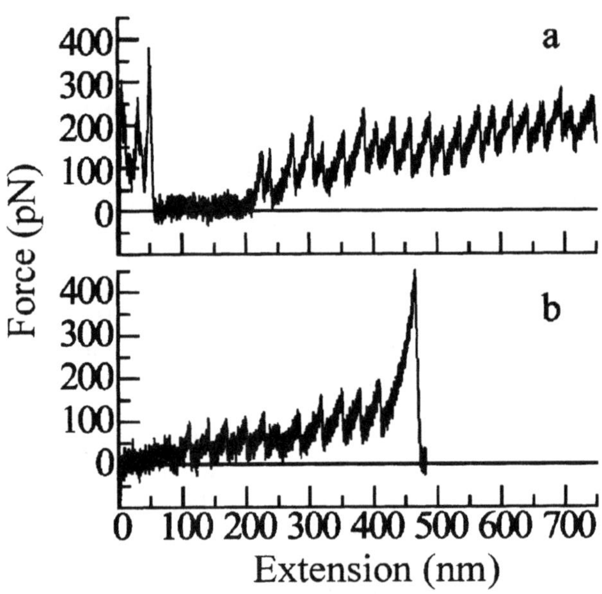

Fig. 4. (a) (b) A force–extension curve of filamin A molecule in aqueous solution measured by AFM at room temperature. Filamin A was stretched at a pulling speed of 0.37 μm/s in 10 mM imidazole–HCl (pH 7.5) buffer containing 0.1 M KCl, and 0.5 mM EGTA. During the extension, the force–extension curves showed a periodic increase and decrease in force; i.e., so-called sawtooth pattern of the force–extension curve (a, b). In some cases such as the (a), this pattern existed to a large extension more than 700 nm. In (b), the sawtooth pattern suddenly disappeared at 480 nm, which is due to the desorption of the filamin A molecule from the gold surface of the mica substrate or the surface of the cantilever. In most cases, at small extensions before the appearance of the sawtooth pattern (at 0–60 nm at (a)), an irregular large attractive force was often observed, which is due to the unspecific weak adsorption of the filamin A with the surfaces of substrate or cantilever and its detachment. Filamin A was purified from human uterine leiomyoma tissue by gel filtration with a 4% agarose column (Ito *et al.*, 1992). The Nanoscope IIIa Multimode AFM (MMAFM) (Digital Instruments, Santa Barbara, CA, USA) equipped with a D scanner (12 μm × 12 μm × 4.4 μm) and a fluid cell were used. For the cantilever, we used an oxide-sharpened Si_3N_4 cantilever (NP-S) with a spring constant of $k = 0.06$ N/m (Digital Instruments).

What kind of physical events occur during this sawtooth pattern in the force–extension curve? The precise periodicity of the sawtooth pattern seems to be related with the 24 tandem repeats of almost the same molecular weight Ig-fold domains of filamin A. As well-analyzed in reports of stretching of titin and its analogue proteins (Rief *et al.*, 1997; Carrion-Vazquez *et al.*, 1999), the abrupt decrease in force in the sawtooth patterns corresponds to the force-induced unfolding of an individual Ig-fold domain of filamin A. It is well known that the unfolding (or unfolding transition) of proteins from native state to unfolded state is induced by temperature or chemical reagents such as guanidine hydrochloride. The force-induced unfolding of proteins is a new kind of unfolding, which was first found in the case of titin (Kellermayer *et al.*, 1997; Rief *et al.*, 1997; Tskhovrevova *et al.*, 1997). The mechanical unfolding of proteins can be described as a nonequilibrium process of transition between two states, i.e., a native state and an unfolded state. The maximum force of each sawtooth in the sawtooth pattern is defined as the unfolding force, F^*, at which the unfolding of Ig-fold domains occurs. As shown in Figure 4, immediately after the abrupt decrease in force, the force again gradually increased with increasing extension. This comes from an entropic force due to the stretching of the unfolded polypeptide chain of filamin A. The force, $F(x)$, vs. extension, x, curve of this part of the sawtooth pattern is described well by the WLC (worm-like chain) model, as follows:

$$F(x) = \frac{k_B T}{p}\left\{\frac{1}{4}\left(1 - \frac{x}{L_c}\right)^{-2} - \frac{1}{4} + \frac{x}{L_c}\right\} \tag{1}$$

where p is persistence length and L_c is contour length. The best fit gave $p = 0.33$ nm and ΔL_c (i.e., the change in contour length between consecutive force peaks) = 31 ± 1 nm. When the force reaches the critical level by extending the unfolded chain, an unfolding transition of another Ig-fold domain of filamin A is induced. This characteristic change in force was repeated several times in the force–extension curve, resulting in the sawtooth pattern. In several cases, we observed sawtooth patterns with irregular periodicity (i.e., wide distribution of the

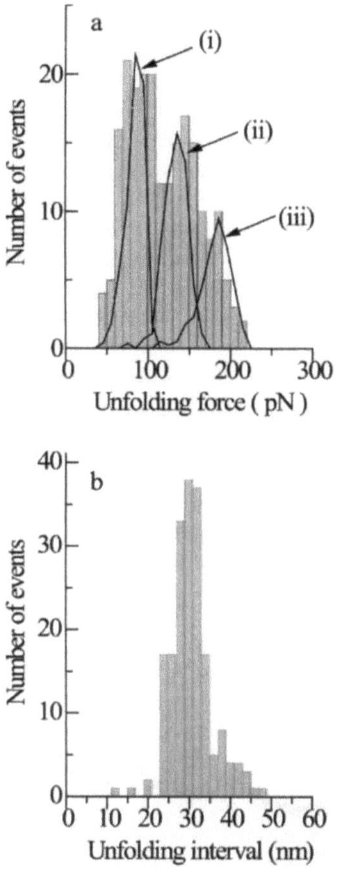

Fig. 5. (a) A histogram of the unfolding force of Ig-fold domains of filamin A at the pulling speed of 0.37 μm/s ($n = 214$ unfolding events from 24 independent experiments). The three curves correspond to the unfolding force obtained from Monte Carlo simulations of 20 Ig-fold domains with different structures, as characterized by two parameters (k_u^0 and x_u). Curve (i), Ig-fold domain A ($x_u = 0.5$ nm and $k_u^0 = 3.0 \times 10^{-4}\ s^{-1}$); curve (ii), Ig-fold domain B ($x_u = 0.4$ nm and $k_u^0 = 3.0 \times 10^{-5}\ s^{-1}$); curve (iii), Ig-fold domain C ($x_u = 0.3$ nm and $k_u^0 = 3.0 \times 10^{-5}\ s^{-1}$). In the Monte Carlo simulation, a protein (i.e., polypeptide chain) is stretched with a constant pulling speed v_c from extension $x = 0$ by a small amount Δx ($= v_c \Delta t$) at each time interval. This stretching increases the force exerted on the protein, which is calculated based on the WLC model (Equation (1)). The probability of observing the unfolding of any Ig-fold domains at this force is then calculated as described in the text, and then random number decision is executed to define their states. We used values of parameters; $v_c = 0.4$ μm/s, $p = 0.4$ nm, the length of the Ig-fold domain in the folded state ($l_f = 4$ nm) and that in the unfolded state ($l_u = 36$ nm). This Monte Carlo simulation produces a force–extension curve ($0 \leq x \leq 500$ nm), showing a sawtooth pattern which is similar to Figure 4. (b) A histogram of the unfolding interval of filamin A, obtained from the same data sets used to construct the unfolding force histogram (a). The unfolding interval was determined by peak to peak spacing (i.e., periodicity) of the sawtooth pattern. This Figure is reprinted from Furuike *et al.* (2001) with permission from Elsevier Science.

unfolding interval) in the force–extension curve, which correspond to data of the stretch more than two polypeptide chains connected with both the tip and the gold surface. In the following analysis, we did not use such sawtooth patterns with irregular periodicity, but only those with regular periodicity.

As shown in Figure 4, in the unfolding of Ig-fold domains of filamin A, there is a wide distribution of the unfolding force, F^*. Figure 5(a) shows a histogram of the unfolding force of filamin A at a pulling speed of 0.37 μm/s when it was pulled until extension, $x = 500$ nm ($n = 214$ unfolding events). The unfolding force F^* ranged from 50 to 220 pN. On the other hand, a histogram of the unfolding interval (i.e., the periodicity of the sawtooth patterns) of filamin A in Figure 5(b) exhibited a narrow distribution around an average value of 30 nm. This indicates that the stretch of only one polypeptide chain of filamin A gave such a wide distribution of the unfolding force (Figure 5a). To investigate the cause of the wide distribution in the unfolding force of filamin A, we performed a Monte Carlo simulation of the unfolding of this protein (Furuike *et al.*, 2001). The rate of transition from the native state to the unfolded state in the absence of the force, k_u^0, depends on activation energy (energy barrier), E_u^0, and is given by the following equation:

$$k_u^0 = A \exp(-E_u^0/k_B T) \quad (2)$$

where k_B is Boltzmann's constant and T is temperature. External force, F, reduces the energy barrier by Fx_u, where x_u is the width of the activation barrier (i.e., the distance to the transition state along the direction of force) (Evans and Ritchie, 1997; Rief *et al.*, 1998), and the unfolding rate $k_u(F)$ is thus expressed as follows:

$$k_u(F) = A \exp(-(E_u^0 - Fx_u)/k_B T) = k_u^0 \exp(Fx_u/k_B T) \quad (3)$$

The transition between the two states is a stochastic process that can be simulated well by the Monte Carlo method (Rief *et al.*, 1998; Carrion-Vazquez *et al.*, 1999). The probability of observing the unfolding of any modules is $P_u = N_f k_u(F) \Delta t$, where N_f is the number of identical folded modules and Δt is the polling interval of the Monte Carlo simulation. During unfolding by force, folding of the unfolded chain does not occur (Carrion-Vazquez *et al.*, 1999). In this model, the probability of mechanical unfolding of protein can be determined using only two parameters related to the protein structure; i.e., the activation energy E_u^0 and the width of the activation barrier, x_u. The k_u^0 and x_u values were determined for proteins consisting of a finite number of identical domains, such as $I27_8$ (the eight tandem repeats of the 27th Ig-fold domain of human cardiac titin prepared using a recombinant DNA technique); $x_u = 0.25$ nm and $k_u^0 = 3.3 \times 10^{-4}\ s^{-1}$. Point mutations in a different amino acid of $I27_8$ affected the mechanical stability significantly: k_u^0 ranged from 8.0×10^{-7} to $3.3 \times 10^{-4}\ s^{-1}$ due to the variation in values of the activation energy E_u^0 in each mutant protein, and the width of the activation barrier, x_u, ranged from 0.25 to 0.60 nm (Li *et al.*, 2000). In native proteins, a similar analysis was reported as follows (Watanabe *et al.*, 2002b); the unfolding force of the recombinant tandem

Ig multidomains at different regions of human titin, i.e., I65-70 and I91-98 domains, were very different, and its analysis indicates that the values of k_u^0 and x_u of these Ig multidomains are different ($k_u^0 = 6.1 \times 10^{-5}$ s^{-1} and $x_u = 0.25$ nm for I91-98 domains and $k_u^0 = 1.0 \times 10^{-5}$ s^{-1} and $x_u = 0.35$ nm for I65-70 domains). Thus, we can reasonably assume that each Ig-fold domain of filamin A has different values of E_u^0 (and, therefore, k_u^0) and x_u, because the individual Ig domains of filamin A have different amino acid sequences and different number of amino acids (Figure 6). To demonstrate the effects of k_u^0 and x_u on the unfolding force of the Ig-fold domain, in Figure 5(a) we plotted three curves obtained from Monte Carlo simulations of 20 identical Ig-fold domains having different structures, as characterized by two parameters (k_u^0 and x_u): curve (i), Ig-fold domain A ($x_u = 0.5$ nm and $k_u^0 = 3.0 \times 10^{-4}$ s^{-1}); curve (ii), Ig-fold domain B ($x_u = 0.4$ nm and $k_u^0 = 3.0 \times 10^{-5}$ s^{-1}); curve (iii), Ig-fold domain C ($x_u = 0.3$ nm and $k_u^0 = 3.0 \times 10^{-5}$ s^{-1}) (Furuike *et al.*, 2001). These curves fit well to the histogram in Figure 5(a). This result clearly indicates that filamin A has various kinds of Ig-fold domains with different mechanical properties due to the different values of E_u^0 and x_u. Such heterogeneity in the unfolding force of filamin A can yield a wide dynamic range in the force needed to extend the molecule, suggesting that, in the mechanical responses of cells, the extension of filamin A may be suitably adjusted by forces of various levels. The unfolding force of filamin A depends on pulling speed. Figure 7 shows dependence of the unfolding force of three kinds of Ig-fold domains (A, B, C) which was obtained by the same Monte Carlo simulations as described above. The unfolding force decreases with a decrease in pulling speed.

As in the case for temperature-induced or chemical substances-induced unfolding of proteins, force-induced unfolding is also a reversible process and refolding to the native state occurs when the applied force is removed. Figure 8 shows a refolding of the unfolded Ig-fold domains of filamin A. After adsorbing the filamin A protein molecules on the gold-coated mica

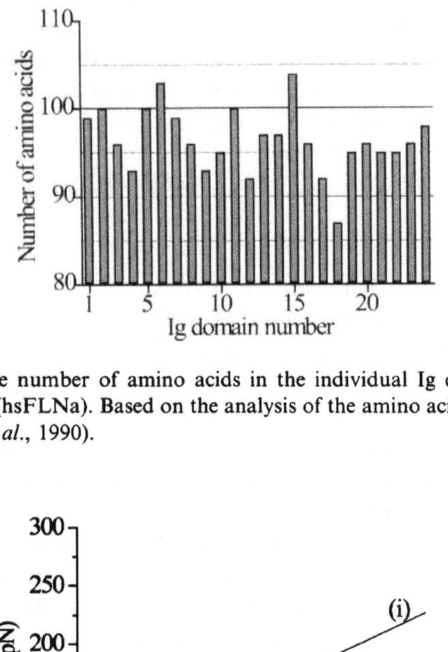

Fig. 6. The number of amino acids in the individual Ig domains of filamin A (hsFLNa). Based on the analysis of the amino acid sequence (Gorlin *et al.*, 1990).

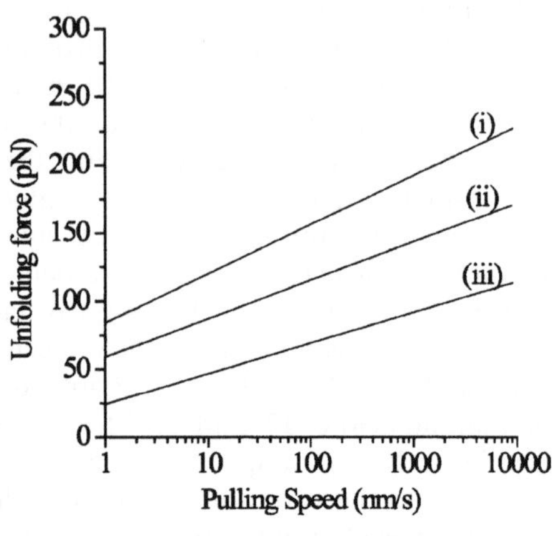

Fig. 7. Dependence of unfolding force on the pulling speed. Curve (i), Ig-fold domain A ($x_u = 0.5$ nm and $k_u^0 = 3.0 \times 10^{-4}$ s^{-1}); curve (ii), Ig-fold domain B ($x_u = 0.4$ nm and $k_u^0 = 3.0 \times 10^{-5}$ s^{-1}); curve (iii), Ig-fold domain C ($x_u = 0.3$ nm and $k_u^0 = 3.0 \times 10^{-5}$ s^{-1}). The method of Monte Carlo simulation is described in the legend of Figure 5.

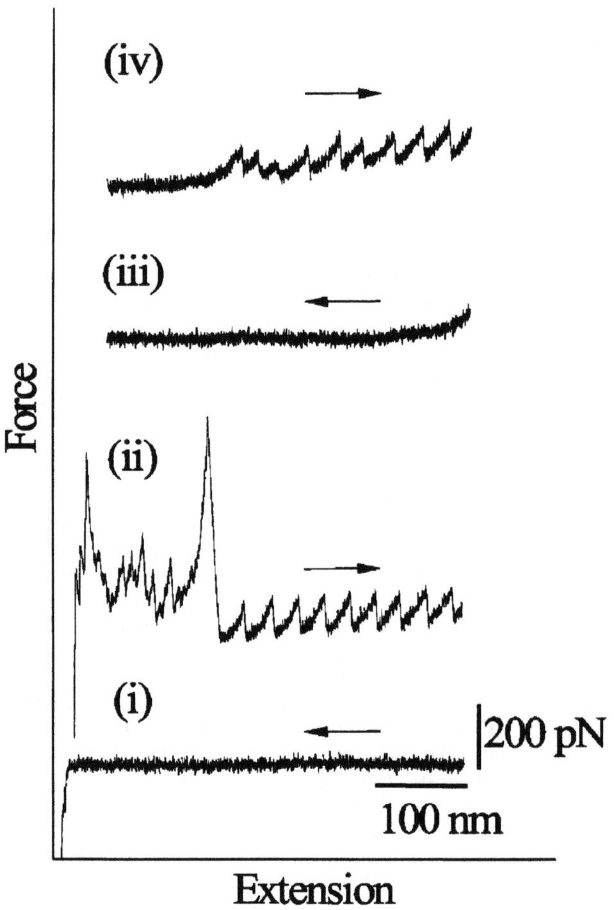

Fig. 8. Refolding of the unfolded filamin A protein. The AFM tip was brought to the gold surface and kept there for 2 s to adsorb the protein onto its tip [curve (i)]. The filamin A was then stretched [curve (ii)], and the characteristic sawtooth pattern appeared. After the extension, the AFM tip was brought to a position 44 nm above the gold surface to reduce the force to near zero [curve (iii)]. After holding there for 10 sec, the protein was stretched again [curve (iv)]. The pulling speed and the approach speed were both 0.40 μm/s. This figure is reprinted from Furuike *et al.* (2001) with permission from Elsevier Science.

substrate in the aqueous solution, the AFM tip was brought to the gold surface and kept there for 2 s to adsorb the segments of the protein onto its tip (Figure 8, step (i)). Then the tip was pulled away and thereby the filamin A molecule was stretched (Figure 8, step (ii)). During the large extension several Ig-fold domains of filamin A were unfolded by the external force (Figure 8, step (ii)). Then the AFM tip was brought to a position 44 nm above the gold surface, to relax the protein to almost its initial length and reduce the force to near zero (Figure 8, step (iii)). After holding the tip at that position for 10 s, the protein was stretched again (the second stretch) (Figure 8, step (iv)). As shown in the figure, most of the sawtooth patterns reappeared in the second stretch, indicating that most of the unfolded domains were refolded by this procedure, i.e., unfolding was reversible. In another similar experiment, after the filamin A was unfolded during the initial large extension, the AFM tip was brought to a position 40 nm above the gold surface and held at that position for 1 s instead of 10 s, after which the protein was stretched again. This time, fewer sawtooth patterns reappeared during the large extension in the second stretch (data not shown), suggesting that it takes time for the unfolded domains to refold.

Hypothesis for the mechanical role of filamin A in the actin cytoskeletons and its biological implication

In previous section, we describe an important feature of single filamin A molecules; Ig-fold domains of filamin A can be unfolded by the critical external force (50–220 pN), and this unfolding is reversible, i.e., the refolding of the unfolded chain of the filamin A occurs when the external force is removed. Due to this reversible unfolding of Ig-fold domains, the filamin A molecule can be stretched to several times the length of its native state in response to the external force. In cells, filamin A binds to several kinds of proteins such as membrane proteins and actin filaments. To consider the mechanical response of the complex of filamin A and several proteins, it is important to compare the unfolding force of Ig-fold domains with the unbinding force between filamin A and other proteins. Here, we consider a model system; a rod-like protein composed of tandem repeats Ig fold domains such as filamin A binds to another protein at the left (Figure 9). The force–extension curve of this complex can be produced by using the

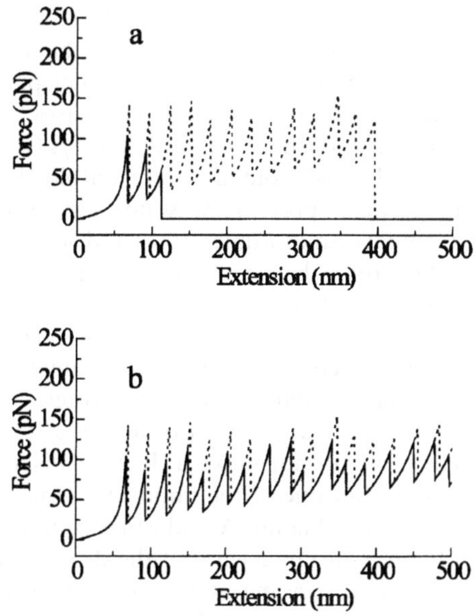

Fig. 10. Comparison between the unbinding of these proteins (see Figure 9) and the unfolding of Ig-fold domains. The force–extension curves were obtained by the Monte Carlo simulation. The dashed line and the solid line show the force-extension curve at pulling speed of 400 and 40 nm/s, respectively. The rate of unbinding between these proteins was obtained by $k_{\text{unbinding}} = k^0_{\text{unbinding}} \exp(Fx_{\text{unbinding}}/k_BT)$. As parameters of Ig-fold domain, Ig-fold domain B ($x_u = 0.4$ nm and $k^0_u = 3.0 \times 10^{-5}$ s^{-1}) was used. The method of Monte Carlo simulation was described in the legend of Figure 5. (a) The rate constant of unbinding of the proteins in the absence of force ($k^0_{\text{unbinding}} = 0.6$ s^{-1}), the width of the activation barrier of unbinding ($x_{\text{unbinding}} = 0.05$ nm). At pulling speed of 400 nm/s, the unbinding is observed at 400 nm, but at pulling speed of 40 nm/s, the unbinding is observed at 110 nm. (b) $k^0_{\text{unbinding}} = 0.06$ s^{-1}. Other parameters are the same as (a). In both cases at pulling speed of 400 and 40 nm/s, more than 16 Ig-fold domains were unfolded and the unbinding between proteins was not observed up to an extension of 500 nm.

Monte Carlo method described above. Figure 10 shows four examples of these force-extension curves. In Figure 10(a), we used the following parameters; rate constant of unbinding of the proteins in the absence of force ($k^0_{\text{unbinding}} = 0.6$ s^{-1}), the width of the activation barrier of unbinding ($x_{\text{unbinding}} = 0.05$ nm), Ig-fold domain B ($x_u = 0.4$ nm and $k^0_u = 3.0 \times 10^{-5}$ s^{-1}) (Furuike *et al.*, in preparation). The dashed line and the solid line show the force–extension curve at pulling speed of 400 and 40 nm/s, respectively. In the case of 400 nm/s, 12 Ig-fold domains were unfolded before the unbinding between proteins, and in the case of 40 nm/s, two Ig-fold domains were unfolded before the unbinding between

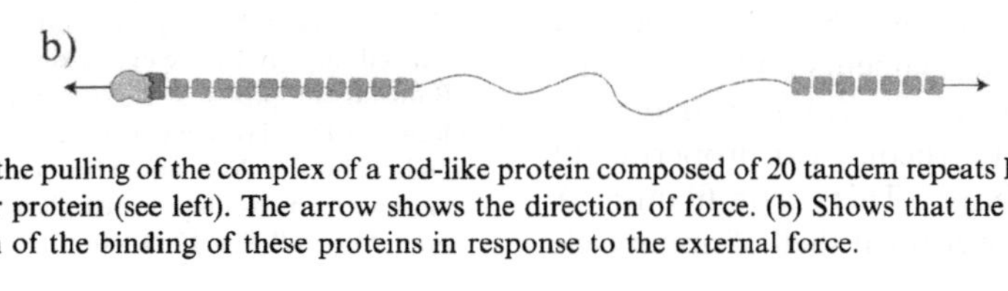

Fig. 9. (a) Schematic model of the pulling of the complex of a rod-like protein composed of 20 tandem repeats Ig-fold domains (as a simple model of filamin A) bound to another protein (see left). The arrow shows the direction of force. (b) Shows that the complex can be stretched to large distance without the disruption of the binding of these proteins in response to the external force.

proteins (Figure 10(a)). In Figure 10(b), as the rate constant of unbinding of the proteins in the absence of force, $k^0_{\text{unbinding}} = 0.06\ \text{s}^{-1}$ is used, but other parameters are the same as in Figure 10(a). In both cases at pulling speed of 400 nm/s (dashed line) and 40 nm/s (solid line), more than 16 Ig-fold domains were unfolded and the unbinding between proteins was not observed up to 500 nm. Comparison of Figure 10(a) and (b) shows that if the binding force between the filamin A and other protein is large the filamin A molecule can be stretched to large extension even at the low pulling speed. These results largely depend on values of parameters which characterize the unbinding of these proteins and the unfolding of Ig-fold domains (Furuike *et al.*, in preparation). Here, we do not make a detailed discussion of this simulation, but one of the important conclusions is that the complex of filamin A and other proteins can be stretched to large distance without the disruption of the binding of these proteins in response to the external force.

In the cells, this characteristic extensibility of the complex of filamin A with proteins may have important physiological roles. For example, filamin A is a linker between the membrane protein and actin cytoskeleton, which can be considered as 'large-extensible linker'. When a large shear strain (or force) is applied to the plasma membrane, the Ig-fold of filamin A unfolds to extend its length, which gives a large deformability to the linker without disruption of the binding of these proteins. After the applied force is removed, the filamin A molecules completely refold, thereby, the native structure of the complex can be rapidly restored. If the binding of these proteins is disrupted, much longer time is necessary for the restoration of the complex.

The mechanical property of single filamin A may also play an important role in the actin/filamin A network. The actin/filamin A network prepared *in vitro* shows the following characteristic properties (Ito *et al.*, in preparation). At low shear strains ($\gamma < 20\%$), it behaves as a linearly deforming gel; at moderate shear strains ($20\% < \gamma < 100\%$), it behaves as a non-linearly deforming gel with deformability that increases with increasing shear strain; at high shear strains ($\gamma > 100\%$), it behaves as a non-Newtonian fluid. In addition, its rheological properties show reversibility; after the removal of a large shear strain, the rheological properties return to their original values. Cross-linking of actin filaments by filamin A may suppress the undulation motion of the actin filaments in the network (Morse, 1998). A relatively large shear force imposed on the network would induce a stepwise unfolding of Ig-fold domains of filamin A, and this unfolding of the cross-linker should bring about a gradual increase in the undulation motion of the filaments. Such a change of the undulation motion of the filaments and the reversibility of the unfolding of the cross-linker may well account for the characteristic rheological properties of the network described above. This actin/filamin A network, which makes up the cortical shell of nonmuscle cells, may play an essential role in the mechanical responses of cells to external deformation. The above-described *in vitro* responses of the actin/filamin A network suggest that cells can stabilize their shapes against a small shear force, and that their deformability increases against a large shear force.

In the case of cardiac titin, under physiological condition at a stretch velocity of 100 nm/s, unfolding of Ig-fold domains is not a likely event, and the PEVK domains and N2B-Us domain play an important role in the elasticity (Watanabe *et al.*, 2002a, b). Thereby, it is important to consider whether the unfolding of Ig-fold domains of filamin A occurs under physiological condition. At present, there is no experimental evidence for the unfolding of filamin A inside cells. However, we can imagine that under certain physiological conditions unfolding does take place. The migration velocity of some kinds of cells such as fibroblasts and melanoma cells is very low (3–8 nm/s) (Friedl *et al.*, 1998). In other cells, under special conditions, the migration velocity becomes very low, and large shear strain is imposed on the plasma membrane and the cortex of the cells. For example, in the inflammatory response, white blood cells must migrate through the narrow space between the capillary endothelial cells from the capillary to the connective tissue. Under these physiological conditions, the unfolding force of Ig-fold domain becomes very low since it depends on the pulling speed (Figure 7), and the large shear force is imposed on the actin/filamin A network and the complex of filamin A and other proteins. In this circumstance, unfolding allows filamin A to function as a large-extensible linker.

To verify our hypothesis on the mechanical role of filamin A, more elaborate experiments are required. We need experimental data on the unbinding force between filamin A and actin filaments or other membrane proteins such as integrin. We would like to develop a new method to observe the unfolding of filamin A in cells.

Acknowledgements

This work was supported in part by a grant-in-aid for General Scientific Research C from the Ministry of Education, Science, and Culture, Japan.

References

Bellanger JM, Astier C, Sardet C, Ohta Y, Stossel TP and Debant A (2000) The Rac1- and RhoG-specific GEF domain of Trio targets filamin to remodel cytoskeletal actin. *Nat Cell Biol* **2:** 888–892.

Browne KA, Johnstone RW, Jans DA and Trapani JA (2000) Filamin (280-kDa actin-binding protein) is a caspase substrate and is also cleaved directly by the cytotoxic T lymphocyte protease granzyme B during apoptosis. *J Biol Chem* **275:** 39,262–39,266.

Calderwood DA, Huttenlocher A, Kiosses WB, Rose DM, Woodside DG, Schwartz MA and Ginsberg MH (2001) Increased filamin binding to beta-integrin cytoplasmic domains inhibits cell migration. *Nat Cell Biol* **12:** 1060–1068.

Carl P, Kwok CH, Manderson G, Speicher DW and Discher DE (2001) Forced unfolding modulated by disulfide bonds in the Ig domains of a cell adhesion molecule. *Proc Natl Acad Sci USA* **98:** 1565–1570.

Carrion-Vazquez M, Oberhauser AF, Fowler SB, Marszalek PE, Broedel SE, Clarke J and Fernandez JM (1999) *Proc Natl Acad Sci USA* **96:** 3694–3699.

Cunningham CC (1995) Actin polymerization and intracellular solvent flow in cell surface blebbing. *J Cell Biol* **129:** 1589–1599.

Cunningham CC, Gorlin JB, Kwiatkowski DJ, Hartwig JH, Janmey PA, Byers HR and Stossel TP (1992) Actin-binding protein requirement for cortical stability and efficient locomotion. *Science* **255:** 325–327.

D'Addario M, Arora PD, Fan J, Ganss B, Ellen RP and McCulloch CAG (2001) Cytoprotection against mechanical forces delivered through β1 integrins requires induction of filamin A. *J Biol Chem* **276:** 31,969–31,977.

Doi M and Edwards SF (1986) The theory of polymer dynamics, Oxford University Press.

Enz R (2002) The actin-binding protein filamin-A interacts with the metabotropic glutamate receptor type 7. *FEBS Lett* **514:** 184–188.

Evans E and Ritchie K (1997) Dynamic strength of molecular adhesion bonds. *Biophys J* **72:** 1541–1555.

Flanagan LA, Chou J, Falet H, Neujahr R, Hartwig JH and Stossel TP (2001) Filamin A, the Arp2/3 complex, and the morphology and function of cortial actin filaments in human melanoma cells. *J Cell Biol* **155:** 511–517.

van der Flier A and Sonnenberg A (2001) Structural and functional aspects of filamins. *Biochim Biophys Acta* **1538:** 99–117.

Florin EL, Moy VT and Gaub HE (1994) Adhesion forces between individual ligand-receptor pairs. *Science* **264:** 415–417.

Fox JEB (1985) Identification of actin-binding proteins as the protein linking the membrane skeleton to glycoproteins on platelet plasma membranes. *J Biol Chem* **260:** 11,970–11,977.

Fox JW, Lamperti ED, Ekçioglu YZ, Hong SE, Feng Y, Graham DA, Scheffer IE, Dobyns WB, Hirsch BA, Radtke RA, Berkovic SF, Huttenlocher PR and Walsh CA (1998) Mutations in filamin 1 prevent migration of cerebral cortical neurons in human periventricular heterotopia. *Neuron* **21:** 1315–1325.

Friedl P, Zänker KS and Bröcker E-V (1998) Cell migration strategies in 3-D extracellular matrix: differences in morphology, cell matrix interactions, and integrin function. *Microscopy Res Tech* **43:** 369–378.

Fucini P, Renner C, Herberhold C, Noegel AA and Holak TA (1997) The repeating segments of the F-actin cross-linking gelation factor (ABP-120) have an immunoglobulin-like fold. *Nature Struct Biol* **4:** 223–231.

Furuike S, Ito T and Yamazaki M (2001) Mechanical unfolding of single filamin A (ABP-280) molecules detected by atomic force microscopy. *FEBS Lett* **498:** 72–75.

Glogauer M, Arora P, Chou D, Janmey PA, Downey GP and McCulloch CAG (1998) The role of actin-binding protein 280 in integrin-dependent mechanoprotection. *J Biol Chem* **273:** 1689–1698.

Gorlin JB, Yamin R, Egan S, Stewart M, Stossel TP, Kwiatkowski DJ and Hartwig JH (1990) Human endothelial actin-binding protein (ABP-280, nonmuscle filamin): a molecular leaf spring. *J Cell Biol* **111:** 1089–1105.

Hartwig JH and Shevlin P (1986) The architecture of actin filaments and the ultrastructural location of actin-binding protein in the periphery of lung macrophages. *J Cell Biol* **103:** 1007–1020.

Hartwig JH and Stossel TP (1981) Structure of macrophage actin-binding protein molecules in solution and interacting with actin filaments. *J Mol Biol* **145:** 563–581.

Hartwig JH and Shevlin P (1986) The architecture of actin filaments and the ultrastructural location of actin-binding protein in the periphery of lung macrophages. *J Cell Biol* **103:** 1007–1020.

Hinterdorfer P, Baumgartner W, Gruber HJ, Schilcher K and Schindler H (1996) Detection and localization of individual antibody-antigen recognition events by atomic force microscopy. *Proc Natl Acad Sci USA* **93:** 3477–3481.

Hjälm G, John MacLeod R, Kifor O, Chattopadhyay N and Brown EM (2001) Filamin-A binds to the carboxyl-terminal tail of the calcium-sensing receptor, an interaction that participates in CaR-mediated activation of mitogen-activated protein kinase. *J Biol Chem* **276:** 34,880–34,887.

Ikai A, Mitsui K, Furutani Y, Hara M, McMurty J and Wong KP (1997) Protein stretching II: results for carbonic anhydrase. *Jpn J Appl Phys* **36:** 3887–3893.

Ito T, Suzuki A and Stossel T (1992) Regulation of water flow by actin-binding protein-induced actin gelatin. *Biophys J* **61:** 1301–1305.

Janmey PA, Hvidt S, Lamb J and Stossel TP (1990) Resemblance of actin-binding protein/actin gels to covalently crosslinked networks. *Nature* **345:** 89–92.

Käs J, Strey H, Tang JX, Finger D, Ezzel R, Sackmann E and Janmey PA (1996) F-actin, a model polymer for semiflexible chains in dilute, semidilute, and liquid crystalline solutions. *Biophys J* **70:** 609–625.

Kellermayer MSZ, Smith SB, Granzier HL and Bustamante C (1997) Folding-unfolding transitions in single titin molecules characterized with laser tweezers. *Science* **276:** 1112–1116.

Krief S, Faivre JF, Robert P, Le Douarin B, Brument-Larignon N, Lefrère I, Bouzyk MM, Anderson KM, Greller LD, Tobin FL, Souchet M and Bril A (1999) Identification and characterization of cvHsp. A novel human small stress protein selectively expressed in cardiovascular and insulin-sensitive tissues. *J Biol Chem* **274:** 36,592–36,600.

Leonardi A, Ellinger-Ziegelbauer H, Franzoso G, Brown K, Siebenlist U (2000) Physical and functional interaction of filamin (actin-binding protein-280) and tumor necrosis factor receptor-associated factor 2. *J Biol Chem* **275:** 271–278.

Li H, Carrion-Vazquez M, Oberhauser AF, Marszalek PE and Fernandez JM (2000) Point mutations alter the mechanical stability of immunoglobulin modules. *Nature Struct Biol* **7:** 1117–1120.

Li M, Bermak JC, Wang ZW and Zhou QY (2000) Modulation of dopamine D2 receptor signaling by actin-binding protein (ABP-280). *Mol Pharmacol* **57:** 446–452.

Lin R, Karpa K, Kabbani N, Goldman-Rakic P and Levenson R (2001) Dopamine D2 and D3 receptors are linked to the actin cytoskeleton via interaction with filamin A. *Proc Natl Acad Sci USA* **98:** 5258–5263.

Loo DT, Kanner SB and Aruffo A (1998) Filamin binds to the cytoplasmic domain of the beta1-integrin. Identification of amino acids responsible for this interaction. *J Biol Chem* **273:** 23,304–23,312.

Marti A, Luo Z, Cunningham C, Ohta Y, Hartwig J, Stossel TP, Kyriakis JM and Avruch J (1997) Actin-binding protein-280 binds the stress-activated protein kinase (SAPK) activator SEK-1 and is required for tumor necrosis factor-alpha activation of SAPK in melanoma cells. *J Biol Chem* **272:** 2620–2628.

Meyer SC, Zuerbig S, Cunningham CC, Hatwig T, Bissel K, Gardner JE and Fox JEB (1997) Identification of the region in actin-binding protein that binds to the cytoplasmic domain of glycoprotein Ibα. *J Biol Chem* **272:** 2914–2919.

Morse DC (1998) Viscoelasticity of concentrated isotropic solutions of semiflexible polymers. 2. Linear response. *Macromolecules* **31:** 7044–7067.

Mullins RD, Heuser JA and Pollard TD (1998) The interaction of Arp2/3 complex with actin: nucleation, high affinity pointed end capping, and formation of branching networks of filaments. *Proc Natl Acad Sci USA* **95:** 6181–6186.

Nikki M, Merilainen J and Lehto VP (2002) FAP52 regulates actin organization via binding to filamin. *J Biol Chem* **277:** 11,432–11,440.

Noegel AA, Rapp S, Lottspeich F, Schleicher M and Stewart M (1989) *J Cell Biol* **109:** 607–618.

Oberhauser AF, Marszalek PE, Erickson HP and Fernandez JM (1998) The molecular elasticity of the extracellular matrix protein tenascin. *Nature* **393:** 181–185.

Ohta Y, Suzuki N, Nakamura S, Hartwig, JH and Stossel TP (1999) The small GTPase RalA targets filamin to induce filopodia. *Proc Natl Acad Sci USA* **96:** 2122–2128.

Petrecca K, Miller DM and Shrier A (2000) Localization and enhanced current density of the Kv4.2 potassium channel by interaction with the actin-binding protein filamin. *J Neuroscience* **20:** 8736–8744.

Pokutta S and Weis WI (2002) The cytoplasmic face of cell contact sites. *Curr Opi Struct Biol* **12:** 255–262.

Rief M, Fernandez JM and Gaub HE (1998) Elastically coupled two-level systems as a model for biopolymer extensibility. *Phys Rev Lett* **81:** 4764–4767.

Rief M, Gautel M, Oesterhelt F, Fernandez JM and Gaub HE (1997) Reversible unfolding of individual titin immunoglobulin domains by AFM. *Science* **276:** 1109–1112.

Sharma CP, Ezzell RM and Arnaout MA (1995) Direct interaction of filamin (ABP-280) with the beta 2-integrin subunit CD18. *J Immunol* **154:** 3461–3470.

Stossel TP, Condeelis J, Cooley L, Hartwig JH, Schleichen M and Shapiro SS (2001) Filamins as integrators of cell mechanics and signaling. *Nature Reviews Mol Cell Biol* **2:** 138–146.

Suzuki A, Yamazaki M and Ito T (1989) Osmoelastic coupling in biological structures: Formation of parallel bundles of actin filaments in a crystalline-like structure caused by osmotic stress. *Biochemistry* **28:** 6513–6518.

Suzuki A, Yamazaki M and Ito T (1996) Polymorphism of F-actin assembly. 1. A quantitative phase diagram of F-actin. *Biochemistry* **35:** 5238–5244.

Tachikawa M, Nakagawa H, Terasaki AG, Mori K and Ohashi K (1997) A 260-kDa filamin/ABP-related protein in chicken gizzard smooth muscle cells is a new component of the dense plaques and dense bodies of smooth muscle. *J Biochem* **122:** 314–321.

Tskhovrebova L, Trinick J, Sleep JA and Simmons RM (1997) Elasticity and unfolding of single molecules of the giant muscle proteins titin. *Nature* **387:** 308–312.

Tsuda Y, Yasutake H, Ishijima A and Yanagida T (1996) Torsional rigidity of single actin filaments and actin–actin bond breaking force under torsion measured directly by *in vitro* micromanipulation. *Proc Natl Acad Sci USA* **93:** 12,937–12,942.

Yamashita Y, Oka M, Tanaka T and Yamazaki M (2002) A new method for the preparation of giant liposomes in high salt concentrations and growth of protein microcrystals in them. *Biochim Biophys Acta* **1561:** 129–134.

Yuan Y and Shen Z (2001) Interaction with BRCA2 suggests a role for filamin-1 (hsFLNa) in DNA damage response. *J Biol Chem* **276:** 48,318–48,324.

Watanabe K, Nair P, Labeit D, Kellermayer MSZ, Greaser M, Labeit S and Granzier H (2002a) *J Biol Chem* **277:** 11,549–11,558.

Watanabe K, Muhle-Goll C, Kellermayer MSZ, Labeit S and Granzier H (2002b) *J Struct Biol* **137:** 1248–1258.

Journal of Muscle Research and Cell Motility 23: 535–540, 2002.

Mechanics of vimentin intermediate filaments

NING WANG[1,*] and DIMITRIJIE STAMENOVIC[2]
[1]*Physiology Program, Harvard School of Public Health, 665 Huntington Ave., Boston MA 02115, USA;* [2]*Department of Biomedical Engineering, Boston University, 44 Cummington St., Boston, MA 02215, USA*

Abstract

It is increasingly evident that the cytoskeleton of living cells plays important roles in mechanical and biological functions of the cells. Here we focus on the contribution of intermediate filaments (IFs) to the mechanical behaviors of living cells. Vimentin, a major structural component of IFs in many cell types, is shown to play an important role in vital mechanical and biological functions such as cell contractility, migration, stiffness, stiffening, and proliferation.

Importance of intermediate filaments in cell functions

Intermediate filaments (IFs) are one of the major structural components of the cytoskeleton (CSK) in addition to actin microfilaments (MFs) and microtubules (MTs) (Amos and Amos, 1991). The CSK is known to play critical roles in controlling and regulating many cell functions including contraction, migration, proliferation, protein synthesis, gene expression, apoptosis, and mechanotransduction (Chicurel *et al.*, 1998; Janmey, 1998; Huang *et al.*, 1998; Beningo *et al.*, 2001; Chen *et al.*, 2001). The relative abundance of IFs in the cell and the fact that IFs span from the nucleus to the cell membrane (Djiabali, 1999) suggest that IFs play important mechanical roles in providing structural stability of the cell and to transmit and/or transduce mechanical signals into biological responses (Coulombe *et al.*, 2000).

Early evidence for vimentin to play minor role in cell functions

Vimentin IFs are expressed abundantly in fibroblasts and endothelial cells (Figure 1). However, the functional role for vimentin has been controversial. There are several lines of evidence that vimentin may not be necessary for some cell functions. For instance, collapsing vimentin networks in fibroblasts with antibodies does not induce any apparent cell shape changes (Klymkowsky, 1981). In addition, microinjection of vimentin-specific antibody does not appear to interfere with cell locomotion and mitosis either (Gawlitta *et al.*, 1981). Furthermore, cell growth and proliferation are not affected in cell lines that lack vimentin network (Venetianer *et al.*, 1983; Hedberg and Chen, 1986). Fibroblasts derived from vimentin null mice do not appear to have abnormal growth and motility (Holwell *et al.*, 1997). At the level of whole animals and tissues, mice lacking vimentin do not exhibit obvious differences in development, breeding, structural or physiological properties of various tissues or organs (Colucci-Guyon *et al.*, 1994). All these studies raise the question whether vimentin IFs are required or not for normal functions of the cells.

*To whom correspondence should be addressed: Tel.: +1-617-432-1686; Fax: +1-617-432-3468; E-mail: nwang@hsph.harvard.edu

Evidence for vimentin's role in biological functions

More recent work provides strong evidence that vimentin IFs are indeed required for many vital cell functions. Some of the earlier work also demonstrates that vimentin expression is associated with motility of various cells (Frankel *et al.*, 1982; Lane *et al.*, 1983; Cochard and Paulin, 1984). It is shown that microinjection of mimetic peptides of IFs into cultured fibroblasts induces rapid disassembly of the IF lattice and profound changes in cell shape, MF and MT networks (Goldman *et al.*, 1996). More recent work shows that when primary fibroblasts derived from vimentin-deficient mouse embryos are compared with those from wild-type mouse embryos, vimentin-deficient cells exhibit decreased motility, chemotactic migration, and delayed wound healing (Eckes *et al.*, 1998). Moreover, the spatial organization of focal adhesions and MFs are altered although there are no apparent differences in cell morphology. This work is important in that it demonstrates that examination of gross cell morphology only is not sufficient to reveal the altered cell functions in these vimentin-deficient cells. After these fibroblasts are synchronized by removing serum from the medium for cell cycle control, cell proliferation is slowed in vimentin-deficient cells due to decreased DNA synthesis (Wang and Stamenovic, 2000). This result does not contradict the earlier finding that mitosis is not changed for those

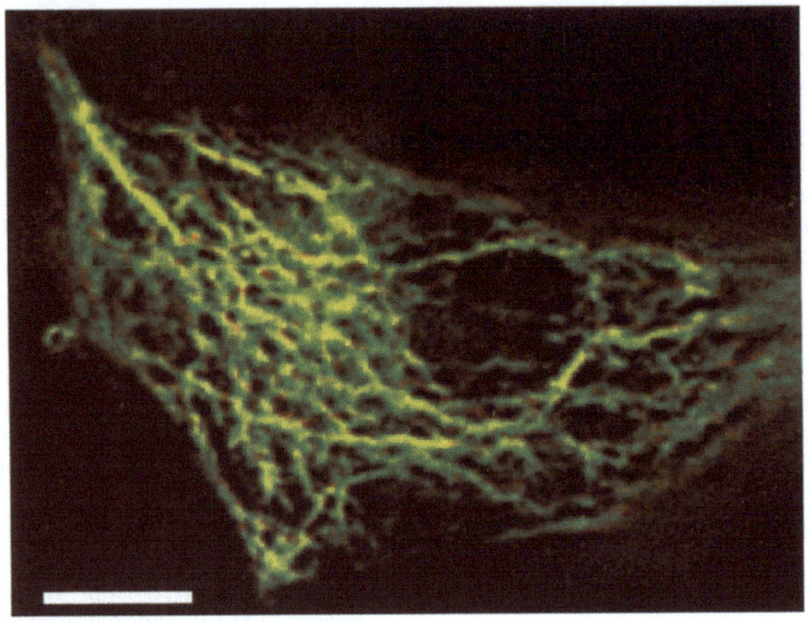

Fig. 1. GFP-vimentin in a living endothelial cell. Bar = 10 μm (Reproduced from Helmke *et al.* (2000) with permission).

unsynchronized cells. The organization of glial fibrillary acidic protein network in astrocytes from vimentin knockout mice is disrupted and later restored after vimentin is transfected back to the cells (Galou *et al.*, 1996). When misdirected vimentin mRNAs are expressed in fibroblasts or SW13 adrenal cells, these cells change their morphology by developing many extremely long processes; they also migrate more slowly in wound healing assays (Morris *et al.*, 2000). In vimentin-deficient mice, it is found that arterial dilation is impaired and arterial remodeling is altered (Henrion *et al.*, 1997; Schiffers *et al.*, 2000).

What might be the underlying mechanism for these significantly altered biological functions? Since vimentin IFs are presumed to play a mechanical role in the CSK of living cells, it would be reasonable to look at the mechanical behavior of the cells when vimentin IFs are disrupted or deleted.

Studies of rheological properties of IF gels

The mechanical roles of IFs in living cells can be inferred from *in vitro* rheological measurements on isolated vimentin and keratin gels. These polymer gels exhibit highly nonlinear stress-strain relationships, characterized by a very low initial stiffness at small strains and relatively high stiffness at large strains (Janmey *et al.*, 1991; Ma *et al.*, 1999). This positive dependence of stiffness on applied stress or strain is known as the strain-hardening or stiffening behavior. Furthermore, mutations in keratins reduce the ability of these IF networks to bundle and to resist large deformations (Ma *et al.*, 2001). These *in vitro* measurements are important in that they predict that the contribution of IFs to cell stiffness would be small during small deformation of the cell and would increase rapidly with increasing cell deformation or distortion.

Vimentin IFs transmit shear stress in living endothelial cells

Recently Peter Davies laboratory has directly measured rapid deformation of the vimentin IFs in living confluent endothelial cells subjected to changes in physiological fluid shear stress (Helmke *et al.*, 2000). By transfecting green fluorescent protein into vimentin IFs, they have quantified spatial and temporal dynamics of vimentin network using time-lapse optical sectioning and deconvolution microscopy. They find that the magnitude and direction of flow-induced vimentin IF displacements are heterogeneous, suggesting a complex force-transmission pattern inside the CSK. This is the first evidence that physiological fluid flow can directly deform cytoskeletal filament systems. They also find that fluid flow results in vimentin displacement over the nucleus and near cell junctions, suggesting mechanical stresses are transmitted to the nucleus and to nearby cells via vimentin IFs (Helmke *et al.*, 2001). Since vimentin IFs may link cell membrane with nuclear lamina, their results suggest that fluid shear stress might directly influence gene expression in living cells.

Mechanics of IFs in living cells

The structural organization of vimentin IFs suggests that they carry tensile stress. When plated within collagen gels, vimentin-deficient cells exhibit reduced contraction of the gels, indicating decreased contractility of these cells, consistent with the prediction that vimentin IFs carry tensile stresses (Eckes *et al.*, 1998). Furthermore, cell–cell physical interactions are dramatically decreased in vimentin-deficient cells, when compared with wild-type cells. These results suggest that vimentin IFs are important for maintaining mechanical interactions between cells, possibly by carrying some of the contractile forces generated among cells.

The role of vimentin IFs in maintaining cell structural strength is examined further by applying a pulling force to the cell surface integrin receptors via a micropipette-displaced fibronectin-coated microbeads. Compared with wild-type cells, similar magnitudes of deformation on the cell surface of vimentin-deficient cells result in tearing of the cytoplasm and rupture of the plasma membrane (Eckes *et al.*, 1998). These results show that vimentin IFs are required for the mechanical strength of the CSK and for protection of the cell structural integrity.

Another important index for cell shape stability is the shear stiffness of the cell. The higher the stiffness, the less deformable the cell is under a given mechanical stress. Using a magnetic twisting cytometry technique, controled rotational mechanical torques can be applied to cell surface integrin receptors via integrin-ligand-coated ferromagnetic microbeads (Wang *et al.*, 1993; Wang and Ingber, 1994). When vimentin-deficient fibroblasts are deformed, they are about 40% less stiff than wild-type cells for the same high amplitude of applied stress under the same culturing conditions (Eckes *et al.*, 1998). Furthermore, disruption of actin MF network with cytochalasin D further decreased the stiffness of both vimentin-deficient and wild-type cells by about 50% from the baseline values. These observations are consistent with earlier findings that stiffness is decreased in endothelial cells when vimentin IFs are disrupted using acrylamide (Wang *et al.*, 1993; Wang, 1998).

An important prediction from the isolated vimentin gels is that vimentin may contribute little to cell stiffness unless the applied stress or strain is large. One way to determine the role of vimentin IFs in the stress-strain relationship of the cells is to vary the amplitudes of stresses that are applied to vimentin-deficient fibroblasts and to measure the resultant strains (Wang and Stamenovic, 2000). When the applied stress is increased from 10 to 80 dyn/cm^2, the stiffness increases nearly linearly in both wild-type and vimentin-deficient cells that have been plated for 4 h (Figure 2B). However, vimentin-deficient cells are less stiff, except at low (10 dyn/cm^2) applied stress, and exhibited less stiffening than wild-type cells. Similar results are obtained from measurements on wild-type fibroblasts and endothelial cells after vimentin IFs are disrupted by acrylamide (Wang and Stamenovic, 2000). All these results in living cells are consistent with the results from the isolated vimentin gels.

Since all cells generate active contractile forces even before external forces are applied, it is important to control for the amount of preexisting cytoskeletal tensile stress (prestress) in the cell and then examine the role of vimentin in cell mechanics. Recently it has been shown that cell traction forces (the contact forces between the cell basal surface and the substrate that arise in response to cell contraction) and thus prestress increase with cell spreading and cell stiffness (Wang and Ingber, 1994; Wang *et al.*, 2001, 2002a; Wang *et al.*, 2002b). After endothelial cells are plated for an extended period of 16 h, stiffness is measured after vimentin IFs are disrupted with acrylamide and is found to be much less than control cells even at lowest applied stress of 10 dyn/cm^2 (Wang and Stamenovic, 2000). These results suggest that for highly spread cells vimentin IFs may be playing an increasingly important role in contributing to the mechanical strength of the cells.

Model prediction for the mechanical role of IFs

To further explore the underlying mechanisms of cell stiffness and stiffening, it is informative to study model systems that may depict fundamental mechanical behavior of living cells. To achieve this, a six-strut tensegrity model that includes the IFs as well as MFs and MTs has been used as a representative of cellular tensegrity. The reason for choosing this particular model was that it could mimic and quantitatively predict many forms of behavior exhibited by living cells (Stamenovic *et al.*, 1996; Coughlin and Stamenovic, 1998; Stamenovic and Coughlin, 2000). It is assumed that the MF cables are elastic and initially under tension, that the IF cables are elastic and initially slack and the MT struts are rigid. The model is stretched uniaxially by pulling apart a pair of parallel struts of MTs (Figure 3A), and its stiffness is calculated using methods of engineering mechanics as described in Wang and Stamenovic (2000). It is interesting that the model predicts several features that are consistent with the data obtained in living cells. First, at a given stress, the stiffness is greater when IFs are present than when they are absent (Figure 3B). This difference is small at low stress, increasing progressively with increasing stress. Second, the stiffening is greater with the presence of IFs (Figure 3B). Third, the stiffening still exists even when the IF elements are absent (Figure 3B), probably due to the effect of kinematic stiffening (the change of orientation and spacing of model elements). Although the model is very simple when compared with the complex structural arrangements in living cells, it may reveal some underlying mechanisms that exist in living cells. Since the presence and absence of IFs in the model exhibit similar changes as those in living cells, it is possible that the mechanisms that determine the model stiffness and stiffening might also be present in the cell, i.e., the cell stiffening is partly due to the rheological properties of IFs and partly due to the kinematic stiffening of the CSK.

Another important mechanical role of IFs is that they stabilize MTs, which buckle under compression during

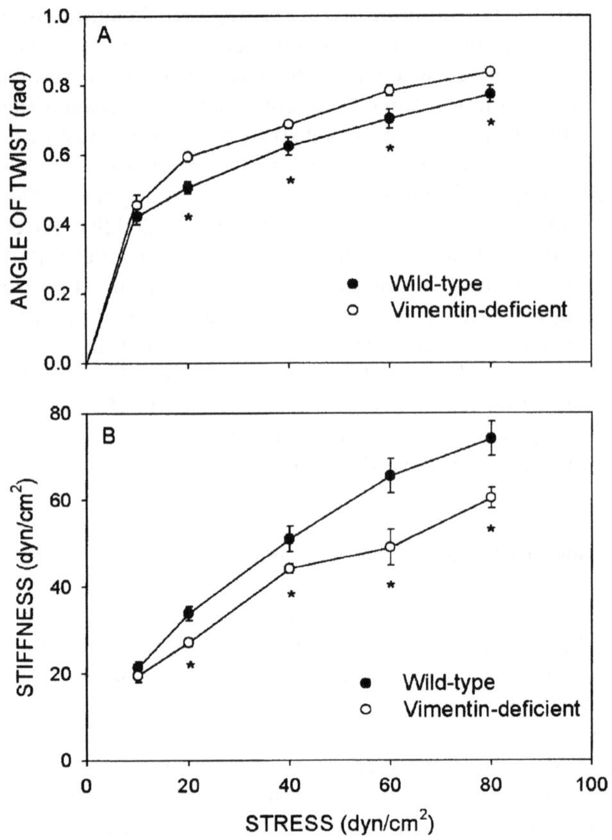

Fig. 2. (A) Angle of twist vs. stress relationships and (B) corresponding stiffness vs. stress relationships for adherent wild-type and vimentin-deficient fibroblasts. Stresses ranging from 10 to 80 dyn/cm^2 were applied trough RGD-coated ferromagnetic beads. Cells were plated on collagen-I coated dishes for 4 h before mechanical measurements. Means – SE ($n = 5$ wells); * = significantly greater values ($P < 0.05$) (Reproduced from Wang and Stamenovic, (2000) with permission).

cell contraction. Brodland and Gordon (1990) proposed that IFs provide a continuous lateral support which in effect reduces critical buckling length of MTs and thus maintain their stability during cell contraction. This model has been recently supported by quantitative experimental data and an energetic analysis. It was shown that in cultured airway smooth muscle cells the energy associated with buckling of MTs that are laterally supported by IFs is quantitatively consistent with the energy transferred from MT's to the substrate upon disruption of MTs (Stamenovic *et al.*, 2002a).

IF interactions with other CSK filament systems

It is now known that plectin molecules that are abundantly expressed in many cell types are cross-linking proteins that physically connect the IFs with MFs and MTs (Seifert *et al.*, 1992; Svitkina *et al.*, 1996; Wiche, 1998). Since shear deformations have to be transmitted through solid structures without exhibiting continuous flow behavior, it is possible that during cell contraction or during externally applied stresses plectins transmit large amounts of forces among the three CSK filament systems. However, it is still not known whether plectin molecules themselves store any elastic energy and/or dissipate any energy during internal contractile force generation or external deformation.

Differences in contribution of IFs to cell mechanics in different cell types

It has been shown that in fibroblasts and endothelial cells the contribution of IFs to cell stiffness is about 20–

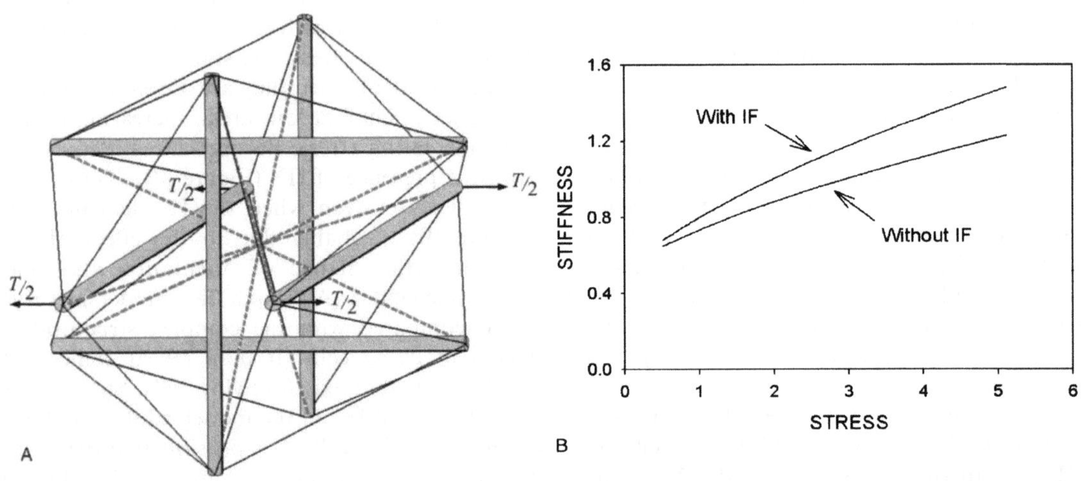

Fig. 3. (A) Six-strut tensegrity model. Microtubule struts: *AA*, *BB*, *CC*; microfilament cables *AB*, *AC*, *BC*; intermediate filament (radial) cables *OA*, *OB*, *OC*. Stretching force *T/2* (thick arrows) is applied at the strut endpoints *A* in the *x*-direction. (B) Stiffness vs. stress relationships predicted by the model for the cases where the IF cables are present and when they are not present in the model; stiffness and stress are given in force per length square. Note that in the presence of IF elements the model exhibit greater stiffening and greater stiffness than when the IF elements are not present. This behavior is consistent with the observations in living cells (Reproduced from Wang and Stamenovic, (2000) with permission).

40%, demonstrated using either vimentin-deficient cells or acrylamide-treated cells (Wang *et al.*, 1993; Eckes *et al.*, 1998; Wang, 1998; Wang and Stamenovic, 2000). On the other hand, it has been recently shown that the rigidity of T lymphocytes primarily depends on vimentin IFs (Brown *et al.*, 2001), whereas MTs play a minor role in T cell rigidity.

It is further speculated that the structural alterations in vimentin IFs in circulating T lymphocytes are important in transendothelial migration of these cells. However, in airway smooth muscle cells, disruption of IFs do not alter the elastic and frictional moduli significantly, suggesting that the IFs do not contribute much to the viscoelastic properties of these cells. However, IFs do play synergistic roles with MFs and MTs together in contributing to the stiffness of the airway smooth muscle cells (Stamenovic *et al.*, 2002b).

Conclusions

Growing experimental evidence shows that IFs play important mechanical roles in living cells. We still have little knowledge on the specific structural and biomechanical interactions between IFs and MFs and MTs. The role of CSK-linking proteins such as plectins and other yet-to-be-discovered proteins in stress distribution and energy dissipation needs to be elucidated. The role of IFs in transmitting and transducing mechanical deformation from the cell surface into the nucleus needs further investigation. We are optimistic that with the emergence of collaboration between biologists and engineers new and unexpected findings on the role of IFs and other CSK filaments in regulation and control of cell functions will occur in the near future.

Acknowledgements

Due to the fact that this is a brief review of mechanics of vimentin intermediate filaments, we regret that some important references related to the topic may have been omitted. We would like to thank J. Chen and Z. Liang for assistance. This work was supported by NASA grant NAG2-1509 and NIH grant HL-33009.

References

Amos LA and Amos WB (1991) *Molecules of the Cytoskeleton*. Guilford Press, New York.

Beningo K, Dembo M, Kaverina I, Small V and Wang Y-L (2001) Nascent focal adhesions are responsible for the generation of strong propulsive forces in migrating fibro0blasts. *J Cell Biol* **153:** 881–888.

Brodland GW and Gordon R (1990) Intermediate filaments may prevent buckling of compressively loaded microtubules. *ASME J Biomech Eng* **112:** 319–321.

Brown MJ, Hallam JA, Colucci-Guyon E and Shaw S (2001) Rigidity of circulating lymphocytes is primarily conferred by vimentin intermediate filaments. *J Immuno* **166:** 6640–6646.

Chen J, Fabry B, Schiffrin EL and Wang N (2001) Twisting integrin receptors increases endothelin-1 gene expression. *Am J Physiol Cell Physiol* **280:** C1475–C1484.

Chicurel ME, Chen CS and Ingber DE (1998) Cellular control lies in the balance of forces. *Curr Opin Cell Biol* **10:** 232–239.

Cochard P and Paulin D (1984) Initial expression of neurofilaments and vimentin in the central and peripheral nervous system of the mouse embryo *in vivo*. *J Neurosci* **4:** 2080–2094.

Colucci-Guyon E, Portier MM, Dunia I, Paulin D, Pournin S and Babinet C (1994) Mice lacking vimentin develop and reproduce without an obvious phenotype. *Cell* **79:** 679–694.

Coughlin MF and Stamenovic D (1998) A tensegrity model of the cytoskeleton in spread and round cells. *ASME J Biomech Eng* **120:** 770–777.

Coulombe PA, Bousquet O, Ma L, Yamada S and Wirtz D (2000) The 'ins' and outs' of intermediate filament organization. *Trends Cell Biol* **10:** 420–428.

Djiabali K (1999) Cytoskeletal proteins connecting intermediate filaments to cytoplasmic and nuclear periphery. *Histol Histopathol* **14:** 501–509.

Eckes B, Dogic D, Colucci-Guyon E, Wang N, Maniotis A, Ingber D, Merckling A, Langa F, Aumailley M, Delouvée A, Koteliansky V, Babinet C and Krieg T (1998) Impaired mechanical stability, migration and contractile capacity in vimentin-deficient fibroblasts. *J Cell Sci* **111:** 1897–1907.

Frankel WW, Grund C, Kuhn C, Jackson BW and Illmensee K (1982) Formation of cytoskeletal elements during mouse embryogenesis. III. Primary mesenchymal cells and the first appearance of vimentin filaments. *Differentiation* **23:** 43–59.

Galou M, Colucci-Guyon E, Ensergueix D, Ridet J-L, Gimenez Y, Ribotta M, Privat A, Babinet C and Dupouey P (1996) Disrupted glial fibrillary acidic protein network in astrocytes from vimentin knockout mice. *J Cell Biol* **133:** 853–863.

Gawlitta W, Osborn M and Weber K (1981) Coiling of intermediate filaments by microinjection of a vimentin specific antibody does not interfere with locomotion and mitosis. *Eur J Cell Biol* **26:** 83–90.

Goldman RD, Khuon S, ChouYH, Opal P and Steinert PM (1996) The function of intermediate filaments in cell shape and cytoskeletal integrity. *J Cell Biol* **134:** 971–983.

Hedberg KK and Chen LB (1986) Absence of intermediate filaments in a human adrenal cortex carcinoma-derived cell line. *Exp Cell Res* **163:** 509–517.

Helmke BP, Goldman RD and Davies PF (2000) Rapid displacement of vimentin intermediate filaments in living endothelial cells exposed to flow. *Circ Res* **86:** 745–752.

Helmke BP, Thakker DB, Goldman RD and Davies PF (2001) Spatiotemporal analysis of flow-induced intermediate filament displacement in living endothelial cells. *Biophys J* **80:** 184–194.

Henrion D, Terzi F, Matrougui K, Duriez M, Boulanger CM, Colucci-Guyon E, Babinet C, Briand P, Friedlander G, Poitevin P and Levy BI (1997) Impaired flow-induced dilation in mesenteric resistance arteries from mice lacking vimentin. *J Clin Invest* **100:** 2909–2914.

Holwell TA, Schweitzer SC and Evans RM (1997) Tetracyclin regulated expression of vimentin in fibroblasts derived from vimentin null mice. *J Cell Sci* **110:** 1947–1956.

Huang S, Chen CS and Ingber DE (1998) Control of Cyclin D1, p27kip1, and cell cycle progression in human capillary endothelial cells by cell shape and cytoskeletal tension. *Mol Biol Cell* **9:** 3179–3193.

Klymkowsky MW (1981) Intermediate filaments in 3T3 cells collapse after intracellular injection of a monoclonal anti-intermediate filament antibody. *Nature* **291:** 249–251.

Janmey PA (1998) The cytoskeleton and cell signaling: component, localization, and mechanical coupling. *Physiol Rev* **78:** 763–781.

Janmey PA, Euteneuer U, Traub P and Schliwa M (1991) Viscoelastic properties of vimentin compared with other filamentous biopolymer networks. *J Cell Biol* **113:** 155–160.

Lane EB, Hogan LM, Kurkinen M and Garrels JI (1983) Co-expression of vimentin and cytokeratin in parietal endoderm cells of early mouse embryo. *Nature* **303:** 701–704.

Ma L, Xu J, Coulombe PA and Wirtz D (1999) Keratin filament suspensions show unique micromechanical properties. *J Biol Chem* **274:** 19145–19151.

Ma L, Yamada S, Wirtz D and Coulombe PA (2001) A 'hot-spot' mutation alters the mechanical properties of keratin filament networks. *Nat Cell Biol* **3:** 503–506.

Morris EJ, Evason K, Wiand C, L'Ecuyer TJ and Fulton AB (2000) Misdirected vimentin messenger RNA alters cell morphology and motility. *J Cell Sci* **113:** 2433–2443.

Schiffers PM, Henrion D, Boulanger CM, Colucci-Guyon E, Langa-Vuves F, van Essen H, Fazzi GE, Levy BI and Mey JG (2000) Altered flow-induced arterial remodeling in vimentin-deficient mice. Arterio., *Throm Vascu Biol* **20:** 611–616.

Seifert GJ, Lawson D and Wiche G (1992) Immunolocalization of the intermediate filament-associated protein plectin at focal contacts and actin stress fibers. *Eur J Cell Biol* **59:** 138–147.

Stamenovic D, Fredberg JJ, Wang N, Butler JP and Ingber DE (1996) A microstructural approach to cytoskeletal mechanics based on tensegrity. *J Theor Biol* **181:** 125–136.

Stamenovic D and Coughlin MF (2000) A quantitative model of cellular elasticity based on tensegrity. *ASME J Biomech Eng* **122:** 39–43.

Stamenovic D, Mijailovich SM, Tolic-Nørrelykke IM, Chen J and Wang N (2002a) Cell prestress. II. Contribution of microtubules. *Am J Physiol Cell Physiol* **282:** C617–C624.

Stamenovic D, Liang Z, Chen J and Wang N (2002b) The effect of the cytoskeletal prestress on the mechanical impedance of cultured airway smooth muscle cells. *J Appl Physiol* **92:** 1443–1450.

Svitkina TM, Verkhovsky AB and Borisy GG (1996) Plectin sidearms mediate interaction of intermediate filaments with microtubules and other components of the cytoskeleton. *J Cell Biol* **135:** 991–1007.

Venetianer A, Schiller DL, Magin T and Franke WW (1983) Cessation of cytokeratin expression in a rat heptoma cell line lacking differentiated functions. *Nature* **305:** 730–733.

Wang N (1998) Mechanical interactions among cytoskeletal filaments. *Hypertension* **32:** 162–165.

Wang N and Ingber DE (1994) Control of cytoskeletal mechanics by extracellular matrix, cell shape, and mechanical tension. *Biophys J* **66:** 2181–2189.

Wang N, Butler JP and Ingber DE (1993) Mechanotransduction across the cell surface and through the cytoskeleton. *Science* **260:** 1124–1127.

Wang N and Stamenovic D (2000) Contribution of intermediate filaments to cell stiffness, stiffening, and growth. *Am J Physiol Cell* **279:** C188–C194.

Wang N, Naruse K, Stamenovic D, Fredberg JJ, Mijailovich SM, Tolic-Nørrelykke IM, Polte T, Mannix R and Ingber DE (2001) Mechanical behavior in living cells consistent with the tensegrity model. *Proc Natl Acad Sci* **98:** 7765–7770.

Wang N, Tolic-Nørrelykke IM, Chen J, Mijailovich SM, Butler JP, Fredberg JJ and Stamenovic D (2002a) Cell prestress. I. Stiffness and pretress are closely associated in adherent contractile cells. *Am J Physiol Cell Physiol* **282:** C606–C616.

Wang N, Ostuni E, Whitesides GM and Ingber DE (2002b) Micropatterning tracitonal forces in living cells. *Cell Motil Cytoskel* **52:** 97–106.

Wiche G (1998) Role of plectin in cytoskeleton organization and dynamics. *J Cell Sci* **111:** 2477–2486.

Extracellular Matrix Proteins

Journal of Muscle Research and Cell Motility **23**: 543–559, 2002.

Mechanics of elastin: molecular mechanism of biological elasticity and its relationship to contraction

D.W. URRY[1,2,*] and T.M. PARKER[2]
[1]*University of Minnesota, Twin Cities Campus, BioTechnology Institute, 1479 Gortner Avenue, St. Paul, MN 55108-6106;* [2]*Bioelastics Research Ltd., 2800 Milan Court, Suite 386, Birmingham, AL, 35211-6918, USA*

Abstract

Description of the mechanics of elastin requires the understanding of two interlinked but distinct physical processes: the development of entropic elastic force and the occurrence of hydrophobic association. Elementary statistical-mechanical analysis of AFM single-chain force–extension data of elastin model molecules identifies damping of internal chain dynamics on extension as a fundamental source of entropic elastic force and eliminates the requirement of random chain networks. For elastin and its models, this simple analysis is substantiated experimentally by the observation of mechanical resonances in the dielectric relaxation and acoustic absorption spectra, and theoretically by the dependence of entropy on frequency of torsion-angle oscillations, and by classical molecular-mechanics and dynamics calculations of relaxed and extended states of the β-spiral description of the elastin repeat, $(GVGVP)_n$. The role of hydrophobic hydration in the mechanics of elastin becomes apparent under conditions of isometric contraction. During force development at constant length, increase in entropic elastic force resulting from decrease in elastomer entropy occurs under conditions of increase in solvent entropy. This eliminates the solvent entropy change as the entropy change that gives rise to entropic elastic force and couples association of hydrophobic domains to the process. Therefore, association of hydrophobic domains within the elastomer at fixed length stretches interconnecting dynamic chain segments and causes an increase in the entropic elastic force due to the resulting damping of internal chain dynamics. Fundamental to the mechanics of elastin is the inverse temperature transition of hydrophobic association that occurs with development of mechanical resonances within fibrous elastin and polymers of repeat elastin sequences, which, with design of truly minimal changes in sequence, demonstrate energy conversions extant in biology and demonstrate the special capacity of bound phosphates to raise the free energy of hydrophobic association.

Introduction

As a brief introduction to the mechanics of elastin, noted are the composition and sequence of bovine elastin, the inverse phase transition behavior of the elastin and related protein components, and the conformational aspects of the most representative repeating sequence of elastin.

Composition and sequence of elastin

The single-residue T_t-based hydrophobicity plots of Figure 1 (Urry, 1997) provide a simple way to gain perspective of the composition and sequence of bovine elastin and to make comparisons with other extracellular proteins, for example human fibronectin. The longest repeating sequence $(GVGVP)_{11}$ is labeled W4 based on the nomenclature of the Sandberg group (Sandberg *et al.*, 1981, 1985). Additional, less extensive repeats are also apparent. Of the nearly 40 lysine (Lys, K) residues, all but two or three are used in forming cross-links. Generally, four Lys residues come together to form a tetrasubstituted pyridinium. By contrast, in the vulcanization of rubber every monomer is a potential cross-link, which allows for cross-link formation whenever two chains are in contact. In elastin only 5% of the residues participate in cross-link formation and these are often distributed along the chain in pairs. Such a situation requires a regular structure in order to bring this limited number of residues into adequate proximity for covalent cross-link formation.

There are only three negatively charged aspartic acid (Asp, D) and no glutamic acid residues in bovine elastin. The more hydrophobic phenylalanine (Phe, F) and tyrosine (Tyr, Y) residues are also apparent. The almost complete absence of charged groups combined with a substantial quantity of hydrophobic residues are responsible for the hydrophobic association between chains that result in parallel-aligned twisted filaments as seen in electron micrographs of negatively stained tropoelastin and α-elastin (Cox *et al.*, 1973, 1974) and in fibrous elastin (Gotte *et al.*, 1974).

By comparison, a sequence that results in a series of globular repeating units of hydrophobically folded β-barrels is seen in the single-residue T_t-based hydrophobicity plots for fibronectin in the lower part of Figure 1 (Kornblihtt *et al.*, 1985). In this case there are

*To whom correspondence should be addressed: E-mail: durry98@aol.com

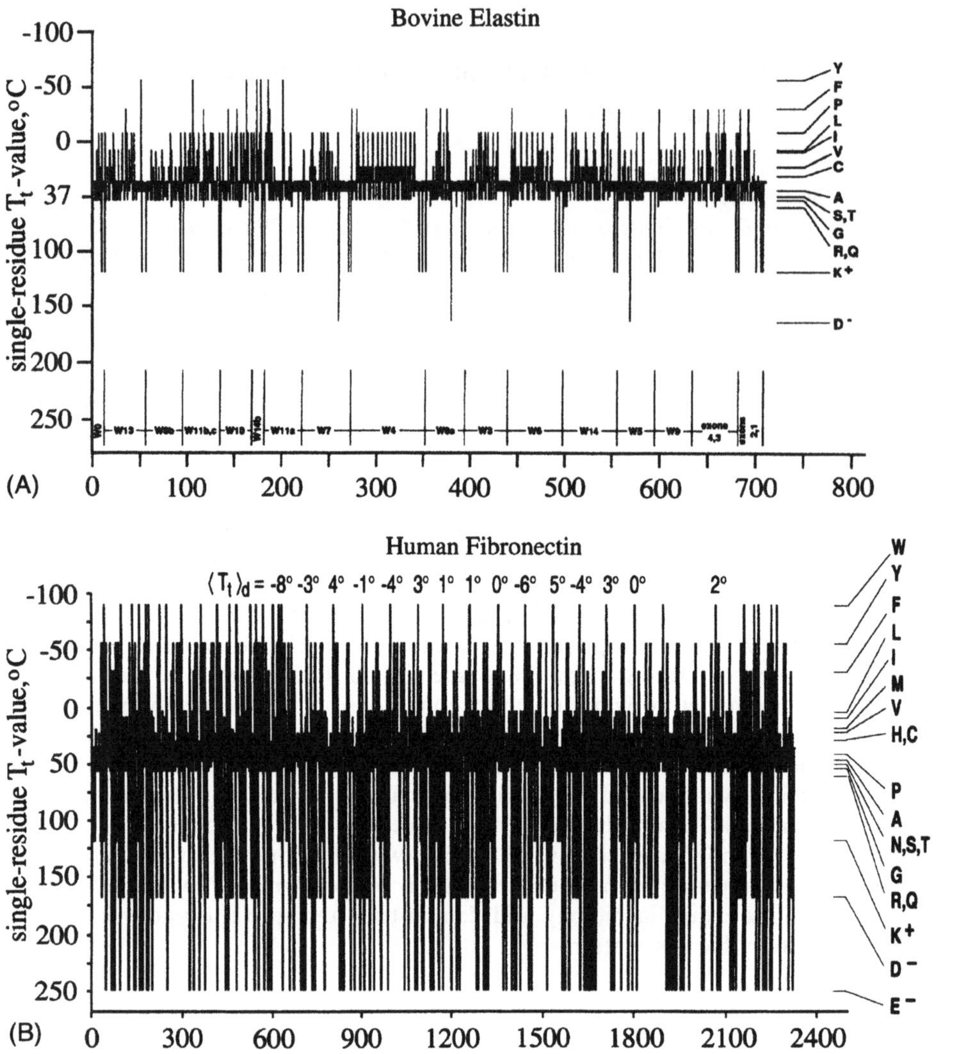

Fig. 1. Single residue T_t-based hydrophobicity plots for bovine elastin (A) and human fibronectin (B) (see text for discussion). Part A reproduced with permission from Urry *et al.* (1995), Part B reproduced with permission from Urry and Luan (1995a).

many more charged Glu, Asp and Lys residues, and the periodicity of the repeating globular units becomes apparent through the repeating tryptophan (Trp, W) residues. This most hydrophobic residue, W, repeating approximately every 90 residues identifies type III domains of fibronectin, which also appear as a repeating unit in titin (connectin). Such repeating globular units give a sawtooth pattern to the single-chain force–extension curves of titin (Rief *et al.*, 1997; Gaub and Fernandez, 1998), which is very different from the smooth single-chain force–extension curves given by elastin models (Urry *et al.*, 2002a, b).

Inverse temperature transition of elastin and component sequences

Increase in order of elastic protein upon raising the temperature

Tropoelastin (precursor protein of fibrous elastin), α-elastin (chemical fragmentation product from fibrous elastin) and high-molecular-weight polymers of repeating sequences of elastin are water soluble at low temperatures, but produce phase separation upon raising the temperature. The initial aggregates of the phase transition, when placed on a carbon-coated EM grid and negatively stained with uranyl acetate/oxalic acid, are observed to form parallel aligned filaments comprised of twisted filaments with interesting periodicities in the optical diffraction patterns of the micrographs (Volpin *et al.*, 1976a, b). Furthermore, cyclic analogues containing pentapeptide and hexapeptide repeats crystallize during temperature increase and redissolve upon lowering the temperature. Clearly, the order in elastin and elastin-based polymers increases with increasing the temperature.

This ordering of protein upon raising the temperature is consistent with the second law of thermodynamics: a structured hydration that surrounds the dissolved hydrophobic groups becomes less ordered bulk water during the phase separation of hydrophobic association (Urry *et al.*, 1997a, b).

This phase transition leading to a fundamental increase in order of the polymer on raising the temperature is called an inverse temperature transition.

Phase diagram

Phase diagrams of the inverse temperature-dependent behavior of elastin-based polymers is seen in Figure 2 (Sciortino *et al.*, 1990, 1993). In the usual polymer phase diagram the polymers are insoluble below and soluble above the binodal (coexistence or coacervate) line. For elastin-based polymers solubility is inverted. With elastin-based polymers solubility occurs below and insolubility occurs above the coexistence line. The shape of the coexistence line is also inverted: while for the usual petroleum-based polymers the line is concave towards the concentration (volume–fraction) axis, for elastin-related polymers it is convex. In case of the elastin-based polymers the coexistence line is also called the T_t-divide. Each point (called T_t) along the T_t-divide corresponds, for a given concentration, to the onset temperature of aggregation.

Elastin-based polymers exhibit a latent heat of transition. Accordingly, within the temperature range of the transition the chemical potential of the polymers in solution and in the phase-separated state are the same. This enables the derivation of the Gibbs free energy of hydrophobic association, ΔG_{HA} (see below). The more hydrophobic elastin-based polymers that associate at lower temperatures have lower values of ΔG_{HA}.

Conformational aspects of the $(GVGVP)_n$ of elastin

The structure of $(GVGVP)_n$, shown in Figure 3, was developed by using methodologies of peptide synthesis, NMR, molecular-mechanics and dynamics calculations

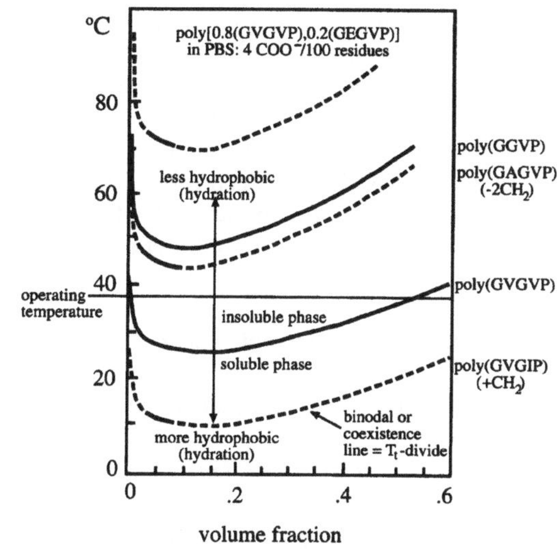

Fig. 2. Schematic representation of phase diagrams for several model elastic proteins based on the elastin repeat, $(GVGVP)_n$ (see text for discussion). Solid curves adapted with permission from Sciortino *et al.* (1990) and (1993).

(Venkatachalam and Urry, 1981; Chang *et al.*, 1989), crystal structure of cyclic analogues (Cook *et al.*, 1980) and the NMR- and computationally characterized relationship between cyclic and linear structures (Urry *et al.*, 1981, 1989), and Raman, absorption and circular dichroism spectroscopies.

The structure involves a series of β-turns, ten-atom hydrogen-bonded rings from the Val^1 C—O to the Val^4 N—H (Figure 3A and B) which, upon raising the temperature, wrap up into a helical structure called the β-spiral (shown schematically in Figure 3C and D and in detail in Figure 3E). It is shown that the β-turns

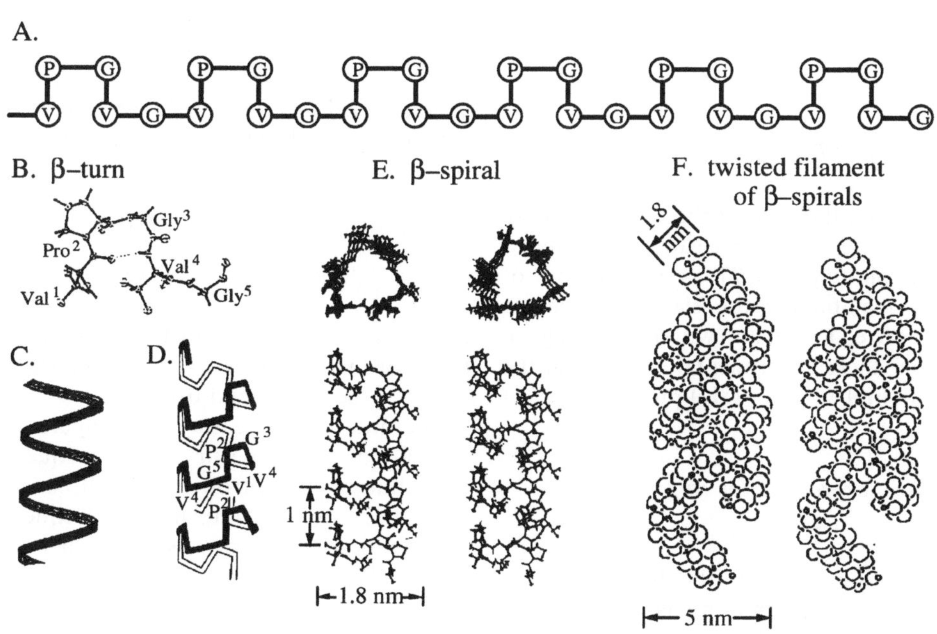

Fig. 3. Representations of the proposed molecular structure of $(GVGVP)_n$ (see text for discussion).

function as hydrophobic spacers between the turns of the β-spiral and that the β-spirals associate hydrophobically to form twisted filaments of the dimensions found in the transmission electron micrographs of negatively stained incipient aggregates as noted above (Volpin *et al.*, 1976a, b). Hydrophobic folding of β-spirals and the hydrophobic association of β-spirals to form twisted filaments occur in a cooperative manner.

General considerations of entropic elastic force

Components of elastic force and their delineation

Internal energy (f_E) and entropy (f_S) components of elastic force

The total elastic force (f) can be thermodynamically described as

$$f = (\partial E/\partial L)_{V,T,n} - T(\partial S/\partial L)_{V,T,n}, \tag{1}$$

where E is the internal energy; S the entropy; T the absolute temperature in °K, and V, T, and n indicate that the change in length (∂L) occurs at constant volume, temperature and composition. Accordingly, the total elastic force comprises two components, the internal energy component (f_E) and the entropic component (f_S):

$$f = f_E + f_S. \tag{2}$$

Experimental delineation of the relative magnitudes of f_E and f_S

Following Flory *et al.* (1960) the total force can also be written as

$$f = (\partial E/\partial L)_{V,T,n} + T(\partial f/\partial T)_{V,L,n} \tag{3}$$

which is equivalent to

$$f_E/f = -T(\partial \ln[f/T]/\partial T)_{V,L,n}. \tag{4}$$

Equation (4) shows that the ratio of the internal energy component of force to the total force can be determined experimentally by plotting $\ln[f/T]$ as a function of temperature while maintaining the elastic element at constant volume, at fixed length and without a change in composition. Under these experimental conditions, the slope of the plot multiplied by ($-T$°K) provides the f_E/f ratio.

Statistical mechanical expression for entropy

The Boltzmann relation

The Boltzmann relation provides the bridge from a statistical mechanical description of molecular structure to experimentally determined thermodynamic quantities. It is an elegant yet simple statement of entropy (S) in terms of thermodynamic probability (W, the number of *a priori* equally probable states accessible to the system) (Eyring *et al.*, 1964), i.e.,

$$S = R \ln W. \tag{5}$$

R (1.987 cal/deg mol) is the gas constant. $R=Nk$, where N is Avogadro's number (6.02×10^{23}/mol) and k Boltzmann's constant (1.38×10^{-16} erg/deg K). W is a volume in phase space, a $2N$-dimensional space, which fundamentally provides a description of a molecule in terms of the momentum (p_i) and coordinate (q_i) of each its i atoms.

Description of entropy using partition functions for different degrees of freedom

In practice, the description of the molecule is achieved by the product of partition functions, which group according to the degrees of freedom of the molecular system. In general, the $2N$ degrees of freedom group as three translational degrees of freedom, three rotational degrees of freedom and the remainder are internal degrees of freedom that comprise vibrations and torsional oscillations (rotations about bonds). In the measurement of elasticity, the single chain molecule or the cross-linked matrix is fixed at both ends. In this case, there are holonomic constraints on the molecular system in that there are neither whole-molecule translational nor rotational degrees of freedom. Thus, we are left only with internal chain dynamics.

Frequency dependence of entropy for the harmonic oscillator representation of internal chain dynamics

The harmonic oscillator partition function (f_v) provides a relatively simple and yet informative representation of internal chain dynamics.

$$f_v = [1 - \exp(-h\nu_i/kT)]^{-1}. \tag{6}$$

When placed in Equation (5) and expressed in terms of the frequency-dependence of entropy (S) we obtain (Dauber *et al.*, 1981),

$$S_i = R\{\ln[1 - \exp(-h\nu_i/kT)]^{-1} + (h\nu_i/kT)[\exp(h\nu_i/kT) - 1]^{-1}\}. \tag{7}$$

Equation (7) is plotted in Figure 4 with S_i on the left-hand ordinate and with the free-energy contribution (TS_i) on the right-hand ordinate.

The interesting features of Figure 4 are that the vibrational modes occur at high frequencies where the contribution to entropy is small, and the torsional oscillations occur at much lower frequencies where the contribution to entropy is much larger. As discussed below, the mechanical resonances exhibited by elastin models and fibrous elastin occur at frequencies near 1 kHz and 5 MHz, i.e., near 3 and 6 on the abscissa of Figure 4. Accordingly, it becomes reasonable to neglect

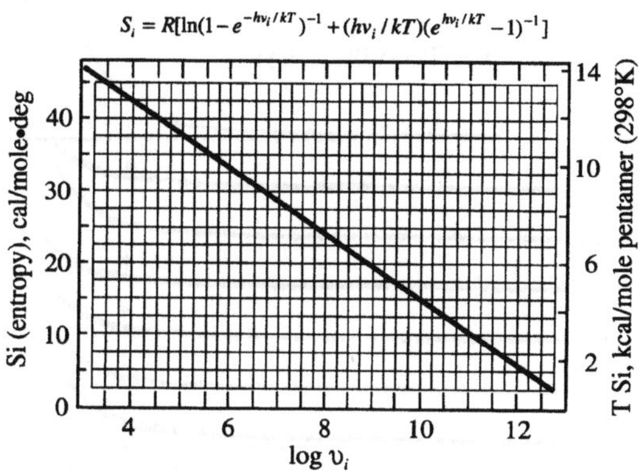

Fig. 4. Plot, based on the harmonic oscillator partition function, of the entropy of the oscillator as a function of log(oscillator frequency). On the left-hand ordinate is plotted the entropy in cal/mol deg (EU, entropy units), and on the right-hand ordinate is plotted the entropic contribution to the Gibbs free energy in kcal/mol pentamer at 298°K. This illustrates the increasing contribution of low frequency oscillations to the entropy and free energy of a protein chain segment. Adapted with permission from Urry *et al.* (1988).

changes in vibrational frequencies on extension and to consider those configurational and dynamic changes due to changes in torsion angles.

Expression for the change in entropy on extension

The change in entropy upon extension is written,

$$\Delta S = (S^{e} - S^{r}) = R\ln(W^{e}/W^{r}), \tag{8}$$

where e and r stand for the extended and relaxed states. With Equation (8) it becomes possible to use both molecular mechanics and dynamics calculations of the molecular structure described in Figure 3 to calculate the contribution of torsional freedom of the structure to entropy in both the relaxed and extended states.

The physical basis for entropic elasticity in elastin

In an ideal or perfect elastomer the energy repeatedly invested in extension is repeatedly and completely recovered during relaxation. Ideality increases as the elastic force results from a decrease in entropy upon extension, because this occurs without stressing bonds to the breaking point. Elastin models and elastin itself in water provide examples of such entropic elastomers with about 90% of the elastic force being entropic, that is, the f_E/f ratio of Equation (4) is about 0.1. This is essential to human life expectancy, because the half-life of elastin in the mammalian elastic fiber is on the order of 70 years. This means that the elastic fibers of the aortic arch and thoracic aorta, where there is twice as much elastin as collagen, will have survived some billion demanding stretch-relaxation cycles by the start of the seventh decade of life. This represents an ultimate in ideal elasticity.

Three main mechanisms proposed for the entropic elasticity of elastin

The classical (random chain network) theory of rubber elasticity

In 1958 Hoeve and Flory reported studies on the nature of elasticity of the bovine fibrous protein, elastin, as representative of the mammalian elastic fiber. Because they were able to determine a very low f_E/f ratio, they concluded that the elastic fiber was without order, that is, it was a random chain network. They re-affirmed it in the spring of 1974 in a Biopolymers paper (Hoeve and Flory, 1974) stating that 'A network of random chains within the elastic fibers, like that in a typical rubber, is clearly indicated.' In order to emphasize this perspective further, the following statement appeared in the legend to Figure 1 of that paper: 'Configurations of chains between cross-linkages are much more tortuous and irregular than shown.' In part because Paul Flory received the Nobel Prize in 1974 for his work on macromolecules, this perspective became firmly set in the minds of the interested scientific community.

The solvent (bulk water ↔ hydrophobic hydration) entropy mechanism

It appeared that the sole purpose of the Hoeve and Flory 1974 paper, which presented neither new experimental data nor theoretical analysis, was to refute the publication of a new mechanism. The new mechanism for the entropic elasticity of elastin was published by Weis-Fogh and Anderson (1970) with the title 'New Molecular Model for the Long-range Elasticity of Elastin.' The proposed new mechanism, stated in the terms of present adherents, was that upon extension hydrophobic side-chains become exposed to the water solvent. The result is formation of low-entropy water of hydrophobic hydration. Here we refer to this as the solvent (bulk water ↔ hydrophobic hydration) mechanism for entropic elasticity. This mechanism has been given credibility by groups using computational methods on $(GVGVP)_n$ in water (Wasserman and Salemme, 1990; Alonso *et al.*, 2001). One of the many arguments marshaled by Hoeve and Flory (1974) against this mechanism was that polymer backbone, not solvent, must bear the force.

The damping of internal chain dynamics on extension

In the hydrophobically folded β-spiral structure (Figure 3), the presence of suspended segments between the β-turns is immediately apparent. With water surrounding the suspended segments and no steric hindrance to limit rotation about backbone bonds of the suspended segments, the peptide moieties would be expected to rock or 'librate'. These suspended segments are probably free to undergo large torsion-angle oscillations, and the amplitude of these torsional oscillations might be damped during extension (Urry, 1982). We devised experimental (physical and chemical) and computational tests of the structure with a focus on the freedom of motion of the suspended segments. The chemical tests (replacement of the Gly residues with D- and L-alanines) resulted in the predicted limited torsional oscillations and elasticity (Urry *et al.*, 1983a, b, 1984, 1991). The principal physical tests involved dielectric and NMR relaxation methodologies (Henze and Urry, 1985; Urry *et al.*, 1985a, b, c, 1986; Buchet *et al.*, 1988). Remarkably, in the dielectric relaxation characterization, where the only dipole moments were those of the backbone peptide moieties, intense localized relaxations developed as the model elastic protein underwent the inverse temperature (phase) transition. Furthermore, molecular mechanics and dynamics calculations demonstrated the decrease in amplitude of torsion-angle oscillations within the suspended segment during extension (Urry, 1982; Chang and Urry, 1989). Indeed, by treating the dynamics of the pentameric unit the oscillating dipole moment reproduced the dielectric relaxation results (Venkatachalam and Urry, 1986). Thus, the concept of the damping of internal chain dynamics by extension as a source of entropic elastic force became established (Chang and Urry, 1989). Key experimental and computational results are treated in more detail below.

Key experimental data relevant to the proposed mechanisms

Atomic force microscopy (AFM) single-chain force–extension results

The remarkable capacity to obtain single-chain force–extension curves provides new insights into the mechanism of entropic elasticity. Figure 5A shows a series of single-chain force–extension curves for a single molecule of Cys-$(GVGVP)_{n\times 251}$-Cys (Urry *et al.*, 2002a). Starting from the bottom, a series of extension–relaxation cycles are displayed. The second trace, during which the molecule was extended from an end-to-end distance of 200–700 nm and then returned to the starting length, demonstrates perfect reversibility at the rate of stretch and within the resolution of the technique. Since all of the energy of deformation was recovered during relaxation, we observe an ideal elastomer! For the third, fourth and sixth traces, a resting time of 30 s near 200 nm was allowed, and these traces demonstrated barely detectable imperfect overlap of extension and relaxation curves. The fifth trace without the 30 s resting time near 200 nm again exhibited perfect reversibility. Finally on the seventh extension of the chain the continuity from substrate to cantilever tip became severed and the force dropped to zero just above 700 nm.

The slight hysteresis of traces one, three, four and six, may be due to backfolding of the chain on itself, interaction of a second chain picked up at low extension or possibly even non-specific adsorption of the single chain to surface at low extension and at a position along the chain of within 100 nm of extended length from the attachment point. Regardless of the source of the barely detectable non-overlap of several of the traces, ideal

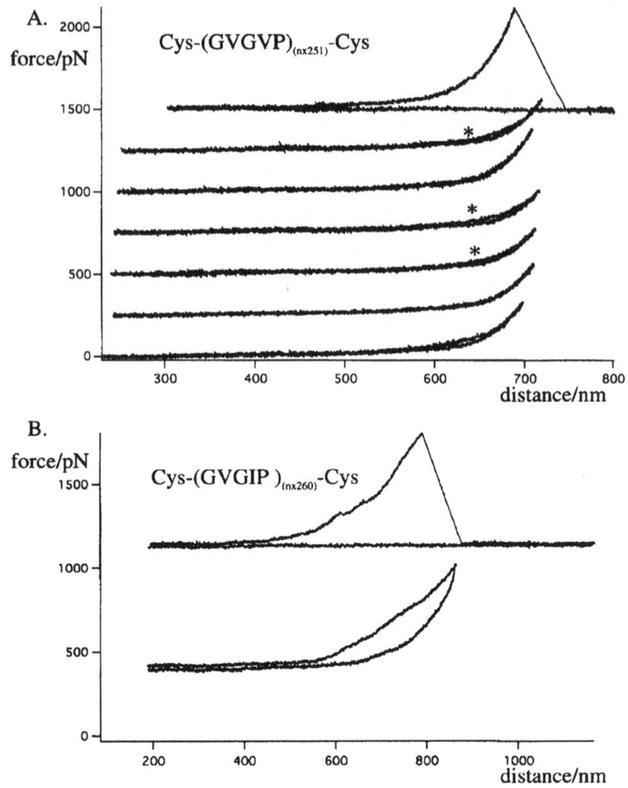

Fig. 5. AFM single-chain force–extension curves of the model elastic proteins based on elastin, Cys-$(GVGVP)_{n\times 251}$-Cys (A) and Cys-$(GVGIP)_{n\times 260}$-Cys (B), showing the ideal elasticity due to perfect reversibility of traces 2 and 5 from bottom of A and marked hysteresis in the lower trace of B. See text for discussion. Reproduced with permission from Urry *et al.* (2002a).

(entropic) elasticity has been observed in a single chain. No longer can the change from a Gaussian distribution of end-to-end chain lengths within a random chain network be insisted upon as the structural representation of entropic elasticity. The ideal single-chain force–extension curves require high dilution, and a higher dilution yet is required for more hydrophobic protein-based polymers where the tendency for aggregation is greater. This and additional data tell us that these perfectly reversible force–extension curves are due to single-chains. A single chain does not constitute a random chain network; and a single chain with ends fixed in space does not comprise a Gaussian distribution of end-to-end chain lengths. It seems that this eliminates the basic tenets of the random chain network theory of entropic elasticity as regards these elastic protein-based polymers. Furthermore, random chains do not exhibit mechanical resonances, e.g., intense localized Debye-type relaxations as observed in the dielectric relaxation spectra near 5 MHz and 3 kHz, and intense localized absorption maxima as observed in the acoustic absorption and loss permittivity spectra. These points constitute only the more apparent elements of the argument.

In addition, on the basis of the elementary statistical mechanical analysis given above, it would seem that the source of entropic elasticity must come from a decrease in internal chain dynamics upon extension.

Reduction of mean solvent-entropy change increases rather than decreases entropic elastic force
Formation of hydrophobic hydration is exothermic (Butler, 1937). Accordingly, when the temperature of the dissolved protein with its hydrophobic hydration is raised from below to above the inverse temperature transition, an endothermic transition due to the conversion of hydrophobic hydration to bulk water occurs. The transition from *hydrophobic hydration to bulk water* represents a positive change in entropy. By finding a suitable solvent that allows the transition but reduces the heat of the transition to near zero, it becomes possible to determine whether a decrease in elastic force occurs due to the loss of the contribution of the negative solvent-entropy change to the entropic elastic force. In other words, if a decrease in solvent entropy occurs as bulk water becomes ordered (in the form of hydrophobic hydration) by exposure of hydrophobic groups during extension, then the extent to which this contributes to entropic elastic force could be measured as a decrease in entropic elastic force.

In our experimental approach (based on Hoeve and Flory (1958)) ethylene glycol was added to suppress the solvent interaction with the protein. In particular, as shown in the differential calorimetry data of Figure 6A increasing the amount of ethylene glycol decreases the heat and lowers the temperature of the transition. By 30% ethylene glycol in water the heat of the transition is minimal. The entropy change attending the transition is calculated by dividing an increment of heat released over the temperature of the increment and summing over all increments of the transition. Accordingly, the decrease in heat of the transition seen in Figure 6A would be expected to result in a decrease in the entropic elastic force due to a decrease in the magnitude of the negative entropy of hydration on extension. In fact, as seen in Figure 6B, the entropic elastic force reached is greater in 30% ethylene–70% water.

The thermoelasticity studies on γ-irradiation-cross-linked poly(GVGVP) of Figure 6B are carried out by stretching the elastomer to a fixed extension of 50% at 37°C. Then held at that fixed length, the temperature is re-equilibrated below that of the onset for the inverse temperature transition, and the temperature is very slowly raised to 55°C. As regards the contribution of solvent entropy change to entropic elastic force, it is expected that the force reached would be less for the ethylene glycol containing solution by a fraction indicative of the contribution of solvent entropy change to the elastic force. The observation that the force actually increased, rather than decreased does not favor the solvent entropy mechanism.

As shown in Equation (4), the value of f_E/f is given by $-T(\partial \ln[f/T]/\partial T)_{V,L,n}$. Using the temperature range from 45 to 55°C, the f_E/f ratio is found to be no more than 0.1. Based on this analysis using Equation (4), the f_S/f value would be 0.9 or greater, and yet this entropic elastic force would appear not to come from a change in solvent entropy. Even more telling is the increase in force at fixed length, $(\partial f/\partial T)_L$, as the temperature is raised through the range of the inverse temperature transition. As will be discussed below, this thermally driven isometric contraction shows an increase in force under conditions of an increase rather than the required decrease in solvent entropy.

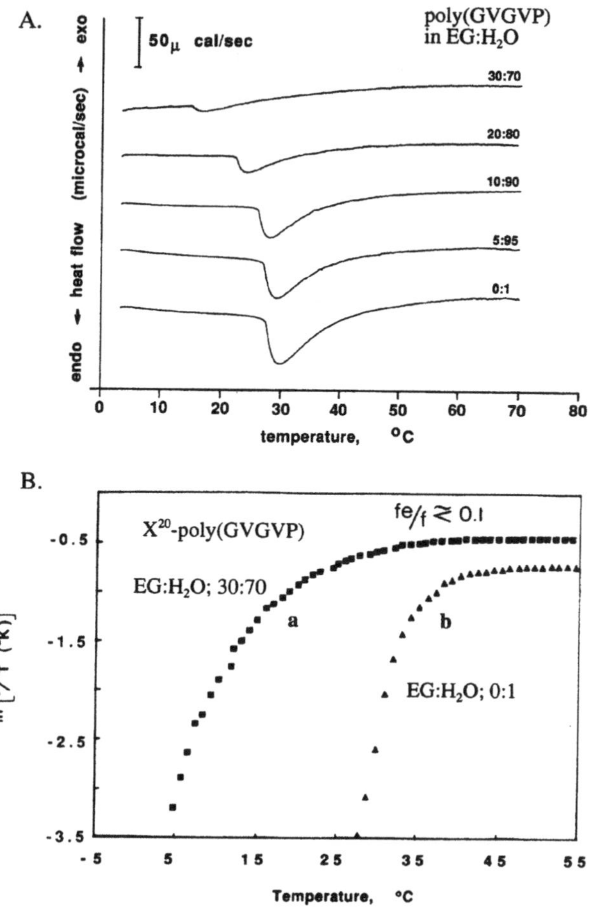

Fig. 6. (A) Differential calorimetry curves as a function of ethylene glycol (EG) in water. Note the decrease in the heat and lowering of the temperature range of the transition as the amount of ethylene glycol is raised to 30%. B. Thermoelasticity curves, plots of log(f/T) vs. T at fixed length, for 20 Mrad γ-irradiation cross-linked poly(GVGVP). The f_E/f ratio, determined from the slope above 45°C for both solvent conditions, is less than or equal to 0.1, that is the entropic component of elastic force is 90% or greater. Significantly, addition of ethylene glycol (EG) results in an increase in entropic elastic force suggesting that solvent entropy change is not a contribution to the entropic elastic force. Reproduced with permission from Luan *et al.* (1989).

Presence of mechanical resonances in elastin models and elastin itself
(i) *Mechanical resonances in the acoustic frequency range*: The elastic models of elastin based on the pentameric repeat (GVGVP) and analogues thereof, such as (GVGIP), exhibit mechanical resonances in the

acoustic frequency range of 100 Hz to 7 kHz. This is seen in Figure 7A for γ-irradiation-cross-linked $(GVGIP)_{260}$, which exhibits a mechanical resonance of increasing absorption intensity as the temperature is raised from below to above the range of the inverse temperature transition (Urry *et al.*, 2002a). By contrast, natural rubber and polyurethane (a particularly good elastomer for sound absorption), exhibit only broad low-intensity absorption as might be expected for a random chain network. For a random chain the frequency for rotation about each polymer backbone bond will be different, whereas for an elastomer with a regular repeating structure such as that in Figure 3 mechanical resonances such as those in Figure 7A can result. For our purposes the observation of a mechanical resonance is evidence for a non-random structure, and therefore challenges the classical theory of rubber elasticity.

Low-frequency dielectric relaxation studies also demonstrate mechanical resonances with absorption in the acoustic frequency range. As seen in Figure 7B, the imaginary part (ε″) of the dielectric permittivity of cross-linked $(GVGIP)_{320}$ exhibits a relaxation that also increases in intensity as the temperature is raised from

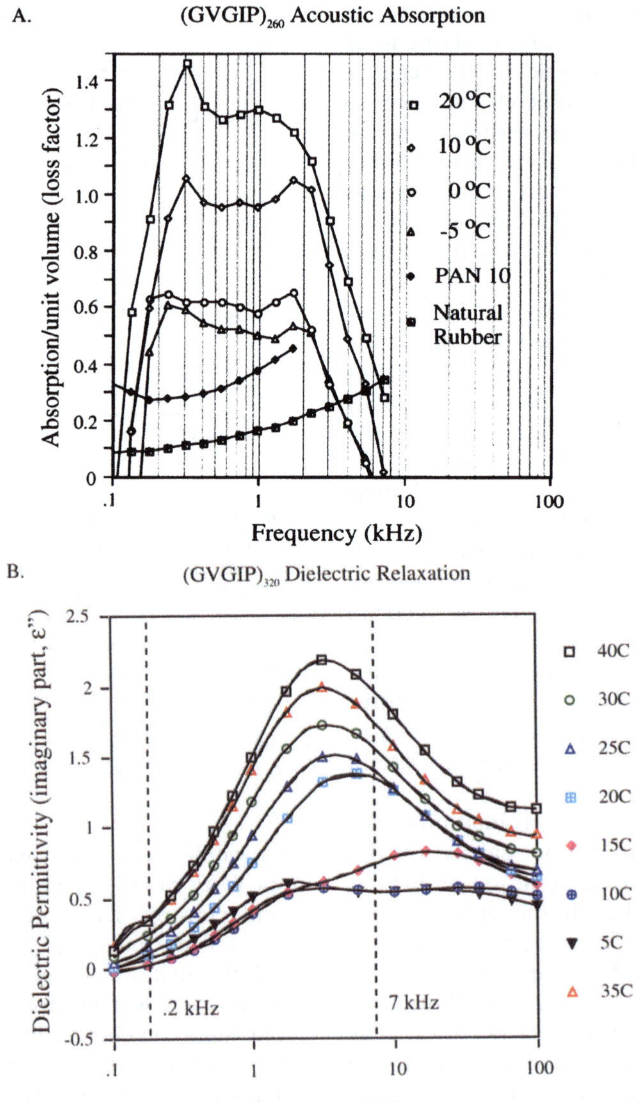

Fig. 7. Mechanical resonances seen in the acoustic absorption frequency range. The repeating pentamers, as they fold into the regularly repeating structure of Figure 3 on raising the temperature, develop a mechanical resonance wherein all pentamers absorb energy over the same frequency range. Although the physical means of exciting the mechanical resonance is different in A and B, an acoustic wave in A and an oscillating electric field in B, the maxima of these low frequency mechanical resonances are only shifted by a few kHz. A second higher frequency mechanical resonance is observed for this same elastic model protein in the dielectric relaxation near 5 MHz, 1000 kHz to higher frequency, as seen in Figure 8 for both the real and imaginary parts of the dielectric permittivity. Part A, acoustic absorption/unit volume (loss factor), reproduced with permission from Urry *et al.* (2002a) and part B, imaginary part of the low frequency dielectric relaxation spectra, reproduced with permission from Urry *et al.* (2002b).

below to above the temperature of the inverse temperature transition (Urry *et al.*, 2002b). As the excitation energy is not that of a compressional acoustic wave, but rather an oscillating electric field, it is interesting to see that the repeating unit of five residues acts as a single unit with an oscillating net dipole that displays mechanical resonance in a similar frequency range. It may be further noted in Figure 7B that the absorption intensity increases by more than 50% upon going from 20 to 40°C. Because of this, the acoustic absorption exhibited by $(GVGIP)_n$ at 20°C in Figure 7A would be expected to increase by an additional 50% upon increasing the temperature further, to 40°C. This would make the absorption per unit volume of $(GVGIP)_n$ much larger than that exhibited by natural rubber or urethane.

(ii) *Mechanical resonances in the MHz range*: Dielectric relaxation studies demonstrate mechanical resonances near 5 MHz for $(GVGVP)_n$, $(GVGIP)_n$, $(GVGLP)_n$ (Buchet *et al.*, 1988), α-elastin and fibrous elastin purified from bovine *ligamentum nuchae*. The data of Figure 8A give the real part of the dielectric permittivity (ε′) for α-elastin, a 70 kDa chemical fragmentation product of natural fibrous elastin (Urry *et al.*, 1988a) and for the elastin model poly(GVGVP) (inset) (Henze and Urry, 1985). ε″ for poly(GVGVP) and fibrous elastin is shown in Figure 8B (Luan *et al.*, 1988).

As seen in Figure 1, there is much more to elastin than poly(GVGVP), of the W4 sequence, although it is the most prominent repeating sequence in bovine elastin. The relaxations for 100% poly(GVGVP) are expected to

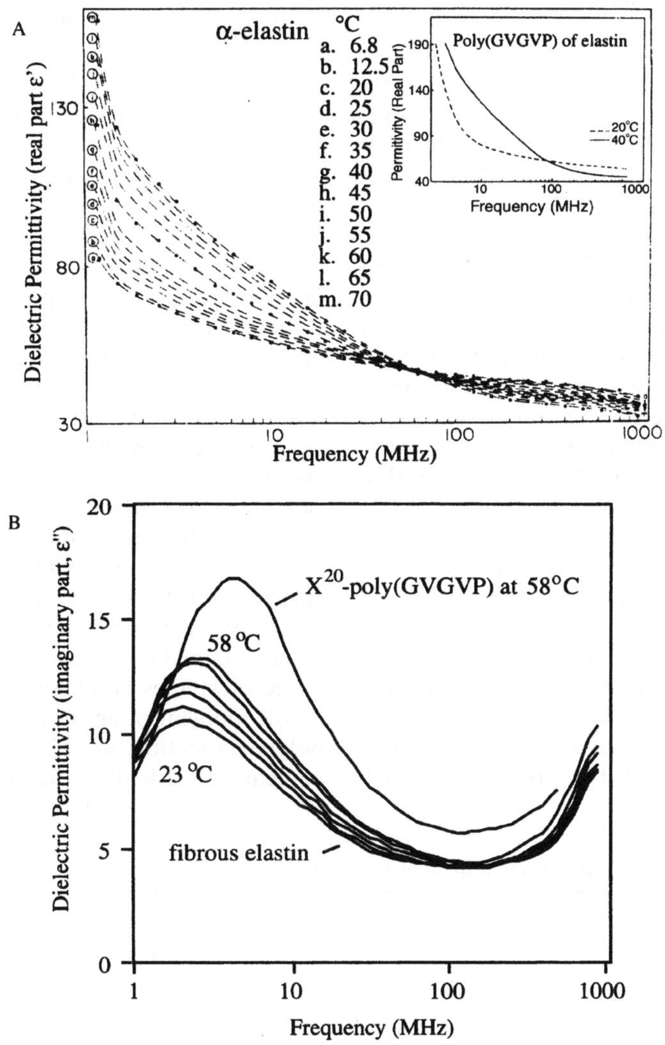

Fig. 8. Dielectric relaxation spectra in the 1–1000 MHz frequency range of α-elastin (part A), the poly(GVGVP) of elastin (inset of part A and part B), and fibrous elastin (part B). In all cases as the temperature is raised there develops an intense mechanical relaxation centered near 5 MHz as repeating elements of these protein systems hydrophobically fold into regular, albeit obviously dynamic, structures. In the case of poly(GVGVP) each pentamer folds into the same conformation, in which the peptide moieties (the only entities with dipole moments) undergo coordinated rocking motions with a resulting oscillating mean dipole moment. The rocking of the mean dipole moments of the pentamers resonate, in this case move, at the same frequency in response to an alternating electric field, that is, they exhibit a mechanical resonance centered near 5 MHz. A similar mechanical resonance is observed near 1 kHz in Figure 7. Part A reproduced with permission from Urry *et al.* (1988a), and part B reproduced with permission from Luan *et al.* (1988).

be more intense than that of α-elastin and fibrous elastin. It is indeed surprising that the frequency overlap and intensity is as close as found. It is apparent that other repeating sequences, or quasi repeats, exhibit dielectric relaxations at slightly lower frequencies.

Importantly, mechanical resonances near 5 MHz and 1 kHz indicate the presence of regularly repeating dynamic structures. Extension of such structures reasonably gives rise to an entropic elastic force upon extension by the damping of internal chain dynamics represented by mechanical resonance. Molecular mechanics and dynamics calculations based on the molecular structure shown in Figure 3 will demonstrate just how effectively the structure can explain the experimental elasticity and relaxation data.

Entropy calculations based on the elastin model, $(GVGVP)_n$

Calculations of structures and the entropies of the structures have used conventional molecular mechanics and dynamics programs such as CHARMM (Karplus and McCammon, 1983; Karplus and Kushick, 1981), AMBER (Weiner and Kollman, 1981), and Scheraga's ECEPP (Momany *et al.*, 1974, 1975). The satisfying feature of these quite distinct computational approaches is that they give essentially identical results whether carried out within our group by different researchers or by different groups. It is important, however, that the constraints and boundary conditions should properly reflect the conditions considered. Paramount among the constraints is adherence to the elasticity experiment in which the ends of the molecular structure are fixed at whatever selected degree of extension.

Molecular mechanics calculations using Scheraga's ECEPP approach for energy surfaces in φ–ψ configuration space

(i) *Calculation of entropy change during extension by an enumeration of states approach for relaxed and extended states*: In Scheraga's approach the internal energy of a chosen chain segment (in our case the pentamer permutation, $V_1P_2G_3V_4G_5$) is calculated as a function of a pair of adjacent torsion angles. Normally, the φ and ψ torsion angles are identical to those in the Ramachandran plot. In our case since the primary sites of motion are the peptide moieties of the suspended segment ($V_4G_5V_1$), such that the two φ–ψ plots, now called lambda plots, become ϕ_5–ψ_4 and ϕ_1–ψ_5.

In the enumeration of states approach, a 5° change in a single torsion angle is counted as a new state, and the number of states is counted for the chosen cut-off energy for both the relaxed and 130% extended structures (Urry *et al.*, 1982; Urry and Venkatachalam, 1983; Urry *et al.*, 1985c; Urry, 1991). A single pentamer is calculated within the relaxed and extended β-spiral structures, and 0.6, 1.0 and 2.0 kcal/mol-pentamer cut-off energies are used as well as an energy weighting using the Boltzmann summation over states. Table 1 gives the number of states where the entropy is simply calculated by Equation (9) using the example of 1 kcal/mol-pentamer,

$$\Delta S = (S^e - S^r) = R\ln(W^e/W^r) = R\ln(58/762)$$
$$= -5.1 \text{ cal/mol-pentamer.} \qquad (9)$$

This, of course, is −1.02 cal/mol-residue as given in Table 1.

(ii) *Calculation of dielectric relaxation data from pentamer molecular mechanics*: With the aid of the Onsager equation for polar liquids (Onsager, 1931; Bottcher, 1973) it becomes possible to use the structures obtained above and to sum the dipole moments of the states of Table 1. In addition, the same can be done for $V_1P_2G_3V_4A'_5$) where the A' stands for the D-Ala amino acid residue (Venkatachalam and Urry 1986 and printers erratum as corrected below; Urry, 1991). Comparison of experimental and calculated values is particularly satisfying. 'In view of this result, it seems reasonable to consider a cutoff energy of 1.5 kcal mol^{-1}. This value of energy cutoff leads to a mean dipole moment change of 3.8 Debye per pentamer in good agreement with the value obtained from dielectric relaxation studies. Similar dielectric studies on $D.Ala^5$-polypentapeptide indicate a dipole moment of about one-third of that found for the polypentapeptide at 37°C... This is in very good agreement with the dipole changes obtained from

Table 1. Perspective of entropy of the poly(GVGVP) β-spiral by enuneration of states

	Number of states		
Cutoff energy (kcal/mol)	Relaxed (r)	Extend (e)	Entropy change per residue
2	1853	162	−0.97
1	762	58	−1.02
0.6	342	24	−1.06
	Using Boltzmann sum over states $f = \sum_i e^{-\varepsilon_i/kT}$		
2	$\Delta S = R\ln(f^e/f^r) + \frac{E^e - E^r}{T}$		−1.01

molecular mechanics calculations presented here' (Venkatachalam and Urry 1986).

The above calculated results utilize the 5 MHz mechanical resonance. They further demonstrate how readily calculations based on the dynamic structure given in Figure 3 calculate experimental dielectric relaxation data with inherent relevance to entropic elastic force. The values for entropy change calculated using the molecular mechanics approach, when used in the second term of Equation (1), provide excess entropic elastic force as expected (see Urry *et al.*, 2002a, b).

Molecular dynamics calculations using Karplus' CHARMm and Kollman's AMBER programs

As the motions of the torsional oscillations are dynamic, it is appropriate to consider molecular dynamics calculations using the Karplus software program adapted by Polygen, Inc. (now Molecular Simulations), called CHARMm, version 20.3. For these Newtonian classical mechanics simulations the potential energy functions and parameters suggested by the Karplus group (Brooks *et al.*, 1983) were used in a software program. The root-mean-square (RMS) fluctuations of the torsion angles, $\Delta\phi_i$ and $\Delta\psi_i$ for the relaxed and extended (130%) of $(GVGVP)_{11}$ as given in Table 2 are used in Equation (10).

$$\Delta S = R \ln[\Pi_i \Delta\phi_i^e \Delta\psi_i^e / \Pi_i \Delta\phi_i^r \Delta\psi_i^r] \tag{10}$$

The calculated change in entropy upon extension of −1.1 cal/mol deg-residue is in remarkable agreement with the previously described molecular-mechanics calculations.

Furthermore, Wasserman and Salemme (1990) using the AMBER software program of the Kollman group (Weiner and Kollman, 1981), but also including a medium of water molecules, obtained essentially identical results. It must be concluded, damping of internal chain dynamics is an abundant source of decrease in chain entropy that is sufficient to account for the entropic elastic force.

As Wasserman and Salemme (1990) included water molecules in their calculations, and an ordering of water molecules was found as the hydrophobic side chains became exposed on chain extension, they considered this decrease in solvent entropy as a possible source of the entropic elastic force. As will be shown below, solvent entropy change does not make a significant contribution to the entropic elastic force development during isometric contractions.

Relationship between hydrophobic association–dissociation, elastic force and energy conversion in elastin mechanics

By introducing glutamate (Glu, E) the ideal elastomer in Figure 5A becomes a chemo-mechanical transducer capable of converting the chemical energy of proton concentration changes into the mechanical energy of isotonic and isometric contractions (See Figure 9A and B). An analysis of this data in combination with that of Figure 6B resolves the issue of the role of changes in solvent entropy in elastic force development and energy conversion.

Table 2. RMS fluctuations of torsion angles (ϕ and ψ) of $(VPGVG)_{11}$ (45 ps of equilibration time and 80 ps of molecular dynamics simulation). Reproduced with permission from Chang and Urry (1989)

	Angle	Relaxed	Extended	Angle	Relaxed	Extended	Angle	Relaxed	Extended
β-turns	ψ_{16}	10.87	14.17	ψ_{26}	27.33	07.64	ψ_{36}	20.19	14.84
	ϕ_{17}	09.86	15.18	ϕ_{27}	11.71	08.33	ϕ_{37}	08.99	10.47
	ψ_{17}	47.59	46.68	ψ_{27}	11.70	13.51	ψ_{37}	21.53	32.96
	ϕ_{18}	61.70	47.41	ϕ_{28}	08.61	10.36	ϕ_{38}	11.15	10.66
	ψ_{18}	09.37	16.05	ψ_{28}	09.33	08.16	ψ_{38}	11.09	27.29
	ϕ_{19}	14.25	08.67	ϕ_{29}	09.70	07.31	ϕ_{39}	12.70	10.24
Suspended	ψ_{19}	44.09	10.99	ψ_{29}	47.32	10.48	ψ_{39}	52.00	12.50
	ϕ_{20}	41.94	09.29	ϕ_{30}	48.57	11.39	ϕ_{40}	55.88	08.37
Segments	ψ_{20}	14.50	11.15	ψ_{30}	42.56	10.62	ψ_{40}	40.67	11.08
	ϕ_{21}	27.13	24.17	ϕ_{31}	11.43	11.38	ϕ_{41}	36.44	19.06
β-turns	ψ_{21}	09.39	22.73	ψ_{31}	12.17	09.21	ψ_{41}	12.97	14.80
	ϕ_{22}	09.94	08.00	ϕ_{32}	09.90	08.93	ϕ_{42}	11.59	07.33
	ψ_{22}	11.58	16.13	ψ_{32}	15.30	10.80	ψ_{42}	11.34	13.17
	ϕ_{23}	16.37	09.33	ϕ_{33}	09.60	07.62	ϕ_{43}	09.23	09.76
	ψ_{23}	14.33	14.25	ψ_{33}	09.88	09.43	ψ_{43}	10.60	12.53
	ϕ_{24}	11.39	29.20	ϕ_{34}	11.86	09.71	ϕ_{44}	11.06	12.82
Suspended	ψ_{24}	19.53	37.87	ψ_{34}	63.80	08.36	ψ_{44}	41.89	35.22
	ϕ_{25}	25.02	23.06	ϕ_{35}	91.70	10.20	ϕ_{45}	48.98	31.31
Segments	ψ_{25}	49.32	32.10	ψ_{35}	15.03	11.51	ψ_{45}	42.05	56.89
	ϕ_{26}	31.43	27.24	ϕ_{36}	21.49	18.66	ϕ_{46}	21.55	30.33

$\Delta S = R \ln \frac{\Pi_i \Delta\phi_i^e - \Delta\psi_i^e}{\Pi_i \Delta\phi_i^r - \Delta\psi_i^r}$.

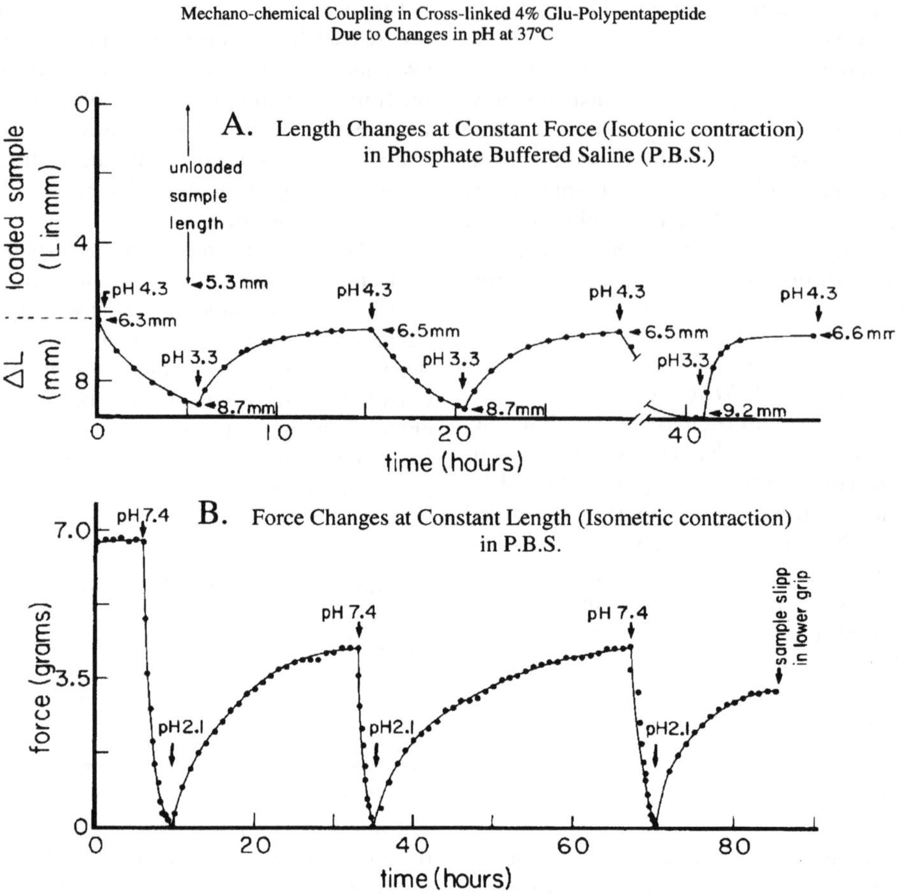

Fig. 9. Mechano-chemical coupling by 20 Mrad γ-irradiation cross-linked of poly[0.8(GVGVP),00.2(GEGVP)] under isotonic (A) and isometric (B) conditions. In part B under isometric contraction conditions there occurs an increase in entropic elastic force resulting from a decrease in entropy of the elastomer while there occurs an increase in the entropy of the solvent as hydrophobic hydration becomes bulk water. Therefore the increase in entropic elastic force cannot be the result of the solvent entropy change. Reproduced with permission from Urry *et al.* (1988b).

Chemically driven isotonic $(\partial L/\partial \mu_H)_f$ and isometric $(\partial f/\partial \mu_H)_L$ contractions

The elastin model protein of interest is poly [x_V-(GVGVP),x_E(GEGVP)], where x_V and x_E are the mole fractions with $x_V + x_E=1$ and for the case where $x_E=$ 0.2. Upon γ-irradiation cross-linking an elastic matrix is formed which can be studied under isotonic and isometric conditions. Results are shown in Figure 9A and B (Urry *et al.*, 1988b). The isotonic contraction $(\partial L/\partial \mu_H)_f$ in Figure 9A is the change in length at constant load (force) resulting from an increase in the concentration of proton, that is, the increase in chemical potential of proton $(\partial \mu_H)$. The isometric contraction $(\partial f/\partial \mu_H)_L$ in Figure 9B is the change in force at constant length due to an increase in proton.

Both of these contractions result from hydrophobic association due to protonation of the carboxylate group. The underlying physical process is the competition for hydration between the charged carboxylate and the hydrophobic side chains of the valine (Val, V) residues (Urry, 1997). This competition, described as an apolar–polar repulsive free energy of hydration, results in increasing positive cooperativity of the acid–base titration curves that correlate with increasingly larger pK_a shifts. As Val residues are replaced by more hydrophobic phenylalanine (Phe, F) residues, the pK_a shifts increase (up to 6 pH units or more) and positive cooperativity increases with Hill coefficients of up to 8 or more (Urry, 1997). The apolar–polar repulsive free energy of hydration is related to the change in Gibbs free energy for hydrophobic hydration, $\Delta G_{HA}(\chi)$, of a phosphate as discussed in relation to Equations (17) and (19) below, but first the source of entropic elastic force is addressed.

Isometric contractions, $(\partial f/\partial T)_L$ and $(\partial f/\partial \mu_H)_L$, confirm basis for entropic elastic force

From Equation (1) the entropic component of the elastic force, f_S, is proportional to $-\Delta S$, that is, an increase in entropic elastic force results from a decrease in entropy. Also the formation of hydrophobic hydration from bulk water constitutes an inherently negative ΔS, and, of course, the loss of hydrophobic hydration constitutes a positive change in entropy. This change in solvent entropy due to formation of hydrophobic hydration has been credited as an important contribution to entropic elastic force (Weis-Fogh and Anderson, 1970; Wasserman and Salemme, 1990, Alonso *et al.*, 2001). Therefore, the immediate question becomes whether experimental results support or eliminate a decrease in solvent entropy as a source of entropic elastic force.

Thermally driven isometric contraction (Figure 6B)

Figure 6B shows the classic thermoelasticity study for determining the f_E/f ratio for γ-irradiation-cross-linked poly(GVGVP). Because extension is fixed, a thermally driven isometric contraction also occurs over the temperature range of the inverse temperature transition of hydrophobic association. During hydrophobic association, hydrophobic hydration becomes bulk water. Therefore, the solvent entropy change is positive. Any contribution to the entropic elastic force due to changes in hydrophobic hydration during this thermally driven isometric contraction would necessarily result in a decrease in force. Yet the experimental result of Figure 6B shows the development of an elastic force that is 90% entropic, while the change in solvent entropy is of the opposite sign. Clearly, the change in solvent entropy is not contributing to the estimated 90% entropic elastic force. Having eliminated solvent entropy change as the source of entropic elastic force during a thermally driven isometric contraction, we now analyze the solvent entropy change attending a chemically driven isometric contraction.

Chemically driven isometric contraction (Figure 9B)

As seen in Figure 9B, at fixed extension of the γ-irradiation-cross-linked elastic matrix comprised of poly[0.8(GVGVP),0.2(GEGVP)], protonation of four carboxylates per 100 residues results in development of elastic force. A thermoelasticity characterization of this matrix at low pH gives the same result of dominantly entropic elasticity as found in curve b of Figure 6B for poly(GVGVP) in the absence of carboxyl moieties.

The process of protonation allows reconstitution of hydrophobic hydration to such an extent that the temperature range for hydrophobic association drops below that of the operating temperature (Urry, 1993, 1997). The result is a contraction due to hydrophobic association. Again, during an isometric contraction (this time chemically driven), hydrophobic hydration becomes less ordered bulk water. The solvent entropy increases during the development of entropic elastic force due to a decrease in entropy.

In addition to eliminating solvent entropy change as a source of entropic elastic force, the isometric contraction results provide the conceptual bridge between the fundamental source of entropic elastic force and contraction by hydrophobic association.

Hydrophobic association effects elastic force development by extending (and thereby damping internal dynamics of) interconnecting chain segments

Clam-shaped globular proteins that open and close due to hydrophobic association

The top part of Figure 10 shows a series of clam-shaped globular proteins strung together near the open end by elastic bands and maintained at fixed extension. In this isometric state, there is depicted an equilibrium between open and closed states. Obviously, shifting the equilibrium toward the closed state would increase the force measured at the force transducer, whereas shifting toward the open state would lower the force on the interconnecting elastic bands. Thus, any process that increased hydrophobic association would increase the elastic force. Now in the lower part of Figure 10, we relate this perspective to the β-spiral representation of Figure 3C.

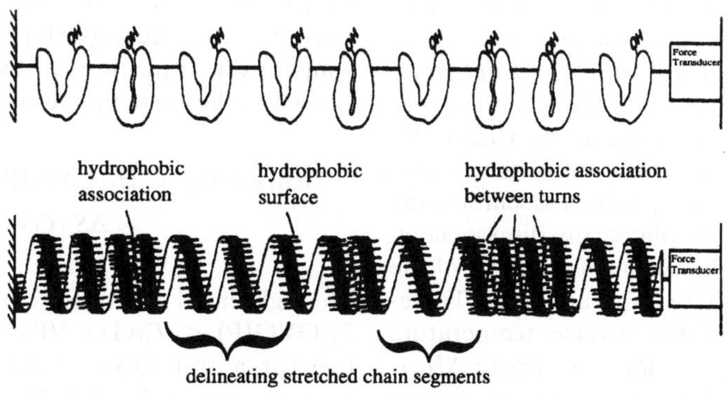

Fig. 10. Cartoon of the relationship between hydrophobic association and entropic elastic force development. Above: A series of clam-shaped globular protein strung together by elastic bands with an equilibrium between open and hydrophobically associated closed states. Clearly, as the equilibrium shifts toward more closed states, the force, sustained by the interconnecting elastic segments, increases. Below: Representation of the β-spiral structure of poly(GVGVP) at an intermediate state of hydrophobic association between turns of the β-spiral. Clearly, as more hydrophobic association occurs between turns, the interconnecting elastic chain segments become further extended with an increase in damping of internal chain dynamics giving rise to greater entropic elastic force. Reproduced with permission from Urry (1990).

Schematic representation of an equilibrium between hydrophobically associated and dissociated turns of a β-spiral held at fixed extension

The bottom half of Figure 10 uses the representation of the β-spiral given in Figure 3C to depict an equilibrium between associated and dissociated turns of the β-spiral. As more of the turns are involved in hydrophobic association, those not involved become extended to a greater extent. Thus, as more hydrophobic association occurs, the interconnecting segments become increasingly extended. By the entropic mechanism of the damping of internal chain dynamics on extension, the entropic elastic force increases.

Opening and closing of the clam-shaped globular proteins of the upper part of Figure 10 (Urry, 1990) would be detectable under conditions of force-clamp atomic force spectroscopy. Such traces have been observed using titin-based elastic constructs comprised of a series of hydrophobically folded β-barrels (Oberhauser *et al.*, 2001). Because of the continuous nature of increases in hydrophobic association between turns of the β-spiral represented in the lower part of Figure 10, such force-clamp traces for poly(GVGVP) would not have such steps, unless sensitivity were sufficient to detect the addition of a repeating unit to hydrophobic association.

Thus, the changes in entropic elastic force and in hydrophobic association are two distinct physical processes that become interlinked when occurring within the same chain system. Hydrophobic association causes the extension of interconnecting dynamic chain segments with the result of the damping of internal chain dynamics within those non-hydrophobically associated interconnecting chain segments.

Derivation of Gibbs free energy for hydrophobic association ΔG_{HA}

The role of hydrophobic association in the development of entropic elastic force has been delineated above. The next step is to derive an expression for the Gibbs free energy of hydrophobic association and to relate it to processes that direct elastic–contractile processes.

(i) *At the temperature of the inverse temperature transition*, $\mu_{HA} = \mu_{HD}$ and $\Delta G = 0$: The chemical potential (μ) is the Gibbs free energy per mole ($\Delta G/\Delta n$). By the Gibbs phase rule, at the temperature of the inverse temperature transition for hydrophobic association the chemical potential of the hydrophobically disassociated (dissolved) state (μ_{HD}) and the chemical potential of the hydrophobically associated state (μ_{HA}) are equal. Therefore, at the temperature of the inverse temperature transition (T_t), $\Delta G_t = \Delta H_t - T_t\Delta S_t = 0$, where ΔH_t is the heat of the transition and ΔS_t is the entropy change for the transition. Accordingly at the value of T_t for a given elastic model protein composition, $\Delta H_t = T_t\Delta S_t$.

By way of example, consider the two different compositions of model elastic proteins, $(GVGVP)_n$ and $(GVGIP)_n$. These two polypentapeptides differ only by a single CH_2 moiety per pentamer. In particular, the R-group for the V residue, $-CH(CH_3)_2$, differs from the R-group of the I residue, $-CH(CH_3)-CH_2-CH_3$, by the addition of a single CH_2 moiety. At the temperature for each of their respective phase transitions where $\mu_{HA}=\mu_{HD}$, we can write,

$$\Delta H_t(GVGVP) = T_t(GVGVP)\Delta S_t(GVGVP) \qquad (11)$$

and

$$\Delta H_t(GVGIP) = T_t(GVGIP)\Delta S_t(GVGIP). \qquad (12)$$

Butler (1937) showed that each addition of a CH_2 moiety in the series of normal alcohols from methanol to *n*-pentanol on dissolution in water resulted in an average exothermic heat $(\Delta H)/CH_2$ of -1.4 kcal/mol-CH_2 and a change of the term, $(-T\Delta S)/CH_2$, of a $+1.7$ kcal/mol-CH_2. While there are many important points to make about this fundamental finding, we use it here simply as a justification to rewrite Equation (12) as

$$\Delta H_t(GVGIP) = \Delta H_t(GVGVP) + \Delta H_t(CH_2), \qquad (13)$$

and

$$T_t(GVGIP)\Delta S_t(GVGIP) = T_t(GVGIP)[\Delta S_t(GVGVP) + \Delta S_t(CH_2)]. \qquad (14)$$

Substituting Equations (13) and (14) into Equation (12) and subtracting Equation (11) gives

$$[\Delta H_t(CH_2) - T_t(GVGIP)\Delta S_t(CH_2)] = [T_t(GVGIP) - T_t(GVGVP)]\Delta S_t(GVGVP). \qquad (15)$$

The left hand side of Equation (15) is recognized as an expression for the change in Gibbs free energy due to addition of the CH_2 moiety. On consideration of the phase transition being analyzed, the left hand side of Equation (5) becomes the change in Gibbs free energy of hydrophobic association due to the addition of a single CH_2 moiety, $\Delta G_{HA}(CH_2)$, that causes the inverse temperature transition of hydrophobic association to occur at a lower temperature for $(GVGIP)_n$.

$$\Delta G_{HA}(CH_2) = [T_t(GVGIP) - T_t(GVGVP)] \times \Delta S_t(GVGVP). \qquad (16)$$

The right hand side of Equation (16) is negative because $T_t(GVGIP) < T_t(GVGVP)$ and $\Delta S_t(GVGVP)$ is positive for a transition whereby hydrophobic hydration becomes less ordered bulk water during hydrophobic association, that is, as insolubility ensues.

The addition of a CH_2 moiety per pentamer is one of very many ways whereby the value of T_t can be changed. A T_t-based hydrophobicity scale has been developed for substitution of each of the naturally occurring amino acid residues and chemical modifications thereof by systematically increasing the mole fraction of pentamers containing the change X and extrapolating to the value of T_t that would occur for poly(GXGVP) (Urry, 1997). Therefore, the value of T_t may be greater than 100°C as for poly(GEGVP) when the glutamic acid residue, E, is ionized as the carboxylate, where T_t=250°C. On the other hand, the value of T_t may be less than 0°C as for poly(GWGVP) when the tryptophan is the residue, where T_t=−90°C. For the amino acids with functional side chains the T_t-value changes upon changing the state of the side chain from carboxyl to carboxylate, or the redox state of an attached redox couple. The most dramatic change in the T_t-value occurs upon phosphorylation of a Ser, Thr, or Tyr residue (Pattanaik *et al.*, 1991). Furthermore, adding salt to the solution of a neutral polymer such as poly(GVGVP) itself lowers the value of T_t (Urry, 1993), but the ΔT_t is 10 times greater when the salt ion-pairs with a charged side chain (Urry and Luan, 1995a; Urry *et al.*, 1997a).

So we generalize to any experimental variable, χ, that alters the value of T_t of a reference model elastic protein and write,

$$\Delta G_{HA}(\chi) = [T_t(\chi) - T_t(\text{reference})]\Delta S_t(\text{reference}), \tag{17}$$

or

$$\Delta G_{HA}(\chi) = \Delta T_t(\chi)\Delta S_t(\text{reference}), \tag{18}$$

where T_t is the value of a reference polymer and state has been changed to $T_t(\chi)$ by the experimental variable χ. All of these energy inputs can now be described in terms of their effect on the free energy of hydrophobic association, $\Delta G_{HA}(\chi)$.

Relevance of Gibbs free energy for hydrophobic association to energizing by phosphorylation

Phosphorylation of poly[30(GVGIP),(RGYSLG)] by the cardiac cyclic AMP-dependent protein kinase at the serine (Ser, S) residue of the polymer goes to an average of 47% completion by the following reaction (Pattanaik *et al.*, 1991; A. Pattanaik, personal communication).

$$\text{ATP} + \text{poly}[30(\text{GVGIP}), (\text{RGYSLG})] = \text{ADP} + \text{poly}[30(\text{GVGIP}), (\text{RGYS}(\text{—OPO}_3^{2-})\text{LG})]. \tag{19}$$

The value of T_t for this polymer before phosphorylation was approximately 20°C and the change in the value of T_t extrapolated as defined to a reference state of one phosphate per GVGIP becomes 860°C, which is represented as $T_t(\text{—OPO}_3^{2-})$. If we now use the experimentally derived value of ΔS_t(GVGIP) = 9.32 cal/mol (pentamer) deg as the ΔS_t(reference) (Luan and Urry, 1999) of Equation (17), the change in Gibbs free energy for hydrophobic association due to phosphorylation, $\Delta G_{HA}(\text{—OPO}_3^{2-})$, is +7.8 kcal/mol. For the reaction of Equation (19) with an equilibrium constant of $K = 47/53 = 0.9 = \exp(-\Delta G/RT)$, the change in Gibbs free energy for the reaction is 0.06 kcal/mol. Thus, the calculated value for the free energy contributed by the terminal phosphate of ATP would be 0.06–7.8 = −7.7 kcal/mol, which is remarkably close to the free energy of hydrolysis of the terminal phosphate of ATP of 7.3 kcal/mol (Voet and Voet, 1995).

Accordingly, phosphorylation raises the free energy of the hydrophobically associated state, that is, it results in hydrophobic disassociation, and phosphate removal drives hydrophobic association, e.g., contraction, more effectively than any other chemical change measured so far. Expectations are that the binding of ATP, and most effectively ADP plus phosphate, would dramatically raise the free energy of the hydrophobically associated state as evidenced by an increase the value of T_t. This $\Delta T_t(\text{—OPO}_3^{2-})$ and $\Delta G_{HA}(\text{—OPO}_3^{2-})$ can be only partially modulated by ion-pairing, for example, with magnesium, of the phosphates at the binding site.

Relationship between entropic elasticity and efficiency of energy conversion

The process of stretching an elastomer constitutes an energy input with the expended mechanical energy $f\Delta L$, where f is force and ΔL is the change in length. Accordingly, the mechanical energy is obtained from the area under the force vs. extension curve. For an ideal elastomer, the plot of force vs. extension is perfectly reversible, as shown by the second and fifth traces of Figure 5A for the single-chain force–extension curve of the entropic elastomer, $(\text{GVGVP})_{n\times 251}$. Thus, the input energy is completely recovered on relaxation for an ideal (entropic) elastomer. Since the energy recovered on relaxation can be viewed in the sense of an energy output, an ideal elastomer could be considered a perfect machine for storing energy of deformation.

Often the force vs. length curve obtained on relaxation falls below the force vs. length curve obtained on extension (as seen in the lower curve of Figure 5B); then the material is said to exhibit a hysteresis. In this case the energy recovered on relaxation is less than that expended on extension. The input deformation energy has been dissipated in some way and to an extent indicated by the difference in the areas below the extension and relaxation curves.

When energy conversion occurs by means of an ideal elastic material, it is possible for the energy conversion to occur at high efficiency, whereas when energy conversion occurs using an elastic material that exhibits hysteresis, the efficiency of energy conversion becomes limited at least to an extent determined by the magnitude of the hysteresis. The elasticity of polymeric materials becomes inextricably intertwined with the efficiency of energy conversion. Thus, the increase in elastic force during the isometric contraction and relaxation of Figure 9B would be exactly reversible for an ideal elastomer as would the isotonic contraction of Figure 9A. This would not be the case for a molecular machine comprised of the elastic polymer functioning in Figure 5B.

An important element, therefore, of an ideal elastomer is to have the energy uptake into the elastomer during extension reside entirely in the backbone modes where it can be recovered on relaxation. Should energy of deformation find its way into side-chain motional modes and into chains not bearing the deformation and irreversibly into solvent, this deformation energy becomes dissipated and unavailable during relaxation, resulting in hysteresis. Comparison of the single-chain force–extension curves of Figure 5 for poly(GVGVP) with poly(GVGIP) provides an example with proposed loss of energy into the side chain motions and interactions of the bulkier isoleucine (I) residue with its added CH_2 moiety into adjacent non-load-bearing chains.

Acknowledgements

The authors wish to acknowledge the support of the Office of Naval Research under contracts, N00014-00-C-0404 and N00014-00-C-0178, and to thank A. Pattanaik for updated details on the phosphorylation study and L. Hayes for assistance in obtaining the hydrophobicity plots and references.

References

Alonso LB, Bennion BJ and Daggett V (2001) Hydrophobic hydration is an important source of elasticity in elastin-based polymers. *J Am Chem Soc* **123:** 11,991–11,998.

Bottcher CJF (1973) *Theory of Electric Polarization.* (vol. 1, p. 178) Elsevier, Amsterdam.

Brooks BR, Bruccoleri RE, Olafson BO, States DT, Swaminathan S and Karplus M (1983) CHARMM: a program for macromolecular energy, minimization, and dynamics calculations. *J Comput Chem* **4:** 187.

Buchet R, Luan C-H, Prasad KU, Harris RD and Urry DW (1988) Dielectric relaxation studies on analogs of the polypentapeptide of elastin. *J Phys Chem* **92:** 511–517.

Butler JAV (1937) The energy and entropy of hydration of organic compounds. *Trans Faraday Society* **33:** 229–238.

Chang DK and Urry DW (1989) Polypentapeptide of elastin: damping of internal chain dynamics on extension. *J Comput Chem* **10:** 850–855.

Chang DK, Venkatachalam CM, Prasad KU and Urry DW (1989) Nuclear overhauser effect and computational characterization of the β-spiral of the polypentapeptide of Elastin. *J Biomol Struct Dynam* **6:** 851–858.

Cook WJ, Einspahr HM, Trapane TL, Urry DW and Bugg CE (1980) Crystal structure and conformation of the cyclic trimer of a repeat pentapeptide of elastin, cyclo-(L-Valyl-L-prolylglycyl-L-valylglycyl)$_3$. *J Am Chem Soc* **102:** 5502–5505.

Cox BA, Starcher BC and Urry DW (1973) Coacervation of α-elastin results in fiber formation. *Biochim Biophys Acta* **317:** 209–213.

Cox BA, Starcher BC and Urry DW (1974) Coacervation of tropoelastin results in fiber formation. *J Biol Chem* **249:** 997–998.

Dauber P, Goodman M, Hagler AT, Osguthorpe D, Sharon R and Stern P (1981) In: Lykos P and Shavitt I (eds) *ACS Symposium Series No. 173. Suypercomputers in Chemistry.* (pp. 161–191) American Chemical Society, Washington, DC.

Eyring H, Henderson D, Stover BJ and Eyring EM (1964) *Statistical Mechanics and Dynamics.* (p. 92) John Wiley & Sons Inc, New York.

Gaub HE and Fernandez JM (1998) The molecular elasticity of individual proteins studied by AFM-related techniques. *AvH-Magazin* **71:** 11–18.

Gotte L, Giro G, Volpin D and Horne RW (1974) The ultrastructural organization of elastin. *J Ultrastruct Res* **46:** 23–33.

Henze R and Urry DW (1985) Dielectric relaxation studies demonstrate a peptide librational mode in the polypentapeptide of elastin. *J Am Chem Soc* **107:** 2991–2993.

Hoeve CAJ and Flory PJ (1958) The elastic properties of elastin. *J Am Chem Soc* **80:** 6523–6526.

Hoeve CAJ and Flory PJ (1974) Elastic properties of elastin. *Biopolymers* **13:** 677–686.

Karplus M and Kushick JN (1981) Method for estimating the configurational entropy of macromolecules. *Macromolecules* **14:** 325–332.

Karplus M and McCammon JA (1983) Dynamics of proteins: elements of function. *Ann Rev Biochem* **53:** 263–300.

Kornblihtt AR, Umezawa K, Vibe-pederson K and Baralle FE (1985) Primary structure of human fibronectin: differential splicing may generate at least 10 polypeptides from a single gene. *EMBO J* **4:** 1755–1759.

Luan C-H, Harris RD and Urry DW (1988) Dielectric relaxation studies on bovine ligamentum nuchae. *Biopolymers* **27:** 1787–1793.

Luan C-H, Jaggard J, Harris RD and Urry DW (1989) On the Source of Entropic Elastomeric Force in Polypeptides and Proteins: Backbone Configurational vs. Side Chain Solvational Entropy. *Int. J. of Quant. Chem.: Quant. Biol. Symp.* **16:** 235–244.

Luan C-H and Urry DW (1999) Elastic, plastic, and hydrogel protein-based polymers. In: Mark JE (ed.) *Polymer Data Handbook*, (pp. 78–89, Table 3A) Oxford University Press, New York, Oxford.

Momany FA, Carruthers LM, McGuire RF and Scheraga HA (1974) Intermolecular potentials from crystal data. III. Determination of empirical potentials and application to the packing configurations and lattice energies in crystals of hydrocarbons, carboxylic acids, amines, and amides. *J Phys Chem* **78:** 1595.

Momany FA, McGuire FF, Burgess AW and Scheraga HA (1975) Energy parameters in polypeptides. VII. Geometric parameters, partial charges, nonbonded interactions, hydrogen bond interactions, and intrinsic torsional potentials for the naturally occurring amino acids. *J Phys Chem* **79:** 2361.

Oberhauser AF, Hansma PK, Carrion-Vasquez M and Fernandez JM (2001) Stepwise unfolding of titin under force-clamp atomic force microscopy. *Proc Natl Acad Sci USA* **98:** 468–472.

Onsager L (1936) Electric moments of molecules in liquids. *J Am Chem Soc* **58:** 1486–1493.

Pattanaik A, Gowda DC and Urry DW (1991) Phosphorylation and dephosphorylation modulation of an inverse temperature transition. *Biochem Biophys Res Comm* **178:** 539–545.

Rief M, Gautel M, Oesterhelt F, Fernandez JM and Gaub HE (1997) Reversible unfolding of individual titin immunoglobulin domains by AFM. *Science* **276:** 1109–1112.

Sandberg LB, Leslie JG, Leach CT, Alvarez VL, Torres AR and Smith DW (1985) Elastin covalent structure as determined by solid phase amino acid sequencing. *Pathol Biol (Paris)* **33:** 266–274.

Sandberg LB, Soskel NT and Leslie JG (1981) Elastin structure, biosynthesis, and relation to disease states. *N Engl J Med* **304:** 566–579.

Sciortino F, Prasad KU, Urry DW and Palma MU (1993) Self-assembly of bioelastomeric structures from solutions: mean field critical behavior and Flory–Huggins free-energy of interaction. *Biopolymers* **33:** 743–752.

Sciortino F, Urry DW, Palma MU and Prasad KU (1990) Self-assembly of a bioelastomeric structure: solution dynamics and the spinodal and coacervation lines. *Biopolymers* **29:** 1401–1407.

Urry DW (1990) Protein folding and assembly: an hydration-mediated free energy driving force. In: Gierasch L and King J (eds) *Protein Folding: Deciphering the Second Half of the Genetic Code*. (pp. 63–71) Am. Assoc. for the Advancement of Sci., Washington, DC.

Urry DW (1991) Thermally driven self-assembly, molecular structuring and entropic mechanisms in elastomeric polypeptides. In: Balaram P and Ramaseshan S (eds) *Mol Conformation and Biol Interactions*. (pp. 555–583) Indian Acad of Sci., Bangalore, India.

Urry DW (1993) Molecular machines: how motion and other functions of living organisms can result from reversible chemical changes, *Angew Chem (German)* **105:** 859–883, 1993; *Angew Chem Int Ed Engl* **32:** 819–841.

Urry DW (1997) Physical chemistry of biological free energy transduction as demonstrated by elastic protein-based polymers (invited FEATURE ARTICLE). *J Phys Chem* **B 101:** 11,007–11,028.

Urry DW, Luan C-H and Peng SQ (1995) Molecular Biophysics of Elastin Structure, Function and Pathology, in Proceedings of The Ciba Foundation Symposium No. 192, The Molecular Biology and Pathology of Elastic Tissues, John Wiley & Sons, Ltd., Sussex, UK, pp. 4–30.

Urry DW and Luan C-H (1995a) A New Hydrophobicity Scale and its Relevance to Protein Folding and Interactions at Interfaces, in Proteins at Interfaces 1994, American Chemical Society Symposium Series, (Thomas A. Horbett and John L. Brash, eds.), pp. 92–110, Washington, D.C.

Urry DW and Luan C-H (1995b) Proteins: structure, folding and function. In: Lenaz G (ed.) *Bioelectrochemistry: Principles and Practice*. (pp. 105–182) Birkhäuser Verlag AG, Basel, Switzerland.

Urry DW and Venkatachalam CM (1983) A librational entropy mechanism for elastomers with repeating peptide sequences in helical array. *Int J Quant Chem: Quant Biol Symp* **10:** 81–93.

Urry DW, Chang DK, Krishna R, Huang DH, Trapane TL and Prasad KU (1989) Two dimensional proton nuclear magnetic resonance studies on poly(VPGVG) and its cyclic conformational correlate, cyclo$(VPGVG)_3$. *Biopolymers* **28:** 819–833.

Urry DW, Harris CM, Luan CX, Luan C-H, Gowda DC, Parker TM, Peng SQ and Xu J (1997a) Transductional protein-based polymers as new controlled release vehicles. In: Park K (ed.) *Controlled Drug Delivery: The Next Generation, Part VI: New Biomaterials for Drug Delivery*, (pp. 405–437) Am Chem Soc Professional Reference Book.

Urry DW, Haynes B, Zhang H, Harris RD and Prasad KU (1988b) Mechanochemical coupling in synthetic polypeptides by modulation of an inverse temperature transition. *Proc Natl Acad Sci USA* **85:** 3407–3411.

Urry DW, Henze R, Redington P, Long MM and Prasad KU (1985a) Temperature dependence of dielectric relaxations in α-elastin coacervate: evidence for a peptide librational mode. *Biochem Biophys Res Commun* **128:** 1000–1006.

Urry DW, Hugel T, Seitz M, Gaub H, Sheiba L, Dea J, Xu J and Parker T (2002a) Elastin: a representative ideal protein elastomer. *Philos Trans Roy Soc London [Biol]* **357:** 169–184.

Urry DW, Hugel T, Seitz M, Gaub H, Sheiba L, Dea J, Xu J, Prochazka F and Parker T (2002b) In: Shewry P & Bailey A (eds) *Ideal Protein Elasticity: The Elastin Model*. Cambridge University Press (in press).

Urry DW, Jaggard J, Prasad KU, Parker T and Harris RD (1991) Poly(Val^1-Pro^2-Ala^3-Val^4-Gly^5): a reversible, inverse thermoplastic. In: Gebelein CG (ed.) *Biotechnology and Polymers* (pp. 265–274) Plenum Press, New York.

Urry DW, Jing N, Trapane TL, Luan C-L and Waller M (1988a) Ion interactions with the gramicidin A transmembrane channel: cesium-133 and calcium-43 NMR studies. In: Agnew W, Claudio T and Sigworth F (eds) *Mol Biol of Ionic Channels* **33:** (vol. pp. 51–90) Academic Press, Inc., New York.

Urry DW, Peng SQ, Xu J and McPherson DT (1997b) Characterization of waters of hydrophobic hydration by microwave dielectric relaxation. *J Am Chem Soc* **119:** 1161–1162.

Urry DW, Peng SQ, Hayes LC, McPherson DT, Xu J, Woods TC, Gowda DC and Pattanaik A (1998) Engineering protein-based machines to emulate key steps of metabolism (biological energy conversion). *Biotechnology and Bioengineering* **58:** 175–190.

Urry DW, Trapane TL, Sugano H and Prasad KU (1981) Sequential polypeptides of elastin: cyclic conformational correlates of the linear polypentapeptide. *J Am Chem Soc* **103:** 2080–2089.

Urry DW, Trapane TL, Long MM and Prasad KU (1983a) Test of the librational entropy mechanism of elasticity of the polypentapeptide of elastin: effect of introducing a methyl group at residue-5. *J Chem Soc, Faraday Trans I* **79:** 853–868.

Urry DW, Trapane TL, Wood SA, Walker JT, Harris RD and Prasad KU (1983b) D-Ala_5 Analog of the elastin polypentapeptide. Physical characterization. *Int J Pept Protein Res* **22:** 164–175.

Urry DW, Trapane TL, Wood SA, Harris RD, Walker JT and Prasad KU (1984) D-Ala_3 Analog of elastin polypentapeptide: an elastomer with an increased Young's modulus. *Int J Pept Protein Res* **23:** 425–434.

Urry DW, Trapane TL, Iqbal M, Venkatachalam CM and Prasad KU (1985b) Carbon-13 NMR Relaxation studies demonstrate an inverse temperature transition in the elastin polypentapeptide. *Biochemistry* **24:** 5182–5189.

Urry DW, Trapane TL, McMichens RB, Iqbal M, Harris RD and Prasad KU (1986) Nitrogen-15 NMR relaxation study of inverse temperature transitions in elastin polypentapeptide and its Cross-Linked Elastomer. *Biopolymers* **25:** S209–S228.

Urry DW, Venkatachalam CM, Wood SA and Prasad KU (1985c) Molecular structures and librational processes in sequential polypeptides: from ion channel mechanisms to bioelastomers. In: Clementi E, Corongiu G, Sarma MH and Sarma RH (eds), *Structure and Motion: Membr, Nucleic Acids and Proteins* (pp. 185–203). Adenine Press, Guilderland, New York.

Urry DW, Venkatachalam CM, Long MM and Prasad KU (1982) Dynamic β-spirals and A librational entropy mechanism of elasticity. In: Srinivasan R and Sarma RH (eds) *Conformation in Biol*. (G.N. Ramachandran Festschrift Volume pp. 11–27). Adenine Press, Guilderland, New York.

Venkatachalam CM and Urry DW (1981) Development of a linear helical conformation from its cyclic correlate. β-spiral model of the elastin poly(pentapeptide), $(VPGVG)_n$. *Macromolecules* **14:** 1225–1229.

Venkatachalam CM and Urry DW (1986) Calculation of dipole moment changes due to peptide librations in the dynamic β-spiral of the polypentapeptide of elastin. *Int J Quant Chem: Quant Biol Symp* **12:** 15–24.

Voet D and Voet JG (1995) *Biochemistry*. 2nd edn. (p. 430, Table 15-3) John Wiley & Sons, Inc., New York.

Volpin D, Urry DW, Cox BA and Gotte L (1976a) Optical diffraction of tropoelastin and α-elastin coacervates. *Biochim Biophys Acta* **439:** 253–258.

Volpin D, Urry DW, Cox BA, Pasquali-Ronchetti I and Gotte L (1976b) Studies by electron microscopy on the structure of coacervates of synthetic polypeptides of tropoelastin. *Micron* **7:** 193–198.

Wasserman ZR and Salemme FR (1990) A molecular dynamics investigation of the elastomeric restoring force in elastin. *Biopolymers* **29:** 1613–1631.

Weiner PK and Kollman PA (1981) AMBER: assisted model building with energy refinement. A general program for modeling molecules and their interactions. *J Comput Chem* **2:** 287–303.

Weis-Fogh T and Andersen SO (1970) New molecular model for the long-range elasticity of elastin. *Nature* **227:** 718–721.

Journal of Muscle Research and Cell Motility **23**: 561–573, 2002.

Molecular basis for the extensibility of elastin

BIN LI and VALERIE DAGGETT*
Department of Medicinal Chemistry, University of Washington, Seattle, Washington 98195-7610, USA

Abstract

Elastin is a cross-linked protein in the extracellular matrix that provides elasticity for many tissues. Its soluble precursor (tropoelastin) has two major types of alternating domains: (1) hydrophilic cross-linked domains rich in Lys and Ala and (2) hydrophobic domains (responsible for elasticity) rich in Val, Pro, and Gly, which often occur in repeats of VPGVG or VGGVG. Since native elastin is large and insoluble, many studies have focused on elastin-based peptides in an effort to elucidate its structure–function relationship. This review focuses on the molecular basis of elastin's conformational properties and associated elasticity. From both experimental and simulation approaches, elastin can be described as a two-phase model consisting of dynamic hydrophobic domains in water. The hydrophobic domain of elastin is best described as a compact amorphous structure with distorted β-strands, fluctuating turns, buried hydrophobic residues, and main-chain polar atoms that form hydrogen bonds with water. Water plays a critical role in determining elastin's conformational behavior, making elastin extremely dynamic in its relaxed state and providing an important source of elasticity.

Introduction

Elastin is responsible for elasticity and resilience in many tissues, and it is well known for its extreme durability (Partridge, 1962). The half-life of elastin is on the order of 70 years, so elastic fibers will have undergone more than 1 billion stretch/relaxation cycles in this time period (Powell, 1990). Calorimetry studies suggested that entropy is the primary source of the remarkable elasticity of elastin (Hoeve and Flory, 1974; Andrady and Mark, 1980). However, despite several decades of almost constant effort, structural information for elastin using circular dichroism (CD) (Mammi *et al.*, 1968, 1970; Urry *et al.*, 1969; Abatangelo *et al.*, 1973; Foster *et al.*, 1976; Tamburro *et al.*, 1977, 1978; Guantieri *et al.*, 1987), Fourier transform infrared (FTIR) (Mammi *et al.*, 1968; Bertoluzza *et al.*, 1989), Raman (Frushour and Koenig, 1975; Prescott *et al.*, 1987), fluorescence (Gosline *et al.*, 1975), and nuclear magnetic resonance (NMR) spectroscopies (Torchia and Piez, 1973; Lyerla and Torchia, 1975; Fleming *et al.*, 1980), as well as X-ray diffraction methods (Gotte *et al.*, 1977), remains severely limited because of the extreme insolubility of the mature cross-linked form of elastin and the high mobility of the elastin backbone. This lack of detailed structural information has prevented the elucidation of the structure–function relationship for this polypeptide.

Elastin is an extremely insoluble molecule due to the extensive cross-linking at Lys residues (Thomas *et al.*, 1963). Its soluble precursor, tropoelastin (Foster *et al.*, 1973), is deposited into the extracellular space and rapidly forms elastic fibers with the help of several microfibril proteins (Gibson *et al.*, 1989, 1991, 1996; Zhang *et al.*, 1994). Tropoelastin biosynthesis, cross-linking and fiber assembly have been covered in previous reviews (Rosenbloom *et al.*, 1993; Vrhovski and Weiss, 1998) and recent research articles (Trask *et al.*, 2000a, b). Tropoelastin genes of human (Indik *et al.*, 1987a, b), chicken (Bressan *et al.*, 1987), bovine (Raju and Anwar, 1987b; Yeh *et al.*, 1989) and rat (Pierce *et al.*, 1990) origin have been completely sequenced, and they have significant homology at both the DNA and amino acid levels (e.g. the C-terminus of tropoelastin is ∼70% conserved among all species tested, Vrhovski and Weiss, 1998). Two major types of alternating domains are found in tropoelastin: (1) hydrophilic cross-linking domains rich in Lys and Ala; and (2) hydrophobic domains (responsible for elasticity), rich in Val, Pro, Ala and Gly, which often occur in repeats of VPGVG or VGGVG. In the last three decades, elastin-based peptides containing these repeats have been studied most extensively by Urry's group (Venkatachalam and Urry, 1981; Urry *et al.*, 1986a, b, 1988, 1989, 1992, 1995; Thomas *et al.*, 1987; Sciortino *et al.*, 1988; Chang *et al.*, 1989; Luan *et al.*, 1992; McPherson *et al.*, 1992; Gowda *et al.*, 1995; Herzog *et al.*, 1997; Lee *et al.*, 2001) and Tamburro' s group (Abatangelo *et al.*, 1973; Tamburro *et al.*, 1978, 1991, 1992, 1995; Tamburro, 1981; Tamburro and Daga Gordini, 1985; Daga Gordini *et al.*, 1989; Castiglione Morelli *et al.*, 1990, 1997; Lelj *et al.*, 1992; Megret *et al.*, 1992, 1993; Villani and Tamburro, 1995, 1999; Broch *et al.*, 1996, 1998; Bisaccia *et al.*, 1998; Lograno *et al.*, 1998; Martino *et al.*, 1998). In this review we focus on a molecular level description of the structure–function relationship of elastin, based primarily on studies of these hydrophobic elastin-based peptides.

*To whom correspondence should be addressed

Structural features of elastin and elastin-based peptides

Elastin structural models

Since high-resolution NMR or X-ray structures are not available for elastin, a variety of phenomenological models have been proposed to explain elastin's behavior. These models can be divided into two major categories: the single and two-phase models. The single-phase model is also known as the random chain model. The two-phase model can be further divided into the liquid-drop, the oiled-coil and the β-spiral models. Also, the β-spiral model is sometimes referred as the fibrillar model.

The single-phase model (Hoeve and Flory, 1974; Dorrington and McCrum, 1977) considers elastin to be like a typical rubber with the polymeric chains in random configurations with any diluent present in the system randomly distributed throughout the swollen material. This model is supported by the following findings: mechanical research suggesting that the structure of elastin is nearly random (Gosline and French, 1979); a glass transition study indicating that elastin is similar to solvent-swollen amorphous polymers (Kakivaya and Hoeve, 1975); a calorimetry study suggesting that entropy is the main source of elasticity (Andrady and Mark, 1980); and NMR studies showing that elastin is composed of highly mobile chains under physiological conditions (Torchia and Piez, 1973; Fleming *et al.*, 1980). However, this model does not agree with the reversible decrease in fluorescence of an elastin-binding dye when it becomes exposed to solvent as elastin is stretched, which instead suggests that the elastin network is a two-phase system of hydrophobic protein globules surrounded by water (Gosline *et al.*, 1975). Also, the single-phase model cannot explain why elastin is not self-lubricating (like rubber) and instead requires water for elasticity (Partridge, 1962).

The two-phase models (Weis-Fogh and Anderson, 1970; Gray *et al.*, 1973) assume that water-swollen elastin is composed of hydrophobic domains of protein (connected by cross-links) and aqueous diluent consigned to the spaces between the domains. Specifically, the liquid-drop model (Weis-Fogh and Anderson, 1970) assumes that the peptide segments between the cross-links are globular domains. The oiled-coil model (Gray *et al.*, 1973) is similar to the liquid-drop model, but it assumes that there are broad coils (the oiled coils) composed of hydrophobic repeats between the cross-linking domains. The Gly residues are on the outside of the coil and hydrophobic Pro and Val residues are inside. Support for these two-phase models is provided by the drop in fluorescence of an elastin-binding dye as elastin is stretched and the dye becomes exposed to solvent (Gosline *et al.*, 1975); NMR studies suggesting that water-swollen elastin is a multi-phase system (Ellis and Packer, 1976), and CD/Raman studies indicating that elastin has more turns and distorted β-strands as the temperature increases and passes through the transition temperature (Urry *et al.*, 1985a; Vrhovski *et al.*, 1997). The main criticism against these two-phase models is that they are not consistent with the experimentally determined high mobility of elastin (Torchia and Piez, 1973; Hoeve and Flory, 1974). Torchia and Piez showed, by ^{13}C NMR measurements, that the rotational relaxation time of the backbone carbon atoms in elastin is about 40 ns, which is only two times that of denatured elastin. In contrast, native collagen shows a rotational relaxation time 10^3 larger than that of denatured collagen (Torchia and Piez, 1973). As denatured elastin and denatured collagen have similar relaxation times, elastin is highly dynamic in its native state (Torchia and Piez, 1973). Therefore, Hoeve and Flory (1974) argued that 'it is highly unlikely that the peptide bonds within the densely packed globular model could undergo isotropic rotation with relaxation times as short as 10^{-8} s'.

Unfortunately, no one of the models described above provide insight into the conformational behavior of elastin at the atomic level – or in other words, the molecular basis for its elastomeric properties. Since (VPGVG) is the most abundant repeating peptide in elastin (Gray *et al.*, 1973; Sandberg *et al.*, 1985), Urry and co-workers carried out a variety of physical–chemical studies using synthetic polypeptide to investigate its structure/function relationship (Urry *et al.*, 1985a, 1995; Urry, 1992). NMR (Urry *et al.*, 1975) and CD (Urry *et al.*, 1985a) spectroscopic studies suggest that poly(VPGVG) adopts β_{II} turns at high temperature. In addition, there is a crystal structure of the cyclo-peptide c(VPGVG)$_3$ in di-methyl sulfoxide (DMSO, Cook *et al.*, 1980). Based on this study, and other, Venkatachalam and Urry (1981) developed a structural model for poly(VPGVG), which they termed a 'β-spiral' (Figure 1). While this model agrees with some experimental studies, it does not appear to agree with the results on water-swollen elastin. For example, Urry's NMR data show that for linear poly(VPGVG), the chemical shift of Val$_4$ NH within a repeat is about 0.3 ppm lower than that of Val$_1$ NH in DMSO, while they are almost the same in H_2O/D_2O (see Figures 3–5 in Urry *et al.*, 1989). Thus, linear poly(VPGVG) appears to adopt different conformations in DMSO and water. NMR studies indicate that elastin is fully mobile (Torchia and Piez, 1973; Ellis and Packer, 1976). Furthermore, birefringence analysis shows that elastin is isotropic (Aaron and Gosline, 1980, 1981), in contrast to the ordered, anisotropic orientation one would expect for the β-spiral. Therefore, given earlier work (Gosline, 1978b; Gosline and French, 1979), a rigid, well-ordered β-spiral model is not an ideal description of the structure of uncross-linked elastin in water. Overall, the various structural models of elastin are supported by some experimental results, but no one of them can explain all of the available evidence.

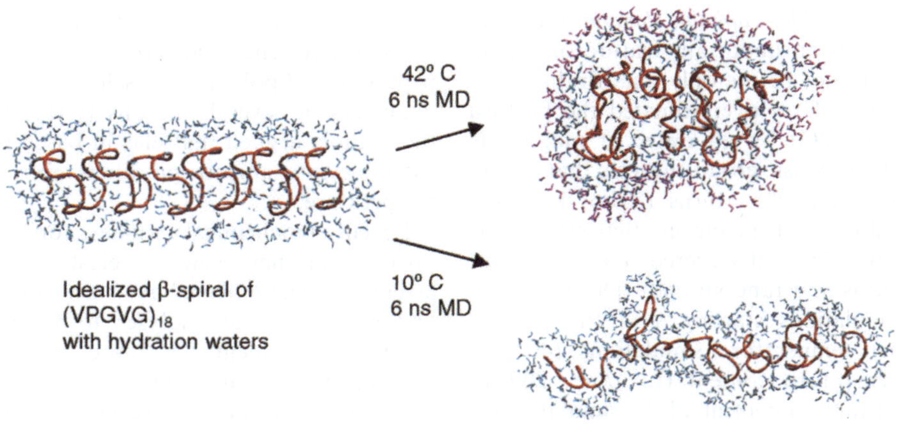

Fig. 1. Temperature-dependent conformational behavior of elastin explored through MD simulations. Elastin undergoes hydrophobic collapse above its transition temperature, expelling waters (shown in magenta) formerly associated with non-polar residues. Only the main chain of elastin is displayed to show the overall fold of the molecule. The simulations began with the β-spiral model, kindly provided by Dr Dan W. Urry.

Recent experimental and simulation studies of elastin structure

Turns in elastin

Various studies of elastin-based repeats suggest that turns, especially the β_{II}-turn, are the main secondary-structural feature of elastin-based peptides. A β-turn is defined as having an $i \rightarrow i + 3$ hydrogen bond, and a β_{II}-turn has approximately $\phi_{i+1} = -60°$, $\phi_{i+1} = 120°$, $\phi_{i+2} = 90°$, $\phi_{i+2} = 0°$ for the two central residues. For VPGVG and its polymer, Urry and co-workers identified turn conformations by CD (Urry *et al.*, 1985a), Raman (Thomas *et al.*, 1987) and NMR (Venkatachalam *et al.*, 1981) experiments. In addition, Tamburro and co-workers also proposed that β_{II}-turns are prevalent in VGGVG and peptides with similar sequences (Castiglione Morelli *et al.*, 1990, 1997; Lelj *et al.*, 1992; Tamburro *et al.*, 1992; Morelli *et al.*, 1993; Broch *et al.*, 1996; Villani and Tamburro, 1999). On the other hand, although both groups agree that β-turns represent the dominant structure in elastin, they provide different explanations of the role of the turns in elasticity. Urry and co-workers built their β-spiral model of elastin as a continuous coil of β-turns interconnected by glycine residues (Figure 1). They suggested that the elastomeric restoring force originates from the reduction of librational entropy in the peptide segments linking the β-turns as the elastin coil is stretched, which they coined the 'librational elasticity mechanism' (Urry and Venkatachalam, 1983). On the other hand, Tamburro's group emphasized that the β-turns in the elastin sequence are very labile and dynamic, and therefore these turns should contribute greatly to the entropy of elastin and elasticity (Broch *et al.*, 1996; Villani and Tamburro, 1999). Since both groups argued that their elastin-based peptides are good models for the description of the structure/function relationship of elastin, Debelle and Alix (1999) questioned whether these peptide models are in fact relevant to those occurring in native elastin.

It is still possible that both VPGVG- and VGGVG-based peptides are relevant to the structure/function of elastin. Arad and Goodman (1990) used depsipeptide analogues of VPGVG-based elastin-repeating sequences for conformational analysis by NMR, CD and IR. The depsipeptides, which replaced valine with its hydroxy acid analogue (S-α-hydroxyisovaleric acid) at position 4 of the VPGVG peptides to eliminate the $i \rightarrow i + 3$ β-turn hydrogen bond, provide a novel way to probe the role of β-turn hydrogen bonds in determining elastin conformation. Their results suggested that an equilibrium exists between a β-turn and a γ-turn (a γ-turn is defined as having an $i \rightarrow i + 2$ hydrogen bond), resulting in a structure that combines flexibility with strong conformational preferences (Arad and Goodman, 1990). Also, we recently employed molecular dynamics simulation methods to study the conformational features of an elastin-based peptide $(VPGVG)_{18}$ and found that the turn portion of the molecule is the most dynamic (Li *et al.*, 2001a). Therefore, VPGVG can contain a highly dynamic turn that contributes to elasticity in a similar manner as that proposed for VGGVG by Tamburro and co-workers (Broch *et al.*, 1996; Villani and Tamburro, 1999). In addition, our simulations began from Urry's idealized β-spiral structure but resulted in amorphous structures at all temperatures investigated (Li *et al.*, 2001a). Simulations above and below elastin's transition temperature (~27°C) indicate that elastin has more turns and distorted β-structure at higher temperatures (Figure 1), consistent with the experimental finding that elastin becomes more ordered above its transition temperature. These results also suggest that a rigid, well-ordered β-spiral is inadequate for describing the structure of water-swollen elastin at physiological temperature (Gosline, 1978a; Gosline and French, 1979; Li *et al.*, 2001a).

The distorted β-sheet and hydrophobic domain of elastin
The β_{II}-turn is indeed an important feature of elastin structure, but it may not be the only one. Brahms and co-workers argue that Urry overestimated the β_{II}-turn content in poly(VPGVG) by an order of magnitude (Brahms *et al.*, 1977). Weiss' group suggests that elastin forms 41% β-sheet and 21% turns (Vrhovski *et al.*, 1997). Debelle and co-workers suggest that elastin has 45% β-strand and 55% disordered conformations (which includes turns and random coil) (Debelle *et al.*, 1995). In our simulations starting with the β-spiral model, the higher temperature simulations (40 and 42°C) have higher β-turn content (Li *et al.*, 2001a). However, the final turn content in all the simulations is low (<15%), which is in agreement with the experimental results described above but in opposition to the β-spiral model (~80% β_{II}-turn). In fact, The Cα root-mean-square deviation (Cα RMSD) from the starting, idealized β-spiral model was quite high for all of our simulations, suggesting that elastin becomes somewhat amorphous in water at all temperatures (Li *et al.*, 2001a). We note that such large shifts from the starting structure would be undesirable in simulations of native proteins, but here it appears to be reasonable because we started with the highly idealized model and the resulting structures from the simulations are in better agreement with the experimental observations than the starting structure. Consequently, one must go beyond β-turns in describing the structure of elastin.

Debelle and co-workers suggest that elastin has about 45% β-strands based on FTIR and Raman data (Debelle *et al.*, 1995). Moreover, they regard the β-strands in elastin as 'very short and/or distorted anti-parallel β-strands' (Debelle *et al.*, 1995). Similarly, our simulations of elastin-based peptides indicate that about 30% of the residues in $(VPGVG)_{18}$ form non-local hydrogen bonds above elastin's transition temperature, corresponding to distorted β-strands (Li *et al.*, 2001a). As implied in Debelle and colleagues' experimental studies, and as directly observed in our simulations, the formation of β-strands occurs concomitantly with hydrophobic collapse when the temperature is raised above elastin's transition temperature. These results support the two-phase model of elastin.

Linear elastin fibers
Previous microscopic studies of elastic fibers suggest the presence of structure, in particular a filamentous or fibrillar structure, in cross-linked elastin. Transmission and scanning electron microscopy of purified elastin reveals fibers with a diameter of 100 nm at low resolution and individual filaments of 3 nm at high resolution (Gotte *et al.*, 1977). Also, a parallel array of primary filaments is seen in undegraded elastin from bovine ligamentum nuchae (Serafini-Fracassini *et al.*, 1976). Interestingly, using electron microscopy Urry's group found that after negative staining, the elastin-based peptide (poly(VPGVG)) showed 5 nm wide filaments with a super-coiled or twisted substructure above its transition temperature (Urry, 1982). Under conditions where chemical cross-linking can occur, this peptide-based polymer can self-assemble into fibers with diameters of several microns (Urry, 1983). Therefore, the authors argued that elastin-based peptides can self-assemble into anisotropic fibers above their transition temperature.

Recent biochemical studies, on the other hand, provide another view of elastin fiber content and assembly. Elastin fibers have been shown by biochemical and ultrastructural analyses to be composed of two distinct components, a more abundant amorphous component (elastin) and a 10–12 nm microfibrillar component (fibrillin and several related proteins) (Rosenbloom *et al.*, 1993). The elastin monomer, tropoelastin, was suggested to have a positively charged pocket with the C-terminal RKRK sequence (Brown *et al.*, 1992). When tropoelastin is deposited into the extracellular space, it is speculated that the pocket could provide a non-covalent binding site for the highly acidic microfibrils. It has been confirmed experimentally that tropoelastin can indeed bind to the microfibrillar protein MAGP-1 (Bashir *et al.*, 1994; Brown-Augsburger *et al.*, 1994). The binding is localized to the C-terminus of tropoelastin and the N-terminal half of MAGP-1 (Brown-Augsburger *et al.*, 1996). Therefore, the microfibrils may serve to align the tropoelastin molecules into the correct orientation for subsequent cross-linking. Indeed, the highly conserved C-terminus of tropoelastin is necessary for correct elastic fiber formation. When it is missing, the fiber integrity is severely compromised, as noted in lamb ductus arteriosis, where a truncated tropoelastin is present and does not form fibers (Hinek and Rabinovitch, 1993). Thus, the linear elastin fibers are a result of microfibril-assisted alignment of tropoelastin molecules, but tropoelastin or just its hydrophobic domains do not have to be linear.

A new view of elastin structure

As mentioned above, elastin fibers are composed of a microfibrillar component (fibrillin and related proteins) and an amorphous component (elastin), and elastin has alternating cross-linking and elastic hydrophobic domains. Biochemical and ultrastructural studies have shown that tropoelastin can be deposited in a suitable position to form linear elastin structure with the assistance of fibrillin proteins (Bashir *et al.*, 1994; Brown-Augsburger *et al.*, 1994). These tropoelastin molecules then link with each other through the cross-linking domains, by forming desmosine (Reiser *et al.*, 1992) or Lys–Lys bridges (Brown-Augsburger *et al.*, 1995). The structural features of the hydrophobic domains of elastin are not so clear, as discussed above for the different models. The hydrophobic domain of elastin undergoes an 'inverse temperature transition' (Figure 1, and Li *et al.*, 2001a) from below to above its transition temperature, in which water is critical, supporting the idea that elastin is basically a two-phase system. Elastin's hydrophobic domain at its physiological state is best described as a compact amorphous structure with distorted β-strands, fluctuating turns, buried hydrophobic residues, and main-chain polar atoms that form hydrogen bonds with water (Li *et al.*, 2001a). In addition, although local and fluctuating β-spiral structure may be present to a certain degree, both experimental (Gosline, 1978b; Gosline and French, 1979) and simulation (Li *et al.*, 2001a) studies suggest, as described above, that a rigid, well-ordered β-spiral model is not a good description of elastin in water.

Molecular basis of elastin's elasticity

Compact but dynamic hydrophobic domain of elastin in its relaxed state

Drawing from our all-atom simulations of elastin in water, we found that the hydrophobic domain of elastin is highly dynamic and therefore quite different from the traditional view of a densely packed globular model (Li *et al.*, 2001a). Elastin's hydrophobic domain displays several types of dynamic behavior. First, the β_{II}-turn portions of elastin rotate through a crankshaft-like motion. This type of motion was also observed in a previous 130-ps elastin simulation (Wasserman and Salemme, 1990). Such a motion results in a reorientation of the central peptide bond with little deformation of the structure on either side of the two rotating bonds, maintaining the overall integrity of the structure. In this way, elastin can have a higher number of turns at elevated temperatures, while maintaining entropy/mobility-based elasticity. Second, the distorted β-strands are short and usually characterized by fluctuating hydrogen bonds during the simulations. This kind of dynamic hydrogen bonding network provides the molecule with the ability to move quickly. In addition, the entire hydrophobic domain undergoes cycles of extension and contraction on the nanosecond time scale (Li *et al.*, 2001a). In this way, elastin is extremely dynamic despite the formation of hydrophobic domains/globules.

To compare the dynamic nature of elastin's hydrophobic domain with a normal, or possibly more dynamic than normal, globular protein, we calculated the average generalized order parameter (S^2) for both main-chain and side-chain atoms at high and low temperatures. We and others have shown that MD-derived S^2 order parameters can be in good agreement with S^2 values derived from NMR relaxation experiments (Wong and Daggett, 1998). S^2 values approaching 1 suggest that the peptide is rigid and those approaching 0 reflect unrestricted motion. The average S^2 values are 0.51 (main chain) and 0.37 (valine side chains) for elastin at 40 and 42°C, compared with values of 0.59 and 0.39 at 7 and 10°C (Li *et al.*, 2001a). For perspective, the average S^2 values for the globular protein barstar are 0.84 and 0.70–0.92 (depending on location, the average value is 0.87) for its main chain and Val side chains, respectively (Wong and Daggett, 1998). Therefore, the S^2 values indicate that elastin is slightly more dynamic above its transition temperature, even as the molecule forms hydrophobic domains, but it is also extremely dynamic even at low temperatures.

The highly dynamic nature of elastin may be due to its high hydrophobic composition. Elastin consists mainly of hydrophobic amino acids and almost all of the lysines in elastin are involved in cross-linking (Rosenbloom *et al.*, 1993). Therefore, it has almost no polar atoms other than main-chain NH/O groups. The primary sequence of a globular protein and its interaction with solvent determine its three-dimensional structure. A globular protein uses side-chain polar groups and some of its main-chain NH/O groups to interact with water, and the bulk of the main-chain polar groups form secondary structural main-chain/main-chain hydrogen bonds. Although elastin can undergo hydrophobic collapse similar to that of globular proteins (Alonso and Daggett, 1998), its main-chain NH/O groups seem to be used mainly for interaction with water and they form few main-chain/main-chain hydrogen bonds. As a result, elastin is extremely dynamic in water, in agreement with experiment (Torchia and Piez, 1973; Hoeve and Flory, 1974). In this way, the hydrophobic domain of elastin is quite different from the traditional view of a densely packed globular domain (Hoeve and Flory, 1974). Instead it is a compact, dynamic hydrophobic domain. Similarly, Tamburro and co-workers suggested a shift in our thinking about elastin from the concept of fixed structure to that of interconverting structures (Martino *et al.*, 1998). The extremely dynamic nature of elastin's hydrophobic domain provides high entropy even in its relaxed state, which may contribute to its elasticity.

Previous models of elasticity

There are three main mechanisms proposed to explain the entropic origins of elastin's elasticity, each based on a corresponding view of elastin structure. Consistent with the single-phase elastin structural model, Hoeve and Flory (1974) suggested that elastin follows the classical theory of rubber elasticity: the collection of chains have a random distribution of end-to-end chain lengths and the displacement from this position of highest entropy provides the source of the elastic restoring force. Several investigators used this model to successfully explain the calorimetry data and mechanical behavior of elastin (Hoeve and Flory, 1974; Dorrington and McCrum, 1977), and optical analysis of single elastin molecules also suggest that elastin is a random network (Aaron and Gosline, 1981). However, this model cannot explain why elastin, unlike rubber, is brittle when it is dry and only water-swollen elastin displays elasticity.

As to the two-phase structural models of elastin, the liquid-drop and oiled-coil models have similar explanations of elasticity. The liquid-drop model considers elastin as being composed of globular domains (liquid drops) connected by crosslinks (Weis-Fogh and Anderson, 1970). When pulling elastin, the hydrophobic domains become exposed to the surrounding water, thus decreasing the entropy of the waters and therefore the system. Elasticity is driven by an increase in entropy for the reverse process as water is expelled. Instead of a liquid drop, the oiled-coil model considers the hydrophobic domain as a broad coil (the oiled coil) consisting of the hydrophobic repeat units (Gray *et al.*, 1973). The Gly residues are on the exterior of the coil and the hydrophobic Val and Pro residues are in the interior. Upon extension, the hydrophobic interior of the coil is exposed to water, leading to a decrease in the entropy of the system. There is an entropic increase that provides the restoring force when released.

In contrast, Urry and co-workers proposed that the elastomeric restoring force originates from the reduction of librational entropy in the peptide segments linking the β-turns as the elastin coil is stretched, which they coined the 'librational elasticity mechanism' (Urry and Venkatachalam, 1983). This librational entropy mechanism depends heavily on the linear β-spiral structural model of elastin. However, recent scanning force microscopy experiments suggest that elastin molecules form a disordered array of beaded filaments containing globular domains in the relaxed state, and this array of globules becomes more expanded but they do not adopt precisely ordered β-spiral conformations in the stretched state (Pasquali Ronchetti *et al.*, 1998). In addition, considering Urry's CD results, Brahms and co-workers argue that their estimate of the β-turn population was exaggerated and instead it is much smaller than that assumed previously (Brahms *et al.*, 1977). Also, as mentioned above, our simulations of elastin starting with the β-spiral conformation result in compact amorphous structures at physiological temperature and the simulations are consistent with experimental results (Li *et al.*, 2001a). Therefore, the β-spiral structure and the librational entropy mechanism may represent some local features of elastin, but they do not appear to provide a good explanation of elasticity.

Molecular dynamics simulations as a tool for studying the molecular basis of elasticity

Previous NMR and CD studies (Urry *et al.*, 1986c) suggest that the entropy change of each single $(VPGVG)_n$ chain, rather than the effects of cross-linking, provides the dominant source of elastomeric force. Gene analysis shows that tropoelastin is divided into several hydrophobic and cross-linking sections (Raju and Anwar, 1987a), and $(VPGVG)_{18}$ appears to nicely represent one of these hydrophobic sections. Also, force–extension studies of single elastin fibers at extensions below 100% indicate a value of approximately 6000–7000 D for the molecular weight of the elastin chain between the cross-linking sections (Aaron and Gosline, 1981), therefore such a length might be treated as a 'unit' of elasticity. These results opened the door for investigating elasticity using relatively small peptides instead of a large cross-linked polymer, but we need to keep in mind that cross-linking may enhance elasticity compared with just the hydrophobic domains (Debelle and Alix, 1999). Here we discuss the molecular basis of elasticity based mainly on molecular dynamics simulations of the elasticity process and compare these results with various experimental studies.

Previous elasticity simulations that support the 'librational elasticity mechanism'

Using simulation methods, Chang and Urry (1989) showed that the 'suspended' part of elastin, the portion between the β-turns in the β-spiral model, had much larger (ϕ, ψ) fluctuations in the relaxed state than in the extended state, but the β-turn portion of elastin was unaffected. As a result, they proposed the 'librational elasticity mechanism' introduced above, which purports that the primary mechanism of elasticity for elastin is the damping of the peptide backbone dihedral angle dynamics upon elongation (Venkatachalam and Urry, 1981). However, their simulation was performed *in vacuo* and was only 100 ps long. Water is known to be a plasticizer of elastin and only water-swollen elastin displays elasticity (Partridge, 1962). Also, dielectric (Henze and Urry, 1985) and NMR (Urry *et al.*, 1985b) relaxation studies have shown that water-swollen elastin has a correlation time from several to tens of nanoseconds. Therefore, one might question whether this simulation can represent elastin in its functional state. Wasserman and Salemme (1990) extended the previous simulation by including water. They stretched their elastin peptide by proportionally translating all atoms out from the center of mass in a direction parallel to the helix axis (similar to Chang and Urry's protocol) over 130 ps, and they obtained larger Gly 5 than Gly 3 (ϕ, ψ) fluctuations. Thus, their simulation supported the librational elasticity mechanism. However, one cannot exclude the possibility that the larger Gly 5 (ϕ, ψ) fluctuations are the results of their method for pulling the molecule. Moreover, they did not discuss elastin's hydration properties and their relationship to elasticity.

Recent MD simulations suggest that hydrophobic hydration is an important source of elasticity

Although elasticity is well known to be linked with entropy (Hoeve and Flory, 1974), it is not clear whether this entropy arises from the conformational entropy of the protein backbone (Urry and Venkatachalam, 1983) or from the change in the hydration of hydrophobic side chains (Gosline, 1978a). The earlier elasticity simulations failed to answer this question since they either were *in vacuo* (Chang and Urry, 1989) or did not analyze water properties (Wasserman and Salemme, 1990). In addition, these elasticity simulations assumed that elastin has a helical structure of 2.7 pentamer units per turn (β-spiral) in both the relaxed and extended states, therefore they simulated the stretching process of elastin by raising the axial distance per turn of the β-spiral. To release the assumption of the helical structure in elastin's extended state, we pulled the ends of elastin (instead of many pairs of atoms), which is closer in spirit to what is done in atomic force microscopy (AFM) experiments, in our recent elasticity simulations (Figures 2 and 3) (Li *et al.*, 2001b). After pulling, we continued the simulation for 3 ns at a particular extension to collect conformations and statistics for elastin in its extended state (we call this the holding process, Figure 2). Moreover, we simulated the elasticity process on the time scale of several nanoseconds, which is on par with the dielectric- and NMR-determined correlation times of water-swollen elastin (Henze and Urry, 1985; Urry *et al.*, 1985b).

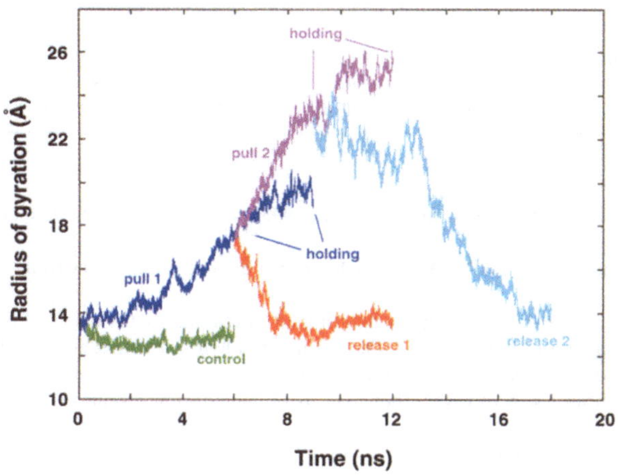

Fig. 2. Elastin's response to force at 42°C (adapted from Li *et al.*, 2001b).

In our simulations, elastin has a large solvent accessible surface area and a low number of hydrophobic contacts in its extended state. When the force is released, the solvent accessible surface area decreases and the intrachain contacts increase. Also, the number of hydration waters decreases upon release as elastin undergoes hydrophobically driven collapse (Table 1). The properties of the simulations after elastin is released are quite similar to those of the corresponding reference simulations at both temperatures in the absence of a pulling force. In comparison, our recent studies on force-induced unfolding of barnase, both by AFM and MD, suggest that non-elastic proteins may not resist force in the same way as elastic ones (Best *et al.*, 2001). These results, and others, suggest that our simulations provide a reasonable representation of the pulling/releasing (elasticity) process of elastin.

Our results indicate that the water orientational entropy for the hydrophobic hydration waters, but not the main-chain polar hydration waters, increases in all relaxed states for all pulling/releasing cycles (Li *et al.*, 2001b) (Table 1). In addition, the structure of the hydration shell does not change much upon extension, but there are more 'ordered' hydration waters in the extended than in the relaxed state (Li *et al.*, 2001b)

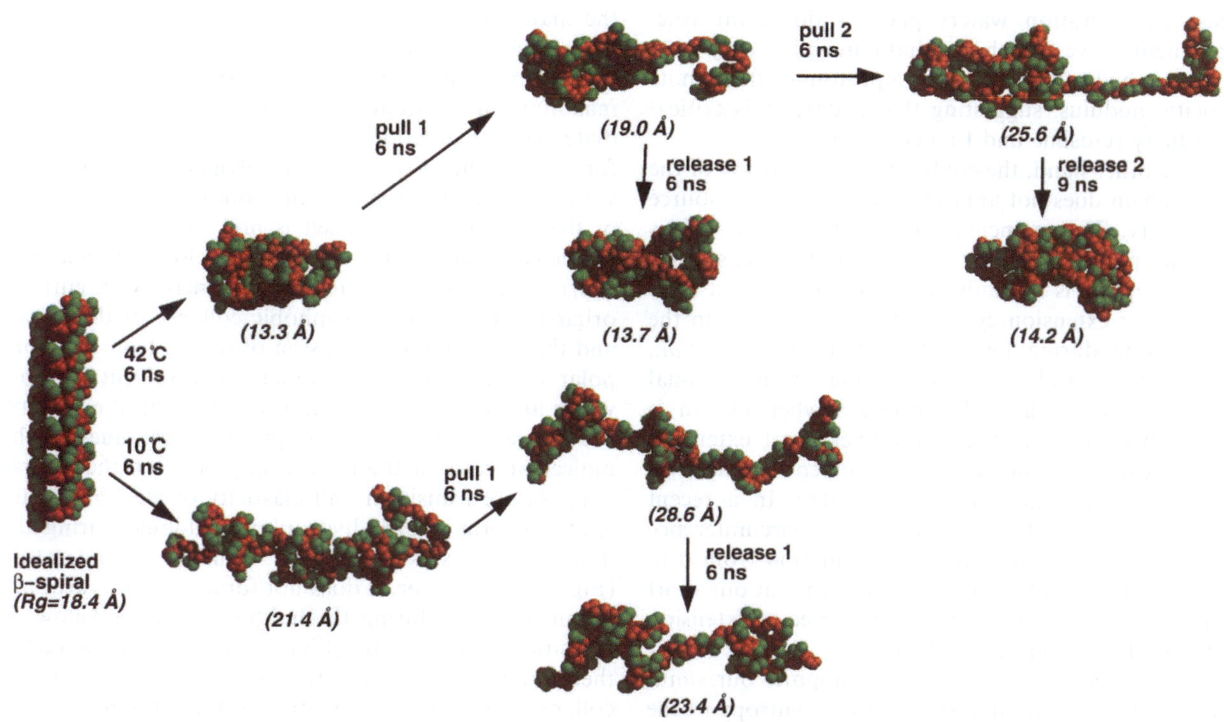

Fig. 3. Conformational behavior of elastin as probed by molecular dynamics simulations in water at 10 and 42°C. The simulations began from the idealized β-spiral structure, which was allowed to adapt to its environment for 6 ns of MD. Then, various cycles of pulling, holding and releasing the molecule were performed. The final structure after each process is shown with the main chain in red and side-chain atoms in green. (adapted from Li *et al.*, 2001b). The Cα-based radius of gyration is given below each structure.

Table 1. Average properties of elasticity simulations for the final nanosecond[a]

Peptide properties					Hydration properties				
					H-bonds[d]		Entropy[e]		
MD name	Rg[b] (Å)	SASA[c] (Å^2)	Non-polar SASA[c] (Å^2)	Side-chain contacts[f]	# protein	Avg. # wat–wat	Side-chain	Main-chain	# Hydration Waters[g]
Elasticity simulations at 42°C									
Control	12.9	5661	4103	53	16	3.01	−22.2	−35.8	376
Pull 1	19.7	6260	4663	40	14	3.03	−31.1	−33.8	449
Release 1	13.8	5625	4082	54	13	3.01	−20.6	−40.8	346
Pull 2	24.9	6736	4898	38	9	3.03	−32.0	−29.3	494
Release 2	14.0	6052	4325	51	14	3.02	−24.0	−33.3	415
Elasticity simulations at 10°C									
Control	21.0	6922	5107	33	4	3.24	−40.6	−41.3	526
Pull 1	28.6	7686	5713	28	0	3.27	−46.6	−39.5	598
Release 1	23.1	7283	5339	35	2	3.26	−43.0	−43.0	553

[a] The data of this table were partly adopted from Li *et al.* (2001b).
[b] Radius of gyration based on the positions of the α-carbons.
[c] All solvent accessible surface areas were calculated using the program NACCESS (Hubbard and Thorton, 1993), which employs the Lee and Richard's algorithm with a probe radius of 1.4 Å (Lee and Richards, 1971).
[d] Hydrogen bonds were tabulated as intact when the hydrogen and the acceptor were within 2.6 Å and the hydrogen bonding angle was within 35° of linearity.
[e] Entropy (Cal mol^{-1} K^{-1}) was calculated over the last nanosecond of the simulations. The entropy is determined by comparing the distributions of water orientations to those expected for a randomly oriented solvent (Daggett and Levitt, 1991). The main-chain hydration including all the water molecules contact to polar and non-polar main-chain atoms (≤4.5 Å); side-chain hydration contact to side-chain non-polar atoms (≤4.5 Å).
[f] Side-chain contacts were counted when two aliphatic carbon atoms were within 5.4 Å of each other. Atoms in neighboring residues were not considered.
[g] Hydration waters were defined as water molecules within 4.5 Å of protein heavy atoms.

(Table 1). These results suggest that hydrophobic hydration of elastin contributes to elasticity and that the release of hydration waters plays a dominant role. Experiments have also shown that removal of only 10% of elastin's hydration waters has a profound effect on its elasticity modulus, suggesting that hydration is critical to elasticity (Gosline and French, 1979).

On the other hand, the conformational entropy of the peptide chain does not appear to be a dominant source of elasticity. The mainchain (ϕ, ψ) fluctuation values and order parameter data from our simulations suggest that the peptide is more dynamic in the extended state during short extension cycles but more dynamic in the relaxed state during long extension cycles (Li *et al.*, 2001b). These results suggest that the conformational entropy of elastin actually decreases when elastin is released from the extended state over short extension cycles. Short extension means ~50% extension based on the radius of gyration (Li *et al.*, 2001b). In a recent mechanical study of elastin, the aortic circumference was assumed to expand 30% upon inflation from 0 to 13.7 K Pa (Lillie and Gosline, 1990), such that our short extension cycles are in line with the degree of extension experienced by elastin physiologically.

Gosline's microcalorimetry study supports our simulation results: the overall conformational entropy of the peptide chain actually increases when elastin is stretched and the polymer becomes more 'random' (Gosline, 1978a). Furthermore, he found that the conformational entropy of elastin does not contribute to elastic force at extensions of less than ~70% (Gosline, 1978a). In the more extreme and non-physiological extension ranges, the main-chain conformational entropy contributes to the elasticity. This is in accord with theoretical work (Flory and Fisk, 1966; Sanchez, 1979) that predicts a maximum in the number of conformations at intermediate radii of gyration. However, that number tails off for very extended or compact polymers. Therefore, the various results suggest that the conformational entropy of the elastin polymer itself is not the main source of elasticity, at least in the case of physiologically relevant, short extension cycles. Instead, the increase in entropy originates from the hydrophobic collapse of the chains and the concomitant expulsion of waters from the non-polar surfaces leading to an increase in solvent entropy, which appears to be the major determinant of elasticity.

These simulations may also provide some clues for the molecular basis of the relationship between the inverse temperature transition and elasticity of elastin. Elastin tends to form a large hydrophobic globule during the holding processes above its transition temperature (Figure 4); however, it does not form similar hydrophobic interactions during the holding processes below its transition temperature (Figure 3). It seems that the thermodynamic features that cause the hydrophobic collapse and represent the driving force for the inverse temperature transition also make elastin form a larger hydrophobic core in the holding state (Figure 4). Moreover, collapse automatically occurs when the pulling force is removed at physiological temperature

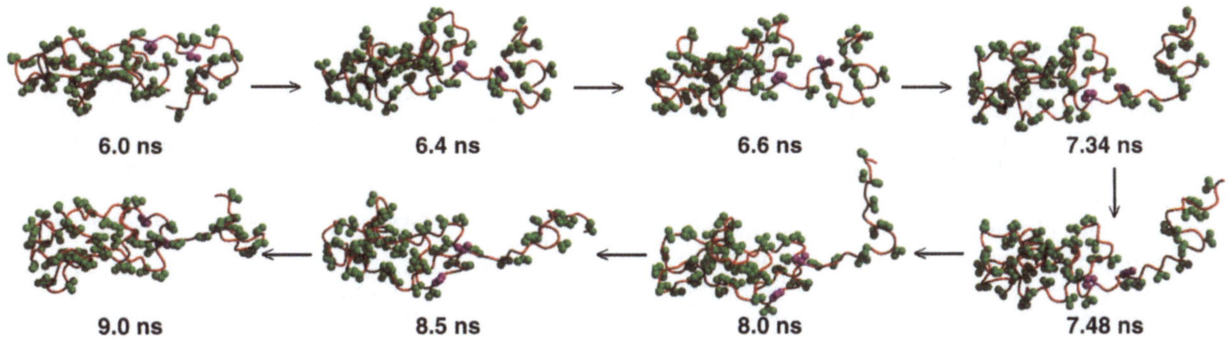

Fig. 4. Hydrophobic collapse in the 42°C *pull 1* holding process – the Pro 67-Val 71 segment joins the main hydrophobic core of elastin. Red: elastin main chain; Green: hydrophobic side-chain atoms; magenta: Pro 67 and Val 71 (adapted from Li *et al.*, 2001b).

(Figure 3). In this way, the hydrophobic hydration mechanism of elasticity we present here is probably most applicable to physiological temperatures. On the other hand, at temperatures below elastin's transition temperature, the hydrophobic side chains of elastin are already 'dissolved' in water and the cross-linking elastin becomes swollen. Therefore, the contribution of hydrophobic hydration to elasticity is lower compared with physiological temperatures (Table 1), and elastin does indeed show lower elasticity below its transition temperature (Luan *et al.*, 1989).

Elastin hydration

The water closely associated with elastin is about 0.9:1 by weight in our simulations (W/W) (Li *et al.*, 2001b), which is in agreement with Urry and co-worker's value of approximately 1:1 (W/W) from coacervation (Urry, 1993) and microwave dielectric relaxation (Urry *et al.*, 1997) studies. In addition, Gosline and French found that a hydrated native elastin sample has a water content of 0.46 g water/g protein at 36°C (Gosline and French, 1979). The dry samples were hydrated under vacuum over water for approximately 7 days, therefore this most likely represents the water content for just hydration waters. Considering that the real elastin sample contains cross-linking sections and is therefore less hydrophobic, this result is in reasonable agreement with the water content in the simulations. Previously, Gosline and French found that elastin has similar elastic character in a thermodynamic open system (swelling equilibrium with water) and in a thermodynamic closed system (constant water content, 0.46 g water/g protein, Gosline and French, 1979). In contrast, removing only 10% of elastin's hydration waters can greatly influence elasticity (Gosline and French, 1979).

Elastin interacts with water in several different ways in the MD simulations. First, a few water molecules are located near the β-turns, and they contribute to the β-turn/γ-turn equilibrium. For example, a water molecule is localized near the Val 69–Val 66 β-turn at 42°C (Li *et al.*, 2001a). The water alternatively formed hydrogen bonds with the main-chain polar groups and disrupted the $i \rightarrow i + 3$ hydrogen bonding of the turn, which seems to be unfavorable to the formation of regular hydrogen bonds and makes the turns the most dynamic portion of the molecule. Second, some of the waters are involved in water/main-chain hydrogen bonds. Elastin forms slightly fewer hydrogen bonds with water as the temperature increases, and the molecule becomes much smaller in this process. In our elastin simulations at 42°C, there are ~120 hydrogen bonds formed between the elastin main-chain and water, compared with ~10 hydrogen bonds between main-chain atoms. For comparison, Baker and Hubbard (1984) found that 43% of the main chain C=O groups and 21% of the main-chain NH groups form hydrogen bonds with water in globular proteins. In addition, elastin's main-chain nitrogen and oxygen (NH/O) groups are generally on the outside of the molecule or on the edges of the cavities (Li *et al.*, 2001a). It seems that the main-chain/water hydrogen bonds of 'wet' elastin replace the main-chain/main-chain hydrogen bonds of dry elastin, thereby contributing to flexibility. Experimentally, elastin needs about 20–29% water (g/100 g) to make it extendable from 10 to 40°C (Kakivaya and Hoeve, 1975). The remainder of the elastin-associated waters are associated with the hydrophobic side chains, since 80% of elastin's solvent accessible surface area is hydrophobic (Li *et al.*, 2001a). Simulation studies of hydrophobic model compounds suggest that the waters tend to align themselves parallel to the non-polar side chains and interact with each other rather than with the protein atoms (Laidig and Daggett, 1996; Beck *et al.*, 2002). This hydrophobic hydration constitutes the driving force for the hydrophobic collapse (Li *et al.*, 2001a) and elasticity (Li *et al.*, 2001b).

Based on our simulations, the hydration waters of elastin play at least the following roles in the elastin–water system: first, the orientational entropy of elastin's hydrophobic hydration layer decreases as the molecule is pulled, but it increases upon release. Then, hydration waters form hydrogen bonds with elastin's main-chain atoms as opposed to elastin forming intramolecular hydrogen bonds like a globular protein, which makes elastin fully dynamic both in the extended and relaxed states. Similarly, AFM studies (Oesterhelt *et al.*, 1999) and MD simulations (Heymann and Grubmuller, 1999) suggest that water drastically affects the elastic properties of poly(ethylene-glycol) (PEG) through its dynamic hydrogen bonding to PEG's main-chain atoms. Finally, the unfavorable interaction between water and elastin's hydrophobic side chains results in hydrophobic collapse with concomitant expulsion of hydration water from non-polar surfaces, which is the source of both elastin's inverse temperature transition and elasticity.

Conclusions

Only water-swollen elastin exhibits elasticity (Partridge, 1962), so both elastin and its hydration shell must be considered in any reasonable model of its behavior. Both experimental data (Gosline, 1978a; Gosline and French, 1979) and our simulation results (Li *et al.*, 2001a, b) suggest that elastin is a two-phase system with dynamic hydrophobic domains in water and that hydrophobic hydration of elastin is an important source of elasticity. While, the proposed β-spiral structure and the librational entropy mechanism by Urry and Venkatachalam (1983) might represent some local features of elastin, they do not to provide a good and complete explanation of elasticity. On the other hand, the random-chain model works well for treating elasticity by considering elastin/water system as a polymer in a poor solvent (Hoeve and Flory, 1974). In fact, water-swollen elastin is highly dynamic and lacks a preferred three-dimensional structure, and the random-chain model provides a good description of elasticity at the macroscopic level (Gosline, 1978b; Gosline and French, 1979), which has now been complemented by the atomic-level information provided by the MD simulations described here. These MD simulations provide a detailed molecular framework for understanding the underlying behavior that gives rise to the fundamental properties of elastic proteins: they must withstand low forces, they must stretch and not break at high forces, and they must recover upon the release of the force.

Acknowledgements

We are grateful for financial support from the Office of Naval Research (N00014-98-1-0477 and N00014-02-1-0373) and for helpful discussions with Dr Jane Clarke.

References

Aaron BB and Gosline JM (1980) Optical properties of single elastin fibres indicate random protein conformation. *Nature* **287:** 865–867.

Aaron BB and Gosline JM (1981) Elastin as a random-network elastomer: a mechanical and optical analysis of single elastin fibers. *Biopolymers* **20:** 1247–1260.

Abatangelo G, Daga-Gordini D and Tamburro AM (1973) Soluble fragments of elastin. Circular dichroism studies. *Int J Pept Protein Res* **5:** 63–68.

Alonso DOV and Daggett V (1998) Molecular dynamics simulations of hydrophobic collapse of ubiquitin. *Protein Sci* **7:** 860–874.

Andrady AL and Mark JE (1980) Thermoelasticity of swollen elastin networks at constant composition. *Biopolymers* **19:** 849–855.

Arad O and Goodman M (1990) Depsipeptide analogues of elastin repeating sequences: conformational analysis. *Biopolymers* **29:** 1652–1668.

Baker EN and Hubbard RE (1984) Hydrogen bonding in globular proteins. *Prog Biophys Mol Biol* **44:** 97–179.

Bashir MM, Abrams WR, Rosenbloom J, Kucich U, Bacarra M, Han MD, Brown-Augsberger P and Mecham R (1994) Microfibril-associated glycoprotein: characterization of the bovine gene and of the recombinantly expressed protein. *Biochemistry* **33:** 593–600.

Beck DAC, Alonso DOV and Daggett V (2002) A microscopic view of the solvation of peptides and proteins. *Biophys Chem*, in press.

Bertoluzza A, Bonora S, Fini G and Morelli MA (1989) Spectroscopic studies of connective tissues – native and hydrated elastin. *Can J Spectrosc* **34:** 13–14.

Best RB, Li B, Steward A, Daggett V and Clarke J (2001) Can non-mechanical proteins withstand force? Stretching barnase by atomic force microscopy and molecular dynamics simulation. *Biophys J* **81:** 2344–2356.

Bisaccia F, Castiglione-Morelli MA, Spisani S, Ostuni A, Serafini-Fracassini A, Bavoso A and Tamburro AM (1998) The amino acid sequence coded by the rarely expressed exon 26A of human elastin contains a stable beta-turn with chemotactic activity for monocytes. *Biochemistry* **37:** 11,128–11,135.

Brahms S, Brahms J, Spach G and Brack A (1977) Identification of beta, beta-turns and unordered conformations in polypeptide chains by vacuum ultraviolet circular dichroism. *Proc Natl Acad Sci USA* **74:** 3208–3212.

Bressan GM, Argos P and Stanley KK (1987) Repeating structure of chick tropoelastin revealed by complementary DNA cloning. *Biochemistry* **26:** 1497–1503.

Broch H, Moulabbi M, Vasilescu D and Tamburro AM (1996) Conformational and electrostatic properties of V–G–G–V–G, a typical sequence of the glycine-rich regions of elastin. An ab initio quantum molecular study. *Int J Pept Protein Res* **47:** 394–404.

Broch H, Moulabbi M, Vasilescu D and Tamburro AM (1998) Quantum molecular modeling of the elastinic tetrapeptide Val–Pro–Gly–Gly. *J Biomol Struct Dyn* **15:** 1073–1091.

Brown PL, Mecham L, Tisdale C and Mecham RP (1992) The cysteine residues in the carboxy terminal domain of tropoelastin form an intrachain disulfide bond that stabilizes a loop structure and positively charged pocket. *Biochem Biophys Res Commun* **186:** 549–555.

Brown-Augsburger P, Broekelmann T, Mecham L, Mercer R, Gibson MA, Cleary EG, Abrams WR, Rosenbloom J and Mecham RP (1994) Microfibril-associated glycoprotein binds to the carboxyl-terminal domain of tropoelastin and is a substrate for transglutaminase. *J Biol Chem* **269:** 28,443–28,449.

Brown-Augsburger P, Broekelmann T, Rosenbloom J and Mecham RP (1996) Functional domains on elastin and microfibril-associated glycoprotein involved in elastic fibre assembly. *Biochem J* **318 (Pt 1):** 149–155.

Brown-Augsburger P, Tisdale C, Broekelmann T, Sloan C and Mecham RP (1995) Identification of an elastin cross-linking domain that joins three peptide chains. Possible role in nucleated assembly. *J Biol Chem* **270:** 17,778–17,783.

Castiglione Morelli MA, Bisaccia F, Spisani S, De Biasi M, Traniello S and Tamburro AM (1997) Structure-activity relationships for some elastin-derived peptide chemoattractants. *J Pept Res* **49:** 492–499.

Castiglione-Morelli A, Scopa A, Tamburro AM and Guantieri V (1990) Spectroscopic studies on elastin-like synthetic polypeptides. *Int J Biol Macromol* **12:** 363–368.

Chang DK and Urry DW (1989) Polypentapeptide of elastin – damping of internal chain dynamics on extension. *J Comput Chem* **10:** 850–855.

Chang DK, Venkatachalam CM, Prasad KU and Urry DW (1989) Nuclear overhauser effect and computational characterization of the beta-spiral of the polypentapeptide of elastin. *J Biomol Struct Dyn* **6:** 851–858.

Cook WJ, Einspahr H, Trapane TL, Urry DW and Bugg CE (1980) Crystal structure and conformation of the cyclic trimer of a repeat pentapeptide of elastin, cyclo-(L-valyl-L-prolyglycyl-L-valylglycyl)3. *J Am Chem Soc* **102:** 5502–5505.

Daga Gordini D, Guantieri V and Tamburro AM (1989) Electron microscopic evidence for elastin-like supramolecular organization in synthetic polytripeptides. *Connect Tissue Res* **19:** 27–34.

Daggett V and Levitt M (1991) A molecular-dynamics simulation of the C-terminal fragment of the L7/L12 ribosomal-protein in solution. *Chem Phys Lett* **158:** 501–512.

Debelle L and Alix AJ (1999) The structures of elastins and their function. *Biochimie* **81:** 981–994.

Debelle L, Alix AJ, Jacob MP, Huvenne JP, Berjot M, Sombret B and Legrand P (1995) Bovine elastin and kappa-elastin secondary structure determination by optical spectroscopies. *J Biol Chem* **270:** 26,099–26,103.

Dorrington KL and McCrum NG (1977) Elastin as a rubber. *Biopolymers* **16:** 1201–1222.

Ellis GE and Packer KJ (1976) Nuclear spin-relaxation studies of hydrated elastin. *Biopolymers* **15:** 813–832.

Fleming WW, Sullivan CE and Torchia DA (1980) Characterization of molecular motions in 13C-labeled aortic elastin by 13C-1H magnetic double resonance. *Biopolymers* **19:** 597–617.

Flory PJ and Fisk S (1966) Effect of volume exclusion on the dimensions of polymer chains. *J Chem Phys* **44:** 2243–2248.

Foster JA, Bruenger E, Gray WR and Sandberg LB (1973) Isolation and amino acid sequences of tropoelastin peptides. *J Biol Chem* **248:** 2876–2879.

Foster JA, Bruenger E, Rubin L, Imberman M, Kagan H, Mecham R and Franzblau C (1976) Circular dichroism studies of an elastin crosslinked peptide. *Biopolymers* **15:** 833–841.

Frushour BG and Koenig JL (1975) Raman scattering of collagen, gelatin, and elastin. *Biopolymers* **14:** 379–391.

Gibson MA, Hatzinikolas G, Kumaratilake JS, Sandberg LB, Nicholl JK, Sutherland GR and Cleary EG (1996) Further characterization of proteins associated with elastic fiber microfibrils including the molecular cloning of MAGP-2 (MP25). *J Biol Chem* **271:** 1096–1103.

Gibson MA, Kumaratilake JS and Cleary EG (1989) The protein components of the 12-nanometer microfibrils of elastic and nonelastic tissues. *J Biol Chem* **264:** 4590–4598.

Gibson MA, Sandberg LB, Grosso LE and Cleary EG (1991) Complementary DNA cloning establishes microfibril-associated glycoprotein (MAGP) to be a discrete component of the elastin-associated microfibrils. *J Biol Chem* **266:** 7596–7601.

Gosline JM (1978a) Hydrophobic interaction and a model for the elasticity of elastin. *Biopolymers* **17:** 677–695.

Gosline JM (1978b) The temperature-dependent swelling of elastin. *Biopolymers* **17:** 697–707.

Gosline JM and French CJ (1979) Dynamic mechanical properties of elastin. *Biopolymers* **18:** 2091–2103.

Gosline JM, Yew FF and Weis-fogh T (1975) Reversible structural changes in a hydrophobic protein, elastin, as indicated by fluorescence probe analysis. *Biopolymers* **14:** 1811–1826.

Gotte L, Mammi M and Pezzin G (1977) Some structural aspects of elastin revealed by X-ray diffraction and other physical methods. *Adv Exp Med Biol* **79:** 236–245.

Gowda DC, Luan CH, Furner RL, Peng SQ, Jing N, Harris CM, Parker TM and Urry DW (1995) Synthesis and characterization of the human elastin W4 sequence. *Int J Pept Protein Res* **46:** 453–463.

Gray WR, Sandberg LB and Foster JA (1973) Molecular model for elastin structure and function. *Nature* **246:** 461–466.

Guantieri V, Tamburro AM, Cabrol D, Broch H and Vasilescu D (1987) Conformational studies on polypeptide models of collagen. Poly(Gly–Pro–Val), poly(Gly–Pro–Met), poly(Gly–Val–Pro) and poly(Gly–Met–Pro). *Int J Pept Protein Res* **29:** 216–230.

Henze R and Urry DW (1985) Dielectric relaxation studies demonstrate a peptide librational mode in the polypentapeptide of elastin. *J Am Chem Soc* **107:** 2991–2993.

Herzog RW, Singh NK, Urry DW and Daniell H (1997) Expression of a synthetic protein-based polymer (elastomer) gene in *Aspergillus nidulans*. *Appl Microbiol Biotechnol* **47:** 368–372.

Heymann B and Grubmuller H (1999) Elastic properties of poly(ethylene-glycol) studied by molecular dynamic stretching simulation. *Chem Phys Lett* **307:** 425–432.

Hinek A and Rabinovitch M (1993) The ductus arteriosus migratory smooth muscle cell phenotype processes tropoelastin to a 52-kDa product associated with impaired assembly of elastic laminae. *J Biol Chem* **268:** 1405–1413.

Hoeve CA and Flory PJ (1974) The elastic properties of elastin. *Biopolymers* **13:** 677–686.

Hubbard SJ and Thorton JM (1993) NACCESS, Computer Program. Department of Biochemistry and Molecular Biology; University College, London.

Indik Z, Yeh H, Ornstein-Goldstein N, Sheppard P, Anderson N, Rosenbloom JC, Peltonen L and Rosenbloom J (1987a) Alternative splicing of human elastin mRNA indicated by sequence analysis of cloned genomic and complementary DNA. *Proc Natl Acad Sci USA* **84:** 5680–5684.

Indik Z, Yoon K, Morrow SD, Cicila G, Rosenbloom J and Ornstein-Goldstein N (1987b) Structure of the 3′ region of the human elastin gene: great abundance of Alu repetitive sequences and few coding sequences. *Connect Tissue Res* **16:** 197–211.

Kakivaya SR and Hoeve CA (1975) The glass point of elastin. *Proc Natl Acad Sci USA* **72:** 3505–3507.

Laidig KE and Daggett V (1996) Testing the modified hydration-shell hydrogen-bond model of hydrophobic effects using molecular dynamics simulation. *J Phys Chem* **100:** 5616–5619.

Lee B and Richards FM (1971) The interpretation of protein structures: estimation of static accessibility. *J Mol Biol* **55:** 379–400.

Lee J, Macosko CW and Urry DW (2001) Elastomeric polypentapeptides cross-linked into matrices and fibers. *Biomacromolecules* **2:** 170–179.

Lelj F, Tamburro AM, Villani V, Grimaldi P and Guantieri V (1992) Molecular dynamics study of the conformational behavior of a representative elastin building block: Boc–Gly–Val–Gly–Gly–Leu–OMe. *Biopolymers* **32:** 161–172.

Li B, Alonso DOV, Bennion BJ and Daggett V (2001b) Hydrophobic hydration is an important source of elasticity in elastin-based biopolymers. *J Am Chem Soc* **123:** 11,991–11,998.

Li B, Alonso DOV and Daggett V (2001a) The molecular basis for the inverse temperature transition of elastin. *J Mol Biol* **305:** 581–592.

Lillie MA and Gosline JM (1990) The effects of hydration on the dynamic mechanical properties of elastin. *Biopolymers* **29:** 1147–1160.

Lograno MD, Bisaccia F, Ostuni A, Daniele E and Tamburro AM (1998) Identification of elastin peptides with vasorelaxant activity on rat thoracic aorta. *Int J Biochem Cell Biol* **30:** 497–503.

Luan C, Jaggard J, Harris D and Urry DW (1989) On the source of entropic elastomeric force in polypeptides and proteins: backbone configurational vs. side-chain solvational entropy. *Int J Quant Chem: Quant Biol Symp* **16:** 235–244.

Luan CH, Parker TM, Gowda DC and Urry DW (1992) Hydrophobicity of amino acid residues: differential scanning calorimetry and synthesis of the aromatic analogues of the polypentapeptide of elastin. *Biopolymers* **32:** 1251–1261.

Lyerla JR Jr and Torchia DA (1975) Molecular mobility and structure of elastin deduced from the solvent and temperature dependence of 13C magnetic resonance relaxation data. *Biochemistry* **14:** 5175–5183.

Mammi M, Gotte L and Pezzin G (1968) Evidence for order in the structure of alpha-elastin. *Nature* **220:** 371–373.

Mammi M, Gotte L and Pezzin G (1970) Comparison of soluble and native elastin conformations by far-ultraviolet circular dichroism. *Nature* **225:** 380–381.

Martino M, Bavoso A, Saviano M, Di Blasio B and Tamburro AM (1998) Structure and dynamics of elastin building blocks. Boc–LG–OEt, Boc–VGG–OH. *J Biomol Struct Dyn* **15:** 861–875.

McPherson DT, Morrow C, Minehan DS, Wu J, Hunter E and Urry DW (1992) Production and purification of a recombinant elastomeric polypeptide, G-(VPGVG)19-VPGV, from *Escherichia coli. Biotechnol Prog* **8:** 347–352.

Megret C, Guantieri V, Lamure A, Pieraggi MT, Lacabanne C and Tamburro AM (1992) Phase transitions and chain dynamics, in the solid state, of a pentapeptide sequence of elastins. *Int J Biol Macromol* **14:** 45–49.

Megret C, Lamure A, Pieraggi MT, Lacabanne C, Guantieri V and Tamburro AM (1993) Solid-state studies on synthetic fragments and analogues of elastin. *Int J Biol Macromol* **15:** 305–312.

Morelli MA, DeBiasi M, DeStradis A and Tamburro AM (1993) An aggregating elastin-like pentapeptide. *J Biomol Struct Dyn* **11:** 181–190.

Oesterhelt F, Rief M and Gaub HE (1999) Single molecule force spectroscopy by AFM indicates helical structure of poly(ethylene-glycol) in water. *New J Phys* **1:** 6.1–6.11.

Partridge SM (1962) Elastin. *Adv Protein Chem* **17:** 227–302.

Pasquali Ronchetti I, Alessandrini A, Baccarani Contri M, Fornieri C, Mori G, Quaglino D Jr and Valdre U (1998) Study of elastic fiber organization by scanning force microscopy. *Matrix Biol* **17:** 75–83.

Pierce RA, Deak SB, Stolle CA and Boyd CD (1990) Heterogeneity of rat tropoelastin mRNA revealed by cDNA cloning. *Biochemistry* **29:** 9677–9683.

Powell JT (1990) In: Greehalgh RM and Marrick JA (eds) *The cause and Management of Aneurysms* (pp. 89–96) W.B. Saunders, London.

Prescott B, Renugopalakrishnan V and Thomas GJ Jr (1987) Raman spectrum and structure of elastin in relation to type-II beta-turns. *Biopolymers* **26:** 934–936.

Raju K and Anwar RA (1987a) Primary structures of bovine elastin a, b, and c deduced from the sequences of cDNA clones. *J Biol Chem* **262:** 5755–5762.

Raju K and Anwar RA (1987b) Primary structures of bovine elastin a, b, and c deduced from the sequences of cDNA clones. *J Biol Chem* **262:** 5755–5762.

Reiser K, McCormick RJ and Rucker RB (1992) Enzymatic and nonenzymatic cross-linking of collagen and elastin. *FASEB J* **6:** 2439–2449.

Rosenbloom J, Abrams WR and Mecham R (1993) Extracellular matrix 4: the elastic fiber. *Faseb J* **7:** 1208–1218.

Sanchez IC (1979) Phase transition behavior of the isolated polymer chain. *Macromolecules* **12:** 980–988.

Sandberg LB, Leslie JG, Leach CT, Alvarez VL, Torres AR and Smith DW (1985) Elastin covalent structure as determined by solid phase amino acid sequencing. *Pathol Biol (Paris)* **33:** 266–274.

Sciortino F, Palma MU, Urry DW and Prasad KU (1988) Nucleation and accretion of bioelastomeric fibers at biological temperatures and low concentrations. *Biochem Biophys Res Commun* **157:** 1061–1066.

Serafini-Fracassini A, Field JM and Spina M (1976) The macromolecular organization of the elastin fibril. *J Mol Biol* **100:** 73–84.

Tamburro AM (1981) Elastin: molecular and supramolecular structure. *Prog Clin Biol Res* **54:** 45–62.

Tamburro AM and Daga Gordini D (1985) Poly(Pro–Nle–Gly): can an amorphus polypeptide take up a supramolecular elastinlike structure? *Biopolymers* **24:** 1853–1861.

Tamburro AM, De Stradis A and D'Alessio L (1995). Fractal aspects of elastin supramolecular organization. *J Biomol Struct Dyn* **12:** 1161–1172.

Tamburro AM, Guantieri V, Daga-Gordini D and Abatangelo G (1977) Conformational transitions of alpha-elastin. *Biochim Biophys Acta* **492:** 370–376.

Tamburro AM, Guantieri V, Daga-Gordini D and Abatangelo G (1978) Concentration-dependent conformational transition of alpha-elastin in aqueous solution. *J Biol Chem* **253:** 2893–2894.

Tamburro AM, Guantieri V and Gordini DD (1992) Synthesis and structural studies of a pentapeptide sequence of elastin. Poly (Val–Gly–Gly–Leu–Gly). *J Biomol Struct Dyn* **10:** 441–454.

Tamburro AM, Guantieri V, Scopa A and Drabble JM (1991) Polypeptide models of elastin: CD and NMR studies on synthetic poly(X–Gly–Gly). *Chirality* **3:** 318–323.

Teeter MM (1984) Water structure of a hydrophobic protein at atomic resolution: pentagon rings of water molecules in crystals of crambin. *Proc Natl Acad Sci USA* **81:** 6014–6018.

Thomas JM, Elsden DF and Partridge SM (1963) Partial structure of two degradation products from the cross-linkages in elastin. *Nature* **200:** 651–652.

Thomas GJ Jr, Prescott B and Urry DW (1987) Raman amide bands of type-II beta-turns in cyclo-(VPGVG)3 and poly-(VPGVG), and implications for protein secondary-structure analysis. *Biopolymers* **26:** 921–934.

Torchia DA and Piez KA (1973) Mobility of elastin chains as determined by 13C nuclear magnetic resonance. *J Mol Biol* **76:** 419–424.

Trask BC, Trask TM, Broekelmann T and Mecham RP (2000a) The microfibrillar proteins MAGP-1 and fibrillin-1 form a ternary complex with the chondroitin sulfate proteoglycan decorin. *Mol Biol Cell* **11:** 1499–1507.

Trask TM, Trask BC, Ritty TM, Abrams WR, Rosenbloom J and Mecham RP (2000b) Interaction of tropoelastin with the amino-terminal domains of fibrillin-1 and fibrillin-2 suggests a role for the fibrillins in elastic fiber assembly. *J Biol Chem* **275:** 24,400–24,406.

Urry DW (1982) In: Cunningham LW and Frederiksen DW, (eds) *Methods in Enzymology* (pp. 673–716) Academic Press, New york.

Urry DW (1983) What is elastin; what is not. *Ultrastruct Pathol* **4:** 227–251.

Urry DW (1992) Free energy transduction in polypeptides and proteins based on inverse temperature transitions. *Prog Biophys Mol Biol* **57:** 23–57.

Urry DW (1993) Molecular machines – how motion and other functions of living organisms can result from reversible chemical-changes. *Angew Chem-Int Edition English* **32:** 819–841.

Urry DW and Venkatachalam CM (1983) A librational entropy mechanism for elastimers with repeating peptide sequences in helical array. *Int J Quant Chem: Quant Biol Symp* **10:** 81–93.

Urry DW, Chang DK, Krishna NR, Huang DH, Trapane TL and Prasad KU (1989) Two-dimensional proton NMR studies on poly(VPGVG) and its cyclic conformational correlate, cyclo(VPGVG)3. *Biopolymers* **28:** 819–833.

Urry DW, Chang DK, Zhang H and Prasad KU (1988) pK shift of functional group in mechanochemical coupling due to hydrophobic effect: evidence for an apolar–polar repulsion free energy in water. *Biochem Biophys Res Commun* **153:** 832–839.

Urry DW, Harris RD, Long MM and Prasad KU (1986a) Polytetrapeptide of elastin. Temperature-correlated elastomeric force and structure development. *Int J Pept Protein Res* **28:** 649–660.

Urry DW, Haynes B and Harris RD (1986b) Temperature dependence of length of elastin and its polypentapeptide. *Biochem Biophys Res Commun* **141:** 749–755.

Urry DW, Luan CH and Peng SQ (1995) Molecular biophysics of elastin structure, function and pathology. *Ciba Found Symp* **192:** 4–22.

Urry DW, Mitchell LW, Ohnishi T and Long MM (1975) Proton and carbon magnetic resonance studies of the synthetic polypentapeptide of elastin. *J Mol Biol* **96:** 101–117.

Urry DW, Peng S, Xu J and McPherson DT (1997) Characterization of waters of hydrophobic hydration by microwave dielectric relaxation. *J Am Chem Soc* **119:** 1161–1162.

Urry DW, Peng SQ and Parker TM (1992) Hydrophobicity-induced pK shifts in elastin protein-based polymers. *Biopolymers* **32:** 373–379.

Urry DW, Shaw RG and Prasad KU (1985a) Polypentapeptide of elastin: temperature dependence of ellipticity and correlation with elastomeric force. *Biochem Biophys Res Commun* **130:** 50–57.

Urry DW, Starcher B and Partridge SM (1969) Coacervation of solubilized elastin effects a notable conformational change. *Nature* **222:** 795–796.

Urry DW, Trapane TL, Iqbal M, Venkatachalam CM and Prasad KU (1985b) Carbon-13 NMR relaxation studies demonstrate an inverse temperature transition in the elastin polypentapeptide. *Biochemistry* **24:** 5182–5189.

Urry DW, Trapane TL, McMichens RB, Iqbal M, Harris RD and Prasad KU (1986c) Nitrogen-15 NMR relaxation study of inverse temperature transitions in elastin polypentapeptide and its cross-linked elastomer. *Biopolymers* **25:** S209–S228.

Urry DW, Trapane TL, Sugano H and Prasad KU (1981) Sequential polypeptides of elastin: cyclic conformational correlates of the linear polypentapeptide. *J Am Chem Soc* **103:** 2080–2089.

Venkatachalam CM and Urry DW (1981) Development of a linear helical conformation from its cyclic correlate. B-spiral model of the elastin poly(pentapeptide) $(VPGVG)_n$. *Macromolecules* **14:** 1225–1229.

Venkatachalam CM, Khaled MA, Sugano KH and Urry DW (1981) Nuclear magnetic resonance and conformational energy calculations of repeat peptides of elastin. Conformational characterization of cyclopentadecapeptide cyclo-(L-Val-L-Pro–Gly–L-Val–Gly)3. *J Am Chem Soc* **103:** 2372–2379.

Villani V and Tamburro AM (1995) Conformational modeling of elastin tetrapeptide Boc–Gly–Leu–Gly–Gly–NMe by molecular dynamics simulations with improvements to the thermalization procedure. *J Biomol Struct Dyn* **12:** 1173–1202.

Villani V and Tamburro AM (1999) Conformational chaos of an elastin-related peptide in aqueous solution. *Ann NY Acad Sci* **879:** 284–287.

Vrhovski B and Weiss AS (1998) Biochemistry of tropoelastin. *Eur J Biochem* **258:** 1–18.

Vrhovski B, Jensen S and Weiss AS (1997) Coacervation characteristics of recombinant human tropoelastin. *Eur J Biochem* **250:** 92–98.

Wasserman ZR and Salemme FR (1990) A molecular dynamics investigation of the elastomeric restoring force in elastin. *Biopolymers* **29:** 1613–1631.

Weis-Fogh T and Anderson SO (1970) New molecular model for the long-range elasticity of elastin. *Nature* **227:** 718–721.

Wong KB and Daggett V (1998) Barstar has a highly dynamic hydrophobic core: evidence from molecular dynamics simulations and nuclear magnetic resonance relaxation data. *Biochemistry* **37:** 11,182–11,192.

Yeh H, Anderson N, Ornstein-Goldstein N, Bashir MM, Rosenbloom JC, Abrams W, Indik Z, Yoon K, Parks W, Mecham R *et al.* (1989) Structure of the bovine elastin gene and S1 nuclease analysis of alternative splicing of elastin mRNA in the bovine nuchal ligament. *Biochemistry* **28:** 2365–2370.

Zhang H, Apfelroth SD, Hu W, Davis EC, Sanguineti C, Bonadio J, Mecham RP and Ramirez F (1994) Structure and expression of fibrillin-2, a novel microfibrillar component preferentially located in elastic matrices. *J Cell Biol* **124:** 855–863.

Journal of Muscle Research and Cell Motility **23**: 575–580, 2002.

Stretching fibronectin

HAROLD P. ERICKSON*
Department of Cell Biology, Box 3079, Duke University Medical Center, Durham, NC 27710, USA

Abstract

Fibronectin (FN) matrix fibrils assembled in cell culture have been observed to stretch in response to cell movements, and when broken relax to 1/3 to 1/4 of their rest length. Two molecular mechanisms have been proposed for the elasticity. One proposes that FN molecules in relaxed fibers are bent and looped into a compact conformation, and stretching pulls the molecules into the extended conformation but domains remain folded. The second proposes that molecules in fibrils are already extended, and stretching is produced by force-induced unfolding of FN type III domains. Experimental observations that may help distinguish these two possibilities are discussed.

Introduction

The definitive extracellular matrix in adult tissues is primarily made of collagen type I fibrils, but this is preceded in embryonic tissue and wound healing by a primitive fibronectin (FN) matrix. Since these developing tissues undergo movements it is reasonable to ask whether the matrix might be pliable and stretchable. This question has only recently been addressed, with experimental evidence showing that the FN matrix is actually very elastic.

Many fibroblasts and other cells secrete FN and assemble it into a matrix in tissue culture, and in the absence of ascorbate (the usual culture condition) FN is the primary constituent of the matrix. These matrices have been studied for many years, primarily by immunofluorescence. In the usual approach the entire culture is fixed before staining with antibodies, so one obtains only a single static image of the matrix. More recently the use of green flourescent protein (gfp) technology has permitted continuous observation of the matrix in living cultures, providing experimental evidence for elasticity.

In an important earlier study of matrix mechanics, Halliday and Tomasek (1995) found that fibroblasts cultured in three-dimensional collagen gels could only assemble a FN matrix when they could generate tension. A recent study extending these results found that there is a cryptic assembly site in or near the first FN-III domain that is exposed in FN matrix fibrils (Zhong *et al.*, 1998). This site is recognized by antibody L8, which stains the FN fibrils in a normal culture. However, in cultures treated with the rho inhibitor C3, which reduces their actin stress fibers and contractility, staining with L8 is substantially lost. The connection between stretching FN and L8 binding was confirmed by showing that FN molecules deposited on an elastic membrane showed a significant increase in binding L8 when the membrane was stretched.

The above two studies suggest that FN matrix fibrils are under tension in normal tissue culture. A simple observation from numerous studies also supports this conclusion. FN matrix fibrils are mostly straight. FN molecules themselves are quite flexible, and the thin matrix fibrils should show bends and kinks if they were relaxed. Since most matrix fibrils are straight this implies that they are under tension.

Characteristic of FN matrix fibrils

FN matrix fibrils are elastic and stretched. Recent observations show that the tension on FN matrix fibrils does more than just straighten them out – it stretches them substantially. These observation were made using FN-gfp (Ohashi *et al.*, 1999), in which a gfp module was inserted between FN-III domains 3 and 4. The matrix assembled by cells transfected with the FN-gfp can be observed multiple times before the gfp bleaches. Figure 1 shows a time course of a culture at 1 h intervals. These cells were transfected with FN-yfp (which appears green in the figure) and also with moesin-cfp (red in the figure), which outlines the cells by staining actin (Ohashi *et al.*, 2002). One cell in the right of the field is almost invisible, apparently expressing only a very low level of moesin-cfp (its nucleus is visible as a dark circle at 1:00 and 2:00, middle right). We will focus our attention on two very prominent FN fibrils, indicated 1 and 2 on the figure. Note that these fibrils are slightly thicker than the finest fibrils visible, meaning that they are slightly thicker than the 200 nm resolution of the microscope. Since extended FN molecules are about 3 nm in diameter, the fibrils are probably several hundred FN molecules thick (although this has never been measured,

* E-mail: h.erickson@cellbio.duke.edu

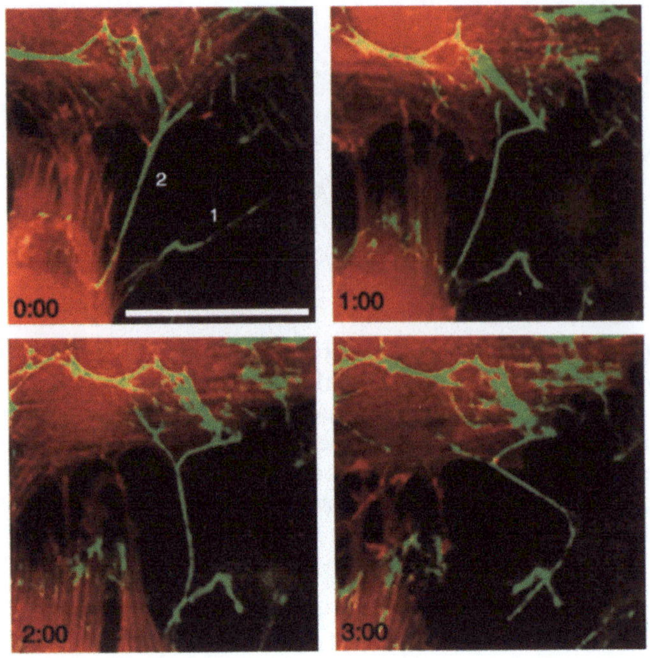

Fig. 1. Movement, stretching and retraction of FN matrix fibrils imaged in real time. FN-yfp is shown as green, and a moesin-cfp construct (red) is used to image the actin cytoskeleton of the cells. A culture of 3T3 cells is shown over a 3-h period, after 72 h of growth. The events are described in the text (reprinted from (Ohashi *et al.*, 2002)).

and we do not know how much water is present between molecules).

Fibril 1 appears to break its attachment to a cell at the upper right and contracts between 0:00 and 1 h. The contraction probably occurred over a time of ~5 min, as we observed in an earlier study (Ohashi *et al.*, 1999). The segment to the right of the bend contracted to about 1/3 of its length at 0:00, similar to the 1/4 contraction measured previously. Note also the increased brightness of the contracted fibril, consistent with a uniform contraction along the entire length. Fibril breakage of this type is not seen very often, but in our several recorded observations the contraction was always to 1/3 to 1/4 of the stretched length.

Fibril 2 does not break, but shows substantial movements and stretching. At 0:00 it seems to be attached to a cell at the top and to another cell at the bottom. At 1:00 the upper cell has moved and the fibril shows a bifurcation that was obscured at 0:00. The cell at the bottom has moved significantly but the fibril has not, suggesting this lower attachment is to the substrate; this is confirmed when the cell moves away entirely in the next two frames. This fibril also seems to be attached to the invisible cell, as a bend is initiated at 2:00. At 3:00 the fibril is pulled strongly to the right, presumably as the invisible cell moves in that direction. As it is pulled the fibril is stretched, its contour length increasing 22% from 1:00 to 2:00, and a total of 65% from 1:00 to 3:00.

Because natural ruptures of FN fibrils are rare, we attempted to create detachments by treating the culture with cytochalasin B to disrupt the actin cytoskeleton (Ohashi *et al.*, 2002). In that study we were looking primarily at newly formed fibrils, aligned on actin stress fibrils. We found that most fibrils showed no change, and concluded that these were attached to the substrate. However, those that did move showed contractions similar to broken fibrils. We conclude that FN fibrils are probably stretched three to four times their rest length even as they are being assembled on the cell surface.

Molecular models for fibril elasticity

A major problem in trying to propose a molecular mechanism for fibril stretching is that we have almost no knowledge about how FN molecules are connected to make up a fibril. There must be at least two types of bonds between molecules. They must be connected longitudinally to make strands that run the length of the fibril, and they must be connected laterally to give the fibrils thickness. For the present discussion I will assume that the longitudinal connections involve bonds, perhaps disulfide bonds, between the N-terminal FN-I domains of adjacent molecules. I will ignore the lateral connections, except to point out that they must exist, and this probably means that laterally connected molecules need to stretch in coordination.

The first model for fibril elasticity is based on the conformational change that the FN molecule undergoes in different solution conditions. In high ionic strength or high pH FN molecules assume an extended conformation, whereas at physiological ionic strength and below the strand-like molecule is bent into a compact conformation (Erickson and Carrell, 1983; Rocco *et al.*, 1983).

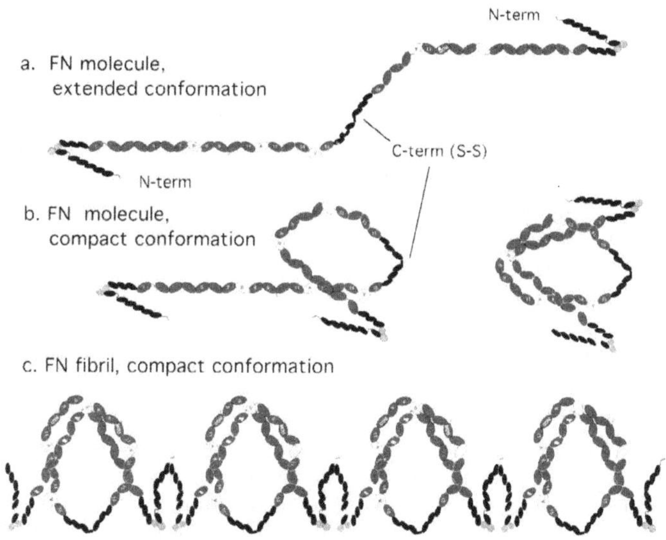

Fig. 2. Model for fibril stretching based on conformational change of the FN molecule. The molecule on top is in the extended conformation. The middle row shows steps of bending the molecule into the compact conformation. The molecule on the left has the FN2-3 domains of one subunit forming an electrostatic bond to domains 12–14 of the other. The molecule on the right has added the same contacts with the other pair of subunits, giving the fully folded compact conformation. The bottom row shows a string of FN molecules bonded to each other at their N-terminal segments, as they might be in a FN fibril (additional lateral contacts would be needed to grow the fibril thicker). In this model stretching the fibril would involve breaking the electrostatic bonds and pulling the FN molecules to the fully extended conformation. Note that the end-to-end length of the extended molecule is about four times that of compact ones.

Studies of proteolytic and recombinant protein fragments indicated that the compact conformation is stabilized by an electrostatic contact between FN-III domains 2–3 of one subunit of the dimer, and domains 12–14 of the other subunit (Johnson *et al.*, 1999).

Figure 2a shows a model of an FN molecule in the extended conformation, and Figure 2b illustrates the formation of the compact conformation by bending first one strand and then the other to make the electrostatic contacts. Figure 2c illustrates how FN molecules in the compact conformation might be linked to make one strand of a fibril. In this model the strand with molecules in the compact conformation would be in the contracted state. Stretching the fibril would involve breaking the electrostatic bonds and pulling some or all molecules into the extended conformation. Importantly, the ends of the extended molecules are 3.5 times farther apart than they are in the compact conformation, in good agreement with the 3–4-fold contraction observed for broken FN fibrils *in vivo*.

The second model assumes that the FN molecules are already fully extended as they are incorporated into the FN fibril, and that stretching involves force-induced unfolding of FN-III domains. The concept that force could unfold FN-III and Ig domains was first proposed as a theoretical model (Erickson, 1994), and has recently been demonstrated experimentally by stretching single molecules of titin (Rief *et al.*, 1997), tenascin (Oberhauser *et al.*, 1998), and FN (Oberhauser *et al.*, 2002) with the atomic force microscope (AFM). The mechanism is explained in detail elsewhere in this volume, but the basic principle is illustrated in Figure 3. Tension on the FN molecule will be supported by the terminal beta

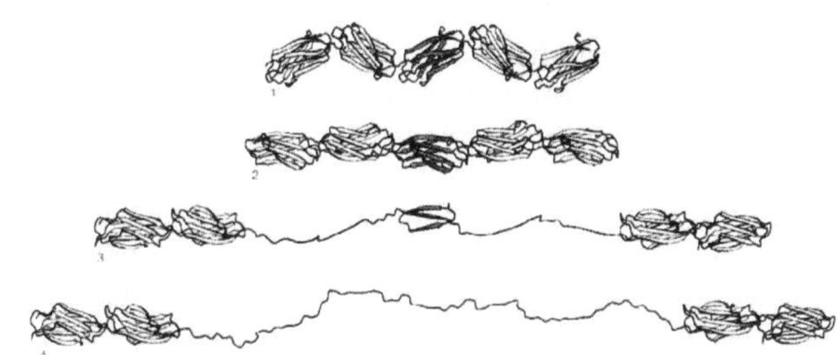

Fig. 3. Model for stretching based on force-induced unfolding of FN-III domains. As tension is applied to the molecule the first effect is to straighten the zigzag between FN-III domains, then tension is applied to the beta strands in each domain. If the force is sufficiently high, one pair of beta strands will be pulled out, and the domain will then be destabilized and completely unfold and unravel (reprinted from (Erickson, 1994)).

strands of each module, and eventually one of them will pull loose. When this happens the rest of the domain is destabilized and it rapidly unfolds, giving an extended polypeptide that is seven times as long as the folded module. The process is reversible, but with a large hysteresis, meaning that refolding only occurs at very low forces.

Forces required for stretching

The forces involved in domain unfolding, the second model, have been explored in considerable detail by stretching molecules with the AFM. The characteristic sawtooth wave, discussed elsewhere in this volume, is generated because one needs a high force to break a few hydrogen bonds and pull the first pair of beta strands a short distance, 2–3 Å. As soon as this is accomplished the domain is greatly weakened and continues unraveling to 300 Å, even as the force drops precipitously. In a simplified picture one can say that the entire free energy of folding the domain, 26 kcal/mol for the titin Ig domain I27 (Li *et al.*, 2000), is dissipated in the first 2.5 Å, after which the domain is destabilized and extends 300 Å with only small force (Lu *et al.*, 1998). The process is reversible but with an enormous hysteresis. At the pulling speeds used in the AFM experiments (~50–5000 nm/s), forces of 100–300 pN are required to initiate the unraveling, but the force needs to be reduced to a few pN to permit the domain to refold. Thus the refolding cannot generate a large contractile force, but it will occur, and will reel in excess polypeptide, when the contractile force has dropped to the range of 1–5 pN.

The forces involved in the first model, disrupting the electrostatic bonds and pulling the molecules from the compact to extended conformation, are more speculative. We can make a good guess by noting that the free energy of the electrostatic bond is probably on the order of 5–10 kcal/mol, and the bond will be dissipated over 5–10 Å, the Debye length in physiological ionic strength. The 2.5–5-fold lower bond energy and the 2–4-fold longer distance suggest that pulling apart the compact conformation would occur at forces 10-fold lower than those required to unfold FN-III domains.

Since forces of ~200 pN force are needed to generate the sawtooth wave in the AFM, one must wonder if domain unraveling could have any significance in biological FN fibrils. An important consideration is that the force needed to unravel a domain depends exponentially on the pulling rate. For a 7-domain segment of tenascin, which goes from 25 to 200 nm when all domains are unfolded, the unfolding force varied from 100 pN at a pulling rate of 10 nm/s, to 180 pN as the pulling rate increased to 5000 nm/s (Oberhauser *et al.*, 1998). The pulling rate on an FN fibril in cell culture would probably be several orders of magnitude slower (it is important here to normalize the extension of the whole fibril to that of a comparable 7-domain segment). At very low pulling rates another phenomenon will dominate the argument. Protein domains, including Ig and FN-III, unfold in the complete absence of force with a time constant of 4×10^{-4} s^{-1} (Clarke *et al.*, 1997; Carrion-Vazquez *et al.*, 1999), or even as fast as 2×10^{-2} s^{-1} (Oberhauser *et al.*, 2002). This means that every domain in FN will spontaneously unfold about once an hour, or at the faster rates once a minute. If there is even a small force, as soon as a domain unfolds it should be rapidly and fully extended. In the absence of force the unfolded domain will rapidly (on the order of 1 s) refold. To estimate the fraction of FN-III domains that are unfolded, the rate of refolding is most important. The question then becomes, what level of force will substantially inhibit refolding?

Carrion-Vazquez *et al.* (1999) studied the refolding of I27 domains in a polyprotein. They first stretched the protein to unravel all five domains, then retracted the AFM tip to a fraction of the extended length. At 30% of the fully stretched length, 30% of the domains refolded in 5 s. Even this relatively relaxed polypeptide will exert a force due to the Brownian spring effect. I have used the formulas and parameters provided in the discussion to their Figure 5 to calculate the forces at different levels of extension, and relate them to refolding rates. A force of 1 pN slows the refolding rate to about half the 1.2 s^{-1} in the absence of force. A force of 5 pN will slow folding to 5%, and 10 pN to 0.3% of the rate in the absence of force. We can therefore predict that at forces greater than 5–10 pN there will be significant domain unfolding, while at 1–5 pN and below domains should refold and be mostly intact.

Similar arguments would suggest that forces 10-fold lower would govern the transition from the compact to extended conformation. If we knew the force per FN molecule in the stretched FN fibrils, we could make a good guess as to which mechanism is involved – the conformational change or domain unfolding. Unfortunately we have no experimental information on the magnitude of the force.

Mechanical hierarchy of FN domains

When the 15 FN-III domains of tenascin were stretched by AFM (Oberhauser *et al.*, 1998), the peaks of the sawtooth wave were all about the same height, 137 ± 12 pN, indicating that they all had about the same mechanical stability. A recent study stretching FN showed a striking contrast (Oberhauser *et al.*, 2002). The force required to unfold domains started at 80 pN and rose steadily to 220 pN. This means that the FN-III domains of FN have a wide range of mechanical stability. This raises the question of how the weak and strong domains are arranged in the molecule. Are the weak domains clustered at one end and the strong ones at the other? Are the important domains, like FN-III-10, mechanically stronger? Oberhauser *et al.* addressed this question by making polyproteins and determining the rupture force for several specific domains.

Surprisingly, the cell adhesion domain FN-III-10 is the weakest domain tested, rupturing at 70 pN. One might have expected the cell adhesion function to be preserved by an especially stable domain. Also, this domain is known to be among the most stable thermodynamically, denaturing only at 100 °C (Litvinovich and Ingham, 1995; Ingham and Litvinovich, 1996). A second surprise is that FN-III-1, which contains the proposed cryptic assembly site, is the strongest domain, rupturing at 220 pN. This contradicts the simplest possibility that the cryptic site is exposed by mechanically unfolding the FN-III-1 domain.

The hierarchy of unfolding forces was already predicted by an elegant computer simulation study (Craig *et al.*, 2001). This study compared FN-III domains 7–10, whose atomic structure is known (Leahy *et al.*, 1996), and concluded that domain 10 was substantially weaker than the others, and domain 7 the strongest. This study, and previous computer simulations referenced therein, provide powerful insights into the molecular pathway of domain unfolding. Preceding the rupture of hydrogen bonds between beta strands, the simulations predicted two states, 'twisted' and 'aligned', corresponding to shifts in the alignment of the two sheets of beta strands. Another prediction of the molecular dynamics simulation is particularly interesting for biological mechanisms. One of the first effects of force in the computer model is that the RGD loop in FN-III-10 is pulled in (Krammer *et al.*, 1999, 2002). This would essentially eliminate the integrin-binding activity. This intriguing prediction still awaits an experimental test.

FRET analysis – a reevaluation

Two recent studies have used fluorescence resonance energy transfer (FRET) to investigate the conformation of FN molecules in FN fibrils (Baneyx *et al.*, 2001, 2002). Although the authors cautiously interpret their measurements as favoring at least some domain unraveling, I will argue here that the data may just as well support the conformational-change model.

The FRET signal is generated when an acceptor fluorophore is within the Förster distance of a donor fluorophore. The Förster distance for a FITC-TMR donor–acceptor pair is 49–54 Å (Wu and Brand, 1994), and it should be very similar for the Oregon green–TMR pair used to label FN. The efficiency of transfer drops off as the sixth power of the distance, so FRET will be maximal for distances below 40 Å, and will fall below 33% for distances greater than 60 Å. To generate a FRET signal the authors attached acceptor TMR to the free cysteines in FN-III domains 7 and 15 (achieving a labeling close to 100%, two acceptors per subunit), and attached donor Oregon green randomly to amines (3.5 fluorophores per subunit). This seems a good strategy for probing the change from the compact to extended conformation, but it will face a problem for probing domain unfolding. The problem is that the FN subunit is 800 Å long (Erickson *et al.*, 1981), so the randomly attached donors will be more than 200 Å apart on average. One can estimate that the labeled FN subunit will have on average one donor within 60 Å of one of the two acceptors. This pair will produce a FRET signal that should be sensitive to domain unfolding. The other acceptor and the 2.5 remaining donors would not contribute to a FRET signal when the FN molecules are extended, and can only contribute noise for probing domain unraveling.

If the FN molecules in fibrils are bent into the compact conformation, additional FRET pairs could be generated by bringing additional donors near the acceptors. This FRET signal should be sensitive to the conformational change, since opening the molecule to the extended conformation will separate the donors and acceptors. Thus this strategy of random labeling may produce molecules that report conformational change with greater sensitivity than they report domain unfolding.

Experimental analysis of the labeled FN in solution showed a pronounced FRET signal that I would argue can be attributed largely to the compact conformation. The FRET signal (ratio of acceptor to donor peak) was 0.85 in physiological salt, dropped to 0.57 in 2 M GuHCl, and to 0.4 in 8 M GuHCl. The 2 M GuHCl is thought to disrupt the electrostatic bonds of the compact conformation but not denature domains, while 8 M GuHCl should completely denature all domains (Khan *et al.*, 1990). Thus 70% of the FRET signal can be attributed to the compact conformation. The remaining 30% may be due to one donor that is randomly within 60 Å of one acceptor, and may be reporting domain unfolding. However, it is also likely that in 2 M GuHCl the FN molecules are relaxed but not fully extended. If the molecules are flexible FRET pairs similar to those in the compact conformation may be formed by random intramolecular contacts. These would be disrupted if the molecule were mechanically pulled into the fully extended conformation. Thus some part of the remaining 30% FRET signal may be due to contacts of the compact conformation.

When the labeled FN was added to cultured cells, some of the FN was diffusely deposited on the cell surface and some was assembled into matrix fibrils. The diffuse FN gave the full FRET signal, consistent with it being in the compact conformation. The matrix fibrils showed a variable reduction in FRET. The average FRET signal from fibrils was 0.54, slightly less than that in 2 M GuHCl (Baneyx *et al.*, 2002). The standard deviation for different fibrils was ±0.1, but the 11 fibrils imaged covered the full range from 0.85 to 0.4. The authors suggest that the fibrils with the lowest FRET probably have some domains unfolded.

Of 11 fibrils reported, five had FRET signals between 0 and 2 M GuHCl, suggesting that they are partially in the compact conformation and partially extended. Six fibrils had FRET signals corresponding to denaturation in >2 M GuHCl. This may indicate domain unfolding, but as argued above these signals could also reflect the molecules being stretched to a fully extended conformation, i.e., more extended than in 2 M GuHCl solution.

Summary

The two models for FN fibril stretching are not yet resolved. As argued above, it appears that much of the loss of FRET signal in FN fibrils is likely due to the conformational change model. This is the first indication that FN molecules in fibrils can exist in the compact conformation. Thus the FRET data actually suggest that the compact to extended conformational change occurs in FN fibrils, and likely accounts for at least some of the stretching. As pointed out in Figure 2 it could account fully for the 3–4-fold extension observed for broken fibrils. Although the present randomly labeled molecules may not be able to report it efficiently, domain unfolding could still occur and contribute to stretching. Perhaps the most important experimental measure would be to determine the force per FN molecule. If it is above 10 pN we would expect domain unfolding to contribute to stretching, but if it is below 1–5 pN the stretching of fibrils may be due entirely to the conformational change.

Although the FRET data were initially interpreted as indicating domain unfolding, the reinterpretation given above favors the conformational change. Clearly more data are needed to resolve the question. For the present both models need to be kept in consideration.

References

Baneyx G, Baugh L and Vogel V (2001) Coexisting conformations of fibronectin in cell culture imaged using fluorescence resonance energy transfer. *Proc Natl Acad Sci USA* **20:** 20.

Baneyx G, Baugh L and Vogel V (2002) Fibronectin extension and unfolding within cell matrix fibrils controlled by cytoskeletal tension. *Proc Natl Acad Sci USA* **99:** 5139–5143.

Carrion-Vazquez M, Oberhauser AF, Fowler SB, Marszalek PE, Broedel SE, Clarke J and Fernandez JM (1999) Mechanical and chemical unfolding of a single protein: a comparison. *Proc Natl Acad Sci USA* **96:** 3694–3699.

Clarke J, Hamill SJ and Johnson CM (1997) Folding and stability of a fibronectin type III domain of human tenascin. *J Mol Biol* **270:** 771–778.

Craig D, Krammer A, Schulten K and Vogel V (2001) Comparison of the early stages of forced unfolding for fibronectin type III modules. *Proc Natl Acad Sci USA* **98:** 5590–5595.

Erickson HP (1994) Reversible unfolding of FN3 domains provides the structural basis for stretch and elasticity of titin and fibronectin. *Proc Natl Acad Sci USA* **91:** 10,114–10,118.

Erickson HP and Carrell NA (1983) Fibronectin in extended and compact conformations. Electron microscopy and sedimentation analysis. *J Biol Chem* **258:** 14,539–14,544.

Erickson HP, Carrell NA and McDonagh J (1981) Fibronectin molecule visualized in electron microscopy: a long, thin, flexible strand. *J Cell Biol* **91:** 673–678.

Halliday NL and Tomasek JJ (1995) Mechanical properties of the extracellular matrix influence fibronectin fibril assembly *in vitro*. *Exp Cell Res* **217:** 109–117.

Ingham KC and Litvinovich SV (1996) Structure and function of fibronectin modules – Comment. *Matrix Biol* **15:** 321.

Johnson KJ, Sage H, Briscoe G and Erickson HP (1999) The compact conformation of fibronectin is determined by intramolecular ionic interactions. *J Biol Chem* **274:** 15,473–15,479.

Khan MY, Medow MS and Newman SA (1990) Unfolding transitions of fibronectin and its domains. Stabilization and structural alteration of the N-terminal domain by heparin. *Biochem J* **270:** 33–38.

Krammer A, Craig D, Thomas WE, Schulten K and Vogel V (2002) A structural model for force regulated integrin binding to fibronectin's RGD-synergy site. *Matrix Biol* **21:** 139–147.

Krammer A, Lu H, Isralewitz B, Schulten K and Vogel V (1999) Forced unfolding of the fibronectin type III module reveals a tensile molecular recognition switch. *Proc Natl Acad Sci USA* **96:** 1351–1356.

Leahy DJ, Aukhil I and Erickson HP (1996) 2.0 Å crystal structure of a four-domain segment of human fibronectin encompassing the RGD loop and synergy region. *Cell* **84:** 155–164.

Li H, Carrion-Vazquez M, Oberhauser AF, Marszalek PE and Fernandez JM (2000) Point mutations alter the mechanical stability of immunoglobulin modules. *Nat Struct Biol* **7:** 1117–1120.

Litvinovich SV and Ingham KC (1995) Interactions between type III domains in the 110 kDa cell-binding fragment of fibronectin. *J Mol Biol* **248:** 611–626.

Lu H, Isralewitz B, Krammer A, Vogel V and Schulten K (1998) Unfolding of titin immunoglobulin domains by steered molecular dynamics simulation. *Biophys J* **75:** 662–671.

Oberhauser AF, Badilla-Fernandez C, Carrion-Vazquez M and Fernandez J (2002) The mechanical heirarchies of fibronectin observed with single molecule AFM. *J Mol Biol* **319:** 433–447.

Oberhauser AF, Marszalek PE, Erickson HP and Fernandez JM (1998) The molecular elasticity of the extracellular matrix protein tenascin. *Nature* **393:** 181–185.

Ohashi T, Kiehart DP and Erickson HP (1999) Dynamics and elasticity of the fibronectin matrix in living cell culture visualized by fibronectin-green fluorescent protein. *Proc Natl Acad Sci USA* **96:** 2153–2158.

Ohashi T, Kiehart DP and Erickson HP (2002) Dual labeling of the fibronectin matrix and actin cytoskeleton with green fluorescent protein variants. *J Cell Sci* **115:** 1221–1229.

Rief M, Gautel M, Oesterhelt F, Fernandez JM and Gaub HE (1997) Reversible unfolding of individual titin immunoglobulin domains by AFM. *Science* **276:** 1109–1112.

Rocco M, Carson M, Hantgan R, McDonagh J and Hermans J (1983) Dependence of the shape of the plasma fibronectin molecule on solvent composition. *J Biol Chem* **258:** 14,545–14,549.

Wu P and Brand L (1994) Resonance energy transfer: methods and applications. *Anal Biochem* **218:** 1–13.

Zhong CL, Chrzanowskawodnicka M, Brown J, Shaub A, Belkin AM and Burridge K (1998) Rho-mediated contractility exposes a cryptic site in fibronectin and induces fibronectin matrix assembly. *J Cell Biol* **141:** 539–551.

Journal of Muscle Research and Cell Motility **23**: 581–596, 2002.

Fibrillin-rich microfibrils: elastic biopolymers of the extracellular matrix

C. M. KIELTY[1,*], T. J. WESS[3], L. HASTON[3], JANE L. ASHWORTH[2], M. J. SHERRATT[2] and C. A. SHUTTLEWORTH[2]
[1]*School of Medicine;* [2]*School of Biological Sciences, University of Manchester, 2.205 Stopford Building, Oxford Road, Manchester M13 9PT, UK;* [3]*School of Biological Sciences, University of Stirling FK9 4LA, UK*

Abstract

Fibrillin-rich microfibrils are evolutionarily ancient macromolecular assemblies of the extracellular matrix. They have unique extensible properties that endow vascular and other tissues with long-range elasticity. Microfibril extensibility supports the low pressure closed circulations of lower organisms such as crustaceans. In higher vertebrates, microfibrils act as a template for elastin deposition and are components of mature elastic fibres. In man, the importance of microfibrils is highlighted by the linkage of mutations in their principal structural component, fibrillin-1, to the heritable disease Marfan syndrome which is characterised by severe cardiovascular, skeletal and ocular defects. When isolated from tissues, fibrillin-rich microfibrils have a complex ultrastructural organisation with a characteristic 'beads-on-a-strong' appearance. X-ray fibre diffraction studies and biomechanical testing have shown that microfibrils are reversibly extensible at tissue extensions of 100%. Ultrastructural analysis and 3D reconstructions of isolated microfibrils using automated electron tomography have revealed new details of how fibrillin molecules are aligned within microfibrils in untensioned and extended states, and delineated the role of calcium in regulating microfibril beaded periodicity, rest length and molecular organisation. The molecular basis of how fibrillin molecules assemble into microfibrils, the central role of cells in regulating this process, and the identity of other molecules that may coassemble into microfibrils are now being elucidated. This information will enhance our understanding of the elastic mechanism of these unique extracellular matrix polymers, and may lead to new microfibril-based strategies for repairing elastic tissues in ageing and disease.

Introduction

Tissue flexibility and elasticity have been essential requirements in the evolution of multicellular organisms. These properties allow transport of fluids, nutrients and oxygen, underpin movement, and endow an ability to respond to changing environment. Tissue elasticity is based on fibrillin-rich microfibrils and elastic fibres, insoluble extracellular matrix assemblies that permit long-range deformability and recoil. These properties are critical to the function of dynamic tissues such as blood vessels, lungs, skin and eyes. The biology of microfibrils has proved difficult to unravel because they are multi-component polymers with a hierarchical assembly and a complex molecular organisation that defines their unique elastomeric properties.

There is currently intense interest in the biology of the unique fibrillin-rich microfibrils of the ECM. This interest is driven by the need to understand the basis of tissue elasticity and genotype-to-phenotype correlations in heritable connective tissue diseases associated with mutations in fibrillin genes, and the possibility of exploiting their biomechanical properties in tissue engineering and wound repair strategies. Here we outline current knowledge of microfibril organisation, composition and elasticity, a molecular folding model of the elastic 'motor', and implications of loss of microfibril integrity in dynamic connective tissues in ageing and disease.

* To whom correspondence should be addressed: Tel.: +44-161-275-5739; Fax: +44-161-2755082; E-mail: cay.kielty@man.ac.uk

Fibrillin-rich microfibrils

Isolated fibrillin-rich microfibrils have a unique and complex repeating beaded structure with an untensioned periodicity of ~56 nm (Sherratt *et al.*, 2001) (Figure 1A). Early indications that they have unique elastic properties were obtained from electron microscopy images that showed both untensioned and stretched regions (Keene *et al.*, 1991) (Figure 1B). Since then, studies of isolated microfibrils and microfibril bundles, and of hydrated tissues such as the ciliary zonules of the eye and crustacean aorta that are based on microfibril bundles, have confirmed and further delineated the elastic properties of microfibrils.

Fibrillins, the principal structural molecules, are encoded by three genes (Pereira *et al.*, 1993; Zhang *et al.*, 1994; Nagase *et al.*, 2001). They are large glycoproteins (~350 kDa) with calcium-binding epidermal growth factor-like (cbEGF) domain arrays that, in the presence of Ca^{2+}, adopt a rod-like arrangement (Downing *et al.*, 1996) (Figure 2A). These domains are interspersed with 8-cysteine motifs (also called 'TB' modules because of their homology to TGFβ binding modules in latent TGF-β binding proteins). Towards the fibrillin-1

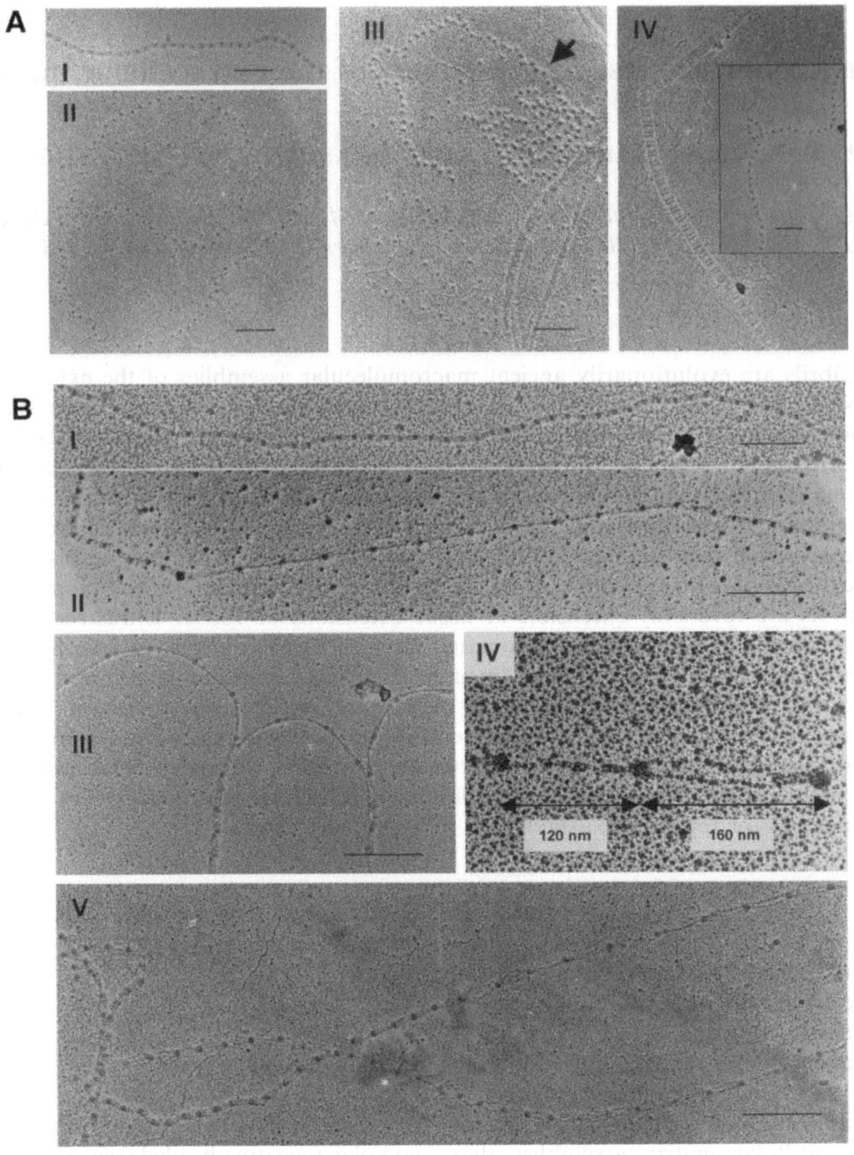

Fig. 1. Rotary shadowing electron micrographs of fibrillin-rich microfibrils isolated from foetal bovine skin or nuchal ligament, showing their characteristic repeating beaded structure. Panel A: (I) Microfibril with ~56 nm periodicity and 'closed' organisation. (II) Microfibril with ~56–60 nm periodicity and 'open' organisation. (III) After treatment with 5 mM EDTA, some microfibrils have reduced periodicity and increased flexibility (Wess *et al.*, 1998a). In some extracts, another population of beaded microfibrils is also apparent that has a more extended appearance. Collagen fibrils are also present. (IV) Invertebrate echinoderm dermis extract, showing a collagen fibril and (inset) a beaded fibrillin-rich microfibril. Panel B: (I) Microfibril with ~56 nm periodicity and 'closed' organisation. (II–V) Microfibrils with stretched regions (periodicities >100 nm). In (IV and V), highly stretched regions appear to comprise defined filaments.

N-terminus is a proline (pro)-rich sequence that may act as a 'hinge'; fibrillin-2 has a glycine (gly)-rich sequence, and fibrillin-3 a pro- and gly-rich sequence. Transglutaminase cross-links between fibrillin peptides have been identified that probably influence microfibril stability and elasticity (Qian and Glanville, 1997). Fibrillin-1 and fibrillin-2 have distinct, but overlapping expression patterns, with fibrillin-2 generally expressed earlier in development than fibrillin-1 (Zhang *et al.*, 1995). It is likely that microfibrils in some tissues may be assembled from different fibrillin isoforms. Determination of the molecular mechanism underlying the remarkable ability of fibrillin molecules, once assembled, to extend and retract is a major goal in microfibril biology (Figure 2B).

Fibrillin molecules and assembled microfibrils are remarkably well conserved throughout multicellular evolution from medusa jellyfish (*Podocoryne carnea* M. Sars), sea cucumber (*Cucumaria frondosa*) and other invertebrates, to man (Pereira *et al.*, 1993; Zhang *et al.*, 1994; Reber-Muller *et al.*, 1995; Thurmond and Trotter, 1996; Nagase *et al.*, 2001) (Figure 1A). In all these organisms, the tissue and developmental distributions of fibrillins correlate closely with the biomechanical needs

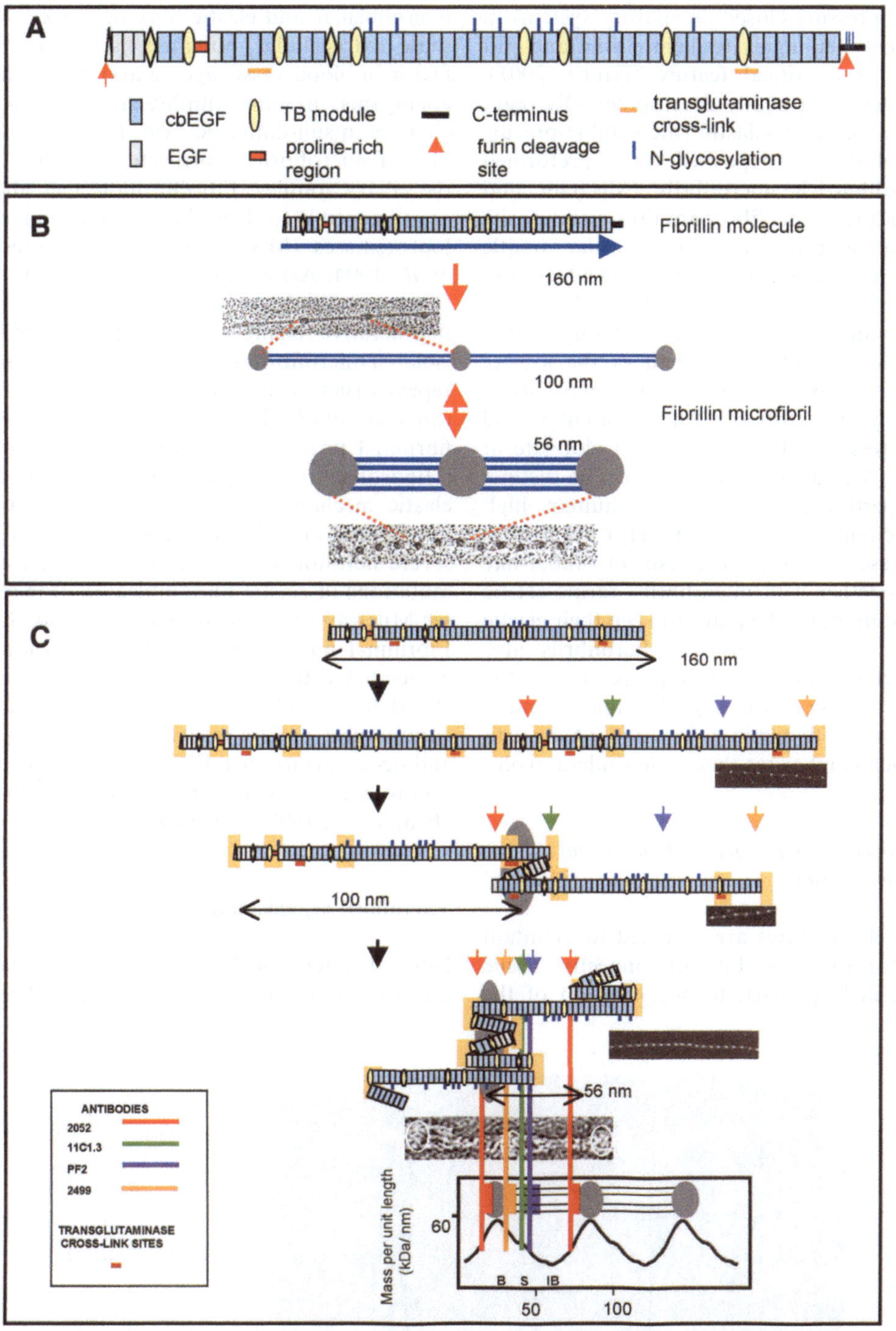

Fig. 2. Schematic representation of the fibrillin-1 molecule and its assembly into extensible microfibrils. Panel A: Domain structure of fibrillin-1, highlighting how contiguous arrays of cbEGF domains are interspersed with TB modules, and the locations of the proline-rich region, N-glycosylation consensus sequences and transglutaminase cross-link sites. Panel B: Schematic diagram outlining that fibrillin-1 molecules assemble to form beaded microfibrils with variable periodicity. Panel C: Schematic diagram depicting a possible folding arrangement of fibrillin molecules in a beaded microfibril (Baldock *et al.*, 2001). The model predicts that two N and C terminally processed molecules associate head-to-tail to give ~160 nm periodicity. Subsequent molecular folding events could generate ~100 nm periodicity, then ~56 nm periodicity. The model is supported by AET, STEM data, antibody localisations, and microfibril extension studies. Whilst predicted fold sites are shown (orange boxes), it should be noted that the precise packing arrangement of folded segments contributing to the bead remains unresolved. Inset are STEM images of microfibrils with periodicities corresponding to those predicted. Observed antibody binding sites are overlaid on the ~56 nm periodic folding arrangement. The axial mass distribution of ~56 nm microfibrils (shown at the bottom) correlates well with the corresponding predicted molecular folding. This figure is reproduced, with minor modifications, from Kielty et al. (2002).

of the organism. Microfibrils, which are considerably more ancient than elastin (a vertebrate protein), have conferred long-range elasticity to connective tissues for at least 550 million years. Elastin evolved more recently to reinforce high pressure closed circulatory systems in which interplay between highly resilient elastin and stiff fibrillar collagen is a critical feature (Faury, 2001). Elastic fibrillogenesis is a highly developmentally regulated process in which tropoelastin (the soluble precursor of mature elastin) is deposited on a preformed template of fibrillin-rich microfibrils (Mecham and Davis, 1994). Mature elastic fibres are thus a composite biomaterial comprising an outer microfibrillar mantle and an inner core of amorphous cross-linked elastin. The association of microfibrils with elastin undoubtedly modulates their biomechanical properties (Faury, 2001). In the low pressure circulatory system of the lobster (arterial pressure 3/1 to 16 mmHg), aortic elasticity is due almost entirely to microfibrils which reorientate and align at lower pressures then deform and elongate at higher pressures (McConnell *et al.*, 1996). The distinct mechanical properties of aorta in the human high pressure closed circulation (120/80 mmHg) are due to the additional presence of elastin, a resilient rubber-like protein which, together with microfibrils, forms elastic fibres (elastin and microfibrils) that can keep their elastic properties up to 140% extension. Microfibrils also contribute important biomechanical properties to tissues that do not express elastin, e.g. the ciliary zonules that hold the lens in dynamic suspension (Ashworth *et al.*, 2000), which emphasises their independent evolutionary function.

Involvement of microfibrils in ageing changes and connective disease diseases

Microfibrils and elastic fibres are designed to maintain elastic function for a lifetime. Loss of connective tissue elasticity is due, in large part, to degeneration of the microfibrillar and elastic fibre network. Degradation of these extracellular polymers is a major contributing factor in blood vessel ageing and development of aortic aneurysms, in lung emphysema, and in degenerative changes in sun-damaged skin. In ageing eyes, degeneration of microfibrils contributes to reduced elasticity of the ciliary zonules. Fibrillin molecules and microfibrils are susceptible to degradation by various matrix metalloproteinases (MMPs) and serine proteinases (Kielty *et al.*, 1994; Ashworth *et al.*, 1999a), and accumulative proteolytic damage is a major mechanism underlying degenerative changes in ageing. MMP treatment of isolated microfibrils can compromise the regular periodic repeat structure and extensible properties of microfibrils (Sherratt *et al.*, 2001) (Figure 3). Some of the major fibrillin-1 proteolytic cleavage sites have been identified (Hindson *et al.*, 1999), and they may provide clues to the elastic mechanism of microfibrils. The importance of microfibrils and elastic fibres is also highlighted by the severe heritable connective diseases that are caused by mutations in elastic fibre molecules (for recent reviews, see Milewicz *et al.*, 2000; Robinson and Godfrey, 2000). Fibrillin-1 mutations cause Marfan syndrome which is associated with severe cardiovascular, ocular and skeletal defects, and fibrillin-2 mutations cause congenital contractural arachnodactyly with overlapping skeletal and ocular symptoms. Elastin mutations cause Williams syndrome, supravalvular stenosis, and cutis laxa (Tassabehji *et al.*, 1998; Milewicz *et al.*, 2000).

Microfibril organisation

Isolated untensioned microfibrils from mammalian tissues such as amnion, skin and ligament have a repeating

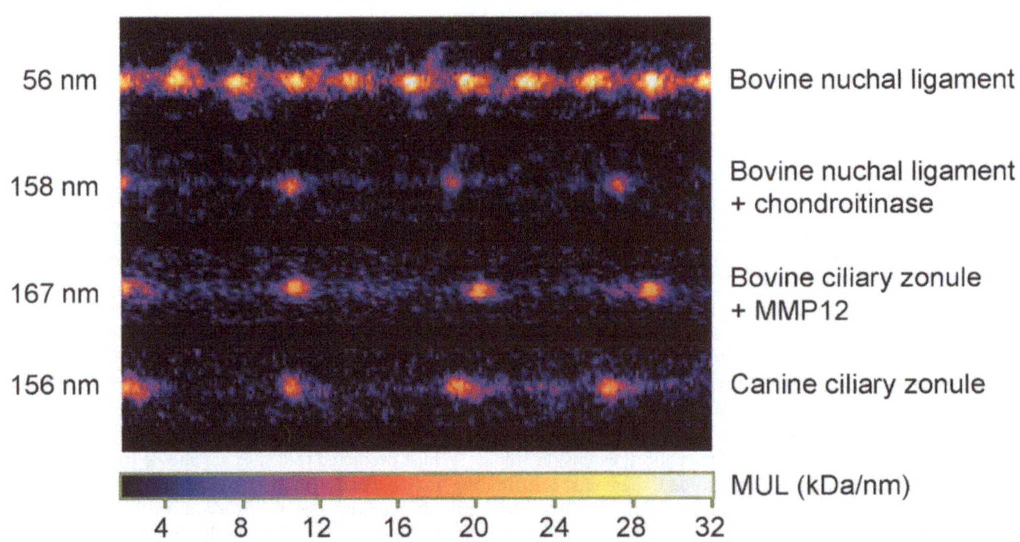

Fig. 3. STEM mass images of isolated fibrillin-rich microfibrils, showing some stretched areas. Bovine nuchal ligament microfibrils were either untreated or pre-incubated with chondroitinase ABC lyase. Bovine ciliary zonule microfibrils were pre-incubated with MMP-12. Zonular filament microfibrils were also examined from a canine ectopia lentis model. In many (but not all) regions of the treated and ectopia lentis microfibrils, highly stretched sections were apparent.

'beads-on-a-string' appearance with an average, but variable, periodicity of 56 nm (Sherratt *et al.*, 2001) (Figure 1). In tissues, microfibrils are generally organised into loose, roughly parallel bundles (Figure 4A). X-ray fibre diffraction of hydrated zonular microfibril bundles has identified staggered 'junctions' that may modulate force transmission (Wess *et al.*, 1998a). In elastic tissues, microfibril bundles form the template for

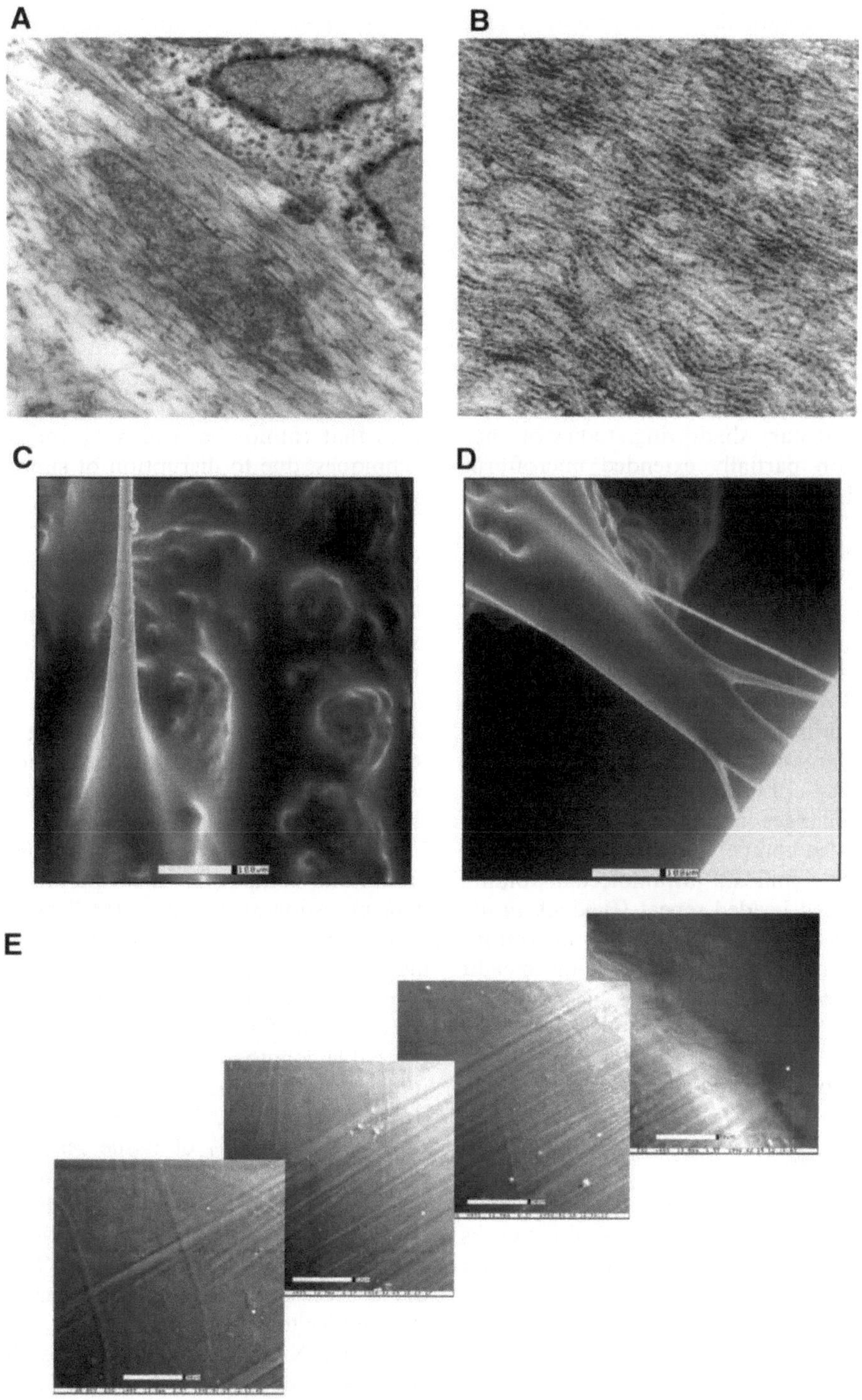

Fig. 4. Tissue microfibril organisation. Panel A: Microfibril bundles are laid down close to the surface of a bovine nuchal ligament fibroblast. Panel B: Microfibrillar array from lobster aorta. This figure is reproduced, with minor modification, from Kielty *et al.* (2002). Panels C and D: Environmental scanning electron microscopy (ESEM) of the posterior aspect of human zonules running from the ciliary process to the lens capsule (68 year old male). Panel C shows the origin of the zonular fibre from the ciliary process. Panel D shows branching of the zonular fibres as they approach the lens capsule. The scale bars = 100 μm. Panel E: Composite ESEM of human zonular filaments aligned on the surface of the lens capsule (68 year old male). The scale bars = 100 μm.

elastin deposition, and in this way influence the orientation and 3D structural architecture of the developing elastic fibre (Mecham and Davis, 1994). Distinct tissues can have profoundly different elastic fibre organisations that reflect specific functions. Thus, elastic fibres are arranged in concentric lamellae in elastic arteries, as networks in deformable auricular cartilage and in lung alveoli that provide elasticity during breathing, and in skin they impart elasticity through a continuous elastic fibre system from the microfibrillar bundles (oxytalan fibres) at the dermal–epidermal junction to the horizontal thick elastic fibres in the reticular dermis. In mature elastic fibres, microfibrils are apparent as an outer mantle, with some microfibrils embedded within the elastin core. The elastic properties of microfibrils may be modified both by inter-microfibrillar interactions within microfibril bundles and by association with elastin in elastic fibres.

Details of the alignment of fibrillin molecules within microfibrils have emerged from ultrastructural studies of isolated microfibrils. Rotary shadowing studies of antibody-binding sites on partially extended microfibrils indicated a parallel head-to-tail alignment of fibrillin molecules in microfibrils (Reinhardt *et al.*, 1996a, b). Automated electron tomography (AET) has provided further insights into the complex organisation of untensioned fibrillin-rich microfibrils, and the alignment of fibrillin molecules in the untensioned state (Baldock *et al.*, 2001). AET-generated 3D reconstructions at 18.6 Å resolution showed that twisting occurs within untensioned microfibrils and revealed new details of bead and interbead organisation. Localisation of antibody and gold-binding epitopes, and mapping of bead and interbead mass changes in untensioned and extended microfibrils provided compelling evidence that fibrillin molecules undergo significant intramolecular folding within each untensioned beaded repeat (Baldock *et al.*, 2001). In addition, STEM analysis defined microfibril mass and its axial distribution, and predicted up to eight fibrillin molecules in cross-section.

Microfibril extensibility

Invertebrate studies

Important insights into microfibril elasticity have emerged from invertebrate studies. Microfibril networks extracted from sea cucumber dermis using guanidine and bacterial collagenase were shown, by tensile testing, to be reversibly extensible up to approximately 300% of their initial length (Thurmond and Trotter, 1996). These networks behaved like viscoelastic solids, with a long-range elastic component as well as a time-dependent viscous component. The strength of the network appeared to be due to non-reducible cross-links, but its elasticity was dependent on disulphide bonds. The modulus of elasticity of lobster (*Homarus americanus*) aorta, a viscoelastic tissue based on microfibril bundles and smooth muscle cells, was found to be similar to that of the rubber-like protein elastin (0.1–1.2 MPa, commonly in range 0.4 MPa), (McConnell *et al.*, 1996). Lobster aorta elastic properties are non-linear; low stiffness allows arterial volume to increase, but then stiffer vessels correspond with arterial volume plateauing. Further assessment of this system revealed that microfibrils alone characterise the stress–strain behaviour of the vessel, when initial reorientation and subsequent deformation are accounted for (McConnell *et al.*, 1997).

X-ray fibre diffraction

Small-angle X-ray fibre diffraction has provided valuable insights into the organisation and elasticity of hydrated ciliary zonules (see Figure 4B for scanning image of zonules). A key advantage of these studies is that a statistically significant number of molecular conformations can be examined in hydrated intact tissues. This technique can thus reveal molecular features that cannot be readily examined by microscopic techniques, due to disruption of supramolecular organisation in sample preparation.

X-ray examination of hydrated bovine ciliary zonules revealed meridional diffraction peaks indexing on a fundamental periodicity of ~56 nm in the relaxed state (Wess *et al.*, 1997, 1998a, b). The strength of the third and sixth order Bragg reflections in the resting state suggested that a specific alignment or staggering of adjacent microfibrils gives rise to electron dense contrast at periodicities of one-third of the axial cell length. Remarkably similar results have now been obtained with zonular filaments from the eyes of pig, roe deer and sheep. The data from ovine eye are presented in Figure 5. All zonules exhibited strong third and sixth orders and comparable axial periodicity (pig (54.9 nm, data not shown), roe deer (56.2 nm, data not shown) and sheep (55.0 nm)) in the resting state, confirming a high degree of inter-species structural homology.

In bovine zonules, the application of a relatively low strain (up to 50% tissue extension) produced a minor reversible lengthening of periodicity by only 4%, as judged by alteration of the D spacing of the principal peaks (Wess *et al.*, 1997, 1998b). In contrast, further extension (to 100% of tissue rest length) produced a higher periodicity, indicating disruption of the supramolecular structure. The fibrillin diffraction image also contained an equatorial diffraction peak that was enhanced upon tissue extension. This peak estimated the nearest approach of microfibrils in the hydrated state under tension to be 28 nm.

Further X-ray diffraction studies have now investigated changes in the supramolecular packing of highly extended ovine zonular filaments (Figure 5). Samples were exposed to X-rays for 12 min in the native state and then in a fully extended state (150% length extension from resting state). Subsequently the tissue was relaxed back to the rest length and re-analysed. The results obtained in this experiment are comparable to

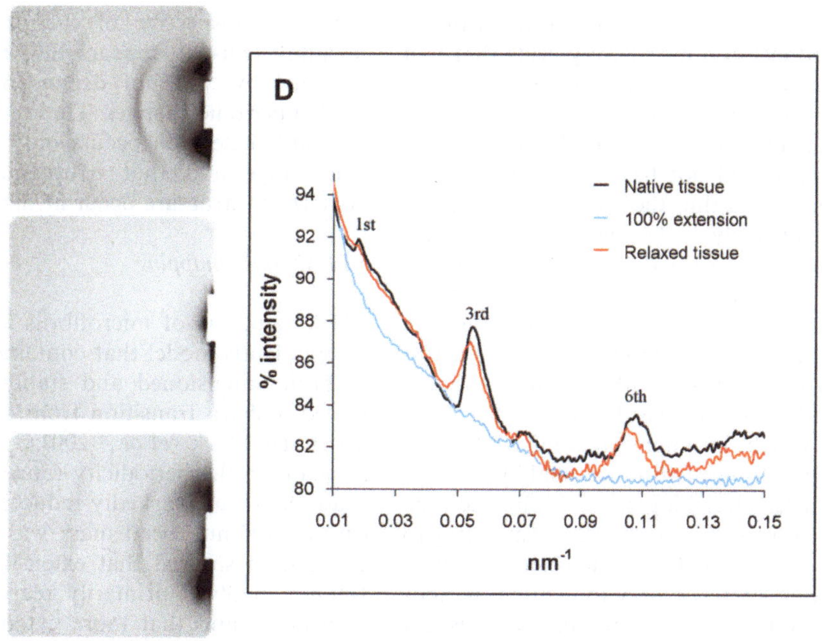

Fig. 5. X-ray fibre diffraction of ovine ciliary zonules before and after extension. X-ray fibre diffraction was carried out using an 8.25 m camera on beamline 2.1 at the CLRC Daresbury laboratory (UK), in order to investigate changes in the supramolecular packing of ovine zonular filaments under extension. Zonular filaments were dissected from ovine eye as previously described for bovine tissue (Wess *et al.*, 1997, 1998a). Briefly, eyes were obtained within 24 h post-mortem from a local abattoir. Dissection of the posterior chamber of the eye produced an intact preparation of ciliary body, lens and vitreous humor. The zonular filaments were mounted securely on small aluminium frames using cyanoacrylate. Samples were kept hydrated using Tris buffered saline (pH 7.2), and were analysed between thin mica sheets. Samples were exposed for 12 min in the native state and then in a fully extended state (150% length extension from resting state). Subsequently the tissue was relaxed back to the rest length and re-analysed. Panel A: native zonules before extension. Panel B: zonules with 100% extension. Panel C: relaxed zonules after extension. Panel D: linear profiles of the integrated small-angle meridional intensity region.

those seen with bovine zonules, with strong third and sixth orders and a weaker fourth clearly visible in the unstretched sample (Figure 5a). Upon stretching up to 50%, no real difference was observed in this diffraction pattern (data not shown). In contrast, tissue extensions of 150% produced a markedly different pattern, with a loss of coherence in the meridional diffraction series (Figure 5b). Relaxation of the tissue re-established the meridional series, indicating that the staggered alignment of adjacent microfibrils is recoverable (Figure 5c), although the relaxed tissue exhibited a slightly longer periodicity, as indicated by a reduction in the reciprocal distance of the third and sixth orders (Figure 5d). No observable difference was seen in the position of the first order, although changes in this peak are difficult to discriminate, even using an 8.25 m camera. Thus, although stretching of the zonular filaments produces an altered supramolecular structure of the component fibrillin-rich microfibrils, these X-ray diffraction studies suggest that the characteristic third order staggered array organisation is recoverable, even beyond what is probably the physiological range of microfibril elasticity (up to ~100% extension).

Mechanical testing of microfibril bundles

Tensile tests and stress-relaxation tests were used on whole ciliary zonular filaments to examine the relationship between their mechanical behaviour and the molecular organisation of the fibrillin-rich microfibrils upon which the tissue is based. Zonular filaments revealed a non-linear (J-shaped) stress–strain curve and appreciable stress-relaxation (Wright *et al.*, 1999), and it was proposed that this behaviour may reflect an initial strain-induced lateral alignment of untensioned microfibrils prior to elastic extension and recoil. A similar non-linear relationship between force and strain was observed for microfibril bundles of less than 1 μm in diameter (Eriksen *et al.*, 2001).

Isolated microfibrils

Early studies of isolated microfibrils revealed that, while most beaded repeats had a regular 56 nm periodicity characteristic of the untensioned state, there were often also seen some highly stretched beaded repeat regions with periodicities of up to ~160 nm (Keene *et al.*, 1991; Sherratt *et al.*, 2001). These highly stretched regions, which probably arose due to tangling in debris during preparation, appeared to be stable in solution, suggesting that they may have been irreversibly extended.

Stretching isolated microfibrils in solution under controlled conditions has proved a challenge, and reproducible approaches have yet to be established. Methods such as centrifugation and aspiring through fine-gauge needles did not reproducibly extend microfibrils

(Sherratt *et al.*, 2001). When microfibrils were adsorbed onto carbon-coated electron microscopy grids that had not been glow-discharged, surface forces were shown to extend beaded periodicity to ~100 nm and to exert a profound effect on their organisation (Baldock *et al.*, 2001). These forces would not have been sufficient to break covalent bonds, and thus the observed 56–100 nm periodic changes are likely to be due to non-covalent forces such as electrostatic or hydrophobic interactions. This observation is consistent with this periodic range being extensible. Incubation of isolated microfibrils with a monoclonal anti-fibrillin-1 antibody (11C1.3) generated extensive microfibril arrays with a double striated pattern corresponding to the beads and to the Mab interbead epitope which, in untensioned microfibrils, is ~41.1% of the bead-to-bead distance (Baldock *et al.*, 2001). However, after stretching these Mab-generated arrays with surface forces, the striation corresponding to the bound antibody was no longer detected except in microfibrils that were stretched only to ~70 nm periodicity. These data imply that the Mab epitope does not have to move until periodicity exceeds 70 nm (at that periodicity, the antibody epitope is 32% of bead-to-bead distance). At higher periodicities, the alignment of the epitope is lost, presumably due to a major structural rearrangement. Thus, microfibrils can stretch at least 10 nm before this epitope has to move.

Effects of calcium on elasticity

Velocity sedimentation and rotary shadowing have revealed a 20–25% decrease in the length of the rod-shaped fibrillin molecule in the presence of 5 mM EDTA (Reinhardt *et al.*, 1997). Bound calcium also strongly influences the organisation of isolated microfibrils (Kielty and Shuttleworth, 1993; Cardy and Handford, 1998; Wess *et al.*, 1998a). X-ray diffraction, dark field STEM and rotary shadowing all showed that chelation of calcium reduces beaded periodicity by 15–30% and produces a diffuse appearance with evidence for increased molecular flexibility. The observed decrease in fibrillin molecule and microfibril rest length is partly accounted for by an increase in the degrees of freedom of the linkages between consecutive cbEGF domains. NMR studies showed that, in the presence of bound calcium, a fibrillin-1 cbEGF domain pair is in a rigid, rod-like arrangement, stabilised by interdomain calcium binding and hydrophobic interactions (Downing *et al.*, 1996).

X-ray diffraction zonular stretching experiments have shown that bound calcium influences microfibril load deformation but is not necessary for high extensibility and elasticity (Wess *et al.*, 1998a, b). The influence of calcium has also been examined in mechanical tests of microfibril bundles using the micro-needle technique (Eriksen *et al.*, 2001). Calcium depletion resulted in a 50% decrease in rest length and a reduction in microfibril stiffness; this effect was reversible upon addition of calcium. However, at high strain, irreversible damage occurred irrespective of the presence or absence of calcium. Thus, it appears that microfibril elasticity is modified by, but not dependent on calcium-induced beaded periodic changes. These studies also showed that reversible extension occurs only within a specific periodic range and that irreversible transitions in the quaternary structure occur at high stretch.

STEM mass mapping

STEM analysis of microfibrils isolated from a canine ectopia lentis model that contained abundant examples of both untensioned and stably stretched regions revealed a sharp transition from 56 to >100 nm periodicities (Baldock *et al.*, 2001). Bead mass remained constant in the periodicity range 56–100 nm but interbead mass was markedly reduced. However, at periodicities >100 nm, bead mass was then rapidly reduced. These data showed that extensibility in the range 56–100 nm involved primarily rearrangements within the interbeads, and that more extended periodicities were accounted for by unravelling of the beads. These data suggested that interbead changes mainly define the reversible elastic range (~56–100 nm), and that bead changes at higher periodicities may be irreversible.

Atomic force microscopy

Air-dried fibrillin-containing microfibrils imaged by tapping mode atomic force microscopy (AFM) and non-contact mode AFM (Hanssen *et al.*, 1998; Sherratt *et al.*, 2001) exhibit the characteristic 56 nm beaded periodicity previously observed by TEM in negatively stained and rotary shadowed and by STEM in unstained microfibrils (Keene *et al.*, 1991; Kielty *et al.*, 2002). Using tapping mode AFM height measurements, we have now also shown that, within the periodic range 44–68 nm, interbead height is reduced at increasing periodicities but bead height remains constant along a 35 nm region (Figure 6). These investigations support the STEM observations and further narrow down the region involved in reversible elastic changes to the centre of the interbead.

AFM single molecule force spectroscopy has proved to be an invaluable tool in the investigation of molecular biomechanics. Using a stiff cantilever, Rief *et al.* (1997) tracked the unfolding of single domains within the giant muscle protein titin with a single amino acid resolution. The large mass (3–4 MDa) and long contour length (1 μm) make titin an ideal subject for single molecule biomechanical studies. The investigation of fibrillin biomechanics by single force spectroscopy AFM is currently underway using recombinantly expressed molecules.

Summary

The lobster aorta studies confirm that microfibrils are responsible for the elasticity of this invertebrate tissue.

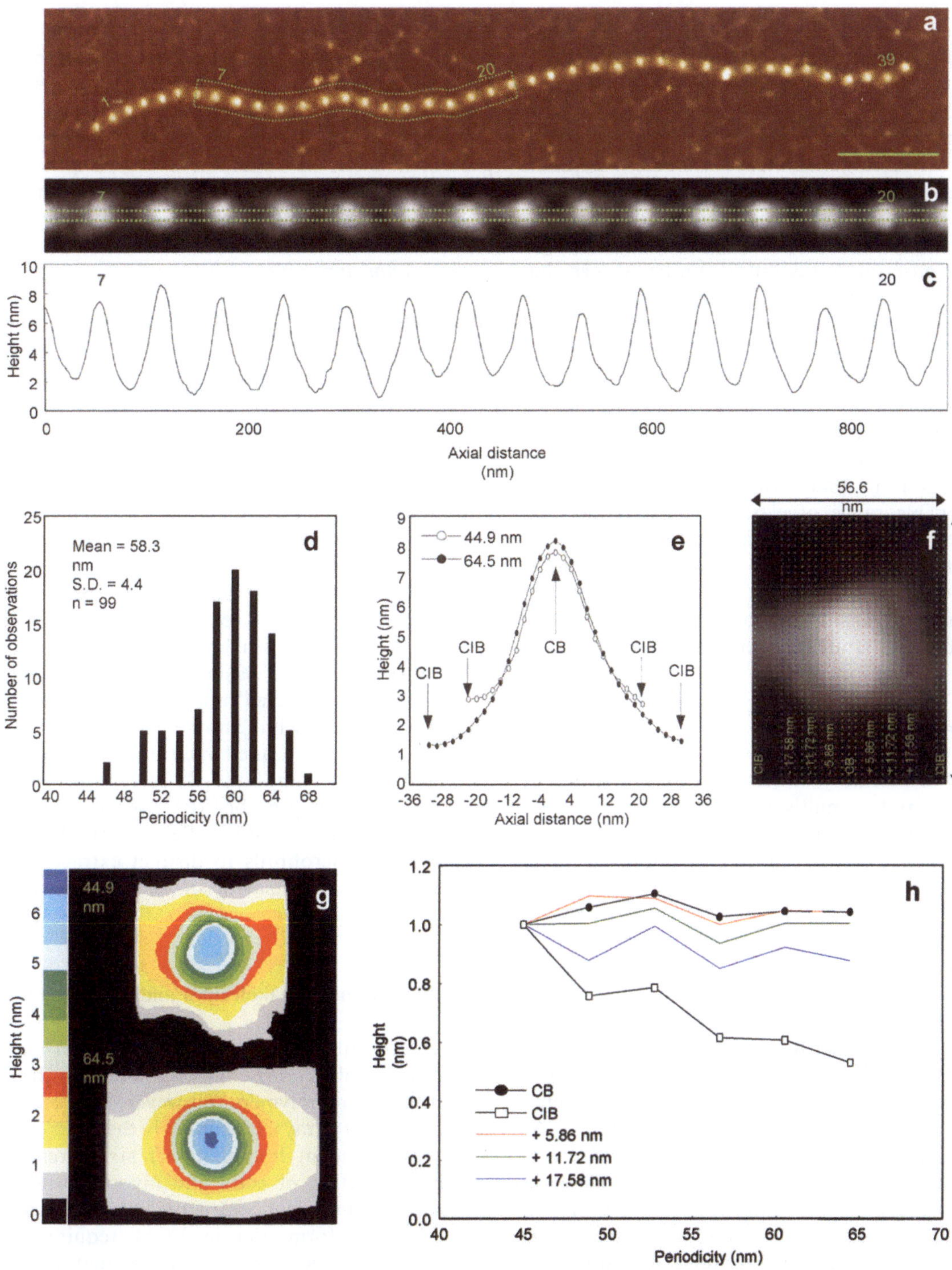

Fig. 6. AFM height measurements of isolated microfibrils. Panel A: AFM tapping mode visualisation of a microfibril isolated from bovine ciliary zonule. Scale bar = 200 nm. Panel B: Fourteen repeats from this microfibril, straightened using a plugin for ImageJ based on an algorithm described in Kocsis *et al.* (1991). Panel C: Mean axial height distributions were calculated at each axial position over 9.8 nm in the transverse direction. Panel D: Mean axial periodicity (determined for 99 repeats) = 58.3 nm. Panel E: Mean axial height distributions were calculated for microfibril repeats in the following range (23, 25, 27, 29, 31, 33 pixels, each ± one pixel, equivalent to 44.9, 48.8, 52.7, 56.6, 60.5, 64.5 ± 1.95 nm). The height of the centre of the bead (CB) is virtually unchanged at periodicities between 44.9 and 64.5 nm. However, the height of the centre of the interbead (CIB) does change. Panel F: In order to assess where height changes occur within the repeat, mean maximum heights were determined for transverse profiles (5.85 nm wide) at the centre of the bead (CB) and centre of the interbead (CIB) and at points 5.86, 11.72 and 17.58 nm to right (+) or left (−) of the bead centre (these correspond to lines on the graph shown in Panel H). Panel G: Mean 3-D height contour maps were determined for repeats in the range 44.9–64.5 nm. The shape and size of the central bead remains unchanged over this period range, but interbead height varies from 2–2.5 nm (44.9 nm) to 1–1.5 nm (64.5 nm). Panel H: Graph showing mean height as a proportion of height at 44.9 nm periodicity. The CB and regions to the plus side of the bead show little change with increasing periodicity. Only the interbead region changes. Over the periodicity range 44.9–68.5 nm, a narrow region at the centre of the interbead undergoes conformation change, but most of the repeat remains topographically unchanged.

The extracted microfibril bundles from sea cucumber infer key roles for transglutaminase cross-links in the provision of strength and disulphide bonds in elastic properties. Perturbing disulphide bonds will have the effect of profoundly altering the secondary and tertiary structure of the contiguous arrays of cbEGF domains and intercalating TB modules in fibrillins, all of which are stabilised by intradomain disulphide bonds, and will also disrupt any inter-molecular disulphide bridges within the microfibril. X-ray diffraction has shown that hydrated physiological microfibril bundles have a reversible range of elasticity (less than 100% stretch), and also that the predicted one-third staggered lateral alignment of microfibrils is recoverable after high stretch. Mechanical tests have shown that untensioned tissue microfibrils undergo initial alignment before extension with strain. The STEM and AFM data support the concept that individual microfibrils also have a reversible range of elasticity based largely on interbead reorganisation. The antibody binding data suggest that a major interbead conformation change occurs at ~70 nm periodicity. Bound calcium can modify microfibril elasticity, probably by altering molecular flexibility at strategic interdomain linkages.

Formation of microfibrils

The process of assembly of elastic fibrillin-rich microfibrils is a complex multistep process that remains poorly understood. Below is outlined our current knowledge of key phases in the assembly of these microfibrils and their deposition in the extracellular matrix.

Cellular regulation

Microfibril assembly is, in part, a cell-regulated process that proceeds independently of tropoelastin. Fibrillin-1 can undergo limited initial assembly in the secretory pathway (Ashworth *et al.*, 1999b; Trask *et al.*, 1999), and in this respect, is similar to other major ECM macromolecules such as collagens, laminins and proteoglycans. Chaperone interactions may be essential in regulating intracellular fibrillin assembly (Ashworth *et al.*, 1999c). Fibrillins have N- and C-terminal cleavage sites for furin convertase; extracellular deposition requires removal of the C-terminus (Raghunath *et al.*, 1999; Ritty *et al.*, 1999), a process influenced by N-glycosylation and calreticulin (Ashworth *et al.*, 1999c).

Cell surface assembly

Microfibrils assemble close to the cell surface, in a process that may require receptors, as shown for fibronectin (Sechler *et al.*, 2001). A role for heparan sulphate proteoglycans (HSPGs) in assembly has been proposed (Tiedemann *et al.*, 2001), and we have evidence that chondroitin sulphate proteoglycans (CSPGs) are involved. Sulphation is needed for microfibril assembly, since chlorate treatment ablates microfibril and elastic fibre formation (Robb *et al.*, 1999). This effect may reflect absence of a proteoglycan or undersulphation of fibrillins or microfibril-associated glycoprotein (MAGP-1). Fibrillin RGD sequences (TB4 module) can also interact with several integrins (Pfaff *et al.*, 1996; Sakamoto *et al.*, 1996).

Extracellular maturation

Extracellular microfibril populations that may differ in macromolecular organisation, composition and stability have been tentatively identified, implying a major role for the extracellular environment in regulating microfibril fate. In human dermal fibroblast cultures, monoclonal antibody11C1.3 (binds beaded microfibrils) did not detect microfibrils until 2 weeks, but a polyclonal antibody (PF2) to a fibrillin-1 pepsin fragment detected abundant microfibrils within 3 days (Baldock *et al.*, 2001). The time-dependent appearance of 11C1.3-reactive microfibrils suggests 'maturation', perhaps due to conformational changes or unmasking of a cryptic epitope. It is not known how such maturation events might affect microfibril elasticity. In developing vascular tissues, 11C1.3 detected microfibrils associated with medial elastic fibres but another monoclonal antibody 12A5.18 (also binds beaded microfibrils, and at the same interbead site as antibody 11C1.3) only recognised microfibrils in collagen fibril rich tissues (Kogake *et al.*, 2002). Interactions with different ECM molecules may commit microfibrils to distinct extracellular fates in association with elastin or in collagen fibre rich tissues.

Microfibril-associated molecules and their interactions

We have described current understanding of microfibril elasticity and fibrillin assembly, implicitly in the context of fibrillin homopolymers. However, like other major supramolecular arrays, it is unlikely that microfibrils are totally independent homopolymer structures. For example, association of actin and myosin is required to generate the functional contractile unit of muscle, whilst collagen fibril formation in ECM requires terminal processing enzymes as well as regulation of fibril diameter and interactions with surrounding ECM by small leucine-rich proteoglycans (e.g. decorin, biglycan) (Danielson *et al.*, 1997; Xu *et al.*, 1998). Microfibrils are probably multimolecular assemblies and they certainly interact with other molecules during elastic fibre formation (Kielty *et al.*, 2002). In fact, a plethora of microfibril-associated molecules has been identified using immunohistochemical and biochemical approaches, and a major current challenge is to determine which molecules are actually structural components of microfibrils and how different associated molecules may modify microfibril elastic properties and integration in the surrounding ECM. The observed time-dependent maturation of microfibrils may well reflect, in part, microfibril associations with other matrix molecules. Thus, it is necessary to consider in some detail the other known molecules of the microfibrillar system, and current understanding of how they may influence fibrillin assembly and microfibril elasticity.

MAGP-1 may also be an integral microfibril molecule (Gibson *et al.*, 1989; Trask *et al.*, 2000a). It is associated with virtually all microfibrils, and widely expressed in mesenchymal and connective tissue cells throughout development (Henderson *et al.*, 1996; Kielty and Shuttleworth, 1997). MAGP-2, the other member of this small microfibrillar protein family, is structurally related to MAGP-1 (Gibson *et al.*, 1998; Segade *et al.*, 2002), and localises to elastin-associated and elastin-free microfibrils in some tissues although it has a more restricted pattern of tissue localisation (Gibson *et al.*, 1998). MAGP-2 is an RGD-containing cell adhesion molecule and may have functions related to cell signalling during microfibril assembly and elastinogenesis (Gibson *et al.*, 1999) (Table 1).

Latent TGFβ binding proteins (LTBPs) are members of the fibrillin superfamily and also contain contiguous arrays of cbEGF domains interspersed with 8-cysteine motifs (also called TB modules due to TGFβ-binding properties) (Sinha *et al.*, 1998; Oklu and Hesketh, 2000). LTBPs form covalent large latent complexes with TGFβ intracellularly that are secreted, transglutaminase cross-linked within the ECM, then released by protolysis (Miyazono *et al.*, 1988; Sinha *et al.*, 1998). LTBP-1 colocalises with microfibrils in skin and in cell layers of cultured osteoblasts and in embryonic long bone, but not in cartilage (Taipale *et al.*, 1996; Raghunath *et al.*, 1998; Dallas *et al.*, 2000). LTBP-1 is unlikely to be an integral microfibril component, but its association implicates microfibrils in TGFβ targeting. LTBP-2 colocalises with fibrillin-rich microfibrils in elastic-fibre-rich tissues (Gibson *et al.*, 1995; Kitahama *et al.*, 2000; Sinha *et al.*, 2002), and is a good candidate for a microfibrillar molecule, although it does not bind TGFβ.

Several other microfibril-associated proteins have been identified immunohistochemically, but little is

Table 1. Reported microfibril and elastic fibre associated molecules

Molecule	Location	References
Fibrillin-1	Microfibrils	Pereira *et al.* (1993) Baldock *et al.* (2001)
Fibrillin-2	Microfibrils	Zhang *et al.* (1994)
Fibrillin-3	? microfibrils	Nagase *et al.* (2002)
MAGP-1	Microfibrils	Brown-Augsberger *et al.* (1996)
MAGP-2	Some microfibrils	Gibson *et al.* (1998)
LTBP-1	Some microfibrils	Miyazono *et al.* (1988) Sinha *et al.* (1998)
LTBP-2	Microfibrils/elastic fibres	Gibson *et al.* (1995) Bashir *et al.* (1996)
LTBP-3	?	Yin *et al.* (1995)
LTBP-4	?	Giltay *et al.* (1997)
Decorin	Microfibrils	Trask *et al.* (2000a)
Biglycan	Elastic fibre core	Reinboth *et al.* (2001)
Versican	Some microfibrils	Isogai *et al.* (2002)
MFAP-1	Some microfibrils	Liu *et al.* (1997)
MFAP-3	Some microfibrils	Abrams *et al.* (1995)
MFAP-4 (MAGP-36)	Some microfibrils	Lausen *et al.* (1999) Hirano *et al.* (2002)
Tropoelastin	Elastic fibre core	Mecham and Davis (1994)
Lysyl oxidase (LOX)	Microfibrils/ tropoelastin	Csiszar (2001)
LOXL	?	Borel *et al.* (2001)
LOXL2	?	Csiszar (2001)
LOXL3	?	Csiszar (2001)
BigH3 (keratoepithelin)	Elastic fibre/collagen interface	Gibson *et al.* (1997)
Fibulin-1	Elastic fibre core	Kostka *et al.* (2001)
Fibulin-2	Elastin–microfibril interface	Tsuda *et al.* (2001)
Fibulin-5	Elastic fibre–cell interface	Nakamura *et al.* (2002) Yanagisawa *et al.* (2002)
Emilin-1	Elastin–microfibril interface	Doliana *et al.* (1999)
Emilin-2	Elastin–microfibril interface	Doliana *et al.* (2001)
Elastin binding protein	Newly secreted tropoelastin	Prody *et al.* (1998)
Vitronectin	Some microfibrils	Dählback *et al.* (1990)
Amyloid	Some microfibrils	Dählback *et al.* (1990)
Collagen VIII (α1(VIII))	Some elastic fibres	Sadawa and Konomi (1991)
Collagen XVI (α1(XVI))	Dermal microfibrils	Grässel *et al.* (1999)
Endostatin (α1(XVIII))	Vascular elastic fibres	Miosge *et al.* (1999)
Collagen VI	Some microfibrils	Finnis and Gibson (1997)

known about whether they are essential microfibrillar components and how they may influence microfibril function. Microfibril-associated protein (MFAP)-1 (or AMP), MFAP-3 and MFAP-4 (or 'MAGP-36') colocalise with elastic fibres in skin and other tissues (Horrigan *et al.*, 1992; Abrams *et al.*, 1995; Liu *et al.*, 1997; Lausen *et al.*, 1999; Toyashima *et al.*, 1999; Hirano *et al.*, 2002). In some tissues, microfibrils associate with amyloid deposits and with adhesive glycoproteins such as vitronectin (Dählback *et al.*, 1990). MP78/70 (also known as beta ig-h3 or keratoepithelin) is another molecule that occasionally appears at elastic fibre interfaces. Originally identified in bovine tissue extracts designed to solubilise microfibrils (Gibson *et al.*, 1989), it localises to collagen fibres in ligament, aorta, lung and mature cornea, to reticular fibres in foetal spleen, and to capsule and tubule basement membranes in kidney (Gibson *et al.*, 1997; Schorderet *et al.*, 2000). In some elastic tissues, MP78/70 is present at the interface between collagen fibres and adjacent elastic fibre microfibrils, which suggests that it has a bridging function. Collagen XVI is expressed by human dermal fibroblasts and keratinocytes and is associated with the microfibrillar apparatus in the upper papillary dermis (Grässel *et al.*, 1999). The angiogenesis inhibitor endostatin is another potential component of elastic fibers in vessel walls (Miosge *et al.*, 1999).

Proteoglycans have important, but still poorly understood structural interactions with microfibrils, and they may contribute to their integration into ECM. Immunocytochemical studies have localised proteoglycans within normal elastic fibres (Baccarani-Contri *et al.*, 1990). Decorin and biglycan, members of the small leucine-rich PG family, have been identified within dermal elastic fibres; biglycan mapped to the elastin core and decorin to microfibrils. Ultrastructural approaches showed that CSPGs associate with microfibril beads (Kielty *et al.*, 1996). Versican, a large chondroitin sulphate PG (CSPG) colocalises with microfibrils in dermis and interacts directly with fibrillin (Zimmermann *et al.*, 1994; Isogai *et al.*, 2002).

Other molecules localise to the elastin–microfibril interface and to the cell surface–elastic fibre interface. These molecules may regulate tropoelastin deposition on microfibrils, and link microfibrils and elastic fibres to cell surfaces. Emilin, a 136 kDa glycoprotein, localises to the elastin microfibril interface (Bressan *et al.*, 1993; Doliana *et al.*, 1999). Four family members have now been identified, emilins-1, -2, -3, and multimerin (emilin-4), but it is not yet known which interact with microfibrils (Colombatti *et al.*, 2000; Doliana *et al.*, 2001). Collagen VIII has been localised to the vascular elastic fibre-cell interface (Sadawa and Konomi, 1991).

Three members of the fibulin family of cbEGF-like domain molecules are implicated in elastic fibre biology. Fibulin-1 is located within the amorphous core of elastic fibres (Kostka *et al.*, 2001). Fibulin-5 has a vascular pericellular localisation and has been shown, in mouse models to be essential for normal elastic fibre formation (Nakamura *et al.*, 2002; Yanagisawa *et al.*, 2002). Fibulin-2 localises preferentially at the interface between microfibrils and the elastin core. It colocalises with fibrillin-1 in skin (except adjacent to the dermal–epithelial junction), perichondrium, elastic intima of blood vessels, and kidney glomerulus, although is apparently not present in ciliary zonules, tendon, and surrounding lung alveoli and kidney tubules (Reinhardt *et al.*, 1996b; Utani *et al.*, 1997; Raghunath *et al.*, 1999; Tsuda *et al.*, 2001). Fibulin may not be needed for microfibril elasticity since it is absent from tissues subject to strong tensional forces (e.g., tendon, ciliary zonule).

Tropoelastin is synthesised as a soluble precursor with a molecular mass of ~70 kDa and alternating hydrophobic and cross-linking domains (Mecham and Davis, 1994; Brown-Augsberger *et al.*, 1995). The 67-kDa enzymatically inactive alternatively spliced variant of beta-galactosidase, which is identical to the elastin/laminin-binding protein, may influence elastin deposition on microfibrils (Prody *et al.*, 1998). Interactions between hydrophobic domains are important in assembly and essential for elasticity (Bellingham *et al.*, 2001; Toonkool *et al.*, 2001). The formation of covalent lysyl-derived desmosine cross-links by lysyl oxidase (Csiszar, 2001), stabilises the polymerised insoluble product (elastin). Five lysyl oxidase-like proteins have now been characterised (LOX, LOXL, LOXL2 [or WS9-14], LOXL3 and LOXC). However, only LOX and LOXL have so far been shown to be able to cross-link insoluble elastin (Borel *et al.*, 2001).

The identification of this plethora of microfibril-associated molecules has now led to an urgent need to define how they interact within functional elastic microfibrils and tissue microfibrillar bundles. Available information is derived from *in vitro* binding studies, so the temporal hierarchy and significance of these interactions in elastic microfibril formation will need to be established.

MAGP-1 interacts with the fibrillin-1 N-terminus (within exons 1–10) in a calcium-dependent manner (Jensen *et al.*, 2001). MAGP-1 and fibrillin-1 are both substrates for transglutaminase and, although only homotypic fibrillin-1 cross-links have been identified to date, MAGP-1 may be cross-linked within microfibrils (Brown-Augsberger *et al.*, 1996; Qian and Glanville, 1997). MAGP-1 and fibrillin-1 both interact with decorin, a sulphated CSPG (Trask *et al.*, 2000a). The fibrillin-1 interacting sequence is within or adjacent to the proline-rich region, and the interaction is with the decorin core protein. Decorin can interact with both fibrillin-1 and MAGP-1 individually, and together they form a ternary complex. However, fibrillin-2 appears not to interact with MAGP-1 or decorin. In a separate study, decorin and biglycan were shown not to bind MAGPs -1 and -2 in solid phase assays, although MAGP-1 in solution interacted with biglycan, but not with decorin (Reinboth *et al.*, 2001).

MAGP-1 also binds tropoelastin; the binding site in MAGP-1 was localised to a tyrosine-rich sequence within its positively charged N-terminal half, which may interact with a negatively charged pocket near the tropoelastin C-terminus (Brown-Augsberger *et al.*, 1996). MAGP-1 may interact first with fibrillin-1 and decorin during microfibril assembly, and then with tropoelastin during elastic fibre formation on the microfibrillar template. Sequences in fibrillins-1 and -2 (within exons 10–16) interact with tropoelastin, but only in solid-phase suggesting that exposure of a cryptic site is needed (Trask *et al.*, 2000b). Decorin and biglycan can both bind tropoelastin, biglycan more avidly than decorin, and the biglycan core protein more strongly than the intact PG (Reinboth *et al.*, 2001). The ability of biglycan to form a ternary complex with tropoelastin and MAGP-1 suggests a role in the elastinogenesis phase of elastic fibre formation.

Molecular folding model of extensible microfibrils

The AET, STEM, X-ray diffraction and immuno-electron microscopy findings described above have formed the basis of a model of fibrillin alignment in extensible microfibrils in which intramolecular folding would act as a molecular 'engine' driving recoil after extension in response to external stretch forces (Baldock *et al.*, 2001; Kielty *et al.*, 2002) (Figure 2C). The model predicts that maturation from initial parallel head-to-tail alignment (~160 nm) to a ~one-third stagger (~100 nm) occurs by folding at the termini and the proline-rich region, which would align known fibrillin-1 transglutaminase cross-link sequences. Microfibril elasticity (in the range 56–100 nm) would require further intramolecular folding at flexible sites which could be links between TB modules and cbEGF domains. This molecular folding model is a novel concept in extracellular matrix biology. The model now needs rigorous, flexible sites within fibrillin molecules need to be determined, and the electrostatic and hydrophobic driving forces of this elastic motor defined. In addition, the multimolecular nature of microfibrils will need to be taken into account.

Future perspectives

We have outlined the evidence that microfibrils are elastic, and current understanding of the molecular mechanism underlying their assembly, and ability to extend and recoil. A priority now will be to assess the proposed molecular folding model, and to identify flexible regions within fibrillin molecules. The precise molecular composition of microfibrils in different tissues urgently needs to be determined using new analytical approaches, so that the implications for elasticity of the presence of other molecules can be evaluated. The molecular basis of microfibril maturation, and the effects of transglutaminase cross-linking on elasticity need to be established. In microfibril bundles, the nature of any inter-microfibrillar cross-links needs defining. This information will enhance our understanding of the elastic mechanism of the unique fibrillin-rich microfibrils of extracellular matrices, and may lead to new strategies for repairing elastic tissues and tissue engineering strategies.

Acknowledgements

Studies from our laboratory that are described here were funded by the Medical Research Council, UK. Studies conducted by CEMB Stirling were supported by BBSRC Grant 98/S15326.

References

Abrams WR, Ma RI, Kucich U, Bashir MM, Decker S, Tsipouras P, McPherson JD, Wasmuth JJ and Rosenbloom J (1995) Molecular cloning of the microfibrillar protein MFAP3 and assignment of the gene to human-chromsosme 5q32-q33.2. *Genomics* **26:** 47–54.

Ashworth JL, Kielty CM and McLeod DM (2000) Fibrillin and the eye. *Brit J Ophthalmol* **84:** 1312–1317.

Ashworth JL, Kelly V, Rock MJ, Shuttleworth CA and Kielty CM (1999c) Regulation of fibrillin carboxy-terminal furin processing by N-glycosylation, and association of amino- and carboxy-terminal sequences. *J Cell Sci* **112:** 4163–4171.

Ashworth JL, Kelly V, Wilson R, Shuttleworth CA and Kielty CM (1999b) Fibrillin assembly: dimer formation mediated by amino-terminal sequences. *J Cell Sci* **112:** 3549–3558.

Ashworth JL, Sherratt MJ, Rock MJ, Murphy G, Shapiro SD, Shuttleworth CA and Kielty CM (1999a) Fibrillin turnover by metalloproteinases: implications for connective tissue remodelling. *Biochem J* **340:** 171–181.

Baccarani-Contri M, Vincenzi D, Cicchetti F, Mori G and Pasquali-Ronchetti I (1990) Immunocytochemical localization of proteoglycans within normal elastic fibres. *Eur J Cell Biol* **53:** 305–312.

Baldock C, Koster AJ, Ziese U, Sherratt MJ, Kadler KE, Shuttleworth CA and Kielty CM (2001) The supramolecular organisation of fibrillin-rich microfibrils. *J Cell Biol* **152:** 1045–1056.

Bellingham CM, Woodhouse KA, Robson P, Rothstein SJ and Keeley FW (2001) Self-aggregation of recombinantly expressed human elastin polypeptides. *Biochim Biophys Acta* **1550:** 6–19.

Borel A, Eichenberger D, Farjanel J, Kessler E, Gleyzal C, Hulmes DJS, Sommer P and Font B (2001) Lysyl oxidase-like protein from bovine aorta. *J Biol Chem* **276:** 48,944–48,949.

Bressan GM, Daga G, Colombatti A, Castellani I, Marigo V and Volpin D (1993) Emilin, a component of elastic fibers preferentially located at the elastin–microfibril interface. *J Cell Biol* **121:** 201–213.

Brown-Augsberger P, Broekelmann T, Mecham L, Mercer R, Gibson MA, Cleary EG, Abrams WR, Rosenbloom J and Mecham RP (1996) Microfibril-associated glycoprotein binds to the carboxyl-terminal domain of tropoelastin and is a substrate for transglutaminase. *J Biol Chem* **269:** 28,423–28,449.

Brown-Augsburger P, Tisdale C, Broekelmann T, Sloan C and Mecham RP (1995) Identification of an elastin cross-linking domain that joins three peptide chains. *J Biol Chem* **270:** 17,778–17,783.

Cardy CM and Handford PA (1998) Metal ion dependency of microfibrils supports a rod-like conformation for fibrillin-1 calcium-binding epidermal growth factor-like domains. *J Mol Biol* **276:** 855–860.

Colombatti A, Doliana R, Bot S, Canton A, Mongiat M, Mungiguerra G, Paron-Cilli S and Spessotto P (2000) The EMILIN protein family. *Matrix Biol* **19:** 289–301.

Csiszar K (2001) Lysyl oxidases: a novel multifunctional amine oxidase family. *Prog Nucleic Acid Res Mol Biol* **70:** 1–32.

Dallas SL, Keene DR, Saharinen J, Sakai LY, Mundy GR and Bonewald LF (2000) Role of the latent transforming growth factor beta binding protein family in fibrillin-containing microfibrils in bone cells in vitro and in vivo. *J Bone Mineral Res* **15:** 68–81.

Dahlbäck K, Ljungquist A, Lofberg H, Dahlbäck B, Engvall E and Sakai LY (1990) Fibrillin immunoreactive fibers constitute a unique network in the human dermis: immunohistochemical comparison of the distributions of fibrillin, vitronectin, amyloid P component, and orcein stainable structures in normal skin and elastosis. *J Invest Dermatol* **94:** 284–291.

Danielson KG, Baribault H, Holmes DF, Graham H, Kadler KE and Iozzo RV (1997) Targeted disruption of decorin leads to abnormal collagen fibril morphology and skin fragility. *J Cell Biol* **136:** 729–743.

Doliana R, Bot S, Mungiguerra G, Canton A, Cilli SP and Colombatti A (2001) Isolation and characterization of EMILIN-2, a new component of the growing EMILINs family and a member of the EMI domain-containing superfamily. *J Biol Chem* **276:** 12,003–12,011.

Doliana R, Bucciotti F, Giacomello E, Deutzmann R, Volpin D, Bressan GM and Colombatti A (1999) EMILIN, a component of the elastic fiber and a new member of the C1q/Tumor necrosis factor superfamily of proteins. *J Biol Chem* **274:** 16,773–16,781.

Downing AK, Knott V, Werner JM, Cardy CM, Campbell ID and Handford PA (1996) Solution structure of a pair of calcium-binding epidermal growth factor-like domains: implications for the Marfan syndrome and other genetic disorders. *Cell* **85:** 597–605.

Eriksen TA, Wright DM, Purslow PP and Duance VC (2001). Role of Ca^{2+} for the mechanical properties of fibrillin. *Proteins: Str Funct Genet* **45:** 90–95.

Faury G (2001) Function–structure relationship of elastic arteries in evolution: from microfibrils to elastin and elastic fibres. *Pathol Biol (Paris)* **49:** 310–325.

Gibson MA, Finnis ML, Kumaratilake JL and Cleary EG (1998) Microfibril-associated glycoprotein-2 (MAGP-2) is specifically associated with fibrillin-containing microfibrils and exhibits more restricted patterns of tissue localisation and developmental expression than its structural relative MAGP-1. *J Histochem Cytochem* **46:** 871–885.

Gibson MA, Hatzinikolas G, Davis EC, Baker E, Sutherland GR and Mecham RP (1995) Bovine latent transforming growth factor β1-binding protein-2: molecular cloning, identification of tissue isoforms, and immunolocalization to elastin-associated microfibrils. *Mol Cell Biol* **15:** 6932–6942.

Gibson MA, Kumaratilake JS and Cleary EG (1989) The protein components of the 12-nanometer microfibrils of elastic and non-elastic tissues. *J Biol Chem* **264:** 4590–4598.

Gibson MA, Kumaratilake JS and Cleary EG (1997) Immunohistochemical and ultrastructural localization of MP 78/70 (beta ig-H3) in extracellular matrix of developing and mature bovine tissues. *J Histochem Cytochem* **45:** 1683–1696.

Gibson MA, Leavesley DI and Ashman LK (1999) Microfibril-associated glycoprotein-2 specifically interacts with a range of bovine and human cell types via alphaVbeta3 integrin. *J Biol Chem* **274:** 13,060–13,065.

Grässel S, Unsold C, Schacke H, Brucner-Tuderman L and Bruckner P (1999) Collagen XVI is expressed by human dermal fibroblasts and keratinocytes and is associated with the microfibrillar apparatus in the upper papillary dermis. *Matrix Biol* **18:** 309–317.

Hanssen E, Franc S and Garrone R (1998) Atomic force microscopy and modelling of natural elastic fibrillin polymers. *Biol Cell* **90:** 223–228.

Henderson M, Polewski R, Fanning JC and Gibson MA (1996) Microfibril-associated glycoprotein-1 (MAGP-1) is specifically located on the beads of the beaded-filament structure for fibrillin-containing microfibrils as visualized by the rotary shadowing technique. *J Histochem Cytochem* **44:** 1389–1397.

Hindson VJ, Ashworth JL, Rock M, Shuttleworth CA and Kielty CM (1999) Fibrillin catabolism by matrix metalloproteinases: characterisation of amino- and carboxy-terminal cleavage sites. *FEBS Lett* **452:** 195–198.

Hirano E, Fujimoto N, Tajima S, Akiyama M, Ishibashi R and Okamoto K (2002) Expression of 36-kDa microfibril-associated glycoprotein (MAGP-36) in human keratinocytes and its localization in skin. *J Dermatol Sci* **28:** 60–67.

Horrigan SK, Rich CB, Streeten BW, Li ZY and Foster JA (1992) Characterization of an associated microfibril protein through recombinant DNA techniques. *J Biol Chem* **267:** 10,087–10,095.

Isogai Z, Aspberg A, Keene DR, Ono RN, Reinhardt DP and Sakai LY (2002) Versican interacts with fibrillin-1 and links extracellular microfibrils to other connective tissue networks. *J Biol Chem* **277:** 4565–4572.

Jensen SA, Reinhardt DP, Gibson MA and Weiss AS (2001) Protein interaction studies of MAGP-1 with tropoelastin and fibrillin-1. *J Biol Chem* **276:** 39,661–39,666.

Keene DR, Maddox BK, Kuo H-J, Sakai LY and Glanville RW (1991) Extraction of extensible beaded structure and their identification as extracellular matrix fibrillin-containing microfibrils. *J Histochem Cytochem* **39:** 441–449.

Kielty CM and Shuttleworth CA (1993) The role of calcium in the organisation of fibrillin microfibrils. *FEBS Lett* **336:** 323–326.

Kielty CM and Shuttleworth CA (1997) Microfibrillar elements of the dermal matrix. *Microsc Res Tech* **38:** 407–427.

Kielty CM, Baldock C, Lee D, Rock MJ, Sherratt MJ, Ashworth JL and Shuttleworth CA (2002) Fibrillin: from microfibril assembly to biomechanical function. *Philos Trans Roy Soc B* **357:** 207–217.

Kielty CM, Whittaker SP and Shuttleworth CA (1996) Fibrillin: evidence that chondroitin sulphate proteoglycans are components of microfibrils and associate with newly synthesised monomers. *FEBS Lett* **386:** 169–173.

Kielty CM, Woolley DE, Whittaker SP and Shuttleworth CA (1994) Catabolism of intact fibrillin microfibrils by neutrophil elastase, chymotrypsin and trypsin. *FEBS Lett* **351:** 85–89.

Kitahama S, Gibson MA, Hatzinikolas G, Hay S, Kuliwaba JL, Evdokiou A, Atkins GJ and Findlay DM (2000) Expression of fibrillins and other microfibril-associated proteins in human bone and osteoblast-like cells. *Bone* **27:** 61–67.

Kocsis E, Trus BL, Steer CJ, Bisher ME and Steven AC (1991). Image averaging of flexible fibrous macromolecules: the clathrin triskelion has an elastic proximal segment. *J Struct Biol* **107:** 6–14.

Kogake S, Hall SM, Kielty CM and Haworth SG (2002) Fibrillin-1 remodelling in the pulmonary arterial wall during development. *Am J Physiol–Lung Cell Mol Physiol* (under review).

Kostka G, Giltay R, Bloch W, Addicks K, Timpl R, Fassler R and Chu M-L (2001) Perinatal lethality and endothelial cell abnormalities in several vessel compartments of fibulin-1-deficient mice. *Mol Cell Biol* **21:** 7025–7034.

Lausen M, Lynch N, Schlosser A, Tornoe I, Sackmose SG, Teisner B, Willis AC, Crouch E, Schwaeble W and Holmskov U (1999) Microfibril-associated protein 4 is present in lung washings and binds to the collagen region of lung surfactant protein D. *J Biol Chem* **274:** 32,234–32,240.

Liu WG, Faraco J, Qian CP and Francke U (1997) The gene for microfibril-associated protein-1 (MFAP1) is located several megabases centromeric to FBN1 and is not mutated in Marfan syndrome. *Human Genet* **99:** 578–584.

McConnell CJ, DeMont ME and Wright GM (1997) Microfibrils provide non-linear behaviour in the abdominal artery of the lobster Homarus americanus. *J Physiol* **499:** 513–526.

McConnell CJ, Wright Gm and DeMont ME (1996) The modulus of elasticity of lobster aorta microfibrils. *Experientia* **52:** 918–921.

Mecham RP and Davis EC (1994) Elastic fiber structure and assembly. In: Yurchenco PD, Birk DE and Mecham RP (eds) *Extracellular*

Matrix Assembly and Structure. (pp. 281–314) Academic Press, New York.

Miyazono K, Hellman U, Wernstedt C and Heldin CH (1988) Latent high molecular-weight complex of transforming growth factor-β1–purification from human-platelets and structural characterization. *J Biol Chem* **263:** 6407–6415.

Milewicz DM, Urban Z and Boyd C (2000) Genetic disorders of the elastic fiber system. *Matrix Biol* **19:** 471–480.

Miosge N, Sasaki T and Timpl R (1999) Angiogenesis inhibitor endostatin is a distinct component of elastic fibers in vessel walls. *FASEB J* **13:** 1743–1750.

Nagase T, Nakayama M, Nakajima D, Kikuno R and Ohara O (2001) Prediction of the coding sequences of unidentified human genes. XX. The complete sequences of 100 new cDNA clones from brain which code for large proteins in vitro. *DNA Res* **8:** 85–95.

Nakamura T, Lozano PR, Ikeda Y, Iwanaga Y, Hinek A, Minamisawa S, Cheng C-F, Kobuke K, Dalton N, Takada Y, Tashiro K, Ross J, Honjo T and Chien KR (2002) Fibulin-5/DANCE is essential for elastogenesis in vivo. *Nature* **415:** 171–175.

Oklu R and Hesketh R (2000) The latent transforming growth factor beta binding protein (LTBP) family. *Biochem J* **352:** 601–610.

Pereira L, D'Alessio M, Ramirez F, Lynch JR, Sykes B, Pangilinan T and Bonadio J (1993) Genomic organization of the sequence coding for fibrillin, the defective gene product in Marfan syndrome. *Hum Mol Genet* **2:** 961–968.

Pfaff M, Reinhardt DP, Sakai LY and Timpl R (1996) Cell adhesion and integrin binding to recombinant human fibrillin-1. *FEBS Lett* **384:** 247–250.

Prody PS, Callahan JW and Hinek A (1998) The 67-kDa enzymatically inactive alternatively spliced variant of beta-galactosidase is identical to the elastin/laminin-binding protein. *J Biol Chem* **273:** 6319–6326.

Qian R and Glanville RW (1997) Alignment of fibrillin molecules in elastic microfibrils is defined by transglutaminase-derived crosslinks. *Biochemistry* **36:** 15,841–15,847.

Raghunath M, Putnam EA, Ritty T, Hamstra D, Park ES, Tschodrich-Rotter M, Peters R, Rehemtulla A and Milewicz DM (1999) Carboxy-terminal conversion of profibrillin to fibrillin at a basic site by PACE/furin-like activity required for incorporation in the matrix. *J Cell Sci* **112:** 1093–1100.

Raghunath M, Tschodrich-Rotter M, Sasaki T, Meuli M, Chu M-L and Timpl R (1999) Confocal laser scanning analysis of the association of fibulin-2 with fibrillin-1 and fibronectin define different stages of skin regeneration. *J Invest Dermatol* **112:** 97–101.

Raghunath M, Unsold C, Kubitscheck U, Bruckner-Tuderman L, Peters R and Meuli M (1998) The cutaneous microfibrillar apparatus contains latent transforming growth factor-beta binding protein-1 (LTBP-1) and is a repository for latent TGF-beta 1. *J Invest Dermatol* **111:** 559–564.

Reber-Muller S, Spissinger T, Schubert P, Spring J and Schmid V (1995) An extracellular matrix protein of jellyfish homologous to mammalian fibrillins forms different fibrils depending on the life stage of the animal. *Dev Biol* **169:** 662–672.

Reinboth B, Hanssen E, Cleary EG and Gibson MA (2001) Molecular interactions of biglycan and decorin with elastic fiber components. Biglycan forms a ternary complex with tropoelastin and MAGP-1. *J Biol Chem* **277:** 3950–3957.

Reinhardt DP, Keene DR, Corson GM, Pöschl E, Bächinger HP, Gambee JE and Sakai LY (1996a) Fibrillin-1: organization in microfibrils and structural properties. *J Mol Biol* **258:** 104–114.

Reinhardt DP, Mechling DE, Boswell BA, Keene DR, Sakai LY and Bächinger HP (1997) Calcium determines the shape of fibrillin. *J Biol Chem* **272:** 7368–7373.

Reinhardt DP, Sasaki T, Dzamba BJ, Keene DR, Chu M-L, Göhring W, Timpl R and Sakai LY (1996b) Fibrillin-1 and fibulin-2 interact and are colocalized in some tissues. *J Biol Chem* **271:** 19,489–19,496.

Rief M, Gautel M, Oesterhelt F, Fernandez JM and Gaub HE (1997) Reversible unfolding of individual titin immunoglobulin domains by AFM. *Science* **276:** 1109–1112.

Ritty TM, Broekelmann T, Tisdale C, Milewicz DM and Mecham RP (1999) Processing of the fibrillin-1 carboxyl-terminal domain. *J Biol Chem* **274:** 8933–8940.

Robb BW, Wachi H, Schaub T, Mecham RP and Davis EC (1999) Characterization of an in vitro model of elastic fiber formation. *Mol Biol Cell* **10:** 3595–3605.

Robinson PN and Godfrey M (2000) The molecular genetics of Marfan Syndrome and related microfibrillopathies. *J Med Genet* **37:** 9–25.

Sadawa H and Konomi H (1991) The alpha 1 chain of type VIII collagen is associated with many but not all microfibrils of elastic fiber system. *Cell Struct Funct* **16:** 455–466.

Sakamoto H, Broekelmann T, Cheresh DA, Ramirez F, Rosenbloom J and Mecham RP (1996) Cell-type specific recognition of RGD- and non-RGD-containing cell binding domains in fibrillin-1. *J Biol Chem* **271:** 4916–4922.

Schorderet DF, Menasche M, Morand S, Bonnel S, Buchillier V, Marchant D, Auderset K, Bonny C, Abitbol M and Munier FL (2000) Genomic characterisation and embryonic expression of the mouse BigH3 (Tgfbi) gene. *Biochim Biophys Acta* **274:** 267–274.

Sechler JL, Rao H, Cumiskey AM, Vega-Colón I, Smith MS, Murata T and Schwarzbauer JE (2001) A novel fibronectin binding site required for fibronectin fibril growth during matrix assembly. *J Cell Biol* **154:** 1081–1088.

Segade F, Trask BC, Broekelmann TJ, Pierce RA and Mecham RP (2002) Identification of a matrix binding domain in the microfibril-associated glycoproteins 1 and 2 (MAGP1 and 2) and intracellular localization of alternative splice forms. *J Biol Chem* (in press).

Sherratt MJ, Wess TJ, Baldock C, Ashworth JL, Purslow PP, Shuttleworth CA and Kielty CM (2001) Fibrillin-rich microfibrils of the extracellular matrix: ultrastructure and assembly. *Micron* **32:** 185–200.

Sinha S, Nevett C, Shuttleworth CA and Kielty CM (1998) Cellular and extracellular biology of the latent transforming growth factor-beta binding proteins. *Matrix Biol* **17:** 529–545.

Sinha S, Shuttleworth CA, Heagerty AM and Kielty CM (2002) Expression of latent TGFbeta binding proteins and association with TGFbeta-1 and fibrillin-1 in the response to arterial injury. *Cardiovasc Res* **53:** 971–983.

Taipale J, Saharinen J, Hedman K and Keski-Oja J (1996) Latent transforming growth-factor-β1 and its binding-protein are components of extracellular-matrix microfibrils. *J Histochem Cytochem* **44:** 875–889.

Tassabehji M, Metcalfe K, Hurst J, Ashcroft GS, Kielty CM, Wilmot C, Donnai D, Read AP and Jones CJP (1998) An elastin gene mutation producing abnormal tropoelastin and abnormal elastic fibres in a patient with autosomal dominant cutis laxa. *Hum Mol Genet* **7:** 1021–1028.

Thurmond FA and Trotter JA (1996) Morphology and biomechanics of the microfibrillar network of sea cucumber dermis. *J Exptl Biol* **199:** 1817–1828.

Tiedemann K, Bätge B, Müller PK and Reinhardt DP (2001) Interactions of fibrillin-1 with heparin/heparan sulphate, implications for microfibrillar assembly. *J Biol Chem* **276:** 36,035–36,042.

Toonkool P, Jensen SA, Maxwell AL and Weiss AS (2001) Hydrophobic domains of human tropoelastin interact in a context-dependent manner. *J Biol Chem* **275:** 44,575–44,580.

Toyashima T, Yamashita K, Shishibori T, Itano T and Kobayashi R (1999) Ultrastructural distribution of 36-kDa microfibril-associated glycoprotein (MAGP-36) in human and bocvine tissues. *J Histichem Cytochem* **47:** 1049–1056.

Trask TM, Ritty TM, Broekelmann T, Tisdale C and Mecham RP (1999) N-terminal domains of fibrillin 1 and fibrillin 2 direct the formation of homodimers: a possible first step in microfibril assembly. *Biochem J* **340:** 693–701.

Trask BC, Trask TM, Broekelmann T and Mecham RP (2000a) The microfibrillar proteins MAGP-1 and fibrillin-1 form a ternary complex with the chondroitin sulfate proteoglycan decorin. *Molec Biol Cell* **11:** 1499–1507.

Trask TM, Trask BC, Ritty TM, Abrams WR, Rosenbloom J and Mecham RP (2000b) Interaction of tropoelastin with the amino-terminal domains of fibrillin-1 and fibrillin-2 suggests a role for the fibrillins in elastic fiber formation. *J Biol Chem* **275:** 24,400–24,406.

Tsuda T, Wang H, Timpl R and Chu M-L (2001) Fibulin-2 expression marks transformed mesenchymal cells in developing cardiac valves, aortic arch vessels, and coronary vessels. *Devel Dynam* **222:** 89–100.

Utani A, Nomizu M and Yamada Y (1997) Fibulin-2 binds to the short arms of laminin-5 and laminin-1 via conserved amino acid sequences. *J Biol Chem* **272:** 2814–2820.

Wess TJ, Purslow P and Kielty CM (1997) Fibrillin-rich microfibrils: an X-ray diffraction study and elastomeric properties. *FEBS Lett* **413:** 424–428.

Wess TJ, Purslow PP, Sherratt MJ, Ashworth JL, Shuttleworth CA and Kielty CM (1998a) Calcium determines the supramolecular organization of fibrillin- rich microfibrils. *J Cell Biol* **141:** 829–837.

Wess TJ, Purslow PP and Kielty CM (1998b) X-ray diffraction studies of fibrillin-rich microfibrils: effects of tissue extension on axial and lateral packing. *J Struct Biol* **122:** 123–127.

Wright DM, Duance VC, Wess TJ, Kielty CM and Purslow PP (1999) The supramolecular organisation of fibrillin-rich microfibrils determines the mechanical properties of bovine zonular filaments. *J Exp Biol* **202:** 3011–3020.

Xu T, Bianco P, Fisher LW, Longenecker G, Smith E, Goldstein S, Bonadio J, Boskey A, Heegaard AM, Sommer B, Satomura K, Dominguez P, Zhao C, Kulkarni AB, Robey PG and Young MF (1998) Targeted disruption of the biglycan gene leads to an osteoporosis-like phenotype in mice. *Nat Genet* **20:** 78–82.

Yanagisawa H, Davis EC, Starcher BC, Ouchi T, Yanagisawa M, Richardson JA and Olsen EN (2002) Fibulin-5 is an elastin-binding protein essential for elastic fibre development *in vivo*. *Nature* **415:** 168–171.

Zhang H, Apfelroth SD, Hu W, Davis EC, Sanguineti C, Bonadio J, Mecham RP and Ramirez F (1994) Structure and expression of fibrillin-2, a novel microfibrillar component preferentially located in elastic matrices. *J Cell Biol* **124:** 855–863.

Zhang H, Hu W and Ramirez F (1995) Developmental expression of fibrillin genes suggests heterogeneity of extracellular microfibrils. *J Cell Biol* **129:** 1165–1176.

Zimmermann DR, Dours-Zimmermann M, Schubert M and Bruckner-Tuderman L (1994) Versican is expressed in the proliferating zone in the epidermis and in association with the elastic network of the dermis. *J Cell Biol* **124:** 817–825.

MIX
Papier aus verantwortungsvollen Quellen
Paper from responsible sources
FSC® C105338

If you have any concerns about our products,
you can contact us on
ProductSafety@springernature.com

In case Publisher is established outside the EU,
the EU authorized representative is:
Springer Nature Customer Service Center GmbH
Europaplatz 3, 69115 Heidelberg, Germany

Printed by Libri Plureos GmbH
in Hamburg, Germany